Natural Small Molecule Drugs from Plants

Guan-Hua Du

Natural Small Molecule Drugs from Plants

PMPH
PEOPLE'S MEDICAL PUBLISHING HOUSE
Springer

Guan-Hua Du
Institute of Material Medica
Chinese Academy of Medical Sciences
Beijing, China

ISBN 978-981-10-8021-0 ISBN 978-981-10-8022-7 (eBook)
https://doi.org/10.1007/978-981-10-8022-7

Library of Congress Control Number: 2018949919

Printed on acid-free paper

This Springer imprint is published by the registered company Springer Nature Singapore Pte Ltd.
The registered company address is: 152 Beach Road, #21-01/04 Gateway East, Singapore 189721, Singapore

Preface

Natural medicines refer to substances that exist in nature and are used as drugs for the prevention and treatment of diseases, including minerals, plants, microorganisms, and animals. Natural medicines are the earliest drugs used for the treatment of human diseases and play an important role in the development of human being. Natural medicines have become important materials used to protect the health of humans after long-term accumulation of application. The formation and development of medicine science is also grounded in natural medicine. Chinese traditional medicine was originated from long-term accumulating experiences of using natural medicine by Chinese people. Up to now, a large number of chemical drugs used in clinic are still from natural products. These small-molecule compounds derived from natural substances are also called natural medicine.

In the long history of natural medicine, it was greatly promoted in different drug applications and need for clinical drugs. To obtain ideal drugs and new natural medicines and to achieve better therapeutic effects, medical scientists had done a lot of researches and modified them, which promoted the development of medicine science. Drug tablets, medicine, and modern formulations in Chinese traditional drugs reflected the development of pharmaceutical science.

The modern pharmaceutical sciences have led to the understanding of composition and mechanisms of natural medicines. With the progress of modern medicinal chemistry, a large number of natural medicines with defined molecular structure were used as clinical drugs. It is common characteristic of drugs to elucidate its compositions and mechanisms at the molecular level. Therefore, natural medicine moves to a new stage. It is the basic requirement of modern drug research to obtain valuable compounds from natural products.

The progress of medical science, the development of life sciences and the breakthrough of biology provide new theories and techniques for drug research. People who understand the role of drugs at the molecular level explore and discover many new drugs. Especially, various types of compounds from natural products including biological macromolecules enriched the resources of drug research and development and provided more new drugs for the treatment of diseases in clinic, which bring a positive effect and good results. Therefore, modern natural medicine mainly refers to drugs derived from natural substances, such as minerals from inorganic compounds and trace elements, antibiotics from microbial microorganisms, biological macromolecules from microorganisms and animals, and small-molecule compounds from plants.

Small-molecule compounds which were derived and isolated from plants are presented in the book. The development history and characteristics of these drugs are reviewed as well. The mechanisms of these small-molecule drugs from natural products are elucidated, which provide support for clinical drugs and provide guidance for the study and development of natural medicine by summing up experiences of research.

Natural Medicines Are a Kind of Important Substances for Prevention and Treatment of Human Diseases

Modern natural drugs are named synthetic chemicals. Natural medicines are drugs which exist in nature. These drugs can be obtained by artificial methods, but since these compounds exist in the natural product, we still call them natural medicine. According to research, modern drugs are divided into macromolecule and small-molecule drugs according to their structure, which showed their basic characteristics and provided guidance for research and development. Therefore, there are also small-molecule and macromolecule drugs in natural medicine, such as hormones (small-molecule drugs) extracted from animals and proteins including insulin (macromolecule drug) obtained from animals.

Structures of macromolecules are complex. They are mostly derived from natural products, such as plants, animals, and microbes. In recent years, a large number of macromolecules were obtained by application and integration of many techniques, especially manual design and biosynthesis, such as antibody drugs, protein drugs, peptide drugs, polysaccharide drugs, DNA and RNA, etc. These drugs brought good results in clinic. With the rapid development of life sciences and biological technology, biological macromolecule drugs will become the most rapid field in research and development of modern drug.

At present, most of chemical drugs used in clinic are small-molecule drugs, which have been studied for more than 100 years. There are detailed mechanisms and a lot of experiences about these drugs, which provided strong support for these drugs in treating diseases.

Up to now, small-molecule drugs used in clinical were divided into three categories according to their sources: (1) those natural small-molecule drugs which were directly separated and purified from natural products, (2) those which were obtained by modifying structure of natural small-molecule drugs that accounted for a large proportion in clinical drugs, and (3) those that are obtained through chemical synthesis. It is obvious that natural products can not only be used as drugs but also provide important structural information in drug research and development.

Therefore, in drug discovery, it is very important to study history of natural small-molecule drugs.

The development and progress of chemical technology provide a strong support for the research of natural products. Scientists can obtain more and more natural compounds with new structure, good activity, and good prospects from natural products. More important drugs would be discovered for diseases by studying these compounds.

Natural Medicine Is the Foundation of Modern Pharmacy

Natural medicines refer to substances which exist in nature. The major sources of natural medicines are minerals, plants, microorganisms, animals, and so on. Obtaining natural medicines from these natural resources is critical for drug development. Currently, natural chemical drugs used in clinic are also derived from many sources, such as minerals, plants, microorganisms, animals, and so on.

Natural Small-Molecule Drugs Derived from Minerals

Major drugs that are used in treating diseases caused by micronutrient deficiencies originate from minerals such as inorganic compounds and some specific elements. Some active minerals in human body are also used as important drugs to treat diseases, such as arsenic trioxide used to treat leukemia, calcium carbonate used to supplement calcium, and iron compounds used to treat iron deficiency anemia. In addition, a large number of chemical elements including zinc, copper, selenium, mercury, manganese, etc., are also used in clinic.

In traditional Chinese medicine, many minerals which play important roles in the development of traditional medicine and in clinic are recorded. Forty-six kinds of mineral drugs, such as realgar, arsenic, gypsum, and so on, are recorded in *Shennong's Classic of Materia Medica* which is a classical book about pharmacy in China. Most of those traditional drugs are still used for treating diseases today.

Natural Small-Molecule Drugs Derived from Microorganisms

In the history of modern chemical drugs, drugs derived from microorganisms play important roles in treating diseases. These drugs which reflected great progress in fighting human diseases occupy a large proportion of natural medicine in type and quantity. In particular, the discovery of antibiotics opened up new areas of developing drugs from microbes. A large number of antibiotics not only are widely used in the treatment of diseases infected by bacteria but also function as antiviral, anti-pathogenic microorganism, and antitumor agents. In addition to a lot of antibiotics, a number of new drugs with extensive roles are also obtained from microorganisms, such as antibiotics treating cancer and some drugs which were developed from fungi including *Ganoderma lucidum*, *Cordyceps*, *Poria*, and ergot and used to regulate physiological functions. These drugs have played very important roles in improving human health.

Microbes have been used as drugs for long time by human being, such as some fungi, bacteria, and even pathogens. Particularly, the research and development of antibiotic drugs was greatly promoted by the discovery of penicillin and its successful application in treating infectious diseases. The discovery of antibiotic prevented death of humans caused by infectious diseases, which is a great progress of medicine. Small-molecule drugs derived from microorganisms, which are one of the important sources of many drugs, not only have diverse chemical structures but also physiological and pharmacological activities.

Natural Small-Molecule Drugs Derived from Animals

Thousands of years ago, humans realized that some substances obtained from animals could be used to treat diseases. These small-molecule drugs, such as adrenaline, glucocorticoid, pituitrin, and dopamine, play important and even indispensable roles in the treatment of serious diseases. In addition, macromolecules in animal body, such as insulin, also play important roles in treating diseases. Using modern molecular and biological techniques, people have developed a large number of artificially manufactured "natural substances," such as proteins, peptides, polysaccharides, nucleotides, antibodies, and so on. These biological drugs have become an important part of modern drug research.

Animals are also one of important drug sources. Sixty-seven drugs derived from animals were recorded in *Shennong's Classic of Materia Medica*. Most of drugs derived from animals are obtained through extraction and preparation, including a substantial portion of macromolecule drugs which are important contents in modern biotechnology drug research. Small-molecule drugs derived from animals are also of important value, especially the discovery of hormonal drugs and neurotransmitter drugs, which not only provides new type of drugs for clinic but also promotes the progress of pharmaceutical theory. The discovery and application of glucocorticoid not only save many lives of patients with serious diseases but also contribute to knowledge of process in hormone regulating organism. Small-molecule drugs derived from animals, represented by epinephrine and acetylcholine, open up a novel field of life science research and improve the effectiveness of many serious diseases, such as hypertension, diabetes, and so on.

Natural Small-Molecule Drugs Derived from Plants

Small-molecule compounds from plants have the characteristics of various types, diversity structures, many activities, and stable resources. Those with strong pharmacological effects and clinical efficacy have been widely used to treat human diseases in clinical. They have attracted great attention of drug researchers, which are the major content described in the book.

Natural small-molecule drugs derived from plants are a large class of drugs. Small-molecule drugs are very abundant for there are numerous species of plants. Therefore, plants are major sources of natural medicine. The book is focused on small-molecule drugs from plants.

The 2015 Nobel Prize of Physiology or Medicine was awarded to the Chinese scientist Prof. Yoyo Tu for her outstanding contributions in the discovery of artemisinin. This award has also attracted attention to the development of small-molecule drugs from plants, which has become an important research field of pharmaceutical science.

Up to now, natural small-molecule drugs that are used in clinic have underwent long processes of research and development, such as artemisinin. Every drug was successfully obtained by continuous research and development, which reflected the development and progress of pharmaceutical science. Therefore, it is important to sum up the mechanisms and research processes of natural small-molecule drugs from plants, which provide experiences for the research and development of new drugs and guidance for the comprehensive evaluation of the characteristics and application prospect of natural medicine.

The Research and Development of Natural Small-Molecule Drugs Embraces the Wisdom of Humans

During the history of natural small-molecule drugs, the experience and wisdom of fighting diseases was recorded by humans. The discovery of every drug underwent a hard process of exploration and even a long process of development. These processes not only provided important references in technology for research of new drug but also had important significance in knowledge and theory of drugs.

The success of natural small-molecule drugs from traditional drugs represented by artemisinin, etc., showed great achievements in the development of traditional Chinese medicine and proved the effectiveness of traditional drugs and the controllability of material basis, which meets the requirement of modern drug. Especially, the research and development process of artemisinin gave us more inspiration. The research and development processes of drugs are hard, and the need for drugs to treat human diseases is eternal. Therefore, exploring and finding new drugs will never end. Previous success and failures are very useful for research and development of new drugs.

Technological progress is an important guarantee for the success of drug development. There are many difficulties in the early research of natural small-molecule drugs, such as its separation and extraction, analysis, identification, and activity evaluation. All these need the support of corresponding technology. Currently, chemical nature and structures of many natural small-molecule drugs used in clinic, such as metformin, paclitaxel, vinblastine, and so on, were elucidated after 10 years or even decades of efforts. Up to now, techniques for modern separation, purification, and analytical technology were greatly improved. It isn't difficult to obtain natural small-molecule drugs any more. However, how to find better drugs is still a big scientific question that we need to consider.

Scientific methods and correct thinking are important conditions for the success of research. Historical experience is of great value to understand and develop natural medicine. For example, the discovery of vitamin C went through a long process of research and development. First of all, it was realized that humans will suffer from scurvy due to lack of ascorbic acid intake by analyzing a large number of seafarers' death. Furthermore, people found that intake of fruits and vegetables could improve symptoms of the disease. In spite of this, people still couldn't avoid forming the wrong guiding ideology in preventing and treating the disease. People even attributed the cause of scurvy to lack of "acid" in the body. Based on this ideology, absurd treatment including taking diluted sulfuric acid or acetic acid to prevent scurvy among seafarers were emerged, which not only caused more death from this disease but also delayed the discovery of ascorbic acid for hundreds of years. Therefore, correct thinking and guidance are vital in the discovery and development of new drugs.

The history of natural small-molecule drug research embraced the wisdom, knowledge, experience, technology, and spirit of medicine science exploration, which is an important reference and significant guide for drug research.

Small-Molecule Drugs Derived from Plants Are the Representatives of Natural Medicines

Plants have been used as drugs by humans to prevent and treat diseases for a long time. Humans had a lot of experiences in using plants to treat diseases before the ancient civilizations, which were recorded in literatures of ancient civilizations in the world. This is mainly due to the fact that plants are much easier to be obtained than other substances and that varieties of plants can have many different activities and become an important source of obtaining drugs.

Small-molecule compounds are secondary metabolites derived from plants. Their production and metabolism reflect the vital processes of life science and have some biological functions, which are reflected by some ways in human body. Some small molecules have an important function in plants and important pharmacological effects in the treatment of diseases.

Small-molecule compounds derived from plants are the foundation of drug development. Small molecules have properties of rich and reliable sources. Both natural collection and planting can be easily achieved. The study of small-molecule drugs derived from plants needs complex techniques, methods, and scientific theories in many disciplines and fields, such as botany, chemistry, biology, life science, etc., which reflected progress of science and technology. Research and development of small-molecule compounds derived from plants can promote the overall scientific and technological level of drug research.

Features and Instructions of the Book

1. Drugs with the following characteristics are presented in the book: ① earliest drugs which were collected in the national pharmacopoeia or national standard and drugs which were used in clinical before and replaced by new alternative medicines now, ② small-molecule drugs directly derived from plants without structural modification or modified drugs without change of major properties, ③ drugs consisting of a variety of small molecules derived from plants, and ④ compounds collected as standard by pharmacopoeia and never directly used as a drug in the form of a monomeric compound.

   This book illustrates that the development of modern pharmacy was promoted by small-molecule drugs derived from plants. The book provides an overview of natural small-molecule drugs derived from plants for researchers in pharmaceutical field.
2. One hundred twenty kinds of drugs are discussed in the book. There are seven chapters that discuss about drugs and their clinical use, such as drugs for cardiovascular diseases, drugs for diseases of metabolism, drugs for neurological and psychiatric system, drugs for immune inflammatory diseases, antitumor drugs, and drugs for parasites and bacterial infection. Vitamins were described in the sixth chapter and the seventh chapter mainly mentions drugs collected only as reference compounds in the pharmacopoeia. The classification is made only for convenient reading. In fact, some drugs have various pharmacological effects, so the classification may not be completely reasonable. However, these drugs are described in detail according to their characteristics. The chapter about vitamins only limits its discussion on those vitamins derived from plants, which does not reflect the whole topic about vitamins. The same class of drugs is sorted according to the first word of their Chinese names. Multiple drugs from the same plant are introduced together to facilitate readers' understanding of the whole situation.

   In addition, for small-molecule drugs, the plant which was firstly found and used as a major source of drugs is introduced.

3. The book introduces every drug from many aspects, such as name, alias, Chinese and English chemical name and structure, physicochemical properties, formulations and indications, source records, development history, pharmacological effects, clinical application, comprehensive content evaluation, references, and so on. Every drug is collected in the pharmacopoeia and its physicochemical properties are briefly described in the book. The development of research and pharmacological activity of every drug are described to help readers' comprehension. The introduction of each drug had no more than ten references, including the earliest and latest published articles, which are most important in the research of drug to help readers understand the whole process.

The book is written by persons who are first-line researchers, teachers, and clinical doctors working in the field of pharmacy and pharmacology. Every drug is described by about 3000 easy-to-understand words, which are accurate in expression of pharmacy knowledge. The book provides guidance and reference for researchers, teachers, and clinical doctors.

The drugs discussed in the book are only small-molecule drugs from plants without natural medicines from other sources. Due to the limited knowledge of authors, there are a lot of omissions and errors in the book. Please don't hesitate to point them out if you find any.

Beijing, China
2017.10

Guan-Hua Du

Contents

Part II Natural Small Molecule Drugs for the Prevention and Treatment of Neuropsychiatric Diseases

Part V Natural Small Molecule Drugs for the Treatment of Parasitic and Bacterial Infections

Part VI Vitamins

Part VII Natural Product Compounds in Pharmacopeia

Part I
Natural Small-Molecule Drugs for the Prevention and Treatment of Cardiovascular Diseases

Overview

All functions and states of the human body are dependent on the functioning of the cardiovascular system, and this is the system most commonly affected by disease. More importantly, cardiovascular disease not only affects the functional status of the body but is also a direct threat to life. Therefore, cardiovascular disease is a disease of particular concern to the population.

Because of the great harm caused by cardiovascular diseases, people have paid special attention to finding drugs for the treatment of these diseases. Both traditional Chinese medicine—which has been used for thousands of years—and modern Western medicine have realized the importance of cardiovascular diseases, and the drugs used for the treatment of cardiovascular disease are therefore also of great concern.

As early as 2000 years ago, in the "伤寒论: On Cold Damage, Translation & Commentaries by Zhang, Zhongjing" the theory of "promoting blood circulation and remove blood stasis" served as an important basis for the treatment of cardiovascular disease in traditional Chinese medicine. "Avicenna Medical Code" and its forerunners have focused on cardiovascular disease, also beginning many years ago. Accordingly, the prevention and treatment of cardiovascular disease with drugs has received considerable attention from medical experts.

The development of modern medicine, especially chemistry and related techniques, has enabled the discovery of compounds with pharmacological effects in plants, animals, and other natural materials. Some of these compounds have been developed as drugs for the treatment of cardiovascular diseases. These natural small-molecule drugs used for the prevention and treatment of cardiovascular diseases represent some of the earliest natural drugs found in plants. From the early cardiac glycosides used for the treatment of heart failure, many drugs have been developed and used clinically for the treatment of a variety of cardiovascular diseases. The discovery of drugs for the treatment of cardiovascular diseases has not only led to progress in the research of natural medicines, but has also increased the understanding of cardiovascular pathology.

Heart failure is a disease with very high mortality. The use of cardiac glycosides has saved the lives of many heart failure patients and has also promoted the study of the pharmacological mechanisms of cardiac drugs. Since the time that cardiac glycosides were first used, a number of natural products with potent actions on the heart have been found, and related compounds with modified structures have been developed; these compounds have been employed for the treatment of heart failure. Among them are many natural small-molecule compounds; for example, digoxin, divaside, digitoxin, and lanatoside. The widespread use of cardiac glycosides has also promoted the clinical study of heart failure. However, the obvious drawbacks of natural cardiac glycosides are their significant toxicity and narrow therapeutic window. The general effective dose is close to the toxic dose, and the clinical application of cardiac glycosides is risky and difficult. For this reason, scientists around the world have conducted lengthy and extensive research on these drugs, but so far have not found ideal cardiac glycosides.

Long-term studies of cardiac glycosides have demonstrated that these natural drugs not only have potent effects on myocardial contractility, but also cause changes in the heart rate. Through studies of the mechanism of glycoside effects on the heart rate, the understanding of the electrophysiological function of cardiac drugs was improved and the study of antiarrhythmic drugs was promoted. Consequently, natural antiarrhythmic drugs such as quinidine and tetrandrine have also been used.

Many natural products that affect cardiovascular function, especially those products that modify chronic pathological changes in the cardiovascular system, have been used clinically, such as the drugs used for the treatment of myocardial ischemia, thrombosis, and other cardiovascular conditions. Natural small-molecule drugs, such as ligustrazine, tanshinone IIA, cyclovirobuxinum D, puerarin, daidzein, ferulic acid, and rutin, have played an important role in the treatment of cardiovascular diseases.

In the study of antihypertensive drugs, the discovery and clinical application of reserpine, ajmaline, and other small-molecule compounds have promoted understanding of both the pathogenesis of hypertension and the mechanism whereby drugs lower blood pressure.

Natural small-molecule drugs used for metabolic regulation, such as metformin, have become important for the treatment of diabetes, and are widely used clinically. The discovery and application of lovastatin has opened up a new area in the development and application of lipid-lowering drugs. The dicoumarin drugs, which have a powerful anticoagulant effect, are important for the prevention of coagulation and thrombosis.

Another important natural small-molecule drug for the treatment of cardiovascular diseases is anisodamine, which improves the microcirculation and is used for the treatment of infection with toxic shock; this drug has saved millions of lives.

These natural small-molecule drugs used in the prevention and treatment of cardiovascular diseases are obtained from different plants worldwide, most of them being isolated from traditional Chinese medicines that have been used for their effects in "activating blood circulation to dissipate blood stasis". The application of these drugs also promotes an understanding of the theories of traditional Chinese medicine.

Many drugs that are now used clinically for the prevention and treatment of cardiovascular diseases represent notable achievements of modern medical science. The breakthroughs made in the development of these drugs are closely related to the development and application of natural medicine; this is particularly seen in modern research investigations of natural medicines used for the prevention and treatment of cardiovascular disease—these investigations provide us with valuable experience and knowledge. Reviewing the history of the development and clinical application of these drugs will promote the development of new drugs.

Ajmaline

Zhi-Hong Yang, Shou-Bao Wang, and Guan-Hua Du

Abstract So far, a total of 91 alkaloids were separated from *Rauwolfia*, and identification of the chemical structures was obtained. Among them, reserpine, ajmaline, and yohimbine have been researched plentifully and thoroughly. Ajmaline, mainly as the class Ia anti-arrhythmic drug, could cause severe hypotension or worsen arrhythmia, so currently ajmaline is almost not applicable in clinic. Many compounds derived from *Rauwolfia* plants have significant and distinct pharmacological effects, just like reserpine as an antihypertensive drug, ajmaline as an anti-arrhythmic drug, and yohimbine as an adrenergic α_2 receptor antagonist. In the foreseeable future, a systematic research on the *Rauwolfia* and its active ingredients will take on an important significance.

Keywords Ajmaline · Anti-arrhythmic · *Rauwolfia*

Alias: Rauwolfine, *Rauwolfia* alkaloid
Origin: *Rauwolfia* (Luo Fu Mu) (Fig. 1)
Chemical name (Fig. 2)

(12R,21R)-Ajmalan-17,21-diol

Molecular formula, $C_{20}H_{26}N_2O_2$; **MW**, 326.43; **CAS**, 4360-12-7

Z.-H. Yang
Institute of Medicinal Plant Development, Chinese Academy of Medical Sciences and Peking Union Medical College, Beijing, China

S.-B. Wang · G.-H. Du (✉)
Beijing Key Laboratory of Drug Targets Identification and Drug Screening, Centre for Pharmaceutical Screening, Institute of Materia Medica, Chinese Academy of Medical Sciences and Peking Union Medical College, Beijing, China
e-mail: dugh@imm.ac.cn

G.-H. Du, *Natural Small Molecule Drugs from Plants*,
https://doi.org/10.1007/978-981-10-8022-7_1

Properties

Appearance: white or yellowish crystal powder. *Melting point*: its anhydride is 205–207 °C. *Optical rotation*: +144° (chloroform). *Solubility*: well soluble in organic solvents, such as methanol, ethanol, etc., and slightly soluble in water. *Storage temperature*: 2–8 °C.

Dosage Form and Indications

Ajmaline is recorded in the *Japanese Pharmacopoeia (17th ed.)*.

Dosage form: tablet (50 mg), injection (50 mg).

Ajmaline has anti-arrhythmic effect for treating atrial premature contraction (APC) or ventricular premature contraction (VPC), paroxysmal supraventricular or ventricular tachycardia, and paroxysmal atrial fibrillation.

Literature

Rauwolfia (Luo Fu Mu) is derived from the plants *Rauvolfia verticillata* (*Lour.*) Baill. and *R. yunnanensis* Tsiang, and the root is usually used as its medicinal part. *R. verticillata* (*Lour.*) *Baill. var. hainanensis Tsiang* and *R. vomitoria Afzel. ex Spreng* have been also used as traditional Chinese herb.

Rauwolfia is one of the important traditional herbs in southern China. Its scope of treatment includes cold, fever, sore throat, headache, and dizziness caused by hypertension, abdominal pain and diarrhea, scabies, and so on. *Rauwolfia* is also used to treat insomniac dizziness, bruises, sprains, and venomous snake bites in folk.

In the "Flora of China," 135 species of *Rauwolfia* were recorded, mostly distributed in the tropics and subtropics, a few in the temperate regions. There are nine species, four varieties, and three cultivated species in China.

History of R&D

Since 1930s, abundant studies have been focused on the *Rauwolfia* plants. So far, a total of 91 alkaloids were separated from *Rauwolfia*, and the identification of the chemical structures was obtained, including 2 non-indole alkaloids (thebaine and papaverine) and 89 indole alkaloids [1]. In accordance with the basic framework, indole alkaloids can be divided into seven classes, such as the ajmaline class, the reserpine class, the yohimbine class, the dehydro-rauwolfine class, the serpentine class, the tetrahydroalstonine class, etc. Among them, reserpine (see No. 95 Reserpine), ajmaline, and yohimbine have been researched plentifully and thoroughly, while studies on other alkaloids are still in progress.

In the early studies of indole alkaloids, three key compounds have been researched and developed for treating cardiovascular diseases.

One of the three key compounds is reserpine. The studies show that reserpine can decrease sympathetic function, blood pressure, and heart rate by reducing the storage of catecholamine and 5-hydroxytryptamine in brain and other tissues, inhibiting the uptake and the synthesis of norepinephrine, a neurotransmitter of sympathetic nerve endings, and promoting the release of norepinephrine. Therefore, reserpine has been developed into a classic drug for hypertension treatment [2]. Although reserpine can be independently used as antihypertensive drug, it is commonly used in combination with diuretics for the treatment of mild to moderate hypertension, because of its weak hypotensive effect and its side effects such as sedation.

The second key compound is ajmaline. Mainly as the class Ia anti-arrhythmic drug, ajmaline also has other pharmacological effects including antivirus, antihypertension, and anti-vasodilation on the coronary artery. The dosage form of ajmaline has been developed into injections, tablets, and other formulations. In the early clinical application, cardiopulmonary resuscitation, temporary pacing, and ajmaline were used together to restore the stability of hemodynamics and terminate arrhythmia. The certain studies have found that ajmaline could cause severe hypotension or worsen arrhythmia, further causing serious ventricular fibrillation or fatal sudden cardiac arrest in patients with the organic heart disease, so currently ajmaline is almost not applicable in clinic.

The last but not the least compound is yohimbine. It is not only a plant-origin drug for treating erectile dysfunction but also a good agent for selective antagonism of epinephrine α_2 receptor [3].

Rauwolfia as the major natural sources of anti-arrhythmic drug (ajmaline) and antihypertensive drug (reserpine) has been a research focus worldwide. Furthermore, the pharmacological studies on *Rauwolfia* are no longer limited to the anti-arrhythmia and antihypertensive effects. A large number of reports on the anti-inflammatory, antibacterial, antiviral, and antitumor effects, inhibition of central nervous system, and reduction in blood glucose and blood lipid have emerged in recent years.

Pharmacology

Anti-arrhythmic

R. serpentina (She Geng Mu) is the first anti-arrhythmic herbal medicine in the psychiatry history. Ajmaline has a potent anti-arrhythmic effect for treating atrial and ventricular arrhythmias and also has a good effect on the treatment of Wolff-Parkinson-White syndrome. Ajmaline plays a main role in decreasing the permeability of sodium ion on the myocardial cell membrane by moderately inhibiting the sodium channel. The electrophysiological effect of ajmaline is similar to that of quinidine, but it is stronger than quinidine.

In addition, ajmaline has a mild anti-sympathetic effect. It affects sympathetic nerve endings to release more sodium, dilate the coronary artery, relax vascular smooth muscle, and then decrease blood pressure. Therefore, it is suitable for clinical treatment of atrial and ventricular premature beats, pre-excitation syndrome with supraventricular tachycardia, but not suitable for the treatment of paroxysmal atrial fibrillation and sinus tachycardia.

Antitumor

Ajmaline and papaverine from *Rauvolfia* plants have a certain degree of inhibition on proliferation of human nasopharyngeal carcinoma KB cells in a concentration-dependent manner. Meanwhile, ajmaline, rescinnamine, and papaverine also have a certain degree of inhibition on the proliferation of human leukemia HL-60 cells in a concentration-dependent manner. It indicates ajmaline maybe has the potential effects on prevention and treatment of leukemia [4].

Pharmacokinetics

Oral absorption of ajmaline is rapid, but with a low bioavailability. It usually takes about 20 min to show efficacy after oral administration and 40–60 min to reach the maximum efficacy. Ajmaline is metabolized mainly in the liver, which is mostly excreted through feces, and a small amount is excreted in the urine.

Drug-Drug Interaction

1. Ajmaline would be increased in blood concentration and prolonged $T_{1/2}$ when administration is combined with quinidine.
2. Ajmaline could slow down the ventricular rate obviously which could cooperate with digitalis to control atrial fibrillation. So both of them have a synergistic effect on atrioventricular conduction [5].

Clinical Application

Ajmaline has the anti-arrhythmic effects for the treatment of atrial or ventricular extrasystoles, paroxysmal supraventricular or ventricular tachycardia, and paroxysmal atrial fibrillation. The lethal dose for human ranged from 100 to 500 mg/kg.

Attention

1. For the treatment of paroxysmal atrial fibrillation, the patient should be under ECG monitoring. When ajmaline is administrated at 50 mg/10 mL at the

speed of 1 mL/min by intravenous injection, the patient's blood pressure should be monitored before and during the drug administration and then continually measured every 15 min after administration. If there is no effect, it can be repeated after 1 h. Furthermore, the patient should stay in hospital for at least 6 h. Once the atrial fibrillation disappears, the treatment should be discontinued.
2. For safety reasons, patients should be hospitalized for medication and given ECG monitoring.
3. Ajmaline should be stored at room temperature away from moisture, heat, and light.

Adverse Reactions

1. Ajmaline could inhibit cardiac conduction, further causing heart block and decreasing myocardial contractility at high doses. It might also cause arrhythmias, coma, and even death. Arrhythmia could accidentally occur even at the normal venous dose.
2. The common side effect events were reported, including vomiting, diarrhea, headache, blurred vision, nystagmus, fatigue, tinnitus, confusion, fall of blood pressure, respiratory depression, lack of granulocytes, hepatotoxicity, etc.

Discussion

Rauwolfia, the traditional Chinese medicine for treating hypertension, have been brought into systematic research of chemistry and pharmacology, finding a number of bioactive compounds and developing different drugs for clinical applications. Researches show that the same plant may contain a variety of compounds with different effects and mechanisms, which may be developed into different drugs, and achieve treatment by the combined pharmacological effects.

Rauwolfia is one of the important folk medicines in the Li national for the treatment of fever, headache, hypertension, traumatic injury, etc. In Hainan province of China, there are five *Rauwolfia* species and one cultivated *Rauwolfia* species, namely, *R. verticillata (Lour.) Baill.*; *R. verticillata (Lour.) Baill. var. hainanensis Tsiang*; *R. tiaolushanensis Tsiang*; *R. serpentina (Linn.) Benth. ex*; *R. teiraphylla Linn.*; and *R. vomitoria Afzel. ex Spreng*. The two *Rauwolfia* species, *R. verticillata (Lour.) Baill. var. hainanensis Tsiang* and *R. tiaolushanensis Tsiang*, are endemically and particularly distributed in Hainan island. Only a few studies on them have been reported. Many medicinal plants in the unusual Hainan's climate and environment have unique pharmacological activities. It is expected that these unique medicinal plants will be a huge and abundant treasure for the research and development of the new drugs.

Many compounds derived from *Rauwolfia* plants have significant and distinct pharmacological effects, just like reserpine as an antihypertensive drug, ajmaline as an anti-arrhythmic drug, and yohimbine as an adrenergic α_2 receptor antagonist. In the foreseeable future, a systematic research on the *Rauwolfia* and its active ingredients will take on an important significance for the further development and reasonable application of these new drugs.

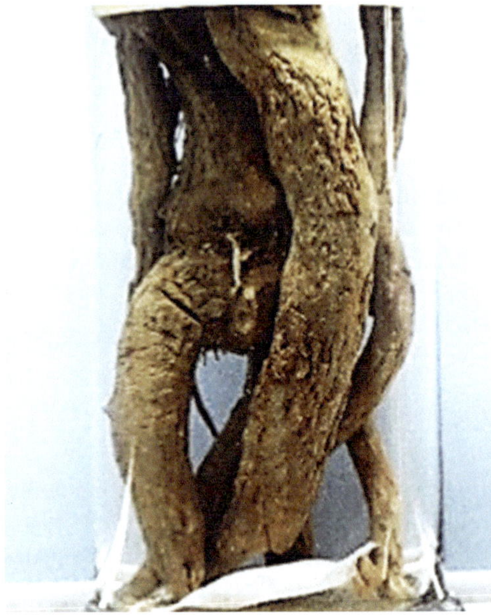

Fig. 1 *Rauwolfia* (Luo Fu Mu)

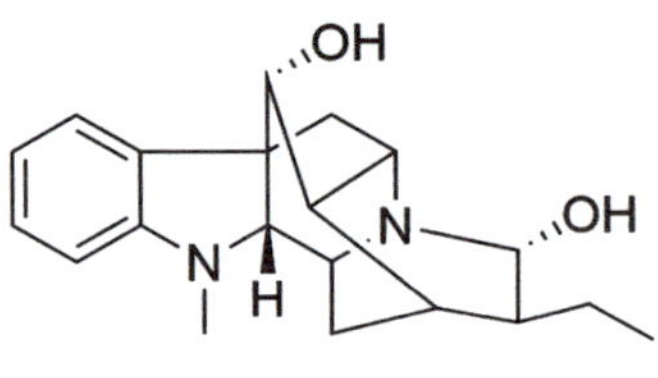

Fig. 2 Chemical structure of ajmaline

References

1. Eurlings M, Lens F, Pakusza C, et al. Forensic identification of Indian snakeroot (Rauvolfia serpentina Benth. Ex Kurz) using DNA barcoding. J Forensic Sci. 2013;58:822–30.
2. Liu YY, Xu QQ, Wang CN, et al. Research situation of pharmacological activity of Nanyao *Rauvolfia*. Chin Pharm J. 2010;45(20):1521–3.
3. Li YJ, Cao FX, Li M, et al. Research progresses of alkaloids pharmacological effects and extraction methods on *Rauvolfia*. Chem Life. 2015;35(2):258–63.
4. Yang XW, Ran FX, Wang RQ, et al. Inhibitory effects of 44 alkaloids compounds against growth of human nasopharyngeal carcinoma cell line KB and human leukemia cell line HL-60 in vitro. Mod Chin Med. 2007;9(1):8–13.
5. Zhang Y. Acute intoxication. Xi'an: The Fourth Military Medical University Press; 2008.

Anisodamine

Hui-Fang Zhang, Lian-Hua Fang, and Guan-Hua Du

Abstract Amisodamine is a kind of alkaloid, which abstracted from anisodus tanguticus. It is a kind of traditional Chinese medicine and used to the treatment of pain. Beijing friendship hospital is the first one use it in clinical practice. With the progress made in clinical reaearch, a lot of pharmacology effects were found, such as cholinolytic effect, calcium antagonist effect, anti-oxygenation effect.

Keywords Amisodamine · Traditional Chinese medicine · Pharmacology effects

Alias: anisodamine hydrobromide; Despun anisodamine hydrpbromide.
Origin: Anisodus tanguticus (Fig. 1)
Chemical name (Fig. 2)

α-(Hydroxymethyl) benzeneacetic acid 6-hydroxy-8-methyl-8-azabicyclo [3.2.1] octan-3-yl ester

Molecular formula, $C_{17}H_{23}NO_4$; **MW**, 305; **CAS**, 17659-49-3

Properties

Appearance: raceanisodamine is a kind of white crystal and crystalline powder. It is odorless and bitter to the taste. *Solubility*: it is easy dissolubility in hydrochloric acid and ethyl alcohol and is soluble in water. *Melting point*: it's melting point is from 103 to 113 °C. The weld spacing is within 6°. *Specific optical rotation*: −9° to +11°.

H.-F. Zhang · L.-H. Fang · G.-H. Du (✉)
Beijing Key Laboratory of Drug Targets Identification and Drug Screening, Centre for Pharmaceutical Screening, Institute of Materia Medica, Chinese Academy of Medical Sciences and Peking Union Medical College, Beijing, China
e-mail: dugh@imm.ac.cn

G.-H. Du, *Natural Small Molecule Drugs from Plants*,
https://doi.org/10.1007/978-981-10-8022-7_2

Derivatives

Anisodamine hydrobromide; Despun anisodamine hydrpbromide.

Dosage Forms

The product is contained in Pharmacopoeia of the People's Republic of China (2015).

Tablet: 5 mg; 10 mg.

Injection: 1 ml, 2 mg; 1 ml,10 mg.

The dosage form contains raceanisodamine tables, raceanisodamine hydrochloride injection, raceanisodamine hydrochloride eye drops, Anisodamine hydrobromide table and Anisodamine hydrobromide injection.

Indications

Anisodamine is a kind of anticholinergic agent. The tablet and injection form are mainly used to relieve smooth muscle spasm, biliary spasm and AOIP in clinical practice. The eye drop is used to treat teenage pseudomyopia; raceanisodamine concomitant with other drugs to treat cluster headache syndrome, renal colic, infantile diarrhea, infantile jaundice hepatitis, infantile purpura, bronchopneumonia, etc.

Literature

Amisodamine is a kind of alkaloid which is mainly living in anisodus tanguticus, which is a M receptor just like Atropine. It is a kind of traditional Chinese medicine which comes from the "National assembly of Chinese Herbal medicine", and it is the alias of zangjia registered in "*Chines herbal medicine summary of shan gan ning qing*". It is a perennial root herb that lives in the bottom slope in the eastern Tibet, Qinghai Province, the Southern GanSu Province, the western SiChuan Province, the northwestward YunNan Province. Since the sources is scare, racemic form of raceanisodamine is the main drug in clinic.

History of R&D

It is the traditional Chinese medicine and the rich folk herbs that create favorable conditions to look for and develop good medicine. Dated from ancient times, anisodus tanguticus used to the treatment of pain. Unfortunately, it can lead to atropine poisoning resulted from large dosage. When found the adverse effects, research made analysis and finally discovered six kinds of ramifications. Besides, they found two types alkaloid anisodamine and anisodine. Compared with atropine, anisodamen has one more hydroxy in tropic ring and named "654".

Beijing friendship hospital is the first one to use it to treat child with fulminant epidemic cerebrospinal meningitis, toxic dysentery, lobar pneumonia, hemorrhagic enteritis and achieved good treatment effect. On account of the limited natural source, the manufacturer synthesis two enantiomer mixture and submitted it to Hangzhou pharmaceutical factory, named 645-2, which has the similar profile action of natural extracts 654-1. Compared with astropine and scopolamine alkaloids, anisodamine has a hydroxy in the six links of parent nucleus hydroxyl radicals, which increases molecular polarity and reduces the fat solubility, which lead it hard to pass BBB, then decrease the central cholinolytic action and increase the peripheral action.

As the usage becomes more and more popular, physicians and researchers find other profile of action, like protection of cells, inhibition of calcium channel, relaxation of smooth muscle and treatment of organophosphorus intoxication. At present, researchers find that it has a good treatment effects in combination with other drugs on nerve headache, infantile dysentery, diarrhea, hepatitis, ureteral calculus.

Pharmacology

1. Cholinolytic effect. Anisodamine can block acetylcholine receptors. We always take it as a merely blocker of M receptor. In recent years, it turns out that anisodamine is also a blocker of N receptor. In the experiment of toad embryo nerve cells and smooth cells, we found that anisodamine can dosage-dependently and reversibility block the transmit between myoneural junctions entirely [1], when used the concentration of 10 mmol/L and continue 15–30 s, which states that anisodamine also can block N receptor.
2. Calcium antagonist effect. Anisodamine will put up Clacium antagonist effect when ischemia, oxygen deficit or other reasons result in irreversible damage and apoptosis. In the research of re-oxygen myocardium damage, before the ischemia reperfusion, ear source intravenous drugs for a 2.213 g/kg, and continue 60 min can restrain the concentration of calcium both free or in organs [2].
3. Increase liquidity in the cell membrane. From the point of molecular structure, anisodamine has position of polar and nonpolar. The polar can combine with fatty acids on the surface of the cell membrane in the way of electrostatic bonding. Nonpolar can be inserted into the membrane phospholipid bilayer, which can increase liquidity in the cell membrane. We used 1,6 diphenyl hexylene as fluorescence probe to detected its effects on red blood cell membrane fluidity. Finally, we found that anisodamine can largely reduce steady-state fluorescence anisotropy and lipid order, when its concentration reached 2 mmol/L [3].

4. Reduce the level of NO. 50 mg/L anisodamine can restrain lipopolysaccharide combined with vascular endothelial cells and promote the production of NO [4] and over-dosage of NO is the pivotal pathologic mechanism of the toxic shock [5]. Pretreatment with anisodamine can reduce the level of NO obviously when cells are damaged and remit toxicity resulted from over-dosage of NO.
5. Anti-oxygenation. anisodamine can restrain myocardial function damage resulted from pyrogallol obviously and can adjust the function of cardioid damaged by ROS [6].

Clinical Application

Anisodamine is a new drug developed by Chinese scientists. It has been widely put into use in clinical practice such as slow reflow phenomenon of ST-elevation acute myocardial infarction patients with percutaneous coronary intervention, infusion leakage [7], neonatal sclerosis. At present, it is mainly used to dilate vessels, remove vasospasm, improve blood circulation, increase tolerance of ischemia hypoxia and reduce the probability of the surrounding tissue necrosis. Besides, 40 mg/L glucose injection can be used to treat bronchial asthma [8], block the vagus nerve, diastole bronchus, decrease respiratory secretion, improve respiratory function and ameliorate bronchial asthma. Combined 654-2 injection and erythromycin can reduce adverse reactions such as the gastrointestinal reaction resulted from respiratory tract infection [9].

Discussion

Anisodamine is a kind of anticholinergic drugs that antagonism peripheral M receptor, and its function is similar with atropine. In clinical, anisodamine are used to treat vascular disease, smooth muscle spasm, neuralgia, toxic shock and fundus oculi disease. At the same time, we also found some features which are superior than atropine and show up its unique advantages. When comes to its adverse effects, they are as follows: lead to constipation by relax gastrointestinal smooth muscle and inhibit the peristalsis; decrease the digestive function by inhibit digestive gland and salivary gland secretion; In addition, some patients appear with the symptom of dry mouth, red face, near vison blurred, near vision blurred, increased heart rate, dysuria.

Fig. 1 Anisodus tanguticus

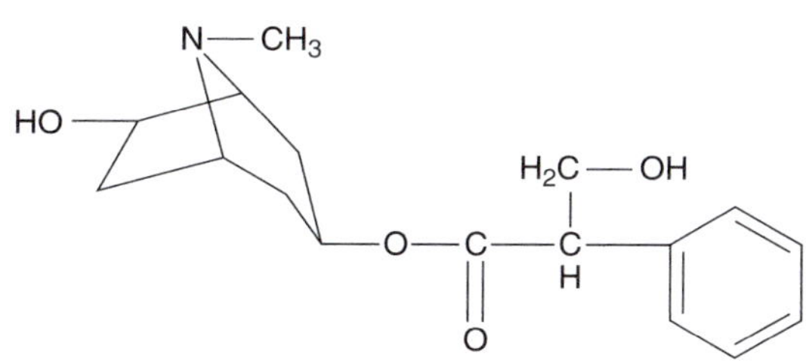

Fig. 2 The structure of anisodamine

References

1. Mei RB, Wang BG, Wang MG. Inhibitory action of aninsodamine on choilnergic transmission of toad sympathetic ganglion in vitro. Chin Pharmacol Bull. 1996;12(5):453–5.
2. Liu XC, Tang WX. An experimental study of the anisodamini hydrobromidum treating rabbit ischemic acute renal failure. Chin J Crit Care Med. 1999;12(6):360–1.
3. Yang GD, Zhou WH, Wang X. Effect of Anisodamine on the fluidity of erythroyte membrane and the intrinsic fluorescence of ghost. Chin J Mod Appl Pharm. 1993;10(1):4–6.
4. Sun J, Yao XJ, Zhao DH. Inhibitory effects of anisodamine on lipopolysaccharides attaching to endothelium and inducing NO release. J Fourth Mil Med Univ. 2001;22(12):1082–4.

5. Tang ZX, Wang XY, Gao H. Changes of nitric oxide metabolism and effects of Anisodamine during Endotoxic shock of goats. Chin J Vet Sci. 1991;19(1):69–72.
6. Esberg LB, Ren J. The oxygen radical generator pyrogallol impairs cardiomyocyte contractile function via a superoxide and p38MAP kinase-dependgdent pathway: protection by anisodamine and tetramethypyrazine. Cardiovasc Toxicol. 2004;4(4):375–84.
7. Fu XH, Fan WZ, Gu XS, et al. Effect of intracoronary administration of Anisodamine on slow reflow phenomenon following primary percutaneous coronary intervention in patients with acute myocardial infatction. Natl Med J China. 2007;120(14):1226–31.
8. He ZY, Kong J. Effect of anisodamine on bronchial asthma. J Clin Exp Med. 2008;7(1):63.
9. Yang B, Yan CM. Anisodamine hydrochloride infection to reduce gastrointestinal reaction of erythromycin, 132 case of clinical analysis. J Ningxia Med Coll. 2006;28(6):534–5.

Caffeic Acid

Wen-Wen Lian and Guan-Hua Du

Abstract Caffeic acid is used for preventing or stopping bleeding during surgery, as well as hemostasis in the departments of medicine, obstetrics, gynecology, etc. It has a wide range of pharmacological effects, including inhibition of platelet aggregation, anti-inflammatory and antioxidant effects, and immune regulation, and it is also an important pharmaceutical product and pharmaceutical intermediate. These make caffeic acid a focus, and scientists are committed to its structural modification trying to find small molecules with better activity for the treatment of oxidative stress-related diseases.

Keywords Caffeic acid · *Solidaginis herba* · Hemostasis

Alias: 3, 4- two hydroxy cinnamic acid
Origin: *Solidaginis herba* (Fig. 1)
Chemical name (Fig. 2)

(2E)-3-(3, 4-Dihydroxyphenyl)-2-propenoic

Molecular formula, $C_9H_8O_4$; **MW**, 180.15; **CAS**, 331-39-5

Properties

Appearance: yellow crystal. The crystal from the concentrated solution does not contain crystal water, and the crystal from the dilute solution contains one molecule crystal water. *Melting point*: 223–225 °C. *Solubility*: It is slightly soluble in cold

W.-W. Lian · G.-H. Du (✉)
Beijing Key Laboratory of Drug Targets Identification and Drug Screening, Centre for Pharmaceutical Screening, Institute of Materia Medica, Chinese Academy of Medical Sciences and Peking Union Medical College, Beijing, China
e-mail: dugh@imm.ac.cn

G.-H. Du, *Natural Small Molecule Drugs from Plants*,
https://doi.org/10.1007/978-981-10-8022-7_3

water but soluble in hot water, cold ethanol, and ethyl acetate. The basic solution is orange-red. Ferric chloride solution was dark green.

Dosage Forms

This product is contained in the *British Pharmacopoeia* (2017) and *European Pharmacopoeia* (9.0th ed.). The dosage form is tablet, with 100 mg per tablet.

Indications

It is used for preventing or stopping bleeding during surgery, as well as hemostasis in the department of medicine, obstetrics and gynecology, etc. It is also used for various causes of neutropenia and thrombocytopenia.

Literature

Caffeic acid is abundant in the whole plant of *Solidago decurrens Lour.* (Yi Zhi Huang Hua) [1], fruit of *Crataegus pinnatifida Bge. var. major N.E.Br.* (Shan Li Hong), *Salix myrtillacea Anderss.* (Po Liu), rhizome of *Cimicifuga foetida L.*, rhizome of Polypodiaceae *Polypodium vulgare L.* (Ou Ya Shui Long Gu), peel of Rutaceae *Citrus limonum* (Ning Meng), the whole plant of Polygonaceae *Polygonum aviculare L.* (Pian Xu), root of *Valeriana officinalis L.* (Xie Cao), the whole plant of *Thymus mongolicus Ronn* (She Xiang), leaves of *Eucommia ulmoides* (Du Zhong), and other herbal plants. It is a kind of polyhydroxy styrene acid, with the general chemical properties of phenolic acid. It is easily oxidized for the reason of the unsaturated double bonds, particularly unstable in alkaline solution.

Caffeic acid has both cis and trans isomers, and the two isomers of caffeic acid have a mutual transformation in plants, which may regulate some important physiological process. Caffeic acid exists in plants in the main form of complexes; free state accounts for a few proportion [2].

History of R&D

Caffeic acid is an important pharmaceutical product and pharmaceutical intermediate and is synthesized mainly through chemical synthesis, biosynthesis, microwave chemical synthesis, and other ways. Its chemical synthesis methods include protocatechuic aldehyde and malondialdehyde condensation, vanillin and malondialdehyde condensation, methyl vanillin and acetic acid condensation, and piperonal and malondialdehyde or acetic acid condensation, and the protocatechuic aldehyde and malondialdehyde condensation is the main method. The synthesis of caffeic acid and its intermediate derivatives (ferulic acid) is mainly accomplished by Knoevenagel-Doebner condensation reaction. Solvent-free microwave irradiation technology works well in this reaction and is the main research of caffeic acid synthesis. This technology-related equipment needs further improvement, limiting its promotion in production. The biosynthesis of caffeic acid includes the hydrolysis of chlorogenic acid and the biotransformation of coumarin. The biosynthesis has the advantages of low cost, high conversion rate, and environmental protection, but the technology needs further research [3].

The bioactivity of caffeic acid and its derivatives make it a focus, and scientists are committed to its structural modification trying to find small molecules with better activity. Considering the importance of the adjacent hydroxyl on the benzene ring of caffeic acid, the current structural modification mainly focuses on the carboxyl, including the synthesis of amine salts, amides, and ester compounds.

Pharmacology

Caffeic acid has a wide range of pharmacological effects, including inhibition of platelet aggregation, anti-inflammatory, antioxidant, and immune regulation. Studies have shown that caffeic acid at 10 mg/kg can significantly inhibit platelet aggregation in rats [4]; caffeic acid at 10 μg/ml can significantly reduce the inflammatory response in LPS-induced endothelial cells [5]; caffeic acid at 100 nM and 1 μM significantly inhibited oxidative stress-induced endothelial injury [6]; caffeic acid at 10 mg/kg can significantly inhibit hyperhomocysteinemia-induced white blood cell rolling and adhesion in mouse brain vein [7]; caffeic acid at 10 mM can significantly inhibit the cell proliferation in cervix cancer [8].

Many of the natural or synthetic derivatives of caffeic acid also have pharmacological effects similar to caffeic acid, such as chlorogenic acid and caffeic acid. Phenethylester is the main lignan in plants, with antioxidant, anti-inflammatory, immunomodulatory, and other activities [9, 10].

Clinical Application

It is used for preventing or stopping bleeding during surgery, as well as hemostasis in the department of medicine, obstetrics and gynecology, etc [11]. It is also used for various causes of neutropenia and thrombocytopenia [12].

Caffeic acid can improve the coagulation function and reduce bleeding, by shrinking blood vessels, improving the function of coagulation factors, and increasing the number of white blood cells and platelets. In the course of chemotherapy, it can increase platelet production, reduce the incidence of thrombocytopenia, increase the generation of white blood cells, synergistic chemotherapy, as well as improve patient quality of life and survival time.

Discussion

Due to the widespread presence of caffeic acid and the ease of structural modification, it is currently modified for the treatment of oxidative stress-related diseases such as anti-inflammatory, neuroprotective, and chemotherapeutic agents, in which the ester compound and amino-based compounds exhibit a series of biological activities. At present, caffeic acid and its derivatives are a series of stabilizers in drug formulation, and they are widely involved in the treatment of many diseases for their extensive biological activities. Small molecules and drug treatment networks developed from caffeic acid may produce new and effective therapeutic agents [13].

Fig. 1 *Solidaginis herba*

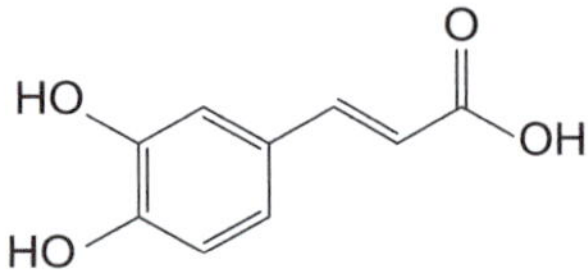

Fig. 2 The structure of caffeic acid

References

1. Wang Y. Analysis of pharmacological action of Solidago canadensis. Chin J Ethnomedicine Ethnopharmacy. 2016;16:150.
2. Macheix JJ, Fleuriet A, Billot J. Fruit phenolic. Boca Raton: CRC Press; 1990.
3. Xin Z, Zhang L. The synthesis study of caffeic acid. Shandong Chem Ind. 2014;43(3):39–43.
4. Anwar J, Spanevello RM, Pimentel VC, et al. Caffeic acid treatment alters the extracellular adenine nucleotide hydrolysis in platelets and lymphocytes of adult rats. Food Chem Toxicol. 2013;56:459–66.
5. Liu M, Song S, Li H, et al. The protective effect of caffeic acid against inflammation injury of primary bovine mammary epithelial cells induced by lipopolysaccharide. J Dairy Sci. 2014;97(5):2856–65.
6. Migliori M, Cantaluppi V, Mannari C, et al. Caffeic acid, a phenol found in white wine, modulates endothelial nitric oxide production and protects from oxidative stress-associated endothelial cell injury. PLoS One. 2015;10(4):e0117530.
7. Zhao HP, Feng J, Sun K, et al. Caffeic acid inhibits acute hyperhomocysteinemia-induced leukocyte rolling and adhesion in mouse cerebral venules. Microcirculation. 2012;19(3):233–44.
8. Chang WC, Hsieh CH, Hsiao MW, et al. Caffeic acid induces apoptosis in human cervical cancer cells through the mitochondrial pathway. Taiwan J Obstet Gynecol. 2010;49(4):419–24.
9. Upadhyay R, Mohan Rao LJ. An outlook on chlorogenic acids-occurrence, chemistry, technology, and biological activities. Crit Rev Food Sci Nutr. 2013;53(9):968–84.
10. Mai FT, Azab SS, Khalifa AE, et al. Caffeic acid phenethyl ester, a promising component of propolis with a plethora of biological activities: a review on its anti-inflammatory, neuroprotective, hepatoprotective, and cardioprotective effects. IUBMB Life. 2013;65(8):699–709.
11. Li P, Li Q. Anti-angina pectoris active ingredient–protocatechuic aldehyde. Chem World. 1991;7:334–5.
12. Wu L. New hemostatic and leukopenic–caffeic acid and caffeine amines. J Nanjing Ins Pharm. 1980;2:24.
13. Silva T, Oliveira C, Borges F. Caffeic acid derivatives, analogs and applications: a patent review (2009–2013). Expert Opin Ther Pat. 2014;24(11):1257–70.

Cyclovirobuxine

Xiao-Li He, Shou-Bao Wang, and Guan-Hua Du

Abstract Cyclovirobuxine is an alkaloid extracted from common boxwood (*Buxus microphylla* Sieb. et Zucc. var. *sinica* Rehd. et Wils) and its genus. Its chemical structure is a derivative of pregnane. Cyclovirobuxine is a new Chinese medicine that is successfully developed for the treatment of cardiovascular diseases in recent years. After decades of clinical observation, cyclovirobuxine has the therapeutic role of anti-myocardial ischemia and antiarrhythmia and protects against acute cerebral ischemia. In addition, cyclovirobuxine can cross the blood-brain barrier and improve brain microcirculation and oxygen supply to treat cerebral arteriosclerosis insufficiency. In all, it has potent curative effects and low risk of side effects. The safety of oral administration of cyclovirobuxine is excellent, while the risk of side effects in long-term use is low, which indicates that this drug has a good application prospect.

Keywords Cyclovirobuxine · Alkaloid · Anticardiovascular · Anticerebrovascular

Alias: Cyclovirobuxine D, bebuxine, Cyclovirobuxin D, CVB-D, NSC 91722
Origin: Common boxwood (Fig. 1)
Chemical name (Fig. 2)

(3β,5α,16α,20S)-4,4,14-Trimethyl-3,20-bis(methylamino)-9,19-cyclopregnan-16-ol

Molecular formula, $C_{26}H_{46}N_2O$; **MW**, 402.36; **CAS**, 860-79-7

X.-L. He
Institute of Medicinal Plant Development, Chinese Academy of Medical Sciences and Peking Union Medical College, Beijing, China

S.-B. Wang · G.-H. Du (✉)
Beijing Key Laboratory of Drug Targets Identification and Drug Screening, Centre for Pharmaceutical Screening, Institute of Materia Medica, Chinese Academy of Medical Sciences and Peking Union Medical College, Beijing, China
e-mail: dugh@imm.ac.cn

G.-H. Du, *Natural Small Molecule Drugs from Plants*,
https://doi.org/10.1007/978-981-10-8022-7_4

Properties

Appearance: colorless needle crystals. *Solubility*: sparingly soluble in acetone; slightly soluble in water; freely soluble in chloroform; soluble in methanol and ethanol. *Melting point*: 219–222 °C.

Derivatives

At present, there are few studies on the structural modification of cyclovirobuxine.

Dosage Forms

Huangyangning tablets are recorded in the *Pharmacopoeia of the People's Republic of China* (2015).

Indications

It is mainly used for the treatment of cardiovascular and cerebrovascular diseases in China, such as the coronary heart disease, arrhythmia, cerebral arteriosclerosis, cerebral embolism and brain vascular accident which are caused by insufficiency of cerebral blood supply.

Literature

Cyclovirobuxine (CVB-D) is an anticardiovascular and anticerebrovascular alkaloid derived from *xiaoyehuangyang* 小叶黄杨 (*Buxus microphylla* Sieb. et Zucc. var. *sinica* Rehd. et Wils) and its genus. Its chemical structure is a derivative of pregnane. Cyclovirobuxine is a potent medicine for the treatment of cardiovascular and cerebrovascular diseases in China.

Firstly, cyclovirobuxine was obtained originally from *Buxus sempervirens* L. (jinshuhuangyang) [1] and then mainly from the *Buxus microphylla* Sieb. et Zucc. var. *sinica* Rehd. et Wils (xiaoyehuangyang) which was widely distributed in Anhui, Zhejiang, Jiangxi and Hubei. It is used in folk medicine for toothache, chest and abdominal bloating, hernia, bruises, hot boils and so on. Nowadays, pharmacological experiments have proved that cyclovirobuxine has significant effects on the heart [2].

History of R&D

Cyclovirobuxine in the treatment of coronary heart disease began in 1969 [3]. The medical teams of Chinese People's Liberation Army No.86489 explored a treatment of coronary heart disease named Guo, prescription composed of *Buxus microphylla* (huangyangmu), *Salvia miltiorrhiza Bge.* (danshen), Ligustici Chuanxiong Rhizoma (chuanxiong), Belamcandae Rhizoma (shegan), Radix Aristolochiae (qingmuxiang) and Asari Radix et Rhizoma (xixin). In the past 10 years, they have developed these six herbs into compound huangyangning tablets to treat some patients with coronary heart disease in Nanjing, Wuhu and other areas and found that these tablets have a better therapeutic effect on rheumatic heart disease and coronary heart disease. Moreover, in 1974, Sun Ruiyuan of Wannan Medical College found a strong heart effect of the water decoction of *Buxus microphylla* (huangyangmu) on frog heart and rabbit heart perfusion, which brought in the systematic research of this plant in heart disease.

In 1977, three kinds of alkaloids have been isolated from *Buxus microphylla* (huangyangmu), i.e. buxine No. 1, 2 and 3. After a series of pharmacological experiments, buxine No.1 was proved as the most effective of the three, which had significant cardiac stimulant action, increased coronary blood flow and promoted anoxia tolerance in animals. Then buxine No.1 was named as the known compound cyclovirobuxine, and its chemical properties, structure identification and pharmacological activity were further studied.

In 1978, the Anhui Huangyang Research Cooperation Group carried out systematic experiments to determine the chemical structure and the separation and identification method of cyclovirobuxine by melting point determination, thin-layer chromatography, infrared spectroscopy, mass spectrometry and NMR [4]. Three studies containing more than 300 cases of coronary heart disease patients treated with cyclovirobuxine showed that this drug can significantly improve angina, chest tightness, arrhythmia and other symptoms caused by coronary heart disease. Fifty-two cases of inpatients treated with this drug or dipyridamole as control group were studied by the double-blind method. The results confirmed that the drug could significantly improve cardiac dysfunction caused by coronary heart disease [5].

Pharmacology

Pharmacological studies have shown that cyclovirobuxine has a positive inotropic effect on the heart that may predominantly be due to the inhibition of cardiomyocyte membrane Na^+-K^+-ATPase activity and promotion of myocardial extracellular Ca^{2+} influx and cardiomyocyte Ca^{2+} release. Moreover, cyclovirobuxine could markedly reduce myocardial oxygen consumption and increase coronary blood flow, suggesting that it has definite anti-myocardial ischemia effect. Experiments also showed that cyclovirobuxine could induce marked inhibition of myocardial ischemia infarction and enhancement of anoxia tolerance in mice [6, 7].

Besides, cyclovirobuxine could have protective effect on the acute experimental cerebral ischemia in mice by bilateral ligation of common carotid arteries, prolong the life span of mice, increase blood flow and decrease the formation of thrombus during cerebral ischemia [8]. In vitro experiments have also shown that cyclovirobuxine has neuroprotective effects on neurons and restrains PC12 cells from excitatory amino acid-induced injury [9].

Clinical Application

After decades of clinical observation, cyclovirobuxine has the therapeutic role of anti-myocardial ischemia and antiarrhythmia and protects against acute cerebral ischemia. In addition, cyclovirobuxine can cross the blood-brain barrier and improve brain microcirculation and oxygen supply to treat cerebral arteriosclerosis insufficiency. In all, it has potent curative effects and low risk of side effects.

Discussion

Regarding the chemical structure, cyclovirobuxine is a derivative of pregane and can be used in a variety of cardiovascular diseases such as arrhythmia, angina pectoris, coronary heart disease, cardiac insufficiency and so on. The safety of oral administration of cyclovirobuxine is excellent, while the risk of side effects in long-term use is low, which indicates that this drug extracted from the natural plant has a good application prospect. Cyclovirobuxine is also included in the *Pharmacopoeia of the People's Republic of China* 2000 edition, 2005 edition, 2010 edition and 2015 edition.

Fig. 1 Common boxwood

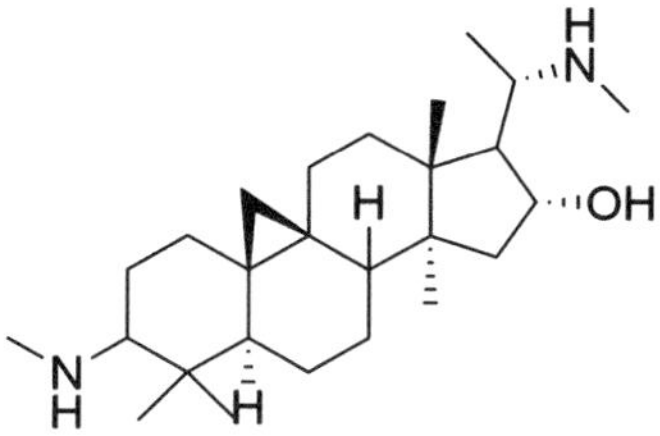

Fig. 2 The structure of cyclovirobuxine

References

1. Brown KS Jr, Morris Kupchan S. Buxus alkaloids. VI. The constitution of Cyclovirobuxine-D. Tetrahedron Lett. 1964;5(39):2895–900.
2. Fang TH, Wu BY, Yao MH, et al. The experimental research of Buxine no.1 on acute myocardial ischemia. Acta Acad Med Wannan. 1979;10:18–34.
3. Wang LZ, Shan WD. The isolation and identification of Cyclovirobuxine D from *Buxus microphylla* Sieb. Zucc. var. *sinica* Rehd. et Wils (Xiaoyehuangyang). Study Chin Tradit Pat Med. 1979;4:24–6.
4. Wang LZ, Shan WD. The isolation and identification of Cyclovirobuxine D from *Buxus microphylla* Sieb. Zucc. var. *sinica* Rehd. et Wils (Xiaoyehuangyang). Jiangsu Med J. 1979;10:14–6.
5. The Anhui Huangyang Research Cooperation Group. The experimental research of Buxine no. 1 on coronary heart disease. Acta Acad Med Wannan. 1984;3:50–3.
6. Liang T, Fang TH, Yao XJ, et al. The pharmacological research progress of Cyclovirobuxine D. Pharm J Chin People's Liberation Army. 2001;17(1):35–8.
7. Yu MQ, Xu L, Fang TH, et al. Pharmacology and toxicology overview of Cyclovirobuxine D. Anhui Med Pharm J. 2008;12(10):885–7.
8. Fang TH, Xu HQ. Protective action of Huangyangning against acute experimental cerebral ischemia in mice. Chin Tradit Herb Drug. 1997;7:413–4.
9. Yuan DP, Long J, Fang TH. Protective effect of CBV-D on Glu-induced injury in cultured PC12 cells. Chin J Integr Med Cardiol-/Cerebrovas Dis. 2004;2(1):37–8.

Daidzein

Zi-Ran Niu, Lian-Hua Fang, Gui-Fen Qiang, and Guan-Hua Du

Abstract Daidzein mainly comes from leguminous plants, the seeds of soybean, red clover grass, or all *Pueraria* roots. Daidzein is a kind of isoflavone compound. It is mainly used for the treatment of hypertension, coronary heart disease, cerebral thrombosis, and vertigo and aids in the treatment of sudden deafness. It can also treat women with menopause syndrome. It can be used to treat osteoporosis of menopause women. It can relax the artery and increase blood flow. It also has certain antihypertensive effects and inhibits the proliferation of endometrial cancer and breast cancer cells. It has good prospective effects in the treatment of hypertension, coronary heart disease, cerebral thrombosis, vertigo, and osteoporosis of menopause women.

Keywords Daidzein · Isoflavone · Hypertension

Origin: Soybean (Fig. 1)
Chemical name (Fig. 2)

4′,7-Dihydroxyisoflavone

Molecular formula, $C_{15}H_{10}O_4$; **MW**, 254.24; **CAS**, 486-66-8
Derivatives (Fig. 3)

Properties

Appearance: pale-yellow prismatic crystal. *Melting point*: 315–323 °C. *Solubility*: soluble in ethanol and ether

Z.-R. Niu · L.-H. Fang · G.-F. Qiang · G.-H. Du (✉)
Beijing Key Laboratory of Drug Targets Identification and Drug Screening, Centre for Pharmaceutical Screening, Institute of Materia Medica, Chinese Academy of Medical Sciences and Peking Union Medical College, Beijing, China
e-mail: dugh@imm.ac.cn

G.-H. Du, *Natural Small Molecule Drugs from Plants*,
https://doi.org/10.1007/978-981-10-8022-7_5

Dosage Forms

Daidzein was included in the *British Pharmacopoeia (*2017) and *European Pharmacopoeia* (9.0th ed.). The dosage form is tablets.

Indications

Daidzein is mainly used for the treatment of hypertension, coronary heart disease, cerebral thrombosis, and vertigo and aids in the treatment of sudden deafness. It can also treat women with menopause syndrome.

Literature

Daidzein mainly comes from leguminous plants, the seeds of soybean (dadou), red clover grass, or all *Pueraria* roots. Its medicinal value was first recorded in Shen Nong Ben Cao Jing 2838 BC. Daidzein, one of the main soy isoflavones, is also the main component of radix puerariae (Gegen) which is the dry root of kudzu leguminous plants. *Radix puerariae* can reduce fever, produce saliva, and relieve diarrhea. Its chemical composition is complicated, including puerarin, xyloside, soy flavonoids, soybean flavonoid glycosides, beta-sitosterol, daidzin, daidzein. The main effective components are puerarin and daidzein [1].

History of R&D

Daidzein is a kind of isoflavone compound, which was first synthetized by researchers in China. It has been widely used in drugs, food supplements, and cosmetics. Because of two phenolic hydroxyl structures, Daidzein has poor water solubility, poor liposolubility, and strong first-pass effect, leading to the low bioavailability of oral absorption, which limits its widely clinical usage.

Ipriflavone is a kind of isoflavone modified from Daidzein, which has been used for the treatment of osteoporosis in Japan and some European countries. The effects of a series of amino alkoxy derivatives of ipriflavones on inhibiting the bone absorption were evaluated. The researchers found that 7-amino alkoxy derivative works best. When Daidzein was alkylated or acylated at 7-hydroxyl selectively, the stability was increased, and thus the proliferation of MCF-7 cell was inhibited [2].

Pharmacology

Daidzein has many kinds of pharmacological effects, such as anticancer, cardiovascular protection, estrogen- and antiestrogen-like effects, antiosteoporosis, antioxidation, improving immunity, and affecting the endocrine system. More attention has been paid by domestic and international pharmaceutical and food industries. Daidzein has obviously antibacterial effect on *Staphylococcus aureus* and *Escherichia coli*. It can also increase the weight of immune organs in mice and has anti-arrhythmic effect. The chemical structure of daidzein is very similar to the endogenous estrogen, so the estrogen-like effect is used to treat menopausal syndrome and increase the levels of osteocalcin (BGP) and also the bone mineral deposits. The clinical efficacy is similar to estrogen replacement therapy (ERT). Daidzein doesn't induce the high expression of estrogen. It has effects on the osteoblast to reduce the bone absorption of osteoclast, thus maintains the dynamic balance of osteoblast and osteoclast, finally reduces the risk of fracture. So it is safe for usage [3]. Daidzein can also increase the bone mineral density (BMD) and bone mineral content (BMC) of the lumbar spine, the number of trabecular bone, and bone volume fraction, improve the bone microstructure, and thus prevent the reduction of femur biomechanics in glucocorticoid-induced osteoporosis in the rat.

Daidzein has an anti-hypoxia effect. The study showed that Daidzein could significantly prolong the survival time of mice in hypoxia tolerance test under normal pressure and after subcutaneous injection of isoproterenol, suggesting that Daidzein has the significant anti-hypoxia effects [4]. Daidzein plays a protective role in myocardial hypertrophy induced by isoproterenol in rat probably by the antioxidative effects [5]. Similarly, Daidzein may protect the ischemia-reperfusion injury in rats by increasing the antioxidative capacity [6]. Daidzein can significantly inhibit the proliferation of two human breast cancer cells (MCF-7 and MDA-MB-231) in vitro with the significant dose-dependent and time-dependent effects. Daidzein can markedly decrease the colony-forming ability, suggesting that Daidzein may have the effect of preventing and treating breast cancer [7].

It was found that Daidzein has the obviously preventive effect on chloroform-induced ventricular fibrillation in mice, therapeutic effect on aconitine-induced arrhythmia in rats, as well as protective effect on the adrenaline-induced arrhythmia in rabbit. Daidzein can significantly reduce the action potential amplitude of sciatic nerve in toad in vitro. All of the above effects were obviously dose-dependent, suggesting Daidzein has the significant anti-arrhythmic effects [8].

Clinical Application

Daidzein can expand the coronary artery, femoral artery, and cerebral artery, increase cerebral blood flow and limb blood circulation, reduce blood viscosity and vascular resistance, decrease myocardial oxygen consumption, improve heart function, increase the microcirculation and blood flow to the tip, lower the blood pressure, and adjust the heart rhythm. Daidzein can be used for the treatment of hypertension, coronary heart disease, angina pectoris, myocardial infarction, cerebral thrombosis, dizziness, and sudden deafness. It can also be used for women's menopause syndrome.

Discussion

Soybean flavin is an important nutrient from soybeans. Similar to endogenous estrogen, it can be used to treat osteoporosis of menopausal women. It can relax the artery and increase blood flow. It also has certain antihypertensive effect and inhibits the proliferation of endometrial cancer and breast cancer cells. It has good prospective effects in the treatment of hypertension, coronary heart disease, cerebral thrombosis, vertigo, and osteoporosis of menopause women.

Fig. 1 Soybean

Fig. 2 Structure of daidzein

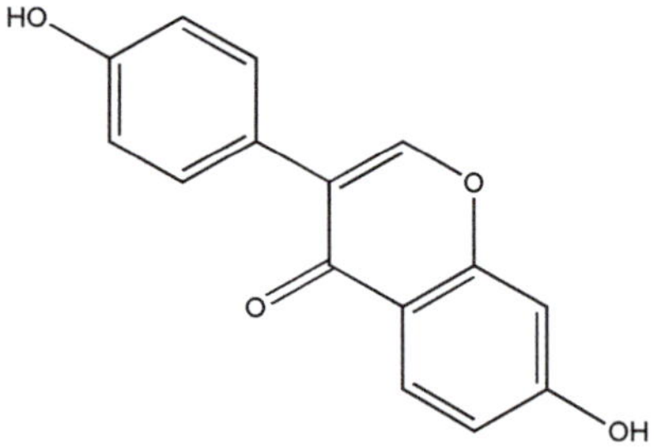

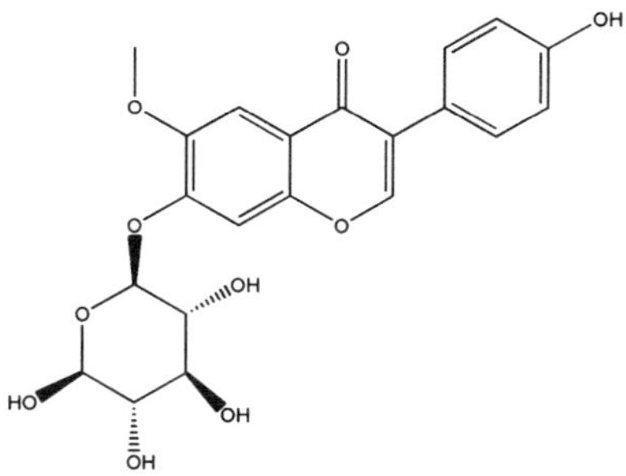

Glycitin

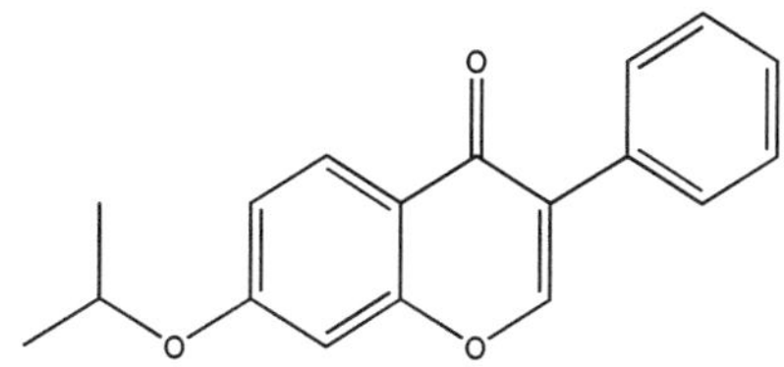

Ipriflavone

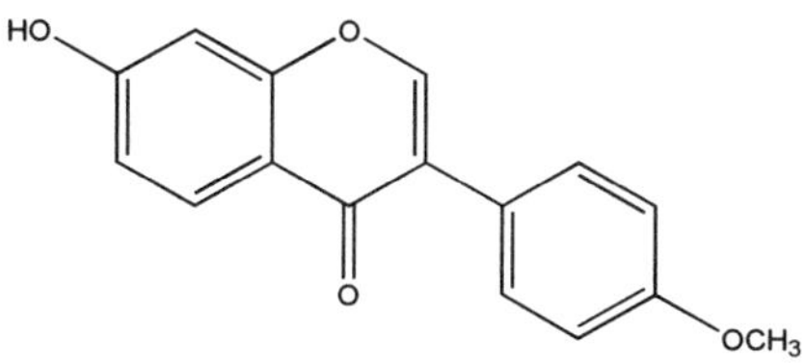

Formononetin

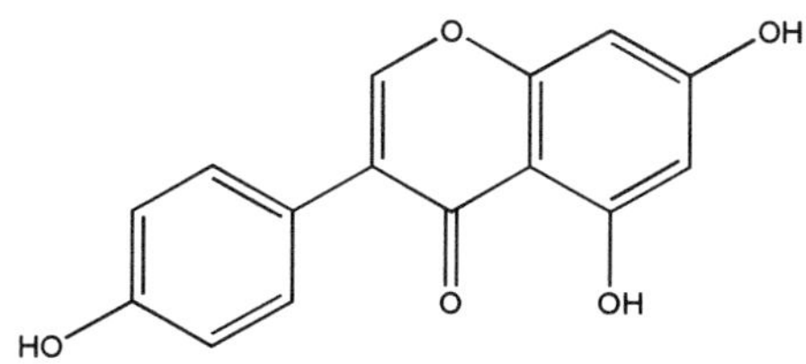

Genistein

Fig. 3 The structure of daidzein derivatives

References

1. Wang SQ, Yao CS, Chen JM. The pharmacokinetic study of the main constituents of gegan. Chin Tradit Herb Drug. 1996;27(11):696.
2. Lu JR, Gui L, Sha L. The synthesis and bioactive study of soybean glycogen derivatives with selective estrogen-receptor modulating activity. Chin J Org Chem. 2011;31(11):1852–63.
3. Gao M, Jin BQ, Zhang H. The effects of soybean side on the ovariate oophora, vitamin D and bone mineral deposits. Chin J Gerontol. 2009;6(29):1480–2.
4. Zeng J. The study of the effects of anoxia of soybean. China J Mod Med. 2004;2(6):453–6.
5. Cao XL. The protective effect of soybean glycoside on cerebral ischemia reperfusion injury. Chin J Pharmacol Toxicol. 2011;9(25):27.
6. Zhong S, Li ZL. The effect of soybean glycogen on the fat and antioxidant properties of myocardial hypertrophy. Chin J Exp Tradit Med Formulae. 2013;19(9):204–6.
7. He FJ. The research on the prevention and treatment of soybean isoflavones. Beijing: Beijing University of Chinese Medicine; 2002.
8. Ye HY, Qiu F, Zeng J. The study of the effect of the anti-cardiac arrhythmia of soybean. China J Chin Mater Med. 2003;28(9):853.

Daphnetin

Lan Sun, Rui Zhao, Xiu-Ying Yang, and Guan-Hua Du

Abstract Daphnetin is mainly extracted from Yu Ruixiang Changbai *Daphne*. Daphnetin has the pharmacological effects on cardiovascular system, central nervous system, and immune system, which made it mainly used for treatment of coronary heart disease, occlusive vascular disease, and Buerger's disease in clinical. In the 1970s, the artificial synthesis of daphnetin, which promoted R&D, makes it a landmark of new drug discovery in China. Daphnetin has extensive pharmacological effects, but pharmacological researches are still based on animal experiments. Pharmacodynamics, pharmacokinetics, and clinical research for other indications are needed as soon as possible.

Keywords Daphnetin · Changbai daphne · Coronary heart disease

Alias: Daphnetin, histidine, daphnetin
Origin: *Daphne aurantiaca* Diels (Fig. 1)
Chemical name (Fig. 2)

7,8-Dihydroxycoumarin

Molecular formula, $C_9H_6O_4$; **MW**, 178.14; **CAS**, 486-35-1

L. Sun · R. Zhao · X.-Y. Yang · G.-H. Du (✉)
Beijing Key Laboratory of Drug Targets Identification and Drug Screening, Centre for Pharmaceutical Screening, Institute of Materia Medica, Chinese Academy of Medical Sciences and Peking Union Medical College, Beijing, China
e-mail: dugh@imm.ac.cn

G.-H. Du, *Natural Small Molecule Drugs from Plants*,
https://doi.org/10.1007/978-981-10-8022-7_6

Properties

Appearance: white or off-white powder, odorless, and tasteless; *Solubility*: slightly soluble in methanol (equilibrium solubility of 485.4 μg/mL), slightly soluble in ethanol (equilibrium solubility of 230.06 μg/mL), and insoluble in water (equilibrium solubility of 176.7 μg/mL) [1]; *Melting point*: 262–264 °C.

Dosage Forms

Capsule

Indications

This product is contained in national standards for chemical drugs, mainly used in Buerger's disease, occlusive vascular disease, and coronary heart disease.

Literature

Daphnetin, mainly extracted from Yu Ruixiang Changbai *Daphne* (Daphne Korean Nakai), *Daphne giraldii* Nitsche bark, and root bark, is the main efficacy component of the plants *D.tangutica Maxim* and *D.retttsa Hemsl.* Changbai daphnes with a distribution in needle broad-leaved forest and coniferous forest between 600 and 1800 m above sea level was firstly recorded in *Xinhua Medica Outline*. The domestic main growth is in the northeast Antu in Jilin province, Liaoning province, and other places. The foreign distribution is in North Korea and the Russian Far East Wusuli River. Changbai daphnes has the pharmacological effects of warming the spleen and stomach, relaxing the muscles and joints, and promoting circulation. In TCM, it's commonly used as medicines for treatment of injuries. Because *Changbai daphne* is a pungent and warm natured herbal medicine, it was used to warm the spleen and stomach by dispelling cold.

History of R&D

Changbai daphne, a Chinese traditional medicine in clinic, is mainly used for treatment of coronary heart disease, rheumatoid arthritis, thromboangiitis obliterans, etc. Changbai daphne belongs to the first-class protective plants in Jilin province in

China, and the rarity restricted the drug development. Scientists tried to find the effective components of Changbai daphne. In 1977, scientists of phytochemistry research group in the Traditional Chinese Medicine research laboratory from Jilin institute of Chinese medicine first successfully isolated the crystalline monomer – daphnetin – from the plant. Pharmacological results show that the pharmacological activities of this crystal are equal to Changbai daphne, which suggested the crystal is the main active ingredient [2]. Through the condensation reaction experiment using equimolecular pyrogallic acid, malic acid, and twice the amount of sulfuric acid, under heating reaction, scientists successfully got pale yellow needle daphnetin crystals. From then on artificially synthesized daphnetin has been successfully implemented [3]. A large-scale industrial production can be acquired. The follow-up researches on the metabolism and pharmacological activity of daphnetin were carried out. In 2009, the State Food and Drug Administration of China approved daphnetin capsule to be used in the clinical.

Pharmacology

Daphnetin pharmacological effect mainly includes:

1. Improvement of the cardiovascular system function [4, 5]:

(a) Inhibits prostaglandin synthesis and platelet aggregation, which reduces experimental arterial thrombosis formation of daphnetin in rats through intravenous injection at the dose of 80 mg/kg
(b) Inhibits the pituitrin-induced acute myocardial ischemia in rabbit by intravenous daphnetin at the dose of 10 mg/kg
(c) Improves egg yolk-caused mice hypercholesteremia at the dose of 800 mg/kg/d
(d) Prolongs beheaded mice mouth action duration and extends the survival time of mice subjected to KCN or $NaNO_2$ poisoning by oral administration of daphnetin at the dose of 300 mg/kg and 600 mg/kg
(e) Antagonizes against the effect of isopropyl adrenaline on oxygen consumption and survival time in the normal pressure hypoxia mice and slows down the body's oxygen consumption rate under the abnormal circumstances by oral administration daphnetin at the dose of 600 mg/kg
(f) Has potential effects on type 2 diabetic rats and atherosclerosis rabbits
(g) Reduces the damage of hippocampal neurons by increasing the expression level of GAP – 43 mRNA and protein in the ischemia-reperfusion rats.

2. Effects on the central nervous system [6]:

(a) Abirritation.
(b) Daphnetin inhibits egg white- and dextran-induced rat voix pedis swelling by oral administration at the dose of 400 mg/kg. The anti-inflammatory mechanisms, which are simulating to dose with salicylic acid, may be through the dynamic balance of pituitary ACTH in hypothalamus neurohumoral or through

the inhibition of ring oxidase and 5-ester oxidase activity, as well as the inhibition of prostaglandin and leukotriene synthesis.

(c) When the intraperitoneal injection of daphnetin is at the dose of 25 mg/kg and 50 mg/kg, the spontaneous activity in mice has not obviously changed; when the dose increased to 100 mg/kg, daphnetin significantly reduces the spontaneous activity in mice; when dose continues to increase to 200 mg/kg, mice appear to be quiet, no activities, and eyelids droop (mice righting reflex disappears); and when the dose increases from 400 to 600 mg/kg, mice finally die of respiratory arrest. Hypnotic dose is close to the lethal dose, so there's no therapeutic significance.

Daphnetin also acts on the NF-*k* B and NFAT signal pathway and results in immune inhibition; it is also used for the treatment of malaria parasite and *Pneumocystis carinii* infection by chelating to Fe^{2+} or by acting on ribonucleotide reductase; daphnetin shows bacteriostatic effects on gold grapes, *Escherichia coli*, *Shigella*'s blessing, and *Pseudomonas aeruginosa* by inhibiting succinic acid oxidase in mitochondria in bacteria; daphnetin reduces the incidences of diarrhea in model mice by inhibiting mice gastrointestinal propulsive movement and has a stronger toxic effect on several important aphids [7].

Plasma protein binding of Daphetin is low. The urinary excretion is higher. The distribution volume is relatively larger. Both of metabolism and excretion are rapid. Drug elimination is mainly through the kidney with a half-life of only 15 min. The tissue distributions after intravenous injection and oral administration are equal. Daphnetin may distribute in the brain through the blood-brain barrier. The metabolites have poor stability and low oral bioavailability. People have studied the metabolic pathways and metabolic product. Daphnetin may be the substrates of MRP2 and BCRP transporters, and thus intestinal absorption is affected; in the plasma of rats, the main metabolite are 7-O and 8-O glucoside substitution products which are mediated by UT1A6 and UT1A9. The anti-inflammatory activity of the metabolites is dependent on daphnetin-7-methyl ether [8–10].

Clinical Application

It has been verified by a large number of clinical cases that oral daphnetin capsule can significantly improve thromboangiitis obliterans; oral daphnetin has a better curative effect than intramuscular injection of Changbai daphne in patients with angina pectoris due to coronary heart disease. Daphnetin shows an obvious analgesic sedative effect on surgical and nonsurgical pain-relieved patients after intravenous drip or injection of daphnetin; oral daphnetin capsule shows significant therapeutic effects on rheumatoid arthritis.

There are no reports about serious adverse reactions. Two cases of whole-body red itch were reported. It sometimes has mild gastrointestinal reaction when continuously used.

Discussion

China's Changbai daphne is a kind of precious plant. In the 1970s, its main active compound daphnetin was discovered and synthesized, which was applied to clinical Buerger's disease, occlusive vascular disease, and coronary heart disease. Daphnetin is the first independent R&D new drug in China, making it a landmark of new drug discovery in China. Daphnetin has extensive pharmacological effects, but pharmacological researches are still based on animal experiments. Pharmacodynamics, pharmacokinetics, and clinical research for other indications are needed as soon as possible.

Fig. 1 *Daphne aurantiaca* Diels

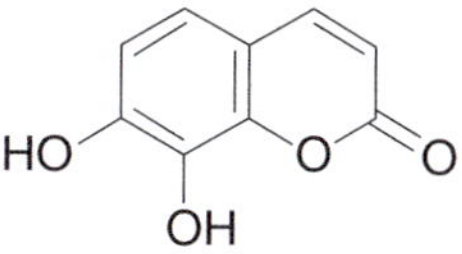

Fig. 2 Daphnetin chemical structure

References

1. Jinjun S, Di Liuqing ZX, et al. Determination of equilibrium solubility and apparent oil/water partition coefficient of Daphnetin. J Nanjing Univ Tradit Chin Med. 2011;27(5):449–50.
2. Plant groups from Traditional Chinese medicine research laboratory, Jilin institute of Chinese medicine. Study of effective component from Changbai daphne – the separation and identification of daphnetin. J Tradit Chin Med. 1977; (4):15–18.
3. Liu Guoqin, Wang Qiujuan, Yang Sixin et al.. Pharmacological researches on Daphnetin. Chinese Herbal Medicine Study. 1977;8(3):21.
4. Plant groups from Traditional Chinese medicine research laboratory, Jilin institute of Chinese medicine. Study of Pharmacological effects on Daphnetin. New Med J.1977;(3):142.
5. Hong L, Weiping L, Xi L, et al. Effect to hemorheology induced by daphnetin and its derivative on type 2 diabetic rats. Jilin Med J. 2010;31(22):3614–6.
6. Xiulian J, Shuyan Q, Pan G, et al. The inhibition of daphnetin on central nervous system. Chin J ChiniseMeteria Med. 1986;11(3):Cover 2.
7. Fang H, Tan L. Screening antimalarial activity of daphnetin derivatives and research on their molecular targets. Int J Med Parasit Dis. 2008;35(6):303–4.
8. Shuyan Q, Yingjie M, Wang Y, et al. Metabolism and pharmacokinetics study of daphnetin. Acta Pharm Sin. 1983;18(7):496–500.
9. Ying J, Yu F. The influence of different transporters on the oral absorption of daphnetin using Caco – 2 cell model. Pharmacol Clin Chin Mater Med. 2012;28(1):52–6.
10. Liang SC, Ge GB, Xia YL, et al. In vitro evaluation of the effect of 7-methyl substitution on glucuronidation of daphnetin: metabolic stability, isoform selectivity, and bioactivity analysis. J Pharm Sci. 2015;104(10):2188–209.

Dicoumarin

Xiao Cheng, Yue-Hua Wang, and Guan-Hua Du

Abstract The plants containing dicoumarol mainly include red carnation grass, rotten alfalfa, rotten white vanilla rhinoceros, and other plants in Leguminosae. Dicoumarin, a coumarin derivative, is an oral anticoagulant drug, and its common mechanism is to inhibit synthesis of the coagulation factor in the liver. There are many coumarin derivatives, and the most common clinical drug is warfarin, which is applicable to long-term maintenance of anticoagulant. The most serious adverse reaction of warfarin is bleeding, which can be against by vitamin K. Up till now, warfarin is more reliable with more clinical experience and will be an important oral anticoagulant in the next few years.

Keywords Dicoumarin · Purple alfalfa · Dicoumarol · Bishydroxycoumarin

Origin: Purple alfalfa (Fig. 1)
Chemical name (Fig. 2)

3,3′-Methylene-bis (4-hydroxy-coumarin)

Molecular formula, $C_{19}H_{12}O_6$; **MW**, 336.30; **CAS**, 66-76-2
Dicoumarin derivatives (Fig. 3)

X. Cheng · Y.-H. Wang · G.-H. Du (✉)
Beijing Key Laboratory of Drug Targets Identification and Drug Screening, Centre for Pharmaceutical Screening, Institute of Materia Medica, Chinese Academy of Medical Sciences and Peking Union Medical College, Beijing, China
e-mail: dugh@imm.ac.cn

G.-H. Du, *Natural Small Molecule Drugs from Plants*,
https://doi.org/10.1007/978-981-10-8022-7_7

Properties

Appearance: white or milky white crystalline powder, slightly fragran; *Solubility*: not dissolved in water, ethanol, and ether, slightly dissolved in chloroform, dissolved in alkali solution; *Melting point*: 287–293 °C. It can be bluer or with purple fluorescence in the ultraviolet light.

Dosage Forms

This product was included in the *Pharmacopoeia of the People's Republic of China* (1995). The dosage forms of current clinical use are tablets, 50 mg/tablet.

Indications

Intravascular thromboembolic diseases include postoperative or postoperative thrombotic phlebitis, pulmonary embolism, myocardial infarction, and atrial fibrillation caused by embolism.

Literature

The plants containing dicoumarol mainly include red carnation grass (*Trifolium pratense* L., hongchezhoucao), rotten alfalfa (*Medicago sativa* L., zimuxu), rotten white vanilla rhinoceros (*Melilotus albus* Desr., baixiangcaomuxi), and other plants in Leguminosae.

History of R&D

In 1940, Karl Paul Link, a fertile scientist from the University of Wisconsin in the United States, first isolated the anticoagulant substance from the moldy alfalfa (*Melilotus*) and determined its structure. It is a kind of dicoumaroloid substance, combined by two molecules of coumarin substances. Since this material was found in the first few years, it has been used as a rodenticide [1]. Over the next few years, people have found several molecular structures similar to this material, all of which have anticoagulant effects. Typical examples are as follows: In 1947, Mark et al. first reacted benzylideneacetone with 4-hydroxycoumarin to form warfarin, which is a more potent anticoagulant. In 1954, warfarin was officially approved by the

FDA. Before the listing of warfarin, the anticoagulant drug for clinical use is heparin, which can only be injected and is very inconvenient for long-term therapy [2]. The emergence of warfarin solved this problem and opened a new chapter in the history of anticoagulant drugs.

In 1979, Conrad et al. reacted p-nitrobenzene ketone with 4-hydroxycoumarin to obtain vinegar coumarin, which is basically the same as warfarin in anticoagulant, but its metabolites also have anticoagulant effect, so the duration of anticoagulation is longer than warfarin. With the further study of coumarin as anticoagulant, a series of coumarin analogues were synthesized and evaluated for their pharmacological properties, such as 4-aminowarfarin, 3-bromowarfarin, 6-hydroxywarfarin, 7-hydroxywarfarin, 6-bromo-4′-nitro-4-bromoferrin, 6-nitro-4′-bromoferrin, and so on. In the 1990s, Manolov et al. studied the pharmacological properties of coumarin derivatives and found that the 4-position hydroxyl esterification in the vinegar coumarin molecule had better anticoagulant effects than warfarin and less side effects [3, 4].

Pharmacology

Dicoumarin is an oral anticoagulant drug and is invalid in vitro [5]. Dicoumarin is a coumarin derivative, and its common mechanism is to inhibit synthesis of the coagulation factor in the liver. The structure of dicoumarin is similar to that of vitamin K and is an antagonist or a competitive inhibitor of vitamin K. It binds to the vitamin K epoxide reductase in the liver, inhibits the conversion of vitamin K from epoxide to hydroquinone, and inhibits the recycling of vitamin K, resulting in that the glutamate side chain of vitamin K-dependent coagulation factors II, VII, IX, and X cannot be carboxylated by γ-carboxy glutamate groups, affecting the binding of coagulation factor with calcium ion, and thereby inhibiting coagulation, reducing platelet adhesion, and prolonging thrombosis time [6]. Dicoumarol drugs have no direct confrontation with synthesized prothrombin and coagulation factor, so it is ineffective in vitro. After withdrawal of dicoumarin, prothrombin and coagulation factors II, IV, IX, and X gradually restore to a certain level, and hence the anticoagulant effect disappear, so its efficacy can be maintained for a long time [4].

Clinical Application

There are many coumarin derivatives, and the most common clinical drug is warfarin (benzalacetone coumarin), the new dicoumarol anticoagulant (vinegar coumarin). Dicoumarins are used clinically for the treatment of stroke with atrial fibrillation (AF) and are not used exclusively for acute antithrombotic therapy. It is applicable to long-term maintenance of anticoagulant. When the rapid anticoagulation is urgently needed in acute arterial occlusion, heparin is the first choice to control risk, and then warfarin is used to continue the treatment. Those who need long-term anticoagulant for the risk of thrombosis because of their hereditary factor should be treated by the combination of this product with heparin for 3–5 days, and then this product is applied to maintain a long-term anticoagulation effect. The most serious adverse reaction of warfarin is bleeding, which can be against by vitamin K, and if necessary, fresh plasma or whole blood can be injected into the body to confront bleeding [6].

Discussion

For half a century, warfarin, as the only oral anticoagulant, has played an important role in the prevention of stroke. As the main anticoagulant in the twenty-first century, its effect is exact, but its plasma concentration is susceptible to other factors, so its clinical application is limited. Its poor compliance is not easy to promote [7]. The clinical application of dicoumarol drugs should be based on their characteristics and should follow the recommendations of the guidelines, and more communication with patients is encouraged to achieve individualized anticoagulant therapy, reducing the prevalence of ischemic stroke in patients with atrial fibrillation and improving their life quality. It is known that some new oral anticoagulants, such as rivaroxaban and dabigatran group, are more convenient and may gradually replace warfarin in the future. But up till now, warfarin is more reliable with more clinical experience and will be an important oral anticoagulant in the next few years.

Fig. 1 Purple alfalfa

Fig. 2 The structure of dicoumarin

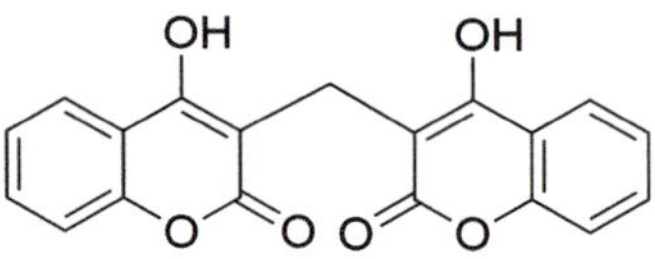

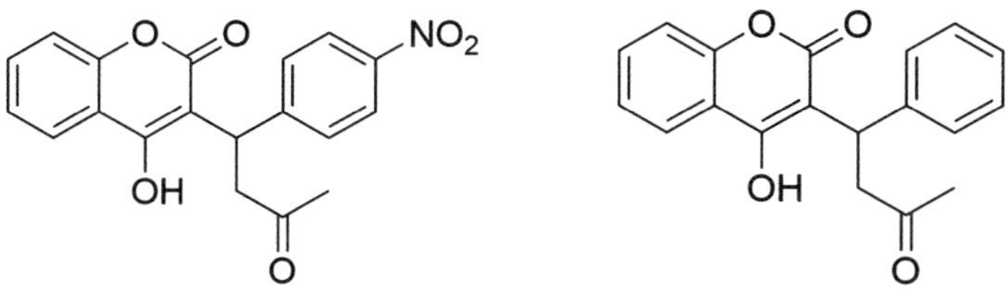

Acenocoumarol 3-(α-acetonylbenzyl)-4-hydroxycoumarin

Fig. 3 The structure of dicoumarin derivatives

References

1. Petit DW, Berne CJ. Dicumarol. Med Prog. 1947;67(1):40–4.
2. Aggeler PM. Heparin and dicumarol-anticoagulants: their prophylactic and therapeutic uses. Cal West Med. 1946;64(2):71–7.
3. Shen D-S, Liu X-F, Yang G-Z. Synthesis of coumarin in anticoagulant and its analogues. J Appl Chem. 2005;22(10):1158–60.
4. Li J, Hou Z, Chen G. Synthesis, antibacterial activities, and theoretical studies of dicoumarols. Org Biomol Chem. 2014;12:5528–35.
5. Dong J, Liu Y. Advances and clinical application of anticoagulant drugs. Chi J Geriatr Heart Brain Vessel Dis. 2013;15(12):1333–5.
6. Liu Y, Jan X. Antithrombotic treatment of atrial fibrillation. Ther Adv Cardiovasc Dis. 2010;31(2):185–9.
7. Wang M, Shi L, Gao M. Research status of anticoagulation therapy in patients with atrial fibrillation. Chin J Clinicians. 2015;9(6):980–4.

Digoxin

Ying Chen, Lan Sun, and Guan-Hua Du

Abstract Cardiac glycosides, including digoxin, digitoxin, and lanatoside digitalis glycosides, are mainly isolated from lanatoside digitalis. They are used in the treatment of chronic congestive heart failure (CHF) and can also be used to treat some arrhythmia, and they show incomparable advantages in clinical practice. Cardiac glycosides have been used in clinical practice for more than two centuries, which is enough to prove its value. Because they are important agents in the treatment of CHF, the issues of narrow dose span and high-risk poisoning make it a meaningful task to find safer cardiac glycosides. Another important topic for future research is the effect of cardiac glycosides on the treatment of other diseases.

Keywords Cardiac glycosides · Digitalis · Digoxin · Digitoxin · Cedilanid

Alias: Digoxin
Origin: Digitalis (Fig. 1)
Chemical name (Fig. 2)

3*β*-[(*O*-2,6-dideoxy-*β*-D-ribo-hexopyranosyl-(1→4)-*O*-2,6-dideoxy-*β*-D-ribo-hexopyranosyl-(1→4)-2,6-dideoxy-*β*-D-ribo-hexopyranosyl]oxy]-12*β*,14-dihydroxy-, 5*β*- card-20(22)-enolide

Molecular formula, $C_{41}H_{64}O_{14}$; **MW**, 780.94; **CAS**, 20830-75-5

Y. Chen · L. Sun · G.-H. Du (✉)
Beijing Key Laboratory of Drug Target Identification and Drug Screening, Centre for Pharmaceutical Screening, Institute of Materia Medica, Chinese Academy of Medical Sciences and Peking Union Medical College, Beijing, China
e-mail: dugh@imm.ac.cn

G.-H. Du, *Natural Small Molecule Drugs from Plants*,
https://doi.org/10.1007/978-981-10-8022-7_8

Properties

Appearance: white crystals or crystalline powder, odorless. *Solubility*: easily dissolved in pyridine, slightly soluble in dilute alcohol, slightly soluble in chloroform, insoluble in water and ethyl ether. *Specific optical rotation*: +9.5 to +12.0°. *Melting point*: 248–250 °C.

Derivatives

Digitoxin, Cedilanid

Dosage Forms

This product is recorded in the Pharmacopoeia of the People's Republic of China (2015), the British Pharmacopoeia (2017), the United States Pharmacopeia (40), the European Pharmacopoeia (9th ed.), the Japanese Pharmacopoeia (17th ed.), and the International Pharmacopoeia (5th ed.).

Tablets, oral solution, and injection.

Indications

Digoxin is used for congestive heart failure (CHF), paroxysmal supraventricular tachycarelia, atrial fibrillation, and atrial flutter.

Digitoxin

Alias: Digitoxinum, Digitox
Chemical name (Fig. 3)

3-[(*O*-2,6-dideoxy-*β*-D-ribo-hexopyranosyl-(1→4)-*O*-2,6-dideoxy-*β*-D-ribo-hexopyranosyl-(1→4)-2,6-dideoxy-*β*-D-robo-hexopyranosyl)oxy]-14-dihydroxy,(3*β*,5*β*)- 20(22)-elollactone

Molecular formula, $C_{41}H_{64}O_{13}$, **MW**, 764.94, **CAS**, 71-63-6

Properties

Appearance: white or white crystalline powder, odorless. *Solubility*: slightly soluble in chloroform, slightly soluble in ethanol or ethyl ether, insoluble in water. *Melting point*: 240 °C.

Dosage Forms

This product is recorded in the Pharmacopoeia of the People's Republic of China (2015), the British Pharmacopoeia (2017), the United States Pharmacopeia (40), the European Pharmacopoeia (9.0th ed.), the Japanese Pharmacopoeia (17th ed.), and the International Pharmacopoeia (5th ed.).

Tablet and injection.

Indications

For the treatment and management of congestive cardiac insufficiency, arrhythmias, and heart failure.

Lanatoside C

Alias: Lanatoside, cedilanide, lanatocid
Chemical name (Fig. 4)

3-[(*O*-*β*-D-Portuguese-hexopyranosyl-(1→4)-*O*-2,6-dideoxy-*β*-D-ribo-hexopyranosyl-(1→4)-2,6-dideoxy-*β*-D-robo-hexopyranosyl)oxy-(1→4)-2,6-dideoxy-*β*-D-robo-hexopyranosyl)oxy]-12,14-dihydroxy-card-20(22)-enolide

Molecular formula, $C_{49}H_{76}O_{20}$, **MW**, 985.25, **CAS**, 17575-22-3

Properties

Appearance: white crystalline powder, odorless, bitter taste, easy to absorb moist. *Solubility:* Slightly soluble in methanol, insoluble in water or chloroform. *Specific optical rotation*: +7 to +9°. *Melting point*: 272–274 °C.

Dosage Forms

This product is recorded in the Pharmacopoeia of the People's Republic of China (1990).

Injection.

Indications

This product is contained in the Pharmacopoeia of the People's Republic of China (2015), the British Pharmacopoeia (2015), the United States Pharmacopeia (36), the Japanese Pharmacopoeia (16th ed.), the European Pharmacopoeia (9th ed.), and the International Pharmacopoeia (4th ed.).

For the treatment and management of chronic heart failure, atrial fibrillation, and paroxysmal ventricular tachycardia.

Literature

Digoxin, digitoxin, and lanatoside digitalis glycosides are mainly isolated from lanatoside digitalis. Digitalis was first recorded for the medical application by Withering in the book of *An Account of the Foxglove and some of its Medical Uses* in 1785. Lanatoside digitalis scrophulariaceae biennial or perennial herbs belong to lanatoside k yellow plants, whose scientific name is *Digitalis lanata* Ehrh. They are perennial or 2-year herbaceous plants. Aliases are narrow Ye Yang rehmannia or digitalis from Greece. Lanatoside digitalis is full of fuzz stem leaves that resemble radix rehmanniae leaves, which is why it is called "foxglove." Because it comes from Europe, it is also called digitalis. Digoxin was first isolated from the foxglove plant, *Digitalis lanata*, in 1930.

History of R&D

Digitalis, a kind of scrophulariaceae biennial or perennial herb, originating in Europe in central and southern mountains, which was isolated from plants of the genus *Digitalis*, has a long history of treating heart disease. Digitalis was also recorded in Chinese Materia Medica and now is largely cultivated in China.

In 1775, Visser (Lin WilliamWithering, 1741–1799), a famous botanist, heard that some peasant woman had applied a secret recipe to the treatment of edema (actually heart exhaustion edema). Visser Lin found that this secret recipe contained more than 20 drugs, and only *Digitalis purpurea* L. really worked. Visser Lin focused on the diuretic effect of this ingredient, but at the same time, he also noticed that digitalis had a cardiotonic effect that had so far not been noticed in other drugs [1]. In fact, these kinds of medicinal plants were used as early as medieval times in sixteenth and seventeenth centuries. They are also mentioned in the published literature in Britain and Germany. In 1785, Visser Lin published the book *An Account of the Foxglove and some of its Medical Uses*. In his book, Visser Lin reported that digitalis plants could serve as an effective treatment of edema and described the use of digitalis derivatives [2]. Thus, he became a world-famous doctor.

Digitalis is unsuitable for direct medical use because of methodological limitations, imprecise dosing, and narrowing of the dose span. Too-high doses could cause toxicity and death. A fundamental way to resolve this problem lies in the extraction of the active components of digitalis. In 1874, the famous pharmacologist Schmid Berg (Oswaldd Schmiedebrg) extracted an analogue of cardiac glycoside from digitalis plants [3].

In clinical practice, the commonly used digitalis so far is known as the cardiac glycoside. (In the twentieth century, this kind of structure was called a glucoside. In the late twentieth century, domestic individual chemical researchers proposed renaming it glycoside, identical to lanatoside k Huang Zhong [extracting the effective ingredients of digoxin, digitoxin, and lanatoside C]). In 1930, a researcher from Burroughs Wellcome pharmaceutical company named Sydney Smith successfully isolated several steroid glycosides, including digoxin. The later production of digoxin by Burroughs Wellcome's successor company GlaxoSmithKline is called LANOXIN, which is also called DIGITEK. Since that time, digoxin has been widely used in clinical practice to treat patients with CHF and atrial arrhythmia. Now digoxin is still the basis of heart failure treatment, and in most cases it is one of the preferred first-line drugs.

In 1954, researchers from the Institute of Materia Medica of the Chinese Academy of Medical Sciences improved the method of extracting digitoxin, resulting in a simpler and 3.5 times higher yield than that of extracting digitoxin in German pharmaceutical [4]. In addition, the purity of domestic digitalis was improved. The injection preparation was then produced and applied in clinical practice.

Recently greatly advances have been made in research on the effects of cardiac glycoside on cancer [5]. Using high-throughput screening methods that have been widely applied to the identification of new cardiac glycosides, researchers found that cardiac glycoside can inhibit transplantation tumor growth in mice [6]. This research also helped in the development of the domestic pharmaceutical industry.

Pharmacology

Digitalis is a kind of cardiac glycoside. It has been used for nearly two centuries to treat CHF and supraventricular arrhythmias due to reentry mechanisms and to control the ventricular rate in the treatment of chronic atrial fibrillation. Digitalis has three key pharmacological mechanisms of action: hemodynamic, neurohormonal, and electrophysiological.

First, it has a positive inotropic action. Digitalis inhibits the sodium–potassium ATPase pump in cardiomyocytes, which is followed by an increase in the intracellular sodium concentration so as to enhance sodium–calcium exchanger activity and increase intracellular calcium concentration. Then it improves the myocardial positive contraction force to lead to a global increase in left ventricular systolic function by interaction between the myocardial contractile proteins.

Second, it has a negative chronotropic action. Digitalis induces vagal activation, leading to a shift in autonomic balance toward parasympathetic dominance. The pathophysiology of CHF includes an initial compensatory and later detrimental neurohormonal activation. Digitalis modulates these neurohormonal abnormalities, reducing plasma norepinephrine.

Third, it affects the electrical activity of the heart. By inhibiting the sodium–potassium ATPase, digitalis increases the slope of phase 4 depolarization, shortens the action potential duration, and then decreases the maximal diastolic potential.

Fourth, it reduces myocardial oxygen consumption. Cardiac glycoside can enhance the CHF myocardial contraction force and increase myocardial oxygen consumption. However, based on the positive inotropic effect, digitalis shortens ejection time, reduces heart indoor residual volume, shrinks ventricular volume, lowers room wall tension, and negatively regulates the frequency of the heart rate; myocardial oxygen consumption always does not increase, which distinguishes cardiac glycosides from catecholamine agents.

Fifth, digitalis affects conducting tissue and cardiac electrophysiological properties. Digitalis can cause a decrease in the autonomy of sinus node, improve the autonomy of the Pukinje fiber, reduce atrioventricular node conductivity, and shorten the atrial and the Pukinje fiber's effective refractory period (ERP).

Sixth, digitalis treatment is associated with changes in electrocardiograms (ecg). Digitalis causes prolongation of the PR and RR interval, shortening of the QTc interval, and depression of the ST-segment and T-wave amplitude. It is the basis of clinical judgment of whether or not cardiac glycoside should be used [7].

In addition, cardiac glycosides can directly act on the nerve-endocrine system, by directly inhibiting sympathetic activity, enhancing vagus nerve activity, and improving the effects of neuroendocrine disorders. Digitalis also has significant diuretic effects in patients with CHF. Digitalis shrinks blood vessels, increases peripheral resistance, elevates blood pressure, and reduces local blood flow.

Digitoxin has a high lipid solubility and good oral absorption. It is metabolized mostly through the liver. Metabolic product is cleared via the renal pathway, and there is also a significant biliary tract discharge and forms enterohepatic circulation, t1/2 for 5–7 days, so the action time is longer, which means digitoxin is classified as a long-term cardiac glycoside drug.

Digoxin, which is isolated from lanatoside k Huang Zhong, is a long-term cardiac glycoside. Digoxin has faster excretion than digitoxin and does not accumulate, and hence it is safer than digitoxin in clinical practice. Oral bioavailability is about 60–80% owing to the particle size and dissolution. About 25% combines with plasma proteins, is distributed to the organs, with the highest concentration being in the kidney. The drug is excreted mostly via the kidneys. Following intravenous administration to healthy volunteers, 50% to 70% of a digoxin dose is excreted unchanged in the urine. Its half-life is 36 h.

Cedilanid has poor and irregular oral absorption. It treatment it is administered intravenously. Its action is quick but short, the serum half-life is about 1.5 days, and it is mainly excreted via the kidneys. Few metabolites combine with proteins, so the therapeutic effects are fast and strong, in favor of clinical rescue.

Clinical Application

Cardiac glycoside is a kind of nucleoside compound, which have strong heart function. In clinical, it is mainly used for the treatment of chronic CHF and can also be used to treat some arrhythmia, and shows incomparable advantages.

Cardiac glycoside is used to treat diseases of the central lobe membrane, congenital heart disease, high blood pressure or heart failure due to atherosclerosis, but it is not suitable for heart failure that is secondary to thyroid hyperfunction, severe anemia, vitamin B1 deficiency, pulmonary heart disease, and active myocarditis. Treatments of heart failure due to narrow pericarditis or severe mitral stenosis are less effective or even ineffective [8]. It has a small dosage window in which it is considered safe. The common gastrointestinal adverse reactions include anorexia, nausea, vomiting, and diarrhea. If these adverse reactions occur, treatment with cardiac glycoside should be reduced or stopped, and simultaneously potassium should be supplemented to avoid severe vomiting, which can lead to loss of potassium and increase cardiac glycoside poisoning. Cardiac toxicity reaction is the most serious adverse reaction. Premonitory symptoms may be the indicators of the occurrence of poisoning. Patients experience an adverse reaction of tachycardia arrhythmia may be treated with intravenous drip potassium or phenytoin sodium when it is serious. In case of extremely severe poisoning, patients should be treated with digoxin antibody Fab fragments by intravenous injection; atropine can be used to rescue slow-type arrhythmia. Rapid pacing should be applied when atropine treatment is invalid.

Discussion

Cardiac glycoside values on clinical application in patients have been fully recognized based upon years of clinical experience and related literature reports. Also, the safety profile of the long-term application of cardiac glycoside has been confirmed. However, heart failure treatment has changed from classical strategies including strengthening of the heart and diuresis to new ones such as inhibition of nerve-endocrine system activation, blocking or slowing the ventricular remodeling process, or protecting the damaged cardiomyocyte. Treatment of heart failure has undergone a fundamental change because of the appearance of the angiotensin-converting enzyme inhibitor (ACEI). Treatment of heart failure with ACEI, beta blockers, and aldosterone receptor inhibitor has formed into a "Golden Triangle" [9].

Cardiac glycosides have been used in clinical practice for more than two centuries, which is enough to prove its value in clinical practice. Because they are important agents in the treatment of CHF, the issues of narrow dose span and high-risk poisoning make it a meaningful task to find safer cardiac glycosides. In addition, another important topic for future research is the effect of cardiac glycosides on the treatment of other diseases.

Fig. 1 Digitalis

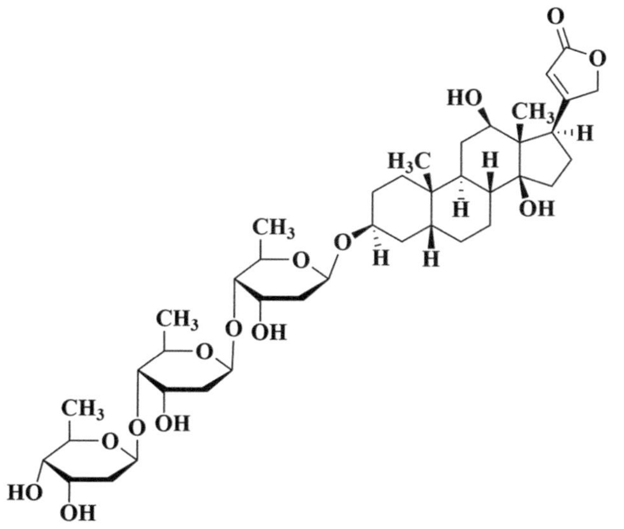

Fig. 2 Chemical structure of Digoxin

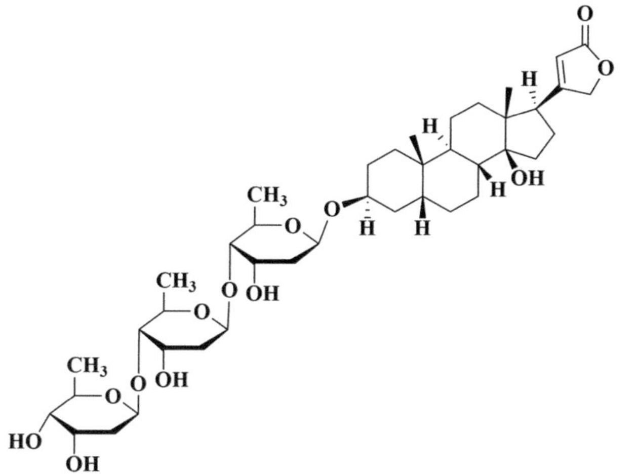

Fig. 3 Chemical structure of digitoxin

Fig. 4 Chemical structure of Cedilanid

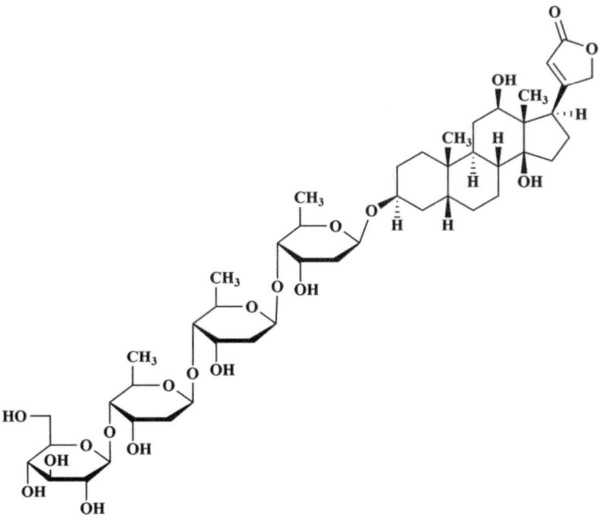

References

1. Wilkins RM, Kendall JM, Wade LO. William Withering and digitalis, 1785 to 1985. Br Med J. 1985;290(6461):7–8.
2. Lüderitz B. Cardiac glycosides: William Withering (1741–1799). J Interv Card Electrophysiol. 2005;14(1):61–2.
3. Hui N. The story of the digitalis. Med People. 2001;5:38.
4. Congratulations on the Chinese academy of sciences institute of medicine for successfully extraction digitoxin. Chin Pharm J. 1954.
5. Yumen J, Wang X, Yang C, et al. Research progress on pharmacokinetics of cardiac glycosides. Chin Tradit Herb Drug. 2014;45(23):3472–7.
6. Calderónmontaño JM, Burgosmorón E, Orta ML, et al. Evaluating the cancer therapeutic potential of cardiac glycosides. Biomed Res Int. 2014;2014(12):85–117.
7. Yang S. Pharmacology. 2nd ed. Beijing: People's Medical Publishing House; 2010. p. 234–8.
8. Shen B. Application of cardiac glycoside: experience for the treatment of 128 cases of heart failure. J Pract Med Tech. 2008;15(1):61–2.
9. Wei YU. Highlights of 2014. Guidelines for diagnosis and treatment of heart failure in China. Chin Clin Dr. 2015;(5):12–14.

Dioscin

Wan Li and Guan-Hua Du

Abstract Dioscin is extracted from *Dioscorea nipponica* Thunb. It has therapeutic effects on reducing blood lipids and protecting the liver, as well as antivirus, anti-inflammatory, and antitumor effects. It can be orally used to treat joint pain, arthritis, malaria, and chronic bronchitis. In addition, it is used externally to treat sore carbuncle swollen. Dioscin is an important predecessor for synthesizing steroid hormones. In China, dioscin tablets have already been used to treat hyperlipidemia in the clinical practice. With the ongoing efforts of evaluating the biological activity, dioscin has a great potential to be fully investigated.

Keywords Dioscin · Blood lipids · Steroid hormones

Alias: Re-saponin III
Origin: *Dioscorea opposite* Thunb. (Fig. 1)
Chemical name (Fig. 2)

(25R)-3β-[2-O,4-O-Bis(α-L-rhamnopyranosyl)-β-D-glucopyranosyloxy] spirosta-5-ene

Molecular formula, $C_{45}H_{72}O_{16}$; **MW**, 869.05; **CAS**, 19057-60-4

W. Li · G.-H. Du (✉)
Beijing Key Laboratory of Drug Targets Identification and Drug Screening, Centre for Pharmaceutical Screening, Institute of Materia Medica, Chinese Academy of Medical Sciences and Peking Union Medical College, Beijing, China
e-mail: dugh@imm.ac.cn

G.-H. Du, *Natural Small Molecule Drugs from Plants*,
https://doi.org/10.1007/978-981-10-8022-7_9

Properties

Appearance: white powder. *Solubility*: insoluble in water, petroleum ether, and benzene; soluble in methanol, ethanol, and acetic acid; slightly soluble in acetone and amyl alcohol. *Melting point*: 294–296 °C. *Specific optical rotation*: −115° ($c = 0.373$, ethanol).

Dosage Forms and Indications

This drug was recorded in the national standards for chemical drugs, volume 5.

Dioscin tablets, containing total dioscin, are used to treat coronary heart disease, coronary insufficiency, as well as control hyperlipidemia.

Literature

Dioscin is extracted from *Dioscorea nipponica*. Rhizoma *Dioscoreae Nipponicae* is the dry rhizome of this plant. It can be orally used to treat joint pain, arthritis, malaria, and chronic bronchitis. In addition, it can also be used externally to treat sore carbuncle swollen. Rhizoma *Dioscoreae Nipponicae* is commonly used in the northeast region, Shanxi province, and Shandong province of China.

History of R&D

Japanese scientists firstly reported and analyzed the structure of dioscin isolated from the saponins of Japanese Dioscoreaceae. V [1]. However, the biological activities of dioscin have not attracted the interests of scientists until the late 1990s and the beginning of this century. It is now believed that dioscin may have therapeutic effects on reducing blood lipids and protecting the liver, as well as antivirus, anti-inflammatory, and antitumor effects. After the implementation of standardization of traditional Chinese medicine, dioscin as the quality control specification of *Dioscorea nipponica* Makino and other traditional Chinese medicine has attracted more attentions and researches. In China, dioscin tablets have been used to treat hyperlipidemia in the clinical practice. However, no corresponding drugs are available on the market in other countries. In addition, dioscin is also an important predecessor for synthesizing steroid hormones [2]. *Dioscorea nipponica* Makino, which is used to extract dioscin, primarily comes from wild plants, whose resources are limited.

Pharmacology

Studies showed that compound preparations containing dioscin can increase coronary blood flow, reduce myocardial oxygen consumption, and protect myocardial ischemia and ischemia-reperfusion injury, which is beneficial for alleviating myocardial ischemia and angina. In addition, dioscin can regulate the lipid metabolism and improve hemodynamics. Dioscin can also prevent atherosclerosis by reducing serum total cholesterol, triglyceride, low-density lipoprotein, and oxidized low-density lipoprotein, reducing the whole blood viscosity and plasma viscosity at high and low shear rates, and relieving arterial wall lipid infiltration and plaque formation. Furthermore, dioscin can reduce liver fat deposition, which may be related to promoting cholesterol secretion metabolism.

It has also been reported that dioscin can induce the apoptosis of a variety of tumor cells, including gastric cancer cells, breast cancer cells, lung cancer cells, oral squamous cell carcinoma, and cervical cancer cells. The mechanisms are related to reducing the oxidative stress and inhibiting some cellular survival signaling pathways [3].

Dioscin can induce antioxidant enzymes, lower blood lipids, improve insulin resistance, inhibit the inflammatory response, reduce the damage of alcoholic hepatitis, and decrease the hepatic injury induced by acetaminophen [4, 5]. These effects may be related to its hydrolysate diosgenin [2]. In addition, the anti-inflammatory effects of dioscin on arthritis and chronic bronchitis are also widely reported [6]. It has been found that dioscin has antitussive and antiasthmatic effects and is used to treat chronic bronchitis, and these effects are related to its active ingredients including total saponin, water-soluble saponin, as well as steroid saponin at high dose. Recent studies also showed that dioscin can alleviate the ischemic-reperfusion damage by inhibiting inflammatory response, suppressing chemotaxis, and reducing inflammatory cell recruitment [7].

Dioscin monomer is rarely used in the clinical practice. In China, dioscin tablets are used to treat coronary insufficiency, coronary heart disease, and hyperlipidemia. It is also used to reduce cholesterol and triglycerides. In China, using dioscin as the quality control specification, *Dioscorea nipponica* Makino is used to treat sputum wheezing, chronic bronchitis, and some arthritis.

Discussion

The dioscin monomer was identified 60 years ago, whose biological activities have not been much studied. However, its anti-inflammatory and antitumor effects are prominent and well characterized, and further studies are being performed.

Dioscorea nipponica Makino has already been used in the clinical practice, whose scientific clinical evaluation criteria and safety assessment can provide important clinical evidences for the biological activity of dioscin. With the ongoing efforts of evaluating the biological activity, dioscin has a great potential to be fully investigated.

Fig. 1 *Dioscorea opposite* Thunb.

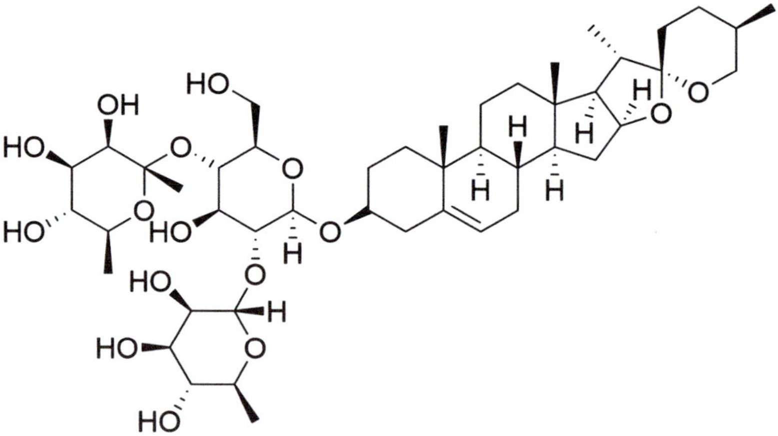

Fig. 2 Chemical structure of dioscin

References

1. Tsukamoto T, Kawasaki T, Yamauchi T. Saponins of Japanese Dioscoreaceae. V. On the structure of dioscin. Pharm Bull. 1956;4(1):35–42.
2. Yue L, et al. Recent advances of diosgenin in its pharmacologic activities and mechanism. Chin J Clin Pharm Ther. 2010;2:233–7.
3. Man S, et al. Chemical study and medical application of saponins as anti-cancer agents. Fitoterapia. 2010;81(7):703–14.
4. Zhao X, et al. Dioscin, a natural steroid saponin, shows remarkable protective effect against acetaminophen-induced liver damage in vitro and in vivo. Toxicol Lett. 2012;214(1):69–80.
5. Kamisako T, Ogawa H. Regulation of biliary cholesterol secretion is associated with abcg5 and abcg8 expressions in the rats: effects of diosgenin and ethinyl estradiol. Hepatol Res. 2003;26(4):348–52.
6. Xu LN, Wei YL, Peng JY. Advances in study of dioscin – a natural product. Zhongguo Zhong Yao Za Zhi. 2015;40(1):36–41.
7. Qi M, et al. Dioscin attenuates renal ischemia/reperfusion injury by inhibiting the TLR4/MyD88 signaling pathway via up-regulation of HSP70. Pharmacol Res. 2015;100:341–52.

Diosmin

Chao Li and Guan-Hua Du

Abstract Diosmin was separated from the root of *Scrophularia nodosa* L., which is applied to treat fever, polydipsia, maculitis, gingivitis, amygdalitis, laryngopharyngitis, acute lymphadenitis, etc. Diosmin is a micronized, purified, flavonoid-structure drug and used as the protecting agent for blood vessel and therapeutic agent for chronic venous disease. It is helpful for lymphatic return and stimulating microcirculation, thus treating hemorrhoids and venous dysfunction by increasing the venous tension. Moreover, the adverse reaction was seldom observed. Thus, it would have great significance in clinical therapy if the application field of diosmin could be enlarged. With further development in its clinical application, the use of diosmin would have a bright future.

Keywords Diosmin · *Scrophularia nodosa* L. · Flavonoid

Origin: *Scrophularia nodosa* L. (Fig. 1)

Alias: 7-Rutinoside
Chemical name (Fig. 2)

7-[[6-*O*-(6-Deoxy-α-L-mannopyranosyl)-β-D-glucopyranosyl]oxy]-5-hydroxy-2-(3-hydroxy-4-methoxyphenyl)-4*H*-1-benzopyran-4-one

Molecular formula, $C_{28}H_{32}O_{15}$; **MW**, 608.54; **CAS**, 520-27-4

C. Li · G.-H. Du (✉)
Beijing Key Laboratory of Drug Targets Identification and Drug Screening, Centre for Pharmaceutical Screening, Institute of Materia Medica, Chinese Academy of Medical Sciences and Peking Union Medical College, Beijing, China
e-mail: dugh@imm.ac.cn

G.-H. Du, *Natural Small Molecule Drugs from Plants*,
https://doi.org/10.1007/978-981-10-8022-7_10

Properties

Appearance: Grayish yellow or light yellow hygroscopic powder. *Solubility*: Practically insoluble in water, soluble in dimethyl sulfoxide, practically insoluble in alcohol. It dissolves in dilute solutions of alkali hydroxides. *Melting point*: 277–278 °C.

Dosage Forms

Diosmin was recorded in the *Pharmacopoeia of the People's Republic of China* (2015), *British Pharmacopoeia* (2017), *United States Pharmacopeia* (40), and *European Pharmacopoeia* (9.0th ed.).

Its main dosage form was tablet (450 mg/tablet).

Indications

The clinical indications of diosmin are those symptoms related to venous-lymphatic dysfunction such as leg heaviness, pains, soreness, and swelling in the morning. Besides, the symptoms related to acute hemorrhoids could also be treated with diosmin.

Literature

Diosmin was separated from the root of *Scrophularia nodosa* L.玄参 in 1925. The root of *Scrophularia nodosa* L. 玄参has the function of nourishing, reducing fever, relieving internal heat, and detoxicating. It was firstly recorded in the *Shen Nong's Classic of Materia Medica*, as a class of middle-grade drug and is currently recorded in the Ch.P 2015, Volume I [1]. The natural resource was abundant in China, mainly provided from Zhejiang, Sichuan, Hubei, Anhui, and Jiangsu provinces [2]. In the clinical practice, *Scrophularia nodosa* L.玄参 was employed to eliminate pathogenic heat from the blood. In the clinical indication, it was mainly applied to treat fever, polydipsia, maculitis, gingivitis, amygdalitis, laryngopharyngitis, acute lymphadenitis, etc. It also has the biological function of antiplatelet aggregation and antitumor. The chemical components include iridoids, phenylpropanoid glycoside, flavonoids, and fatty acids [3].

History of R&D

Diosmin was firstly introduced as a medicine in 1969. It was launched in France in 1987 with the product name "Alvenor." As the protecting agent for blood vessel and therapeutic agent for chronic venous disease, diosmin has been used for over 30 years in Europe.

Diosmin is a typical flavonoid. It could be prepared from natural resources or semisynthesized from the natural product hesperidin by dehydrogenation reaction.

Pharmacology [4–6]

Diosmin is a micronized, purified, flavonoid-structure drug. It is helpful for lymphatic return and stimulating microcirculation, treating hemorrhoids and venous dysfunction by increasing the venous tension.

1. Increasing the venous tension. Diosmin could strengthen the tensile force of the venous wall even under high temperature. Diosmin's action in venous constriction is stronger than other drugs like rutin. Even under the acidic toxicity in the body, it could still raise the venous tension. Diosmin has special affinity to the venous other than the arterial system.
2. Improving the microcirculation. Diosmin could effectively reduce adhesion and migration of leukocyte and vascular endothelial cells. It could also release inflammatory substances such as histamine, bradykinin, alexin, leukotriene, prostaglandin, and surplus free radicals. Hence, the penetration of capillary blood vessel was reduced, and the tension was enhanced. Besides, diosmin could decrease blood viscosity, accelerate erythrocyte flow rate, and finally reduce the chances of microcirculatory stasis.
3. Stimulating the lymphatic return. Diosmin could effectively increase the speed of lymphatic drainage and the lymphatic contraction, accelerating the circumfluence of interstitial fluid, improving lymphatic return, and relieving edema.

Clinical Application

Diosmin was mainly used to treat diseases like chronic venous dysfunction, haemorrhoids, lymphedema, phlebeurysm [7], etc.

Discussion

Diosmin has a special affinity to the venous system. In contrast, it has no apparent influence to the arterial system. The adverse reaction was seldom observed. The safety could be guaranteed and suitable for long-term administration. The local or global swell-related diseases caused by venous or lymphatic obstruction were epidemic. Thus, it would have great significance in clinical therapy if the application field of diosmin could be enlarged. With further development in its clinical application, the use of diosmin would have a bright future.

Fig. 1 *Scrophularia nodosa* L.

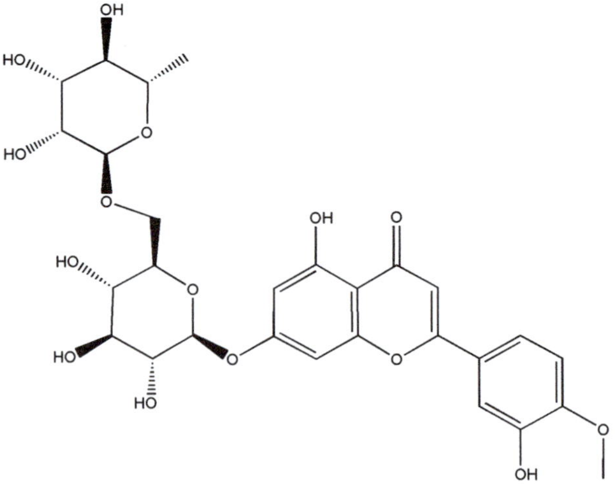

Fig. 2 Chemical structure

References

1. Chinese Pharmacopoeia Commission. Pharmacopoeia of the People's Republic of China 2010 English edition, volume I. Beijing: Chemical Industry Press; 2015.
2. Li JS. Authentication of traditional Chinese medicine. Shanghai: China Science and Technology Press; 1996.
3. Chen C. The genealogy, phylogeography of *Scrophularia ningpoensis* Hemsley and phylogeny of East Asian Scrophularia study. Zhejiang University. 2011.
4. Dieterling P. Contribution of anorectal manometry to the treatment of hemorrhoids. Phlebologie. 1992;21(Suppl 2):57–60.
5. Ho YH, Foo CL, Seow-Choen F, et al. Prospective randomized controlled trial of a micronized flavonidic fraction to reduce bleeding after haemorrhoidectomy. Br J Surg. 1995;82(8):1034–5.
6. Buckshee K, Takkar D, Aggarwal N. Micronized flavonoid therapy in internal hemorrhoids of pregnancy. Int J Gynaecol Obstet Off Organ Int Fed Gynaecol Obstet. 1997;57(2):145–51.
7. Li FM, Tian XD, Hu GW, et al. The advances in clinical application of Diosmin. Mod J Integr Tradit Chin West Med. 2012;21(6):680–1.

Divasidum

Kun Hu, Li-Da Du, and Yang Lu

Abstract Divasidum is a cardiac glycoside compound found in the seeds of Apocynaceae plant *Strophanthus divaricatus*. In China, divasidum has been developed into injection for the treatment of heart failure. The current clinical application of the divasidum preparation is divasidum injection, which is a mixture of divasidum and divostroside. The cardiotonic action of divasidum is similar to strophanthidin. Another study shows that divasidum has a diuretic effect and sedative effect, and it is similar to strophanthin K.

Keywords Divasidum · Cardiac glycoside · Anti-heart failure

Alias: Divaricoside, divasid
Origin: *Strophanthus divaricatus* (Lour.) Hook. et Arn (Fig. 1)
Chemical name (Fig. 2)

3-[(3S,5R,8R,9S,10S,11R,13R,14S,17R)-11,14-Dihydroxy-3-[(2R,4S,5S,6S)-5-hydroxy-4-methoxy-6-methyloxan-2-yl]oxy-10,13-dimethyl-1,2,3,4,5,6,7,8,9,11,12,15,16,17-tetradecahydrocyclopenta[a]phenanthren-17-yl]-2H-furan-5-one

Molecular formula, $C_{30}H_{46}O_8$; **MW**, 534.68; **CAS**, 508-84-9

K. Hu · Y. Lu (✉)
Beijing Key Laboratory of Polymorphic Drugs, Research Center of Polymorphic Drugs, Institute of Materia Medica, Chinese Academy of Medical Sciences and Peking Union Medical College, Beijing, China
e-mail: luy@imm.ac.cn

L.-D. Du
School of Biomedical Sciences, Chinese University of Hong Kong, Hong Kong, China

G.-H. Du, *Natural Small Molecule Drugs from Plants*,
https://doi.org/10.1007/978-981-10-8022-7_11

Properties

Appearance: small white crystalline powder, odorless, bitter taste. *Solubility*: soluble in water, chloroform, or ether.

Dosage Forms and Indications

This product is included in the *Pharmacopoeia of the People's Republic of China* (1963).

Divasidum injection can treat congestive heart failure and myocardial infarction.

Literature

Divasidum is a cardiac glycoside compound found in the seeds of Apocynaceae plant *Strophanthus divaricatus*. *Strophanthus divaricatus* is also known as *Gelsemium elegans* and *Morinda umbellata L.* This plant was recorded in *Flora of China* [1]. "Bencaoqiuyuan" suggested that it can be used for treating itching and tinea capitis. "Lu Chuan Materia Medica" and "Lingnan herbal medicine" also suggest that it is useful for treating sore, congestion, malignant ulcers, and bruises. It grows in Guizhou, Yunnan, Guangxi, Guangdong, and Fujian provinces in China. It is also distributed in Vietnam and Laos. Divasidum, decogenin, sarmentogenin, sarmutogenin, bipindogenin, and other types of cardiac glycosides are found in the roots, stems, leaves, and other parts of *Strophanthus divaricatus*. In China, divasidum has been developed into injection for the treatment of heart failure.

History of R&D

A white amorphous saponin was first extracted from the seeds of *Strophanthus divaricatus* in 1940, and the content is 1.8%. Strophanthiline A, B, and C were then identified. Its effect is about two-thirds of that of strophanthin K. Then, divaricoside (content 0.46%) and caudoside (content 0.19%) were identified in 1953 and 1954 [2].

Pharmacology

The cardiotonic action of divasidum is similar to strophanthidin [3]. The positive inotropic action of strophanthidin at the high concentration is achieved by inhibiting Na^+-K^+-ATPase activity. Divasidum increases the calcium ion concentration in ventricular myocytes, and this effect is dependent on the T-type calcium channels, L-type calcium channels, and Na^+-Ca^{2+} exchange [4]. Another study shows that divasidum has a diuretic effect and sedative effect, and it is similar to strophanthin K [5]. Studies have also shown that the diuretic mechanism of divasidum is the direct effect of the cardiac glycoside on the kidney, so that water and chlorine reabsorption in the renal tubules decreased, while the nitrogen excretion increased. In addition, similar as strophanthin K, divasidum is able to excite uterine smooth muscle. Six new cardiac glycosides that were isolated from the ethanol extract of *Strophanthus divaricatus (Roupellina boivinii)* grown in Africa inhibit the proliferation of human ovarian cancer cell line A2780 [6]. One study in cats suggested that oral administration of divasidum did not induce drug accumulation after 5 days.

Clinical Application

It was observed that among the 22 patients with congestive heart failure, divasidum treatment displayed good therapeutic effect in 14 patients. In another report, divasidum treatment of congestive heart failure in 58 cases, 39 cases have good results. However, fast injection induces side effects [7]. The current clinical application of the divasidum preparation is divasidum injection, which is a mixture of divasidum and divostroside.

Discussion

Divasidum, as an alternative to strophanthin K found in the 1950s, is effective for treating heart failure and has been clinically used. Further studies are needed to confirm their effect for treating diuresis, sedation, and cancer.

Fig. 1 *Strophanthus divaricatus* (Lour.) Hook. et Arn

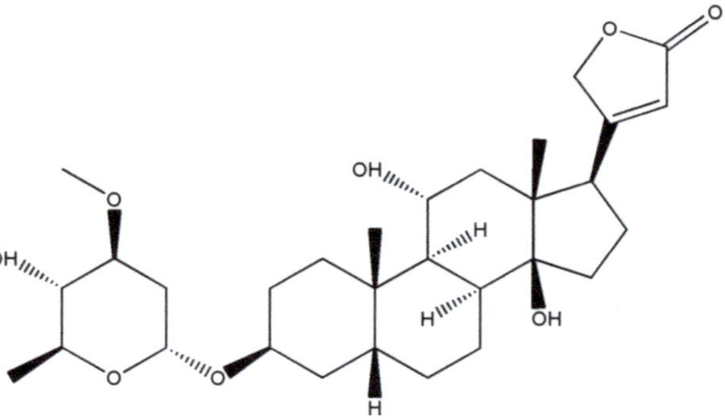

Fig. 2 Chemical structure of divasidum

References

1. "Chinese Flora" (1977 edition);63:152.
2. Kun S. Pharmacology and clinical application of divasidum. Tianjin Medical Journal. 1962;(07): 396-401.
3. Jiang M, Li Z. Comparison of the effects of divasidum and strophanthin K on the heart failure caused by metabolic inhibitors. Acta Phys Sin. 1958;(04).
4. Qiu Y, et al. Effects of divaricoside (Div) on [Ca2+] of guinea pig ventricular myocytes in different extra cellular solution. Chin J Pharmacol Toxicol. 2007;21(5):381–4.
5. Zheng S. The diuretic and sedative action of divasid a gardiac glycoside preparation from Yang. Chio-Ou (Strophanthus divari.catus (Lour.)Hook,et arn.). Acta Pharm Sin. 1959;7(5):161–5.
6. Karkare S, et al. Cytotoxic cardenolide glycosides of Roupellina (Strophanthus) boivinii from the Madagascar rainforest. J Nat Prod. 2007;70:1766–70.
7. Li Y et al. A preliminary observation on the clinical efficacy of divasidum preparation in the treatment of 55 cases of heart failure. J Wuhan Med Coll. 1958;(2):117–22.

Ferulic Acid

Ping Wu, Li Li, and Guan-Hua Du

Abstract Ferulic acid is widely found in plants, especially in artichoke, eggplant and corn bran. This product is mainly used for the treatment of atherosclerosis, coronary heart disease and ischemic cerebrovascular disease. Recently, ferulic acid has attracted attention in the research and treatment of neurodegenerative diseases, nephropathy and cancer. The new pharmacological effects and possible indications of ferulic acid are being widely shown. Therefore, how to improve the efficacy of ferulic acid and its derivatives as well as the best indications is of great significance.

Keywords Ferulic acid · Atherosclerosis · Coronary heart disease · Neurodegenerative diseases

Alias: Hydroxy-3-methoxycinnamic acid
Origin: *Resina ferulae* (Fig. 1)
Chemical name (Fig. 2)

(E)-3-(4-Hydroxy-3-methoxy-phenyl)prop-2-enoic acid

Molecular formula, $C_{10}H_{10}O_4$; **MW**, 194.2; **CAS**, 1135-24-6

Derivatives (Fig. 3)

P. Wu · L. Li · G.-H. Du (✉)
Beijing Key Laboratory of Drug Targets Identification and Drug Screening, Centre for Pharmaceutical Screening, Institute of Materia Medica, Chinese Academy of Medical Sciences and Peking Union Medical College, Beijing, China
e-mail: dugh@imm.ac.cn

G.-H. Du, *Natural Small Molecule Drugs from Plants*,
https://doi.org/10.1007/978-981-10-8022-7_12

Properties

Appearance: light yellow crystalline powder. *Solubility*: slightly soluble in cold water; soluble in hot water, with poor stability in aqueous solution; easily decomposed when encounter light; soluble in ethanol and ethyl acetate; slightly soluble in ether; insoluble in benzene and petroleum ether. *Melting point*: 170–173 °C.

Dosage Forms

This product is recorded in *Pharmacopoeia of the People's Republic of China* (2015) as sodium ferulate and ferulic acid piperazine and recorded as ferulic acid in the *British Pharmacopoeia* (2017) and *European Pharmacopoeia* (8.7th ed.). Its dosage forms are tablets and injections.

Indications

This product is mainly used for the treatment of atherosclerosis, coronary heart disease and ischemic cerebrovascular disease.

Literature

Ferulic acid is widely found in plants, especially in artichoke, eggplant and corn bran. In addition, it is also present in a variety of Chinese herbal medicines, such as angelica, dome, motherwort, snow ganoderma lucidum and so on.

History of R&D

Ferulic acid was first isolated from the medicinal plants ferulic in 1866. The biological activity of ferulic acid was not revealed until 1957 when the pioneering study of Preziosi P in Italy showed for the first time the efficacy of ferulic acid in regulating blood lipids and diuretic [1]. In 1979, Lin Mao and others isolated ferulic acid from the Chinese medicine angelica and reported that ferulic acid had the inhibitory effect on platelet aggregation [2]. Since then, more and more medicinal efficacy of ferulic acid has gradually been recognized.

Pharmacology

In vitro studies have shown that ferulic acid has anti-inflammatory and antioxidant biological properties, suggesting that it could be used for the treatment of some chronic diseases. Ferulic acid was found to significantly reduce Aβ-induced neurological inflammation in mice and improve the memory function of mice. Ferulic acid was also found to prevent neuron apoptosis; therefore, it could be used in the treatment of Parkinson's syndrome characterized by neuronal loss [3].

Ferulic acid can also increase coronary blood flow and protect the myocardium from ischemic injury. Due to its blocking effect on the α-receptor, ferulic acid can inhibit the aortic smooth muscle contraction and prevent the blood pressure-boosting effect induced by methachlor and phenylephrine β epinephrine. Sodium ferulate is conducive to improving the oxygen supply and demand imbalance in the myocardium [4].

Ferulic acid also has some protective effects on atherosclerosis. Ferulic acid can prevent lipid peroxidation due to its antioxidant properties; it can restrain cholesterol deposition in the blood vessel wall and inhibit thrombosis caused by platelet aggregation. Therefore, ferulic acid can be used to treat coronary heart disease and angina caused by atherosclerosis [4, 5].

In vitro and in vivo studies have shown that ferulic acid had good anti-radiation effects, effectively inhibiting radiation-induced cell damage, improving cell viability and preventing cell apoptosis. The mechanism was related to the activation of Nrf2-ARE signalling pathway, which can reduce the expression of adhesion molecules, inhibit granulocyte colony-stimulating factor and promote the production of erythropoietin. Thus, ferulic acid can promote the recovery of peripheral blood and the growth of haematopoietic progenitor cell colonies in mice after irradiation [5].

Clinical Application

At present, there are sodium ferulate tablets and ferulic acid injection used in clinic. Sodium ferulate tablets are mainly used for the adjuvant therapy of atherosclerosis, coronary heart disease, cerebrovascular disease, glomerular disease, pulmonary hypertension, diabetic vascular disease, vasculitis and other vascular disorders. Ferulic acid can also be used for the treatment of migraine headache and vascular headache. Ferulic acid injection is mainly used for the treatment of ischemic cardiovascular and cerebrovascular disease. In addition, sodium ferulate combined with atorvastatin can be used for the treatment of pulmonary hypertension, diabetic nephropathy and chronic glomerulonephritis in clinic [6].

Ferulic acid is also used in combination with other drugs to treat other diseases. For example, the ferulic acid piperazine tablets are useful in the treatment of glomerular diseases associated with microscopic haematuria and hypercoagulable conditions such as nephritis, chronic nephritis, nephrotic syndrome, early uraemia and the adjuvant therapy on coronary heart disease and cerebral infarction [7].

Discussion

Ferulic acid has a wide range of pharmacological activities. Clinically, ferulic acid has been used to treat a variety of diseases, such as atherosclerosis, coronary heart disease and cerebrovascular disease. Recently, ferulic acid and its derivatives have attracted attention in the research and treatment of neurodegenerative diseases, nephropathy and cancer [8]. The new pharmacological effects and possible indications of ferulic acid are being widely shown. Therefore, how to improve the efficacy of ferulic acid and its derivatives as well as the best indications is of great significance.

Fig. 1 *Resina ferulae*

Fig. 2 Chemical structure of ferulic acid

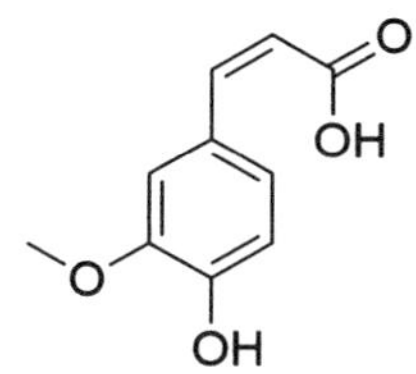

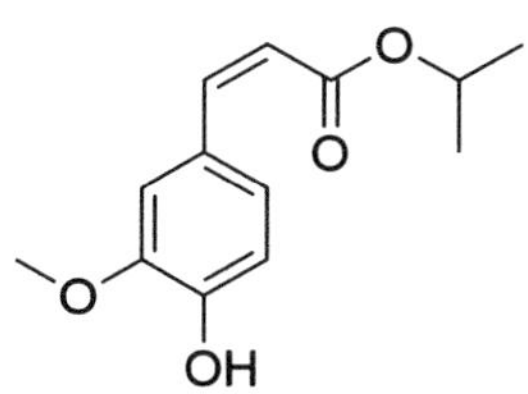

Isopropyl ferulate

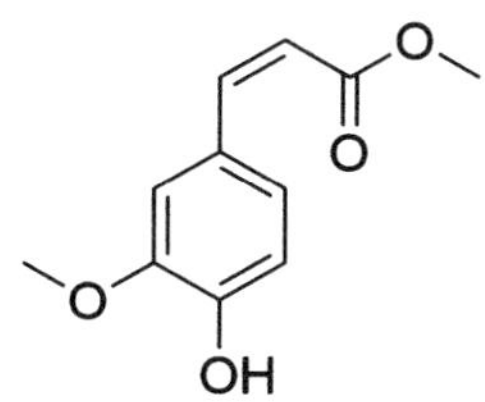

Ferulic acid methyl ester

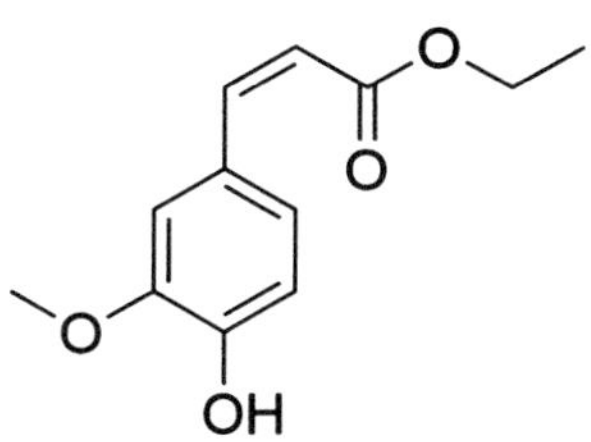

Ethyl ferulate

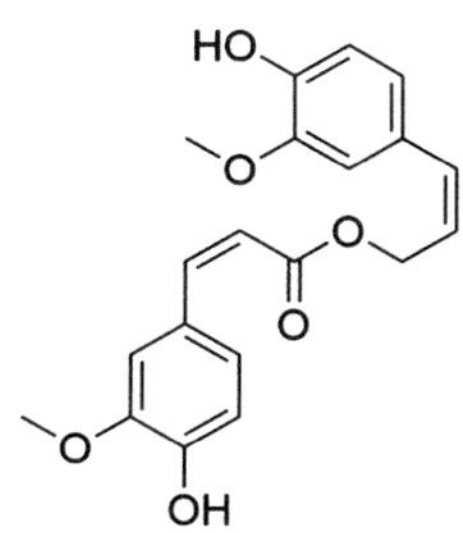

Coniferyl ferulate

Fig. 3 Chemical structure of ferulic acid derivatives

References

1. Preziosi P, Loscalzo B. Pharmacodynamic research on the active principle of Cynara scolymus(1, 4-dicaffeiylquinic acid): effect on blood cholesterol values & on triton-induced hypercholesterolemia. Boll Soc Ital Biol Sper. 1957;33:679–82.
2. Lin M, Zhu Z, Qingmin S, et al. Studies on chemical constituents of Angelica sinensis. Chin J Pharm Sci. 1979;14(9):529–33.
3. Sgarbossa A, Giacomazza D, Di Carlo M. Ferulic acid: a hope for Alzheimer's disease therapy from plants. Forum Nutr. 2015;7(7):5764–82.
4. Dong-ping H, Wen-yu Y, Xing-fu C. Progress in ferulic acid research. J Tradit Chin Med. 2008;19(8):1839–41.
5. Barone E, Calabrese V, Mancuso C. Ferulic acid and its therapeutic potential as a hormetin for age-related diseases. Biogerontology. 2009;10(2):97–108.

6. Ma F, Jiaming L, Chuanrun L, et al. Ferulic acid derivatives and their biological activities. Prog Pharm. 2008;32(8):345–50.
7. Yingchun X, Xiaohui Q. Treatment of chronic glomerulonephritis with atorvastatin combined with sodium ferulate. West Med. 2011;23(2):250–2.
8. Huixing L, Haiying Y. Clinical study of atorvastatin combined with sodium ferulate on renal interstitial fibrosis in diabetic nephropathy. Chin J Biochem Pharm. 2014;4:043.

Hesperidin

Jiao Chen, Zhen-Zhen Wang, Ling-Lei Kong, and Nai-Hong Chen

Abstract Hesperidin, as a derivative of dihydroflavonoids, widely exists in legume, birch, lip flower, butterfly flower, Rutaceae, and citrus plants. Hesperidin has the function of vitamin P, controlling blood pressure, antiallergic, reducing bone mineral density and cholesterol, improving enzyme activity and microcirculation, antibacterial, anti-inflammatory, anti-hepatitis B, and antitumor. As auxiliary materials, hesperidin is widely used to aid in the treatment of the cardiovascular system. Hesperidin, as a natural plant extract and nontoxic side effect to the human body, will have a broad market prospect.

Keywords Hesperidin · Dihydroflavonoids · Vitamin P

Alias: Orange peel glucoside; dihydroflavonoside; hesperidium glucoside
Origin: Citrus plants (Fig. 1)
Chemical name (Fig. 2)

7-[[6-O-(6-deoxy-α-L-mannopyranosyl)-β-D-glucopyranosyl]oxy]-2,3-dihydro-5-hydroxy-2-(3-hydroxy-4-methoxyphenyl)

Molecular formula, $C_{27}H_{32}O_{14}$; **MW**, 610.56; **CAS**, 520-26-3

J. Chen · Z.-Z. Wang · N.-H. Chen (✉)
Institute of Materia Medica, Chinese Academy of Medical Sciences and Peking Union Medical College, Beijing, China
e-mail: chennh@imm.ac.cn

L.-L. Kong
Beijing Key Laboratory of Drug Targets Identification and Drug Screening, Centre for Pharmaceutical Screening, Institute of Materia Medica, Chinese Academy of Medical Sciences and Peking Union Medical College, Beijing, China

G.-H. Du, *Natural Small Molecule Drugs from Plants*,
https://doi.org/10.1007/978-981-10-8022-7_13

Properties

Appearance: fine dendritic crystal (precipitation at pH 6–7), odorless, tasteless. *Melting point*: 258–262 °C (softening at 250 °C). *Solubility*: 1 g of hesperidin can be soluble in 50 l of water. Hesperidin dissolves in dimethyl formamide and formamide at 60 °C. It is slightly soluble in methanol and hot ice acetic acid and hardly soluble in acetone, benzene, and chloroform and is soluble in dilute alkali and pyridine.

Dosage Forms

Injection

Indications

Hesperidin can be used for cardiovascular disease prevention and treatment, blood sugar and blood lipid and blood pressure regulation, circulatory system regulation, and body regulation, and it can also be used as an antibacterial, anti-inflammatory, and antiviral.

Literature

Hesperidin exists mainly in the peel of lemon, orange, and Seville orange flower, which belong to facilitating medicine. These Chinese medicines are warm and fragrant with the function of eliminating depression and knots. They cure abdominal distension, belching swallow acid pain, nausea and vomiting, diarrhea or constipation caused by the spleen, and stomach qi zhi; and they also cure depression, hernia, breast pain, and menoxenia caused by liver qi; moreover, they cure chest pain, cough, and asthma, which are caused by lung qi.

Modern research has shown that li qi medicines have an extensive effect, such as regulatory effect on the digestive system, and they control the bronchial smooth muscle, the uterus smooth muscle, and the cardiovascular system. The base of the effect of li qi medicines in inverse, anti-nausea, antidiarrheal, and analgesia pharmacological effects is its effect in inhibiting gastric bowel movement; its exciting gastrointestinal movement is the foundation to eliminate swelling; its effect in relaxation of the bronchial smooth muscle is the foundation of pharmacological effects in anti-nausea. The intravenous injection to treat shock effect is the new development of li qi medicine.

History of R&D

Hesperidin is the glycoside in the form of hesperidin and rubiose and is a derivative of dihydroflavonoids. It widely exists in legume, birch, lip flower, butterfly flower, Rutaceae, and citrus plants [1]. Hesperidin is an important composition of citrus pulp and peel; most of hesperidin exists in citrus processing waste such as skin and fruit bag. Mature skin and tissue have the highest content of hesperidin (30–50% in endocarp; 30–50% in orange collaterals, nuclear, and pulp; and 10–20% in exocarp); the content of hesperidin is relatively low in juice and orange bag, which is about 1–5%. The crude extracts of hesperidin was first discovered in 1827 by Lebreton. Then the Hungarian scholar Albert Szent-Gyorgi discovered that the flavonoids have a protective microvascular effect in 1936, which is similar to that of vitamin P [2]. Preparation of vitamin P was made in 1938 [3]. It was not until 1949 that it was discovered that vitamin P was made up of two flavonoids, luteolin and hesperidin, which are believed to be vitamin active. This substance, which was later named as vitamin P, was designed to reduce blood vessel permeability and brittleness, as well as alleviate bad blood and vitamin C deficiency. It was later discovered that the substance had an antioxidant effect, so the name of vitamin P was abandoned. Due to the widespread distribution of hesperidin in plant medicine, the research and development have been widely followed.

At present the main methods of extracting hesperidin include [4] methanol Soxhlet extraction process, solvent ultrasonic extraction, reflux extraction, column chromatography, methanol ultrasonic extraction, methanol dilution centrifugation, and so on. Hesperidin extraction is to use its two main phenolic hydroxyls in alkaline conditions, and the solution of sodium ion reaction between sodium and dissolution, acidification, and cooling, makes its precipitation from a solution. The hot extraction method and the soaking extraction method are not ideal for extracting hesperidin from citrus peel. In recent years, the research on the effective components of ultrasonic extraction of natural plants has been widely carried out, and some progress has been made. Hesperidin exists in open loop and dissolves under the condition of alkaline and precipitations in acid condition, so increasing dosage of alkali can reduce the dosage of ethanol, but the basic dosage should not be too large; otherwise, hesperidin is susceptible to oxidative damage.

Pharmacology

The pharmacological effect of hesperidin is widespread, and people thought it was vitamin P in the early days, but in recent years, people found that it has other functions such as controlling blood pressure, antiallergic, reducing bone mineral density and cholesterol, improving enzyme activity and microcirculation, antibacterial, anti-inflammatory, anti-hepatitis B, antitumor, and other pharmacological effects.

Hesperidin has the function of vitamin P, which can reduce capillary permeability and prevent microvascular hemorrhage. Intraperitoneal injection of hesperidin at 175–250 mg/kg in mice could increase permeability of blood vessels by antihistamine and inhibiting hemolytic lecithin. Hesperidin has antiviral and antimicrobial effect, and preincubation with hesperidin at 200 mg/ml protects the cells from viruses. One to 10 μg/ml of hesperidin effectively inhibits the growth of the fungus. It has the effect of maintaining the normal osmotic pressure of the blood vessels, reducing the shortness of blood vessels, shortening bleeding time, reducing blood fat, and preventing atherosclerosis; hesperidin has an effect on the gastrointestinal tract, which can excite the smooth muscle transiently and then inhibit it, and it is a major component of the diet drug; hesperidin has an effect of anti-lipid peroxidation and scavenging hydroxyl radical. Hesperidin is a newly discovered flavonoid compound which has an effect in the central nervous system; it has a sedative effect. At the same time, hesperidin has the effect of lowering cholesterol, curing rheumatism, and inhibiting skin pigmentation. Hesperidin is a strong affinity for estrogen receptors, which can be used in estrogen receptors to prevent bone loss and reduce the number of osteoblasts. Hesperidin has a significant inhibitory effect on human lung cancer, colorectal cancer, kidney cancer, and human breast cancer cells, which can be used for cancer prevention [5].

Clinical Application

Hesperidin has the effect to maintain osmotic pressure, strengthen the capillary toughness, shorten the bleeding time, lower cholesterol, and so on. Although hesperidin cannot be used as independent medication, it is recorded in the pharmacopoeia that hesperidin, as auxiliary materials, is widely used to aid in the treatment of cardiovascular system; it can be configured as a variety of drugs to prevent hardening of the arteries and myocardial infarction. It is one of the main raw materials of medicine "pulse." Hesperidin is used as auxiliary materials for the treatment of vascular brittleness, bedsore, rheumatoid arthritis, vitamin C deficiency disease, trauma, obstetric disease, gum inflammation, edema, and gastrointestinal tract disease in the world [6]. Hesperidin can be used to produce an anticancer drug called diosmin. Natural antioxidant is available in the food industry [7]. It is also used in the cosmetics industry.

Discussion

Studies at home and abroad show that hesperidin has broad application prospects in health care, functional food, food additives, and cosmetics. Hesperidin is rich in resources and is widely available in natural green plants. Citrus, lemon, and Buddhist plants are all good sources of Hesperian, which provides a good resource base for the development and utilization of hesperidin. Hesperidin has a great amount of physiological effects, and it is a natural plant extract and has nontoxic side effects to the human body. All these indicate a wide development prospect in the field of health food. With the progress of science and technology and the gradual improvement in people's health consciousness, the research and development about hesperidin will be deeper. Hesperidin, as the main component of functional food, will have a broad market prospect.

Fig. 1 *Citrus aurantium*

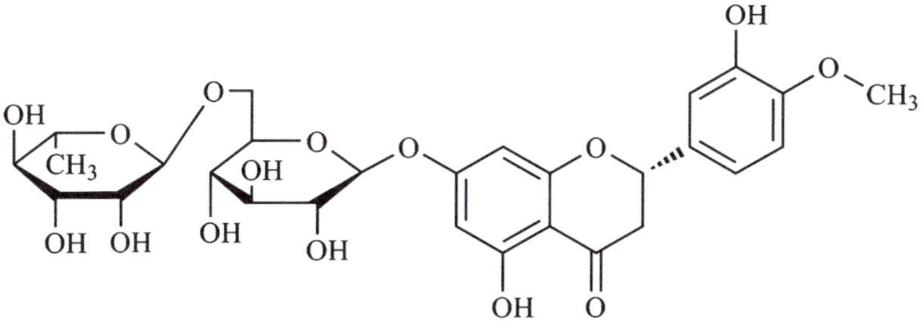

Fig. 2 The structure of hesperidin

References

1. Barthe GA. Radioimmunoassay for the quantitative determination of hesperidin and analysis of its distribution in *Citrus sinensis*. Phytothem. 1988;27(1):249.
2. Armentano L, Bentsath A, Beres T, Rusznyak S, Szent-Gyorgi A. Uber den Einfluss von substanzen der flavon gruppe auf die permeabilitat der kapillaren. Vitamin P Dtsch Med Wochenschr. 1936;62:1325–8.
3. Szent-Gyorgi A. Preparation of citrin. Physiol Chem. 1938;225:126–31.
4. Ma QY, Chen CT, Jing XY, Yuan WX. Studies on extraction of hesperidin and other active components from citrus peels and their antimicrobial effects. Food Sci. 2004;259(12):112–5.
5. Li YS. Advances in research on hesperidin. Sci Technol Rev. 2009;27(22):108–16.
6. Garg A, Garg S, Zaneveld LJD, Singla AK. Chemistry and pharmacology of the citrus bioflavonoid hesperidin. Phytother Res. 2001;15:655–69.
7. Saija A, Tomaino A, Loascio R, et al. In vitro antioxidant activity and in vivo photo protective effect of orange extract. Int J Cosm Sci. 1998;20(6):331–3.

Ligustrazine

Guo-Rong He, Shou-bao Wang, and Guan-Hua Du

Abstract Ligustrazine is extracted from *chuanxiong* 川芎 (*Ligusticum chuanxiong Hort; L. wallichii*), which is the most important alkaloid in *chuanxiong*. Studies have demonstrated that ligustrazine is effective at promoting blood flow, the circulation of Qi, and the removal of blood stasis, so it has been extensively used to treat ischemic cardiovascular and cerebrovascular diseases, such as atherosclerosis, hypertension, and stroke. Ligustrazine has also used to treat children with viral myocarditis, chronic renal failure, hepatic fibrosis, portal hypertension, tumors, and diabetic nephropathy. Ligustrazine can relieve symptoms rapidly, and its effect is stable and lasting. It has good spread of the value in clinic. To decide the specificity of ligustrazine used in clinic, we carried out a large-sample, multicenter, randomized, double-blind, placebo-controlled trial, which provides evidences for the extensive use of ligustrazine in clinic.

Keywords Ligustrazine · Alkaloid · Antiapoptosis · Promoting blood circulation

Alias: 2,3,5,6-tetramethylprazine; Tetramethylpyrazine; Pyrazine
Origin: *Ligusticum chuanxiong Hort* (Fig. 1)
Chemical name (Fig. 2)

2,3,5,6-tetramethylprazine

Molecular formula, $C_8H_{12}N_2$; **MW**, 136.20; **CAS**, 1124-11-4

G.-R. He · S.-b. Wang · G.-H. Du (✉)
Beijing Key Laboratory of Drug Targets Identification and Drug Screening, Centre for Pharmaceutical Screening, Institute of Materia Medica, Chinese Academy of Medical Sciences and Peking Union Medical College, Beijing, China
e-mail: dugh@imm.ac.cn

G.-H. Du, *Natural Small Molecule Drugs from Plants*,
https://doi.org/10.1007/978-981-10-8022-7_14

Properties

Appearance: colorless needle crystal. *Solubility*: soluble in hot water, petroleum ether, chloroform, dilute hydrochloric acid, slightly soluble in ether, insoluble in cold water. *Melting point*: 77–80 °C. *Boiling point*: 190 °C. *Flash point*: 128–130 °C/200 mm. Ligustrazine has a particular odor, moisture absorption, and sublimation.

Dosage Forms

Ligustrazine hydrochloride was included in the Pharmacopoeia of the People's Republic of China (2015).

Ligustrazine hydrochloride injection; ligustrazine phosphate tablets; ligustrazine phosphate injection; ligustrazine phosphate pills.

Indication

Ligustrazine is used for occlusive vascular diseases, cerebral thrombosis, vasculitis, coronary heart disease, angina pectoris, and others.

Literature

Ligustrazine is an alkaloid derived from *chuanxiong* 川芎 (*Ligusticum chuanxiong Hort; L. wallichii*). *Ligusticum wallichii* Franchat (or *Ligusticum chuanxiong Hort; L. wallichii*), a member of the Umbelliferaceae family, is a Chinese medicinal herb that is a common constituent in polypharmaceutical herbal drugs. It has been widely used in Asian countries to increase coronary blood flow and systemic circulation and to relieve stasis.

The earliest Chinese materia medica monograph "Shennong's classic of materia medica" contained Xiong, which is listed in the top grade. Since ancient times its original plants have been found in several species, with different habitats and different plant names. Therefore, geographical names are added to show the difference. In the 2005 edition of the Pharmacopoeia of the People's Republic of China, among 564 prescriptions and single-herb preparations collected, 85 prescriptions used chuanxiong, which accounted for about 15% of the total amount of Chinese patent medicines contained in the pharmacopoeia.

Ligusticum walliichi has the effects of promoting Qi and nourishing blood. *Ligusticum walliichi*, *angelica*, and *peony* were used as medicinal herbs, and the commonly used Yangxue Siwu decoction was proposed for clinical practice. The combination of gypsum and *Ligusticum walliichi* can clear lung heat and eliminate cough and asthma. It is a common drug for clearing away heat and toxic substances and purging fire. The compatibility of *Ligusticum walliichi* and *Radix Linderae* can regulate the blood and Qi and treat women's headaches and colds and retrograde and other related pains.

Ligusticum walliichi is warm, acrid, and bitter. It can regulate the blood and conduct Qi, which is commonly used in oral administration. *Ligusticum walliichi* is used to treat apoplexy, headache, cold pain, the upper reaches of the wind, tears clogging of nose by cold, and so on. At present, the study of Chuanxiong in China and abroad has occurred at the molecular level [1].

History of R&D

Ligusticum walliichi has a long medicinal history in China and is used in the treatment of headache, promoting blood and Qi circulation, curing chills and alleviating pain". Clinically, it is mainly used for the treatment of cardiovascular and cerebrovascular diseases. Ligustrazine (2,3,5,6-tetramethylpyrazine) is the main active ingredient of *Ligusticum walliich.* It was first isolated in 1962 [2, 3].

In recent years, studies have found that ligustrazine can inhibit the activity of the cerebral cortex and medullary respiratory center, excite vasomotors, induce the expansion of peripheral blood vessels, promote the blood flow of the coronary artery and lower limbs, and reduce blood pressure. Ligustrazine can also paralyze the nerve center and has sedative and analgesic effects.

Studies have suggested that ligustrazine has a variety of new effects, including improving leptomeningeal and peripheral microcirculation and increasing cerebral blood flow, and its effect is better than that of papaverine. Ligustrazine can also protect against experimental cerebral ischemia and alleviate brain edema and the deposition of fibrin in microvessels. Ligustrazine is also used in the treatment of pulmonary heart disease, heart failure, dilated heart disease, portal hypertension, type II diabetes mellitus, tumor and coronary heart disease, and restenosis after stent implantation [4, 5].

Pharmacology

Traditional Chinese medicine has some advantages in comprehensive treatment under multisite, multitarget conditions and overall regulation. Ligustrazine can inhibit phosphodiesterase activity and inhibit platelet aggregation. It has the typical characteristics of a calcium antagonist and can adjust the release of various vasoactive substances against the vasoconstrictive action of the sympathetic nerve, improve microcirculation, reduce red blood cell aggregation, shorten the erythrocyte electrophoresis time, and thereby reduce blood viscosity [6].

Ligustrazine can inhibit the proliferation of vascular endothelial cells induced by vascular endothelial growth factor, which may be one of the mechanisms by which angiogenesis is inhibited. The mechanism by which ligustrazine inhibits the proliferation of vascular endothelial cells may be through calcium ion retardation, inhibition of cell adhesion molecules, or through the antagonistic or synergistic effects of some growth factors involved in the proliferation and angiogenesis of vascular endothelial cells.

Studies have shown that ligustrazine plays a role in resistance to ischemia-reperfusion injury and antiapoptotic effects via the phosphatidylinositol 3- kinase/protein kinase B (PI3K/Akt) pathway. The phosphorylation of endothelial nitric oxide synthase (eNOS) and generation of nitric oxide (NO) are important downstream effectors, which significantly enhance the myocardial protective effects of ligustrazine.

In addition, it has been reported that ligustrazine can protect against rat myocardial cells infected with coxsackievirus 3 (CVB3), and the mechanism may be related to the decrease of lactate dehydrogenase (LDH) activity and the expression of NF-kappa B. Ligustrazine can inhibit the proliferation of rat cardiac fibroblasts induced by angiotensin II and reduce the secretion and synthesis of type I collagen, which is one of the mechanisms of resistance to fibrosis [1].

Clinical Application

The main clinical formulation of ligustrazine is ligustrazine hydrochloride injection. Ligustrazine hydrochloride injection was used to prevent restenosis following coronary stent implantation and to treat acute coronary syndrome and ischemic stroke. Ligustrazine hydrochloride injection is helpful in alleviating pulmonary hypertension in patients with pulmonary heart disease. The therapeutic effect of ligustrazine hydrochloride injection on severe cholestatic liver cirrhosis caused by chronic heart failure of rheumatic heart disease is remarkable. In addition, ligustrazine is widely used in the treatment of vertigo syndrome, vertebrobasilar insufficiency, tension headache, and so on. It has a good symptomatic-treatment effect [7–9].

Ligustrazine is also reported for the treatment of children with viral myocarditis, chronic renal failure, hepatic fibrosis, portal hypertension, tumors, and diabetic nephropathy. Ligustrazine can relieve symptoms rapidly, and its effect is stable and lasting. It has good clinical promotion value [10].

Discussion

Clinical reports on ligustrazine mainly focus on small-sample clinical observations and lack advanced evidence-based support, thereby hindering ligustrazine's further application. To decide the specificity of ligustrazine used in clinic, we carried out a large-sample, multicenter, randomized, double-blind, placebo-controlled trial, which provides the basis for the clinical promotion of ligustrazine.

Fig. 1 *Ligusticum chuanxiong Hort*

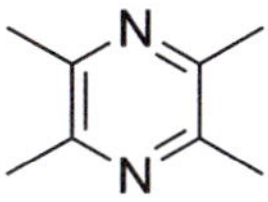

Fig. 2 The structure of Ligustrazine

References

1. Jia L, Sun X. Research progress of Chinese herbal medicine Ligusticum Wallichii. Heilongjiang Sci Technol Inform. 2009;11:146–7.
2. Kosuge T, Kamiya H. Discovery of a pyrazine in a natural product: tetramethylpyrazine from cultures of a strain of Bacillus subtilis. Nature. 1962;193:776.
3. Chen K, Qian Z, Zhang W, et al. Efficacy of Ligustrazine in the treatment of acute occlusive cerebrovascular disease. Report on coordination meeting of ligustrazine in Beijing 1975.
4. Jiang F, Qian J, Chen S, et al. Ligustrazine improves atherosclerosis in rat via attenuation of oxidative stress. Pharm Biol. 2011;49(8):856–63.
5. Tan F, Fu W, Cheng N, et al. Ligustrazine reduces blood-brain barrier permeability in a rat model of focal cerebral ischemia and reperfusion. Exp Ther Med. 2015;9(5):1757–62.
6. Jiang Y, Chen K. Pharmacological effects and clinical application of ligustrazine on cardiovascular and cerebrovascular diseases. Chin J Integr Tradit West Med. 2013;33(5):707–11.
7. Wu W. The clinical curative effect of Ligustrazine and citicoline sodium combined with diphenhydramine hydrochloride in treatment of vertigo syndrome. China Mod Med. 2011;18(4):51–5.
8. Li X, Chen Y. Efficacy of ligustrazine in the treatment of tension headache. Chin J Pract Nerv Dis. 2010;13(18):64–5.
9. Dong M, Li Y, Ge N, et al. Curative efficacy of Qingxuan decoction combined with ligustrazine injection in the treatment of vertebrobasilar insufficiency. Beijing J Tradit Chin Med. 2011;30(1):8–10.
10. Peng X. Efficacy of ligustrazine injection in treatment of viral myocarditis in children. J Youjiang Med Univ Natl. 2010;3:376–7.

Lovastatin

Jian Ying, Li-Da Du, and Guan-Hua Du

Abstract Lovastatin exists in red yeast rice, post-fermented pu-erh tea, oyster mushroom, and different varieties of fungi. It is used to treat dyslipidemia, prevent cardiovascular disease, or combined with dietary interventions. However, it is not effective in treatment of receptor-negative homozygous familial hypercholesterolemia. Lovastatin was originally known as an HMG-CoA reductase inhibitor, thus reducing cholesterol biosynthesis; more complicated mechanism of actions were discovered later. Pharmaceutical preparation strategies are needed to improve the absorption and bioavailability of lovastatin. Lovastatin can lead to drug-drug interactions and several adverse reactions though. Chemical structure modifications led to series of new statins, one of which has been list in the best-selling drugs in the history of pharmaceuticals.

Keywords Lovastatin · Red yeast rice · Cholesterol · HMG-CoA reductase · Dyslipidemia

Alias: Mevinolin, Mevacor, Monacolin K, Altoprev, Altocor, Lovalord, Mevinacor
Origin: Pu-erh tea (Fig. 1)
Chemical name (Fig. 2)

[(1S,3R,7S,8S,8aR)-8-[2-[(2R,4R)-4-Hydroxy-6-oxooxan-2-yl]ethyl]-3,7-dimethyl-1,2,3,7,8,8a-hexahydronaphthalen-1-yl] (2S)-2-methylbutanoate

J. Ying
Beijing Key Laboratory of Nutrition Health and Food Safety, COFCO Nutrition and Health Research Institute, Beijing, China

L.-D Du
School of Biomedical Sciences, Chinese University of Hong Kong, Hong Kong, China

G.-H. Du (✉)
Beijing Key Laboratory of Drug Targets Identification and Drug Screening, Centre for Pharmaceutical Screening, Institute of Materia Medica, Chinese Academy of Medical Sciences and Peking Union Medical College, Beijing, China
e-mail: dugh@imm.ac.cn

G.-H. Du, *Natural Small Molecule Drugs from Plants*,
https://doi.org/10.1007/978-981-10-8022-7_15

Molecular formula, $C_{24}H_{36}O_5$; **MW**, 404.54; **CAS**, 75,330-75-5

Properties

Appearance: White, nonhygroscopic crystalline powder. *Solubility*: Insoluble in water (0.0004 mg/mL); sparingly soluble in ethanol, methanol, isobutanol, isopropanol, acetonitrile, n-propanol; soluble in acetone, *N*-dimethylformamide; freely soluble in chloroform. *Melting point*: 174.5 °C. *Specific optical rotation*: 25 °C for D (sodium) line, +323° (concentration 0.5 g in 100 ml acetonitrile). *Stability*: Lovastatin is sensitive to light. Following exposure to extreme light conditions, the drug is stable for 24 h or 1 month when exposed to UV (approximately 3230 lux) or fluorescent (approximately 10,764 lux) light, respectively, at 28 °C in air.

Derivatives

Simvastatin; pravastatin; fluvastatin; atorvastatin; cerivastatin; rosuvastatin

Dosage Forms

This product is contained in *Pharmacopoeia of the People's Republic of China* (2015), *British Pharmacopoeia* (2017), *United States Pharmacopeia* (40), and *European Pharmacopoeia* (9.0th ed.).

Tablet, capsule, granule.

Indications

Hypercholesterolemia, combined hyperlipidemia.

Literature

Lovastatin is a natural product at low level in some foods, such as red yeast rice [1], pu-erh tea [2], and oyster mushroom [3]. It is one of the secondary metabolites produced by fungi.

Red yeast rice has been used in flavoring, coloring, and preserving foods for a long time in Asian countries, especially in China. In southeast region of China, people use red yeast rice as fermentation starter to produce yellow wine. It is also used in traditional Chinese medicine, known as *hongqu*. Red yeast rice is made by fermenting white rice with the yeast *Monascus purpureus* and other related fungi. At least 13 monacolins have been isolated from red yeast rice, one of which is monacolin K, i.e., lovastatin [1].

Pu-erh tea is a type of post-fermented dark tea in China. It is manufactured from leaves of *Camellia sinensis (Linn.) var. assamica (Masters) Kitamura*, which grows in Yunnan Province in south China. Raw pu-erh tea is turned into ripe pu-erh after natural aging for years or by industrial fermentation at higher temperature and higher humidity. During the post-fermentation process, *Aspergillus*, *Penicillium*, and *Trichoderma* produce lovastatin, usually ranging from tens up to hundreds of nanograms per gram of dried tea leaves [2].

Pleurotus ostreatus (Pleurotaceae) (Jacq. ex. Fr, Kummer) is an edible mushroom that is grown all over the world and common in culinary use. The content of lovastatin in *Pleurotus ostreatus* ranged from hundreds of micrograms up to milligrams per gram of dried mushroom [3].

Although lovastatin exists in varieties of foods, it was firstly isolated from cultured *Aspergillus terreus*. The organisms that have been reported to produce lovastatin as a secondary metabolite include *Aspergillus terreus*, *Monascus* sp., *Aspergillus niger*, *Aspergillus flavus*, *Penicillium purpurogenum*, *Pleurotus* sp., *Trichoderma viride*, and *Penicillium* sp. For commercial purpose, *Penicillium* sp., *Monascus ruber*, and *Aspergillus terreus* are the most often used [4].

History of R&D

Statins are the most extensively used class of lipid-lowering medication. Lovastatin is the second statin discovered by scientists. The first one is compactin isolated from *Penicillium citrinum* by Sankyo Research Laboratories and identified as a HMG-CoA inhibitor [5–7].

In the 1970s, under a confidentiality agreement with Sankyo, Merck Research Laboratories confirmed the finding of compactin and isolated another statin from the fermentation broth of *Aspergillus terreus*. They named their discovery mevinolin [8]. Akira Endo from Sankyo Co. independently isolated lovastatin from *Monascus rube*r and got the patent 4 months earlier than Merck. He named the chemical monacolin K [9]. In the fall of the same year, it was confirmed that monacolin K and mevinolin are the same compound.

Later, the official name lovastatin was established. The activity of lovastatin is much better than compactin. In early 1981, Brown and Goldstein reported that lovastatin could upregulate liver LDL receptors in dogs and significantly lower plasma LDL levels [10]. Later, LDL cholesterol-lowering effects of lovastatin were confirmed in a clinical study [11].

In July 1982, lovastatin showed dramatic effects on lowering LDL cholesterol in patients with severe hypercholesterolemia who were unresponsive to the existed medicines, with very few adverse reactions [12, 13]. Merck confirmed the effects of lovastatin in a large-scale clinical trial. A long-term toxicity study showed no tumorigenesis in dogs.

In 1986, Merck applied for regulatory approval of lovastatin. The FDA approval was obtained on 31 August 1987. However, although compactin was highly effective in hens, dogs, and monkeys, it resulted in lymphoma in dogs at extremely high dose in a 2-year study. The development of compactin was canceled afterward [14].

Chemical structure modifications of compactin and lovastatin lead to a series of new drugs. Simvastatin was approved for marketing in Sweden in 1988 and subsequently worldwide followed by pravastatin in 1991, fluvastatin in 1994, atorvastatin in 1997, cerivastatin in 1998, and rosuvastatin in 2003. Simvastatin is a semisynthetic derivative of lovastatin. Pravastatin is derived from the natural product compactin by biotransformation. Other HMG-CoA reductase inhibitors are totally synthetic products. Atorvastatin, under the brand name Lipitor, has become the best-selling drug in the history of pharmaceuticals.

Pharmacology

Lovastatin is an inhibitor of 3-hydroxy-3-methylglutaryl-coenzyme A reductase (HMG-CoA reductase), an enzyme that catalyzes the conversion of HMG-CoA to mevalonate [15], a rate-determining step of cholesterol biosynthesis. Lovastatin is metabolized as a prodrug into an active form, lovastatin acid. Lovastatin acid is a reversible competitive inhibitor for HMG-CoA.

However, the reduction in plasma cholesterol by statins is not only due to reduction in cholesterol biosynthesis. As an example, lovastatin also lowers human plasma cholesterol by increasing the uptake of LDL via LDL receptor [8] and decreases the production of apolipoprotein B-containing lipoproteins by the liver [16].

In addition to their lipid-lowering properties, statins produce several nonlipid-related properties, include decreasing levels of high-sensitivity C-reactive protein (hsCRP), improving endothelial function, reducing inflammation at the site of the atherosclerotic plaque, inhibiting platelet aggregation, anticoagulant, etc. [17, 18].

Almost 30% of lovastatin can be absorbed after oral administration, but only 5% can reach the systemic circulation due to an extensive first-pass effect. Hepatic metabolism hydrolyzes lovastatin into its active form. Its time to peak blood concentration is 2–4 h; and the half-life time is 1.1–1.7 h. 80% to 85% of lovastatin is excreted in feces, while 10% excreted in urine. Advanced pharmaceutical technologies are developed to improve the bioavailability of lovastatin. Recently, it was found that a rod-shaped lovastatin nanocrystals exhibited superior *in vitro* dissolution and *in vivo* bioavailability over spherical nanocrystals [19]. A mesoporous nanomatrix-supported lipid bilayer was designed to deliver amorphous lovastatin, thus improving the dissolution behavior and *in vivo* absorption [19]. Lovastatin can inhibit CYP3A4 in the liver and P-gp in the intestine, which contribute to drug-drug interactions.

Clinical Application

The primary uses of lovastatin are the treatment of dyslipidemia and the prevention of cardiovascular disease. It is recommended to be used only when other approaches such as diet, exercise, and weight reduction have not improved the cholesterol profile ("Lovastatin". The American Society of Health-System Pharmacists. Retrieved 3 April 2011.). Lovastatin is useful in treating hypercholesterolemia and combined hyperlipidemia. However, lovastatin is not effective in treatment of receptor-negative homozygous familial hypercholesterolemia [20].

Lovastatin is usually well tolerated. Its common side effects are listed in approximately descending order of occurrence frequency: creatine phosphokinase elevation, flatulence, abdominal pain, constipation, diarrhea, muscle aches or pains, nausea, indigestion, weakness, blurred vision, rash, dizziness, and muscle cramps. As with all statin drugs, it can rarely cause myopathy, hepatotoxicity (liver damage), dermatomyositis, or rhabdomyolysis. These can be life-threatening if not recognized or treated in time, so any unexplained muscle pain or weakness should be reported to the prescribing doctor while using lovastatin.

Drug interaction happens with warfarin, gemfibrozil. Other medications can affect the removal of lovastatin from your body, which may affect how lovastatin works. These medications include certain azole antifungals (such as itraconazole, ketoconazole, posaconazole, voriconazole), cobicistat, colchicine, cyclosporine, delavirdine, macrolide antibiotics (such as clarithromycin, erythromycin), nefazodone, HIV protease inhibitors (such as lopinavir, ritonavir), hepatitis C virus protease inhibitors (such as boceprevir, telaprevir), telithromycin, etc.

Red yeast rice products are not recommended to take together with lovastatin, since the total amount of lovastatin can increase to an uncontrolled amount.

Discussion

Lovastatin is a natural-derived drug which exists at low level in varieties of foods. The primary uses of lovastatin are to treat dyslipidemia and to prevent cardiovascular disease. However, the medication should be combined with dietary interventions. Lovastatin is insoluble in water. The absorption and bioavailability of lovastatin can be improved by pharmaceutical preparation strategies. Although lovastatin is generally safe, it can lead to drug-drug interactions and several adverse reactions. It is not effective in treatment of receptor-negative homozygous familial hypercholesterolemia.

Fig. 1 Pu-erh tea

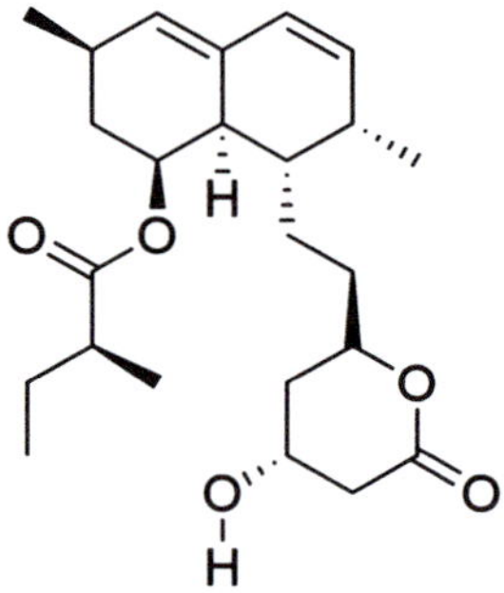

Fig. 2 The structure of lovastatin

References

1. Nguyen T, Karl M, Santini A. Red yeast rice. Foods. 2017;6(3):E19.
2. Zhao ZJ, Pan YZ, Liu QJ, et al. Exposure assessment of lovastatin in Pu-erh tea. Int J Food Microbiol. 2013;164(1):26–31.
3. Gunde-Cimerman N, Cimerman A. Pleurotus fruiting bodies contain the inhibitor of 3-hydroxy-3-methylglutaryl-coenzyme A reductase-lovastatin. Exp Mycol. 1995;19(1):1–6.
4. Goswami SVA, Bhunia B, Mandal T. A review on lovastatin and its production. J Biochem Technol. 4(1):581–7.
5. Betteridge DJ, Krone W, Reckless JP, et al. Compactin inhibits cholesterol synthesis in lymphocytes and intestinal mucosa from patients with familial hypercholesterolaemia. Lancet. 1978;2(8104–5):1342–3.
6. Endo A, Kuroda M, Tanzawa K. Competitive inhibition of 3-hydroxy-3-methylglutaryl coenzyme A reductase by ML-236A and ML-236B fungal metabolites, having hypocholesterolemic activity. FEBS Lett. 1976;72(2):323–6.
7. Brown AG, Smale TC, King TJ, et al. Crystal and molecular structure of compactin, a new antifungal metabolite from Penicillium brevicompactum. J Chem Soc Perkin. 1976;1(11):1165–70.

8. Ma PT, Gil G, Sudhof TC, et al. Mevinolin, an inhibitor of cholesterol synthesis, induces mRNA for low density lipoprotein receptor in livers of hamsters and rabbits. Proc Natl Acad Sci U S A. 1986;83(21):8370–4.
9. Endo A. Monacolin K, a new hypocholesterolemic agent that specifically inhibits 3-hydroxy-3-methylglutaryl coenzyme A reductase. J Antibiot (Tokyo). 1980;33(3):334–6.
10. Brown MS, Dana SE, Goldstein JL. Regulation of 3-hydroxy-3-methylglutaryl coenzyme A reductase activity in human fibroblasts by lipoproteins. Proc Natl Acad Sci U S A. 1973;70(7):2162–6.
11. Brown MS, Dana SE, Goldstein JL. Regulation of 3-hydroxy-3-methylglutaryl coenzyme A reductase activity in cultured human fibroblasts. Comparison of cells from a normal subject and from a patient with homozygous familial hypercholesterolemia. J Biol Chem. 1974;249(3):789–96.
12. Vagelos PR. Are prescription drug prices high? Science. 1991;252(5009):1080–4.
13. Liu J, Wang M, Peng S, et al. Effect of green tea catechins on the postprandial glycemic response to starches differing in amylose content. J Agric Food Chem. 2011;59(9):4582–8.
14. Endo A. A historical perspective on the discovery of statins. Proc Jpn Acad Ser B Phys Biol Sci. 2010;86(5):484–93.
15. Alberts AW. Discovery, biochemistry and biology of lovastatin. Am J Cardiol. 1988;62(15):10J–5J.
16. Arad Y, Ramakrishnan R, Ginsberg HN. Effects of lovastatin therapy on very-low-density lipoprotein triglyceride metabolism in subjects with combined hyperlipidemia: evidence for reduced assembly and secretion of triglyceride-rich lipoproteins. Metabolism. 1992;41(5):487–93.
17. De Denus S, Spinler SA. Early statin therapy for acute coronary syndromes. Ann Pharmacother. 2002;36(11):1749–58.
18. Ray KK, Cannon CP. Optimal goals for statin use in patients with coronary artery disease. Curr Opin Cardiol. 2005;20(6):525–9.
19. Zhang Y, Zhang H, Che E, et al. Development of novel mesoporous nanomatrix-supported lipid bilayers for oral sustained delivery of the water-insoluble drug, lovastatin. Colloids Surf B Biointerfaces. 2015;128:77–85.
20. Uauy R, Vega GL, Grundy SM, et al. Lovastatin therapy in receptor-negative homozygous familial hypercholesterolemia: lack of effect on low-density lipoprotein concentrations or turnover. J Pediatr. 1988;113(2):387–92.

Metformin

Xiu-Ying Yang and Guan-Hua Du

Abstract Metformin is a biguanide compound originated from the extraction of goat's rue plants, also known as *Galega officinalis*. The structure of metformin was identified by British scholars in the early 1920s, and its hypoglycemic effect was found through the study of galegine in 1950s. Metformin is a first-line drug for the treatment of type 2 diabetes, mainly used as an oral hypoglycemic drug in clinical practice, especially for obese patients with normal renal function. Metformin is able to reduce the levels of blood glucose and lipid as well as capable of regulating cell growth, anti-inflammation, antiaging, etc. The main pharmacological mechanisms involve the inhibition of hepatic gluconeogenesis, activation of AMPK, and regulation of mitochondrial function. With more and more novel pharmacological effects discovered besides hypoglycemic, the application prospect of metformin would be more optimistic.

Keywords Metformin · Glucophage · Diabetes · *Galega officinalis*

Alias: Dimethylbiguanide; dimethylguanylguanidine; glucophage; difrex
Origin: *Galega officinalis* (Fig. 1)
Chemical name (Fig. 2)

N,N-dimethylimidodicarbonimidic diamide

Molecular formula, $C_4H_{11}N_5$; **MW**, 129.16; **CAS**, 657-24-9
Derivatives (Fig. 3)

X.-Y. Yang · G.-H. Du (✉)
Beijing Key Laboratory of Drug Targets Identification and Drug Screening, Centre for Pharmaceutical Screening, Institute of Materia Medica, Chinese Academy of Medical Sciences and Peking Union Medical College, Beijing, China
e-mail: dugh@imm.ac.cn

G.-H. Du, *Natural Small Molecule Drugs from Plants*,
https://doi.org/10.1007/978-981-10-8022-7_16

Properties

Appearance: white crystalline or crystalline powder, odorless. *Solubility*: freely soluble in water, soluble in methanol, slightly soluble in ethanol, and insoluble in chloroform or ether. *Melting point*: 223–226 °C.

Dosage Forms and Indications

This product was collected in the *Pharmacopoeia of the People's Republic of China* (2015), *British Pharmacopoeia* (2017), *United States Pharmacopeia* (40), the *Japanese Pharmacopoeia* (17th ed.), and *European Pharmacopoeia* (9.0th ed.).

The main clinical dosage forms are oral preparations, including ordinary tablets (capsules), enteric-coated tablets (capsules), and sustained-release tablets (capsules). Metformin is primarily used for type 2 diabetes, especially for the patients ineffective in diet therapy and physical exercise. In addition, combined with insulin, metformin can reduce the dosage of insulin to prevent the occurrence of hypoglycemia. It can also be used in combination with sulfonylurea hypoglycemic agents for synergistic lowering blood glucose.

Literature

The study of metformin and its hypoglycemic effects originated from the study of goat's rue plants, also known as *Galega officinalis* 山羊豆 (French lilac). Goat's rues are native plants in the Middle East and introduced to Europe later and have been used as forage and ornamental plants throughout the world, including China. As early as in the Middle Ages in Europe, it was found that goat's rues could ease polyuria, which is one of the typical symptoms of diabetes. While goat's rues were used to treat a variety of other diseases in the Middle Ages, it was found to cause poisoning symptoms in livestock. Goat's rues are still used as medical plants at present, mainly for diabetes, diuretic, hepatoprotection, aiding in digestion and promoting lactation, etc. In China, goat's rues were recorded first in the dictionary of Chinese seed plants and mainly used for the treatment of diabetes. However, because of high toxicity, it is rarely used in traditional Chinese medicines at present.

History of R&D

Metformin is a biguanide compound which originated from the extraction of goat's rue plants. The structure of metformin was identified by British scholars in the early 1920s. In 1922, Werner and Bell et al. first synthesized metformin in 31 institutes in Dublin, Ireland [1]. In 1929, Slotta and Tschesche found metformin's hypoglycemic action [2]. However, because of other potent antidiabetic drugs such as insulin which were widely used in clinical practice, the pharmacological effects of metformin didn't receive much attention.

Until the 1950s, a French diabetic scientist Jean Sterne found the hypoglycemic effect of metformin through the study of galegine. Then the drug was used in diabetic patients for the first time, and the results were published in 1957 [3]. In 1958, metformin was collected in the *British National Formulary* and began selling. At the same time, the more potent hypoglycemic drugs phenformin and buformin also came in the market. Since the hypoglycemic effect is relatively weaker compared with phenformin and buformin, metformin was less competitive. Later in clinical practice, phenformin and buformin showed a higher risk of lactic acidosis and a higher mortality rate. By the end of the 1970s, phenformin was almost completely out of the market. Since metformin belongs to biguanide family, metformin was also affected and once proposed to be out of the market.

In the late 1970s, a prospective diabetic study was carried out in the UK (UKPDS). This study led metformin to be the first-line drug for the treatment of type 2 diabetes eventually. UKPDS, which began from 1977 and ended in 1997 and was then followed up for 10 years, is the longest in the history of clinical trials and has a significant impact on practice and guidelines for prevention and treatment of diabetes mellitus. In this trial, metformin was found to reduce the risk of diabetic complications by 32%. In addition, it was proved for the first time that metformin can reduce blood glucose and protect against cardiovascular function, especially in obese patients [4]. In 1994, metformin was approved by the US FDA for type 2 diabetes treatment. Currently, metformin has become the world's most widely used antidiabetic drug.

Aiming at improving the stability of the absorption of metformin, chemists have also carried out a series of structural renovation and modification. Metformin activates with carbonyl, esters, chlorides, and aldehydes to form triazine compounds, with 1,3-diketone to produce pyrimidine compounds, and with disulfides to produce C-S coupling products, etc.

Pharmacology

As a traditional antidiabetic drug, metformin can reduce the levels of blood glucose and lipid, as well as regulating cell growth, anti-inflammation, antiaging, etc. The main pharmacological mechanisms are inhibition of hepatic gluconeogenesis, the activation of AMP-activated protein kinase (AMPK), and the regulation of mitochondrial function.

Improvement of Insulin Resistance and Decrease of Blood Glucose Levels The main pharmacological effects of biguanide drugs are to reduce blood glucose output and improve peripheral insulin resistance. Metformin inhibits hyperglycemia mainly by inhibiting hepatic glucose production (hepatic gluconeogenesis), increasing muscle glucose uptake. Hepatic gluconeogenesis in type 2 diabetic patients was three times faster than healthy people; metformin can effectively inhibit the process. The hypoglycemic molecular mechanisms of metformin is involved in the inhibition of mitochondrial respiratory chain complex I, activation of AMPK, glucagon suppression induced by cyclic adenosine monophosphate (cAMP), activation of protein kinase A (PKA), inhibition of mitochondrial glycerol-3-phosphate dehydrogenase (mGPD), promoting the binding of insulin and insulin receptor, induction of GLUT4-enhanced phosphorylation [5] etc. Polycystic ovary syndrome has a close relationship with insulin resistance; metformin treatment can restore ovarian function, increase ovulation, adjust the menstrual cycle, reduce androgen levels, and thereby improve the clinical symptoms of polycystic ovary syndrome.

Regulation of Lipid Metabolism and Reducing Body Weight Metformin can reduce the blood triglycerides in circulation, improve liver steatosis, promote the oxidation of brown adipose tissue and VLDL-fatty acid triglyceride uptake, and inhibit fat formation; the process may be related with the activation of AMPK [6]. AMPK can phosphorylate acetyl coenzyme A carboxylase (ACC) and inhibit the conversion of acetyl coenzyme A into malonyl-CoA. Malonyl-CoA is a precursor of fatty acid production and is an allosteric inhibitor of fatty acid transporting to mitochondria.

Prevention and Treatment of Tumors Epidemiological studies have shown that metformin reduced the risk of multiple types of tumors in type 2 diabetes and non-diabetes patients and reduced tumor-related mortality [7]. Metformin also has a therapeutic effect on various tumors. The antitumor effect of metformin may be through the reduction of serum insulin and insulin-like growth factor-1 (IGF-1) levels or activation of LKB1/AMPK, thereby blocking the mammalian target of rapamycin-sensitive complex 1 (mTORC1) signaling pathway.

Antiaging Antiaging effect is a major discovery in the new role of metformin [4]. FDA approved the phase 4 of clinical trial of metformin for antiaging effect, and the trial is ongoing [8]. The study suggests that the antiaging effect of metformin is closely related to the mitochondria, through the regulation of mitochondrial function to activate AMPK and inhibit mTOR, thereby reducing energy consumption. Furthermore, metformin can regulate oxidative stress, reduce tissue inflammation, and reduce the growth factor level and cell proliferation. The above effects and the combinational effect result in the improvement of health and achieving longevity.

Other Effects Studies have shown that metformin helps to prevent type 2 diabetes. For young people with a high body mass index, metformin is more effective than lifestyle control. In addition, metformin can improve and prevent vascular disease, improve mitochondrial function, and serve as anti-inflammatory agent and antioxidant.

Metformin is hydrophilic and distributed in tissues in an active manner. After oral administration, metformin is absorbed quickly and rapidly distributed into various tissues, mainly in the liver, gastrointestinal tract, and kidney.

Clinical Applications

Metformin is mainly used as an oral hypoglycemic drug in clinical practice. It is a first-line drug for the treatment of type 2 diabetes [9], especially for obese patients with normal renal function. Metformin is also used to treat other diseases associated with insulin resistance, such as polycystic ovary syndrome (PCOS), nonalcoholic fatty liver disease, cancer, obesity, osteoporosis, inflammation, etc.

Metformin was rarely associated with adverse reactions. The most common adverse event is gastrointestinal discomfort. There may be lactic acidosis in the case of overdose or in the presence of contraindications.

Discussion

The application of metformin has a relatively long history. It is safe and effective in clinical application that makes its clinical demand increasing. With more and more new pharmacological roles discovered apart from its hypoglycemic activity, the application prospect of metformin will be more optimistic.

Fig. 1 *Galega officinalis*

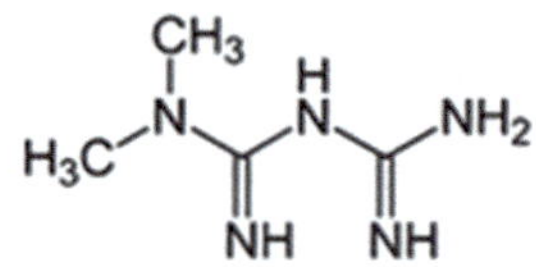

Fig. 2 Chemical structure of metformin

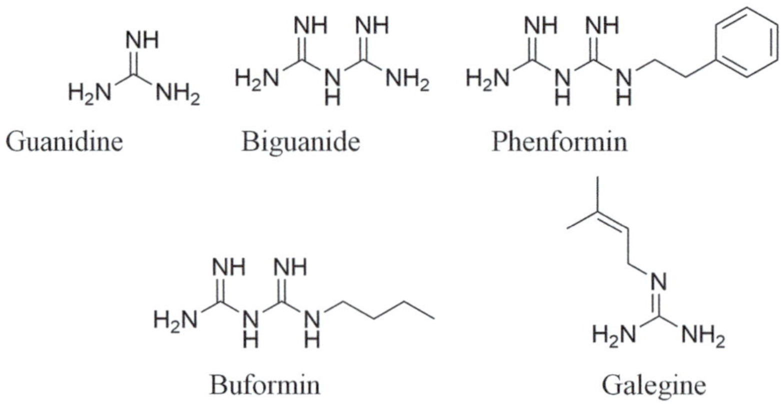

Fig. 3 Chemical structure of metformin derivatives

References

1. Werner A, Bell J. CCX1V. The preparation of methylguanidine and of ââ-dimethylguanidine by the interaction of dicyanodiamide and methylammonium and dimethylammonium chlorides respectively. J Chem Soc Trans. 1922;121:1790–4.
2. Slotta KH, Tschesche R. Uber Biguanide. II. Die Blutzuckersenkende Wirkung der Biguanides. Ber Dtsch Chem Ges B: Abh. 1929;62:1398–405.
3. Bailey CJ, Day C. Metformin: its botanical background. Pract Diabetes Int. 2004;21(3):115–7.
4. Hall SS. A trial for the ages. Science. 2015;349(6254):1274–8.
5. Madiraju AK, Erion DM, Rahimi Y, et al. Metformin suppresses gluconeogenesis by inhibiting mitochondrial glycerophosphate dehydrogenase. Nature. 2014;510(7506):542–6.
6. Geerling JJ, Boon MR, van der Zon GC, et al. Metformin lowers plasma triglycerides by promoting VLDL-triglyceride clearance by brown adipose tissue in mice. Diabetes. 2014;63(3):880–91.
7. Morales DR, Morris AD. Metformin in cancer treatment and prevention. Annu Rev Med. 2015;66:17–29.
8. ClinicalTrials.gov [Internet]. Identifier NCT00287391, Metformin in Longevity Study (MILES). 2015.
9. Inzucchi SE, Bergenstal RM, Buse JB, et al. Management of hyperglycaemia in type 2 diabetes, 2015: a patient-centred approach. Update to a position statement of the American Diabetes Association and the European Association for the Study of Diabetes. Diabetologia. 2015;58(3):429–42.

Papaverine

De Kang, Gui-Fen Qiang, Li-Da Du, and Guan-Hua Du

Abstract Papaverine is mainly extracted from the mature capsule shell of Papaveraceae poppy. As a vasodilator, papaverine is a nonspecific antispasmodic drug of smooth muscle, used for the treatment of ischemia caused by cerebral, cardiac, and peripheral vascular spasm, as well as visceral spasm of the kidney, gallbladder, or gastrointestinal tract. However, high dose of papaverine can cause hypotension and tachycardia without effect on the central nervous system. As an old variety recorded in pharmacopoeia, papaverine hydrochloride has demonstrated its clinical efficacy and safety after its clinical usage for decades. To avoid the serious adverse reaction caused by intravenous injection, the improved dosage forms were developed for the rational administration of papaverine hydrochloride to benefit the patients.

Keywords Papaverine · Papaveraceae poppy · Vasodilator · Ischemia

Alias: Papaverin; Robaxapap; Pavabid; Cerespan
Origin: Opium poppy (Fig. 1)
Chemical name (Fig. 2)

1-(3,4-dimethoxybenzyl)-6,7-dimethoxyisoquinoline

Molecular formula, $C_{20}H_{21}NO_4$; **MW**, 339.39; **CAS**, 61-25-6

D. Kang · G.-F. Qiang · G.-H. Du (✉)
Beijing Key Laboratory of Drug Targets Identification and Drug Screening, Centre for Pharmaceutical Screening, Institute of Materia Medica, Chinese Academy of Medical Sciences and Peking Union Medical College, Beijing, China
e-mail: dugh@imm.ac.cn

L.-D. Du
School of Biomedical Sciences, Chinese University of Hong Kong, Hong Kong, China

G.-H. Du, *Natural Small Molecule Drugs from Plants*,
https://doi.org/10.1007/978-981-10-8022-7_17

Properties

Appearance: colorless prismatic or acicular crystal. *Melting point*: 147–148 °C. *Relative density*: 1.337 (20/4 °C). *Solubility*: very soluble in benzene, acetone, hot ethanol, and glacial acetic acid; soluble in concentrated sulfuric acid; slightly soluble in ether, chloroform, and carbon tetrachloride; and insoluble in water. The common crystalline salt is hydrochloride salt, papaverine hydrochloride.

Papaverine Derivatives (Fig. 3)

Dosage Forms

Papaverine is included in *Pharmacopoeia of the People's Republic of China* (2015), *British Pharmacopoeia* (2017), *US Pharmacopeia* (40), the *Japanese Pharmacopoeia* (17th ed.), *European Pharmacopoeia* (9.0th ed.), and the *International Pharmacopoeia* (5th ed.). The current dosage forms include papaverine hydrochloride tablets and papaverine hydrochloride injection.

Indications

Papaverine is mainly used for the treatment of ischemia caused by cerebral, cardiac, and peripheral vascular spasm, as well as visceral spasm of the kidney, gallbladder, or gastrointestinal tract.

Literature

Papaverine mainly exists in the mature capsule shell of Papaveraceae poppy (*Papaver somniferum* L). Opium poppy shell is proved to be poisonous and contains more than 30 kinds of alkaloids including morphine, codeine, narcotine, and papaverine, the content of which varies greatly between different habitats and batches. The main effects of papaverine are astringing the lung and intestine and relieving the pain. The usage of opium poppy shell as medicine was recorded in the *Compendium of Materia Medica*, *Ben Cao Jing Shu*, and *Materia Medica of Seeking Truth*. Opium poppy shell can also be used for the treatment of chronic cough, diarrhea, archoptosis, and abdominal pain, but the long-term usage turns out to be addictive easily.

As an annual or biennial herb, the cultivation of opium poppy requires certain climate and regional conditions, generally in the altitude of 900–1300 m. Opium poppy is native in Asia Minor, India, Armenia, and Iran, and Afghanistan is the largest origin of poppy flower in the world. In China, there is only a small amount of cultivation of opium poppy for medicinal research in the herbal plantation.

Considering the low content and the low production rate of papaverine in the opium poppy, the supply of papaverine by extracting from opium poppy can hardly meet the clinical demand. To this end, the artificial synthesis of papaverine was explored since 1909 [1], and Chinese established the synthetic method of papaverine hydrochloride by using guaiacol as starting material in the 1980s [2].

History of R&D

Human beings have a long history of using opium poppy as food and narcotics, which can be dated back to the Neolithic age. The opium, which is the dried product of the opium poppy juice, is widely used as analgesics in the old and new civilizations of the old continent except the Chinese civilization.

In 1848, German chemist Georg Merck successfully isolated a new alkaloid from the opium liquor for the first time and named it as "papaverine" [1, 3]. He also determined the correct molecular formula of papaverine as $C_{20}H_{21}NO_4$ and prepared the papaverine hydrochloride and nitrate by recrystallization method. Then, Goldschmiedt et al. put forward the molecular structure of papaverine by studying its oxidation products in 1883 and determined the isoquinoline ring as the core structure of papaverine in 1888 [4] (see Fig. 1).

Since the content of papaverine in plants is very low (less than 1%) and can hardly meet the clinical needs, so Pictet and Gams proposed a synthetic method of papaverine in 1909, which makes the large-scale industrialized production of papaverine possible [1].

Study on the pharmacology of papaverine was first published in 1914 by Professor Pal of Vienna, who found that papaverine possessed smooth muscle relaxation effect without causing smooth muscle paralysis and can be used for the treatment of hypertension, angina, and acute uremia [5]. The foreign clinical records of papaverine could be dated back to 1934, when Denk used papaverine hydrochloride to treat ten patients with acute arterial embolism. Among them, the blood circulation of six patients was fully recovered [6]. In the article published in JAMA by Meyer in 1965, angiography showed that intravenous injection of 64 mg papaverine hydrochloride can significantly increase the effective oxygen content of brain in patients with acute stroke [7]. The study of papaverine also resulted in the discovery of several papaverine derivatives, which are effective drugs in the clinical treatment of cardiovascular diseases, such as verapamil (Fig. 2).

In the 1980s, Hunan Pharmaceutical Industry Research Institute and Lianyungang Pharmaceutical Factory produced papaverine by using guaiacol as the raw material to meet the clinical needs. The pharmacological studies indicated that the hypotensive effect and the smooth muscle spasm relieving effect of the synthetic papaverine were similar to those natural analogues. After clinical validation of 200 cases, the synthetic papaverine hydrochloride has shown the same curative effect as its natural analogue without any obvious side effects [2].

Pharmacology

As a vasodilator, papaverine is a nonspecific antispasmodic drug of smooth muscle, which is especially effective on pulmonary artery, coronary artery, and great vessels, producing systemic and nonspecific hypotensive and smooth muscle relaxation effect. Papaverine has a direct effect on the smooth muscle cells by inhibiting phosphodiesterase, thus increasing the intracellular concentration of cyclic adenosine monophosphate (cAMP), which further removes the catalase calcium from cytoplasmic of vascular smooth muscle, leading to the direct smooth muscle relaxation without nerve. Papaverine can also inhibit the cardiac conduction, directly act on myocardial cells, and prolong the refractory period. No effect of papaverine on the central nervous system has been found. High dose of papaverine can cause hypotension and tachycardia.

Clinical Application

Papaverine hydrochloride is used for the treatment of ischemia induced by brain, heart, and peripheral vascular spasm, as well as visceral spasm of the kidney, bladder, or gastrointestinal tract. Besides, papaverine is also suitable for angina and arterial embolism, occasionally used for the treatment of erectile dysfunction. In recent years, papaverine hydrochloride also combines with other drugs like nimodipine to treat the vasospasm and crisis triggered by subarachnoid hemorrhage and calculi-induced acute renal colic. The main adverse effects of papaverine include liver function damage, exaggerated respiration induced by rapid parenteral administration, facial flushing, heart rate acceleration, as well as hypotension with vertigo. In addition, overdose of papaverine can lead to blurred vision, diplopia, lethargy, and weakness.

Discussion

As an old variety recorded in pharmacopoeia, papaverine hydrochloride has demonstrated its clinical efficacy and safety after its clinical usage for decades. In order to meet the clinical needs, the improved dosage forms such as sodium chloride injection of papaverine hydrochloride (100 ml:30 mg) were developed to avoid the serious adverse reaction caused by intravenous injection of papaverine hydrochloride at high concentration. Cong et al. confirmed that papaverine hydrochloride is not addictive by analyzing 70 clinical cases [8]. The rational administration of papaverine hydrochloride to patients with clinical indications can not only relieve the pains but also benefit the patients.

Fig. 1 Opium poppy

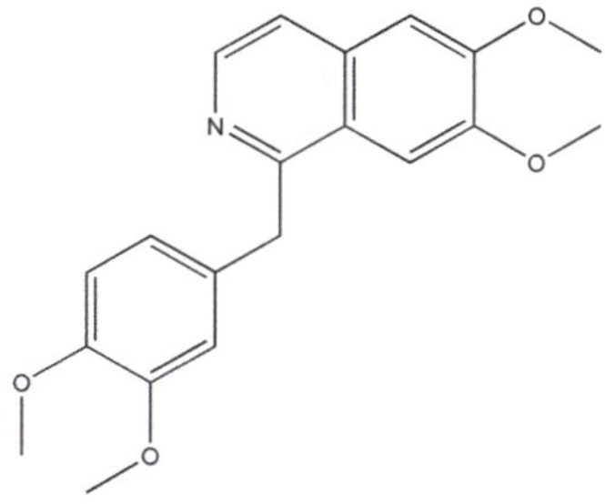

Fig. 2 Chemical structure of papaverine

Fig. 3 Chemical structure of papaverine derivatives

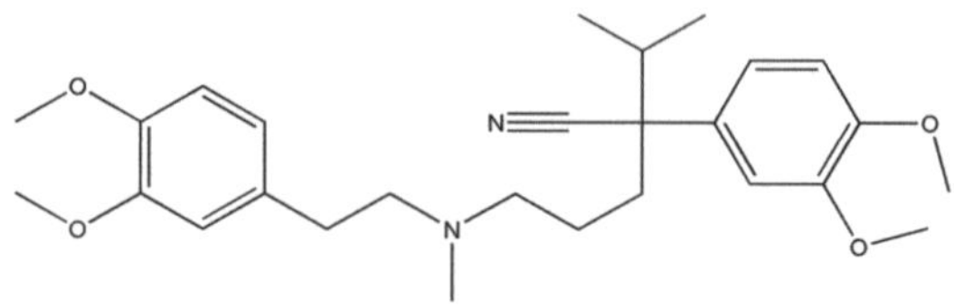

Verapamil

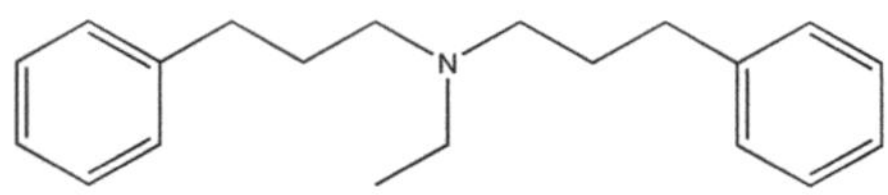

Alverine

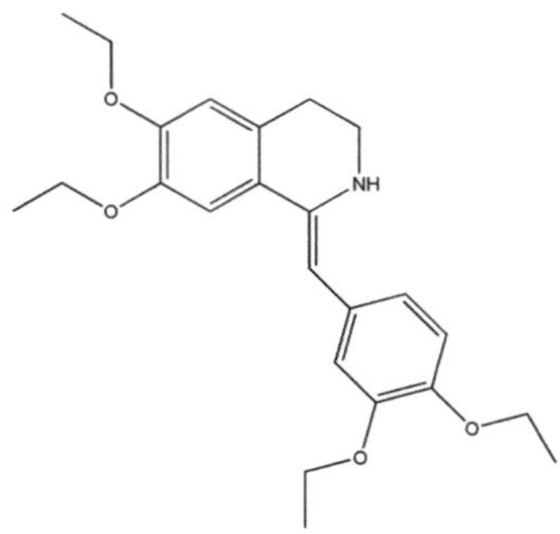

Drotaverine

References

1. Elad D, Ginsburg D. The synthesis of papaverine. Bull Narc (UNDOC). 1952;3:27–34.
2. Mao HK. Papaverine hydrochloride (synthesis) appraisal meeting. Pharm Ind. 1981;3:45.
3. Merck & Co. Merck's 1907 Index.
4. Goldschmiedt. Monatsch. 1883;4:704.
5. Pal J. Das Papaverine als Gefässmittel und Anaestheticum. Dtsch Med Wochenschr. 1914;40:164.
6. Macht DI. A pharmacologic and clinical study of papaverine. Ann Int Med. 1916;17:786.
7. Meyer JS, Gotoh F, Gilroy J, Nara N. Improvement in brain oxygenation and clinical improvement in patients with strokes treated with papaverine hydrochloride. JAMA. 1965;194(9):957–61.
8. Wang C, Yuan WJ, Wang YX. Analysis on 70 cases about no addiction of papaverine hydrochloride. Chin Hosp Pharm J. 1982;2:17–9.

Protocatechualdehyde

Yue-Hua Wang, Wei-Han Li, and Guan-Hua Du

Abstract Protocatechualdehyde is a kind of phenolic acid compound extracted from roots of the common traditional Chinese medicine *Salvia miltiorrhiza*. Protocatechualdehyde is an important pharmaceutical intermediate, which could be used to synthesize many kinds of antibiotics and anti-inflammatory drugs. Protocatechualdehyde has wide pharmacological activities including anti-atherosclerosis, myocardial protection, antithrombosis, neuroprotection, antisepsis, antiviral, and anti-fibrosis. In clinical therapy, it is used to treat coronary heart disease, chest tightness, angina, and so on.

Keywords Protocatechualdehyde · Catechol formaldehyde · Coronary heart disease · Chest tightness · Angina

Origin: *Ilex chinensis Sims* (Fig. 1)
Chemical name (Fig. 2)

3,4-Dihydroxy-benzaldehyde

Molecular formula, $C_7H_6O_3$; **MW**, 138.12; **CAS**, 139-85-5

Y.-H. Wang · W.-H. Li · G.-H. Du (✉)
Beijing Key Laboratory of Drug Targets Identification and Drug Screening, Centre for Pharmaceutical Screening, Institute of Materia Medica, Chinese Academy of Medical Sciences and Peking Union Medical College, Beijing, China
e-mail: dugh@imm.ac.cn

G.-H. Du, *Natural Small Molecule Drugs from Plants*,
https://doi.org/10.1007/978-981-10-8022-7_18

Properties

Appearance: pale beige acicular crystal (in water or methylbenzene) or off-white powder, crystal twin shape. *Solubility*: easily soluble in ethanol, acetone, ethyl acetate, ethyl ether, and hot water; soluble in cold water; insoluble in benzene and chloroform. *Melting point*: 150–153 °C. *Specific optical rotation*: easily oxidized to benzoquinone and changes color, and it is instable in water.

Dosage Forms and Indications

Protocatechualdehyde tablet was used to treat angina and coronary heart disease. Perhexiline injection containing protocatechualdehyde 100 mg/2 mL is used to treat angina and coronary heart disease.

Literature

Protocatechualdehyde is extracted from roots of the common traditional Chinese medicine *Salvia miltiorrhiza Bge*, which was harvested in autumn for better quality. Take root and remove stems, leaves and fibrous roots, then dry it. Most of them are wide, and they are mainly cultivated in recent years. South *Salvia* and Gansu *Salvia* are also widely used. Besides, there are a variety of sibling plant roots used as *Salvia miltiorrhiza* in Yunnan. Now, the existence of protocatechualdehyde has been found in variety of herbs, for example, in the leaf of *Stenoloma Chusanum (L.) Ching* and *Ilex chinensis Sims*.

History of R&D

In the 1940s, the research was conducted overseas to obtain protocatechualdehyde form herbs. In 1972, the chemical group of Chinese herbal medicine in Nanjing College of Pharmacy systematically studied chemical composition in purple flower holly leaf and separated six monomers including protocatechualdehyde [1]. Then they compared the effect of protocatechualdehyde, *Salvia miltiorrhiza* injection and hairy holly root injection, in which protocatechualdehyde is the most effective in increasing coronary sinus flow [2]. Further experiments examined that protocatechualdehyde can increase coronary flow and improve coronary circulation, so it was called after perhexiline [3]. Further clinical observation indicated that protocatechualdehyde has effect on coronary heart disease. The graduates of Nanjing College of Pharmacy in 1975 extracted, separated, isolated, and identified protocatechualdehyde in *Salvia miltiorrhiza* during the internship in Nanjing Municipal Hospital and studied its distribution, excretion, and toxicity in animals for clinical rational drug usage [4, 5].

Pharmacology

Protection of Myocardial Cells and Myocardial Ischemia

Reducing Ca^{2+} concentration in adult erythrocyte cytosolic to protect the myocardium, dilating the coronary arteriae, promoting collateral circulation, increasing myocardial oxygen supply coronary blood flow, improving collateral circulation without increasing the ventricular and myocardial oxygen consumption, decreasing heart rate, inhibiting myocardial contractility, expanding peripheral vascular, and ultimately reducing cardiac load and myocardial oxygen consumption. Used for the treatment of coronary heart disease and angina pectoris.

Effects on Atherosclerosis

Protocatechualdehyde improves atherosclerosis from the effect of inhibition of inflammation, apoptosis, and leukocyte chemotaxis.

Antithrombotic Effects

1. Platelet aggregation inhibition: Administration of protocatechualdehyde in vitro and in vivo has apparently effect on inhibiting platelet aggregation induced by ADP. Protocatechualdehyde 0.625, 1.25, and 2.5 mg/ml can decrease the degree and slow down the rate of platelet aggregation and promote the aggregation of platelet [6, 7].
2. Improving microcirculation: Protocatechualdehyde can increase the blood flow of microcirculation, accelerate the flow of blood to improve the oxygen supply of cells, and reduce the number of normal red cells to spiny red cells and the abnormal shape of spiny red cells [8].

Antioxidation Effects

Protocatechualdehyde contains phenolic hydroxyl in ortho-position, which is material base for its antioxidation. It has good activities for scavenging free radical. So it treats cardiovascular and cerebrovascular diseases induced by active oxygen [9].

Protection of Nerve Cells and Cerebral Ischemia Injury

Protection of cerebral ischemia injury might be concerned with reducing the production of TXA2 in brain tissue, inhibiting the release of excitatory amino acids in brain tissue, and improving the microcirculation of brain tissue.

Repair of Damaged Venous Valve and Treatment of Phlebeurysm

Protocatechualdehyde reduced the fibrosis of tissues and organs effectively, dissolved fibrin, promoted the regeneration of fibrotic cells, and repaired damaged venous valves to prevent flowing back of blood and cure phlebeurysm effectively.

Other Effects

Protecting the liver and promoting repair and regeneration of liver tissue; antifibrosis of liver and promoting the healing of fracture and wounds; antibacterial, anti-inflammatory, antiviral, anti-sepsis, and preventing pigmentation.

Clinical Application

Perhexiline injection containing protocatechualdehyde 100 mg/2 mL by intravenous infusion or intramuscular injection was used to treat coronary heart disease, chest tightness, angina, and myocardial infarction.

Injection treats ischemic stroke and improves both symptoms and signs of patients.
Treating chronic hepatitis and early cirrhosis: relieving the symptoms, promoting the recovery of liver function, and hepatosplenomegaly.
It can significantly improve the blood rheology index for patients with acute exacerbation of chronic pulmonary heart disease.
Treatment of peptic ulcer: a certain effect.
Others: it has effect on many diseases like viral myocarditis, embolism of central retinal artery, thromboangiitis obliterans, scleredema neonatorum, scleroderma, psoriasis, nerve deafness, and toxemia of pregnancy.

Discussion

Protocatechualdehyde is one of the main degrading products from the water-soluble components of common traditional Chinese medicine salvianolic acid B. It is an important pharmaceutical intermediate, which could be used to synthesize many kinds of antibiotics and anti-inflammatory drugs. It has wide pharmacological activities including anti-atherosclerosis, myocardial protection, antithrombosis, neuroprotection, antisepsis, antiviral, and anti-fibrosis, but it has cardiotoxicity on some degree [10].

Fig. 1 *Ilex chinensis Sims*

Fig. 2 The structure of protocatechualdehyde

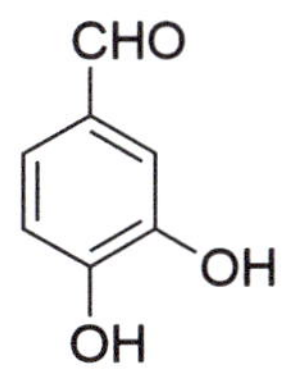

References

1. Yang YL. Study on Sijiqing. Jiangsu Med. 1976;2:59–61.
2. Jiangsu New Medical College of pharmacology group and Chinese herbal medicine group. Preliminary analysis of Sijiqing, Salvia miltiorrhiza, Mao Dongqing, et al. on the coronary sinus flow, myocardial oxygen consumption. Jiangsu Med. 1977;4:19–21.
3. Zhang KZ, Zou YB, Lu WJ. Guanxinning (protocatechuic aldehyde) treatment of angina pectoris of coronary heart disease: a clinical observation of 27 cases. Jiangsu Med. 1978;1:8–9.
4. Yao JY, Chen ZJ, Shao ZG, et al. Isolation and identification of the effective components of protocatechuic aldehyde in Salvia miltiorrhiza Bunge. J Nanjing Col Pharm. 1979;1:74–6.
5. Yang YL, Cao JY, Xu XQ, et al. Study on distribution, excretion and toxicity of protocatechuic aldehyde in animals. Jiangsu Med. 197(10):16–9.
6. Yang YL, Wu YN, Duan WY. Preliminary study on the antiplatelet effect of protocatechuic aldehyde. J Nantong Med Coll. 1981;2:15–6.
7. Shi L, Wu CQ, Yang YL, et al. Effects of protocatechuic aldehyde on platelet aggregation and platelet cAMP content. J Suzhou Med Coll. 1982;2:1–6.
8. Shen LH, Wang BY, Wang CQ, et al. Experimental study on the effect of protocatechuic aldehyde on microcirculation in rats. J Microcirc. 2002;12(2):12–4.
9. Liu M, Xia XH, Zhang ZM, et al. Comparative study on antioxidant activity of Danshensu, protocatechuic aldehyde, caffeic acid and salvianolic acid B in vitro. Chin Herb Med. 2009;32(2):265–7.
10. Zhang CY, Guo LL, Wang Z. Pharmacological research progress of protocatechuic aldehyde. Chin J Exp Tradit Med Formul. 2013;19(23):338–42.

Puerarin

Yue-Rong Zhao, Li-Da Du, Li Zhang, and Guan-Hua Du

Abstract Puerarin is an isoflavone component extracted from *Pueraria lobata* (Willd.) Ohwi and has widely protecting functions on the cardiovascular system, nervous system, glucose and lipid metabolism, osteoporosis, liver injury and inflammation. Puerarin has a great value of utilization due to its less adverse reactions and wide security range. The most widely used medicinal formulations of puerarin in the clinical area are puerarin injection. However, puerarin is difficult to absorb orally. In the long run, modifying the structure of puerarin and developing new types of dosage forms to improve bioavailability and selectivity will be effective to enhance its pharmacological activities, so as to expand the clinical application of puerarin and its derivatives.

Keywords Puerarin · Isoflavone · Pharmacological effects

Alias: Daidzein-8-C-glucoside**,** 7,4′-Dihydroxy-8-C-dlucosylisoflavone
Origin: *Pueraria lobata* (Willd.) Ohwi (Fig. 1)
Chemical name (Fig. 2)

7-Hydroxy-3-(4-hydroxyphenyl)-8-[(3R,4R,5S,6R)-3,4,5-trihydroxy-6-(hydroxymethyl) oxan-2-yl]chromen-4-one

Molecular formula, $C_{21}H_{20}O_{10}$; **MW**, 416.38; **CAS**, 3681-99-0

Y.-R. Zhao · L. Zhang · G.-H. Du (✉)
Beijing Key Laboratory of Drug Targets Identification and Drug Screening, Centre for Pharmaceutical Screening, Institute of Materia Medica, Chinese Academy of Medical Sciences and Peking Union Medical College, Beijing, China
e-mail: dugh@imm.ac.cn

L.-D. Du
School of Biomedical Sciences, Chinese University of Hong Kong, Hong Kong, China

G.-H. Du, *Natural Small Molecule Drugs from Plants*,
https://doi.org/10.1007/978-981-10-8022-7_19

Properties

Appearance: White to light-yellow crystalline powder. *Solubility*: Soluble in methanol, freely soluble in ethanol, slightly soluble in water, insoluble in chloroform and ether. *Melting point*: 187–189 °C.

Dosage Forms

This product is recorded in the *Pharmacopoeia of the People's Republic of China* (2015), *British Pharmacopoeia* (2017), *European Pharmacopoeia* (8.7th ed.).

Puerarin injection, Puerarin eye drops.

Indications

At present the most widely used medicinal formulation of puerarin in clinical application is puerarin injection, with more than 90 enterprises producing it. The injection was mainly used for the treatment of coronary heart disease, angina pectoris, myocardial infarction, retinal movement, vein occlusion, sudden deafness and other diseases. Another dosage form of puerarin is eye drops for the treatment of primary open-angle glaucoma, ocular hypertension, primary angle closure glaucoma and secondary glaucoma. In addition, there are puerarin tablets and puerarin capsules, belonging to the health-care product.

Literature

Puerarin, also known as pueraria flavonoids, is a kind of flavonoid glycoside extracted from the roots of *Pueraria alba* or *Pueraria thomsonii* and is also one of the main effective ingredients of *Pueraria lobata*. That *Pueraria lobate* was used to treat diseases has already been recorded in China's ancient medical books such as *Shen Nong's Materia Medica*, *Treatise on Miscellaneous Diseases* and *Medical Dictionary*.

Pueraria lobate was firstly recorded in *Shen Nong's Herbal Classic* in the Tang dynasty. In the Ming dynasty, "the grass roots of the grass" recorded *Pueraria lobate* for "clearing wind and coldness, clearing surface evil, solving muscle fever, stopping vexed thirst"; *Compendium of Materia Medica* records the effect of *Pueraria lobate's* cold, sweet, decreasing internal heat and detoxification.

The Chinese pharmacopoeia stipulates that kudzu vine is a dry root of the *Pueraria lobata* (Wild.) Ohwi. The *Pueraria lobata* (Wild.) Ohwi was collected by the pharmacopoeia of the calendar as it contains the highest active ingredient, puerarin. And *Radix Pueraria thomsonii* was taken in the 2005 and 2010 versions of *Pharmacopoeia of the People's Republic of China* for containing the second highest puerarin. Kanzu root is widely distributed in our country and of rich resources. It has been reported that puerarin could be extracted from *Pueraria lobata* (Wild.) Ohwi, *Radix Pueraria thomsonii*, *Pueraria omeiensiswanget Tang*, *Pueraria edulis Pamp* and *Pueraria phaseoloides*, but the content of puerarin differs.

Pueraria has a great value for nutrition and medicine and was considered the south ginseng of China. Puerarin has been widespread concerned over our country for its use of food and medicine in recent years.

History of R&D

In 1959, Japanese chemist Shibata Shoji first studied the chemical production of pueraria root, indicating that isoflavones are the main active ingredients of pueraria root, including puerarin, daidzin and daidzein [1]. Then in 2003, David Lee first total synthesized analogical puerarin [2]. And it was in 1974 that puerarin was first extracted by Fang Qicheng and other chemists in China [3]. Due to its strong activity of anti-cardiovascular and anti-cerebrovascular ischemia and hypoxia, expanding the coronary artery and cerebrovascular, reducing myocardial oxygen consumption, and improving myocardial systolic and microcirculation function, puerarin was approved for clinical use by the Ministry of Health in 1993.

As a new drug to improve cardiovascular circulation, puerarin has significant clinical application value as small toxicity and wide range of safety and efficacy. But in the process of research and development, it was found that the oral dosage of this drug was difficult to be absorbed in the gastrointestinal tract, with a lower concentration through the blood-brain barrier into the brain, which was difficult to play a role in the physiological activity of the cerebrovascular system. Therefore, large amounts of experimental research work were carried out, such as modifying the structure to change its space structure and improving the puerarin water-soluble and fat-soluble properties, so as to enhance its pharmacological activity and efficacy.

For instance, by organic synthesis reactions, such as alkylation, acylation or basic salt-forming reaction, and directional synthesis technology, researchers could modify the structure of the main active site of puerarin to prepare a series of puerarin derivatives with some fat-soluble and water-soluble properties. In recent years, researchers have synthesized a series of puerarin derivatives, and active research showed that the part of the puerarin derivatives has a good biological activity and bioavailability. For example, the derivatives of puerarin phenolic hydroxyl and C8 alcohol hydroxyl are both fat-soluble enhanced and water-soluble decreased, indicating higher bioavailability compared with puerarin [4]. In a new dosage form research, researchers have developed several types of dosage forms such as quick-release solid dispersions, solid from microemulsion preparation, etc., which greatly increase the solubility of puerarin, thus improving bioavailability. In recent years, the research on crystal drug has opened up new ways for the application of puerarin.

Pharmacology

Contemporary pharmacological research has demonstrated that puerarin has wide pharmacological activities on the cardiovascular system, nervous system, liver impairments, osteoporosis and hangover [5].

The clinical and experimental researches have shown that puerarin has a significant role in the prevention of cerebral vascular disease, which can expand the coronary artery and improve the metabolism of an ischemic heart muscle, improving brain circulation, increasing cerebral blood flow and reducing blood volume and heart rate. There is good control of puerarin on myocardial ischemia and cerebral ischemic disease.

A large number of experiments have proved that puerarin can promote the glucose and lipid metabolism and reduce blood sugar and blood lipid. The protection of puerarin on diabetes was attributed to its functions on inhibiting apoptosis and oxidative stress.

Puerarin have protection efficacy on liver damage induced by chemical substances, alcohol, surgery and other experimental damage. This effect is closely related to the antioxidant effect, resisting lipid peroxidation, inhibiting platelet aggregation and improving circulation.

Puerarin can also improve the eye microcirculation obviously and reduce the pressure of the eyes, thus acting obvious therapeutic effect on glaucoma. In addition, it has a variety of preparations, such as liquors, which are still under development.

Clinical Application

Currently, puerarin is used clinically for the treatment of hypertension, coronary heart disease, angina pectoris, arrhythmia, myocardial infarction, ischemic cerebrovascular disease, retinal arteriovenous obstruction, sudden deafness, diabetes complications, dizziness and other diseases. Besides, it also is used to treat chronic pharyngitis, cerebral infarction and Parkinson's syndrome.

However, puerarin also has adverse reactions in clinical therapy. Fever is the main symptom. In addition, allergic dermatitis, anaphylactic shock, laryngeal oedema, increased transaminase, gastrointestinal bleeding, haemolysis phenomenon and kidney damage occurred occasionally, and these symptoms disappeared after withdrawal of the medication.

Discussion

Puerarin has a great value of utilization due to its less adverse reactions and wide security range. With the development of modern clinical pharmacology research, many pharmacological effects and clinical efficacy of puerarin are gradually confirmed, and the treatment of the disease has achieved gratifying effect, especially in the treatment of coronary heart disease. However, the research on the mechanism of the pharmacological action, the discovery and verification of the new pharmacological action, and the new pharmacological action are also to be continued. Enhancing its pharmacological activities by modifying the structure of puerarin and improving bioavailability and selectivity will expand the clinical application of puerarin and its derivatives.

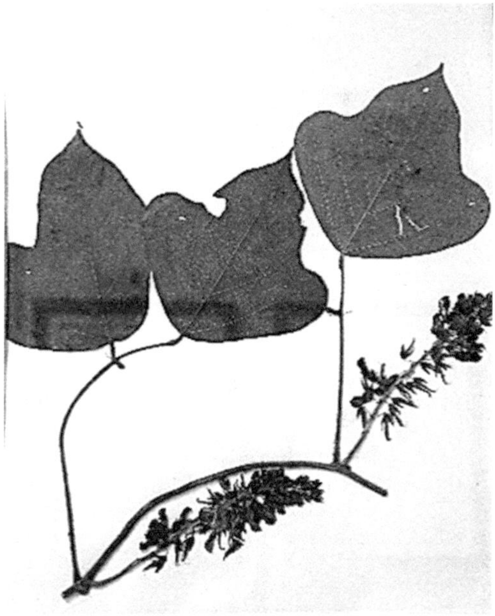

Fig. 1 *Pueraria lobata* (Willd.) Ohwi

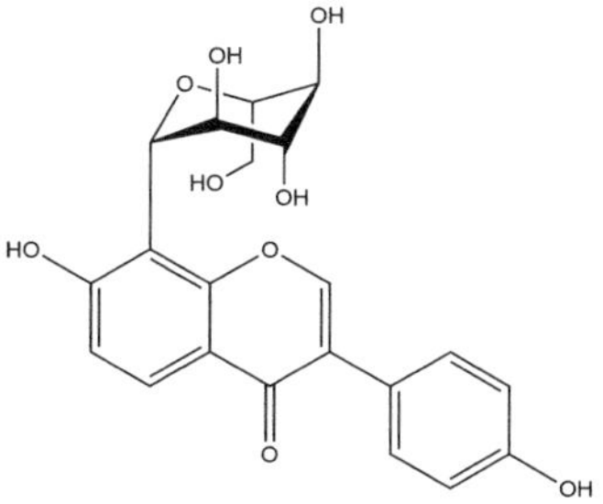

Fig. 2 Chemical structure of puerarin

References

1. Shibata S, Murakami T, Nishikawa Y, et al. The constituents of Pueraria root. Chem Pharm Bull. 1959;7(1):134–6.
2. Lee David YW, Zhang WY, Karnati Vishnu Vardhan R. Total synthesis of puerarin, an isoflavone C-glycoside. Tetrahedron Lett. 2003;44(36):6857–9.
3. Fang QC, Lin M, Sun QM, et al. The research of kudflavone. Chin Med J. 1974;54(5):271.
4. Liu YJ, Lan ZP, Yu Y, et al. Research of the synthesis and its clinical applications of puerarin derivatives. J Guiyang Coll Tradit Chin Med. 2012;34(5):31–3.
5. Wei SY, Yi C, Xiao XU. Progress on the pharmacological research of puerarin: a review. Chin J Nat Med. 2014;12(6):407–14.

Quinidine

Jun-Ke Song, Gui-Fen Qiang, and Guan-Hua Du

Abstract Cinchona bark is an ancient herb with magic medical effects. Quinidine is one of the alkaloids in the cinchona bark, also known as the isomer of antimalarial drug quinine. Quinidine acts as a class I antiarrhythmic agent (Ia) in the heart. The main mechanism of quinidine is to inhibit the fast inward sodium current (I_{Na}). It also blocks the slow inward calcium current (I_{Ca}), the rapid (I_{Kr}) and slow (I_{Ks}) components of the delayed potassium rectifier current, and the inward potassium rectifier current (I_{KI}). It directly acts on the cell membrane to prolong the cellular action potential and decrease automaticity. Currently, quinidine has been replaced by new types of drugs because of many adverse effects.

Keywords Quinidine · Alkaloid · Antiarrhythmic agent · Cinchona bark

Alias: (+)-Quinidine; Chinidin
Origin: The cinchona tree (Fig. 1)
Chemical name (Fig. 2)

(S)-[(2R,4S,5R)-5-Ethenyl-1-azabicyclo[2.2.2]octan-2-yl]-(6-methoxyquinolin-4-yl)methanol

Molecular formula, $C_{20}H_{24}N_2O_2$; **MW**, 324.42; **CAS**, 56–54-2

J.-K. Song · G.-F. Qiang · G.-H. Du (✉)
Beijing Key Laboratory of Drug Targets Identification and Drug Screening, Centre for Pharmaceutical Screening, Institute of Materia Medica, Chinese Academy of Medical Sciences and Peking Union Medical College, Beijing, China
e-mail: dugh@imm.ac.cn

G.-H. Du, *Natural Small Molecule Drugs from Plants*,
https://doi.org/10.1007/978-981-10-8022-7_20

Properties

Appearance: Quinidine is commonly used in its sulfate form with white needle-like crystal and bitter smell. It changes color easily when exposed to light. *Solubility*: It was soluble in ethanol and chloroform. Its water solubility is 0.05 g/100 mL (20 °C). *Specific optical rotation*: 256° (c = 1, EtOH). *Melting point*: 168–172 °C.

Dosage Forms

Quinidine has been listed in the *Pharmacopoeia of the People's Republic of China* (2015), the *Japanese Pharmacopoeia* (17th ed.), *British Pharmacopoeia* (2017), *United States Pharmacopeia* (40), *European Pharmacopoeia* (9th ed.), and the *International Pharmacopoeia* (5th ed.). Its main dosage form is quinidine sulfate tablets.

Indications

Quinidine acts as a class I antiarrhythmic agent (Ia) in the heart. It was clinically applicable to the treatment of recurrent, documented, life-threatening ventricular arrhythmias [1, 2].

Literature

Cinchona bark is an ancient herb with magic medical effects. Quinidine is one of the alkaloids in it, also known as the isomer of antimalarial drug quinine. Quinine was used to augment digitalis therapy in the nineteenth century. The effects of quinidine on cardiac arrhythmias were described in 1914. Then, it was reported as one of the most effective cinchona alkaloids in controlling arrhythmias.

History of R&D

In 1820, the French chemists Pierre Pelletier and Joseph Caventou extracted some alkaloids from the cinchona bark, including quinine and quinidine. Subsequently, quinine was demonstrated to play a very important role in the treatment of malaria after a number of scientific researches. Quinidine is the dextroisomer of quinine and has the similar pharmacological properties as quinine, but quinidine's effects are five to ten times stronger on the heart than quinine.

Pharmacology

Quinidine, a classical sodium channel blocker, belongs to the class I antiarrhythmic drugs. The main mechanism of quinidine is to inhibit the fast inward sodium current (I_{Na}). It also blocks the slow inward calcium current (I_{Ca}), the rapid (I_{Kr}) and slow (I_{Ks}) components of the delayed potassium rectifier current, and the inward potassium rectifier current (I_{KI}). It directly acts on the cell membrane to prolong the cellular action potential and decrease automaticity.

Quinidine is absorbed rapidly after oral administration, but its bioavailability varies between individuals. Quinidine has high protein affinity and wide distribution in various human organs and tissues.

The onset time of quinidine is around 30 min after oral administration, and it takes 1–3 h to achieve the maximum efficacy. The efficacy lasts for about 6 h. The effective blood concentration of quinidine is 3–6 μg/mL, and toxic blood concentration is 8 μg/mL. The half-life of quinidine in adult is 6–8 h and child 2.5–6.7 h.

In vivo metabolism of quinidine occurs mainly in the liver by cytochrome P450 system. Hepatic enzyme inducer can increase its metabolism. Some of its metabolites also have pharmacological activities. About 18.4% of quinidine is excreted by urine, 5% by feces, and a small amount by milk and saliva.

Clinical Application

Quinidine acts as a class I antiarrhythmic agent (Ia) in the heart. It is clinically used for the treatment of recurrent, documented, life-threatening ventricular arrhythmias and should not be used for ventricular arrhythmias of lesser severity. It is useful for atrial fibrillation or atrial flutter after the maintenance of electrical therapy [3]. Currently, quinidine has been replaced by new types of drugs, because of many adverse effects and narrow therapeutic index, even though it is effective to atrial premature beats, paroxysmal supraventricular tachycardia, supraventricular tachycardia and pre-excitation syndromes, ventricular premature beats, ventricular tachycardia, as well as atrial fibrillation or atrial flutter.

Discussion

Quinidine showed effects on atrial fibrillation and ventricular arrhythmia as sodium channel blockers. However, in clinical applications, quinidine can increase the blood concentrations of lidocaine and some other drugs as an inhibitor of the cytochrome P450 enzyme 2D6. It can also cause thrombocytopenia, granulomatous hepatitis, and myasthenia gravis limiting its clinical application [4–7]. Above all, the drug R&D process of quinidine and its clinical application are helpful for us to understand the new drug development process.

Fig. 1 The cinchona tree

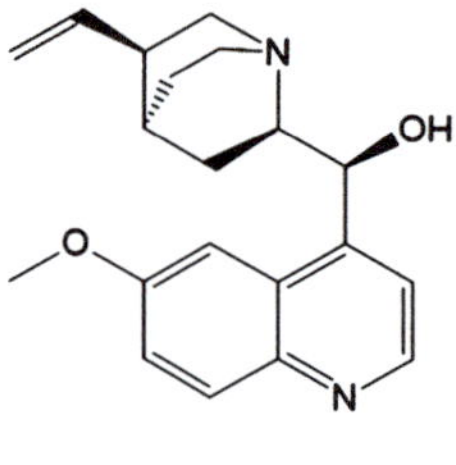

Fig. 2 The chemical structure of quinidine

References

1. Roden DM. Antiarrhythmic drugs. Goodman and Gillman's: the pharmacological basis of therapeutics. New York: McGraw-Hill; 2006.
2. Schwaab B, Katalinic A, Boge UM, et al. Quinidine for pharmacological cardioversion of atrial fibrillation: a retrospective analysis in 501 consecutive patients. Ann Noninvasive Electrocardiol. 2009;14(2):128–36.
3. Fang MC, Stafford RS, Ruskin JN, et al. National trends in antiarrhythmic and antithrombotic medication use in atrial fibrillation. Arch Intern Med. 2004;164:55–60.
4. Zhang WB, Zhang ZM, Liu XH, et al. The revaluation and selection of Class I and II anti-arrhythmic drugs. J Binzhou Med Univ. 2006;29(2):113–7.
5. Chen HZ, Zong P, Pu SY. Clinical experience of 20 patients with quinidine syncope. Natl Med J China. 1996;76(8):624–5.
6. Shan HL, Pan ZW, Feng TM, et al. Mechanism of proarrhythmic effect of quinidine. Chin J Endemiol. 2007;26(1):33–5.
7. Dai GZ. Proarrhythmia effects of antiarrhythmic drugs. Chin J Postgrad Med. 1995;8(2):7–8.

Reserpine

Ying Chen, Lan Sun, and Guan-Hua Du

Abstract Reserpine is extracted from indole alkaloid, which is one of the most important alkaloids in *Rauvolfia*. Reserpine has a medium antihypertensive effect, and its initial effect is slow but lasts long. However, due to the insufficient clinical data and the adverse effect of reserpine, such as depression, Parkinson's disease, dizziness, nightmare, nasal congestion, headache, diarrhea, abdominal pain, inattention, and suicidal tendencies, reserpine has already been eliminated in foreign countries, and in China, it is only used as a second-line antihypertensive drug. In the history of reserpine R&D, large scales of studies about Chinese *Rauvolfia* as well as the numerous participants promote the rapid progress of natural medicine. It can be taken as the earliest study in plant medicine since the founding of China, providing valuable experience to our later studies.

Keywords Reserpine · *Rauvolfia* · Antihypertension

Alias: Serpasil, serpentine
Origin: *Rauvolfia verticillata* (Fig. 1)
Chemical name (Fig. 2)

Methyl(3S,16S,17R,18R,20R)-11,17-dimethoxy-18-(3,4,5-trimethoxy-benzoyloxy)yohimban-16-carboxylate

Molecular formula, $C_{33}H_{40}N_2O_9$; **MW**, 608.68; **CAS**, 50-55-5

Y. Chen · L. Sun · G.-H. Du (✉)
Beijing Key Laboratory of Drug Targets Identification and Drug Screening, Centre for Pharmaceutical Screening, Institute of Materia Medica, Chinese Academy of Medical Sciences and Peking Union Medical College, Beijing, China
e-mail: dugh@imm.ac.cn

G.-H. Du, *Natural Small Molecule Drugs from Plants*,
https://doi.org/10.1007/978-981-10-8022-7_21

Properties

Appearance: crystalline powder, colorless to yellowish brown, darker in case of light. *Solubility*: soluble in chloroform, slightly soluble in acetone, and almost insoluble in water, methanol, ethanol, or ether. *Melting point*: 264–265 °C. *Specific optical rotation*: −117.7°.

Dosage Forms and Indications

This product is recorded in the *Pharmacopoeia of the People's Republic of China* (2015), the *Japanese Pharmacopoeia* (17th ed.), *British Pharmacopoeia* (2017), *United States Pharmacopeia* (40), *European Pharmacopoeia* (9.0th ed.), and *The International Pharmacopoeia* (5th ed.)

At present, the clinical dosage forms of reserpine include tablets and injections. It is widely used in the treatment of mild and moderate hypertension and also used as ataractics.

Literature

Reserpine is an indole alkaloid, consisted in various species *Rauvolfia*. About *Rauvolfia*, there is a very ancient legend. It is said that *Rauvolfia* has been called serpentine and Chandrh. Chandrh represents both the moon and the treatment of "moon" disease and mental disorders. This species may have been mentioned by Rbeede who suggested using its roots to resist snake bites and scorpion stings.

The dried roots of *Rauvolfia* have been on the market in India for at least 20 years before it was received as a drug by Western medicine. Sen and Bose reported that it was of great value and safety in the treatment of hypertension in 1931 and its main active ingredient was dubbed the chemical name "reserpine" in 1942. Deb put forward that the effect of *Rauvolfia* in inhibiting nerve excitability was much better than barbiturates in 1943. In 1946, the extract from *Rauvolfia* was included in the *Indian Pharmacopoeia* [1].

History of R&D

In 1931, Indian scholar Sen discovered Indian *Rauvolfia* have the antihypertensive and antipsychotic effects. The following studies in medicinal chemistry and pharmacology found that the main active ingredient is reserpine and clarified the mechanism of lowering blood pressure. In 1952, reserpine was first isolated and won an important status in treating hypertension and neurological and psychiatric disorders because of its remarkable physiological attributes. The structure analysis of reserpine peaked in 1955.The total synthesis of reserpine is completed in 1956 [2].

There are no effective antihypertensive drugs in clinic at the initial stage in our country, and the reserpine imported from India was scarce and expensive, which could not meet the urgent needs of patients. China nutrition and chronic disease status report *(2015)* shows that the prevalence rate of hypertension in 2012 of Chinese residents aged 18 years and above is 25.2%, higher in male than in female and higher in cities than in rural areas. It is estimated that adult patients with hypertension in China now are about 260 million and the prevalence rate of hypertension increased significantly compared with 2002 [3]. Hypertension is a common chronic disease and an important risk factor in cardiovascular disease. The complications of hypertension cause serious consumption of medical and social resources, resulting in a heavy burden to the families and the countries. Therefore, the central health research institute at that time conducted a comprehensive study of Chinese *Rauvolfia* and proved that there were abundant resources in Guangdong, Guangxi, and Yunnan. The research shows that different preparations of *Rauvolfia* could depress the blood pressure of hypertensive dogs and at the same time cause a series of typical symptoms of reserpine, including quietness, ptosis, myosis, diarrhea, and so on, suggesting that reserpine might be the main antihypertensive effective component which was confirmed by phytochemistry and pharmacology researches. In 1958, the Bureau of Drug Administration of Ministry of Health presided over the work of identifying the total alkaloids in *Rauvolfia* and approved the first Chinese antihypertensive drug commercially named "Verticil". Large scales of studies about Chinese *Rauvolfia* as well as the numerous participants promote the rapid progress of natural medicine. It can be taken as the earliest study in plant medicine since the founding of China, providing valuable experience to our later studies [4].

Pharmacology

Reserpine has an antihypertensive effect by reducing cardiac output, reducing the peripheral resistance, and inhibiting cardiovascular reflection. Aside from the effect of antihypertension, reserpine can also decrease heart rate, deplete adrenaline in the sympathetic nerve ending, and reduce the catecholamine and 5-serotonin in the heart, brain, or other tissues.

The mechanism of the hypotensive effect of reserpine: combined with amine pump vesicles (the $Mg2^{+}$-ATP-dependent amine transporter system), (a) it inhibits amine reuptake, (b) inhibits uptake of dopamine, (c) leads to the synthesis disorder of norepinephrine (NA), and (d) induces neurotransmitter depletion by vesicle reduction. The transmitter depletion in sympathetic nerve system would decrease blood pressure. By inhibiting the central nervous system, reserpine plays a sedative effect through relieving the tension and anxiety of patients with hypertension. New research shows that reserpine administration can lead to the fibromyalgia of mice via the loss of monoamine [5]. Some studies showed that reserpine (1 mg/kg) can suppress the pharmacological effects of vesicular monoamine transporter, resulting in the changes of behavior and neurochemistry, which can be reversed by withdrawal for a period of time. The neurochemical change is considered to be an indicator of reduced activity of dopaminergic neurons [6].

If taken orally, the antihypertensive effect of reserpine is slow and gentle, and the disappearance of this effect after discontinuation is also slow. Reserpine has been widely applied to the therapy of light and moderate hypertension and has a good effect in conjunction with thiazide diuretics. Reserpine is one of the most commonly used antihypertensive drugs, and its common oral dose is 0.25–0.5 mg daily. The antihypertensive effect of reserpine would occur about 30–60 min after intramuscular injection or intravenous injection of 0.5–1.0 mg, peaking in 3 h, so reserpine should be repeatedly injected in 3 h.

Clinical Application

Reserpine has a medium effect in antihypertension, and its initial effect is slow but lasts long. Reserpine can be used for severe and advanced or acute hypertension in combination with other antihypertensive drugs or be used for the psychotic symptoms of mania. Due to the insufficient clinical data and the adverse effect of reserpine, such as depression, Parkinson's disease, dizziness, nightmare, nasal congestion, headache, diarrhea, abdominal pain, inattention, and suicidal tendencies, reserpine has already been eliminated in foreign countries, and in China, it is only used as a second-line antihypertensive drug. Various preparations, such as reserpine tablets and compound reserpine tablets, have been widely used in clinical therapy. In recent years, with the emergence of new antihypertensive drugs, reserpine has been gradually faded out of clinical therapy [7] and has been deleted from the basic drugs directory by the WHO expert committee [8].

Discussion

Because of its mild effect and low price, reserpine is often used in combined preparation in the clinical treatment of hypertension and plays an important role. The lack of large scales of randomized trial studies, contrast test, and safety evaluation makes it a challenge for clinical usage. The research and development of reserpine have promoted the development of antihypertensive drugs, which provides valuable theories and experiments in the progress of new drug research.

Fig. 1 *Rauvolfia verticillata*

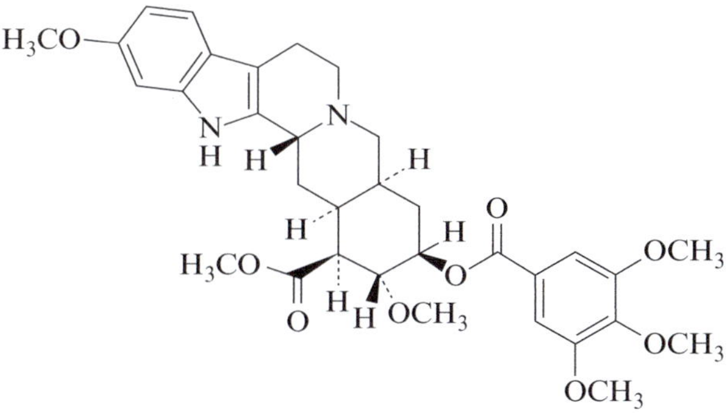

Fig. 2 The structure of reserpine

References

1. Monachino J. Rauvolfia serpentina – its history, botany and medical use. Econ Bot. 1954;8:349–65.
2. Woodward RB, Bader FE, Bickel H, et al. The total synthesis of reserpine. Tetrahedron. 1958;2(1):1–57.
3. Liu-Lisheng. Guidelines for prevention and treatment of hypertension in China, 2010. Guidelines on the rational use of drugs in hypertension. Chin J Front Med Sci (Electronic Version). 2015;7(6):22–3.
4. Institute of Materia Medica. Chinese academy of medical science and Peking Union Medical College. Modern research of Chinese herbal medicine. Beijing: Beijing Medical University and Peking Union Medical College Associated Press; 1995. p. 306–25.
5. Blasco-Serra A. Depressive-like symptoms in a reserpine-induced model of fibromyalgia in rats. Physiol Behav. 2015;151:456–62.
6. Catiuscia Molz de Freitas. Behavioral and neurochemical effects induced by reserpine in mice. Psychopharmacology. 2016;233(3):457–67.
7. Liu-xiu. Application of reserpine in perioperative period of hypertension. Shandong Med J. 2015;9:103–4.
8. Yang-Yue. Taking antihypertensive drugs as an example for systematic review of WHO for essential drugs. China Pharm. 2011;22(32):2990–3.

Rhomotoxin

Cheng Xing, Li-Da Du, and Yang Lu

Abstract Rhomotoxin is a white needle crystal that is extracted from the dried fruit of Rhododendron molle (Bl.) G. Don (yáng zhí zhú). Its functions as an analgesic or hypotensive or in kidney protection are the mainly pharmacological properties of rhomotoxin. Rhomotoxin acts as an antihypertensive drug, and it is able to lower blood pressure in patients with severe hypertension. Rhomotoxin is quite stable. However, further research is needed to understand how it works .

Keywords Rhomotoxin · Antihypertensive · Kidney protection

Alias: Andromedotoxin
Origin: Rhomotoxin (Fig. 1)
Chemical name (Fig. 2)

(6R) -2,2,6,10- tetramethyl -4a, 7- methanol heptyl [6, 7] azuleno [2,3-b] epoxy -2a, 3,6,10,11- pentyl alkyl

Molecular formula, $C_{20}H_{32}O_6$; **MW**, 368; **CAS**, 4720-09-6

Properties

Appearance: white crystalline or powder. *Solubility*: soluble in water, acetone, and chloroform, almost insoluble in diethyl ether or petroleum ether. *Melting point*: 265–268 °C. *Specific optical rotation*: −50 to −60°.

C. Xing · Y. Lu (✉)
Beijing Key Laboratory of Polymorphic Drugs, Research Center of Polymorphic Drugs, Institute of Materia Medica, Chinese Academy of Medical Sciences and Peking Union Medical College, Beijing, China
e-mail: luy@imm.ac.cn

L.-D. Du
School of Biomedical Sciences, Chinese University of Hong Kong, Hong Kong, China

G.-H. Du, *Natural Small Molecule Drugs from Plants*,
https://doi.org/10.1007/978-981-10-8022-7_22

Dosage Forms and Indications

This product is available in the Pharmacopoeia of the People's Republic of China (1977).

Injection (1 mg/mL) and tablets (25 mg). Rhomotoxin is a potent antihypertensive drug that can significantly slow down the heart rate and lower blood pressure for a variety of supraventricular tachycardia and hypertension conditions.

Literature

Fructus Rhododendri (ba li ma) is the dried fruit of Rhododendron molle (Bl.) G. Don (yáng zhí zhú), mainly distributed in the Jiangsu, Hubei, and Jiangxi Provinces of China [1].

"Compendium of Materia Medica" recorded that Rhododendron molle (Bl.) G. Don (yáng zhí zhú) can be used to treat waist pain, arm pain, and poisoning. It is effective at lowering blood pressure and slowing the heart rate, but a large dosage can cause death [2]. Rhomotoxin's formulations include injection and tablet. Side effects include a burning sensation, numbness, dizziness, dry mouth, nausea, vomiting, vertigo, chest tightness, slow heartbeat, and low blood pressure. In addition, pregnant women should use it with caution.

History of R&D

Because rhomotoxin can slow down heart rhythm and blood pressure, Deng Dao ji et al. from the First Affiliated Hospital of Wuhan Medical College researched *Fructus Rhododendri* in the early 1970s. They extracted the crystallization from fruits and obtained preliminary results by animal testing and clinical application [2].

Rhomotoxin is a white needle crystal that is extracted from the dried fruit of Rhododendron molle (Bl.) G. Don (yáng zhí zhú). Its biological activity is closely related to three-dimensional specificity and hydrophobicity. 5β-hydroxyl, 6β-hydroxyl, and, especially, the 2,3-epoxy groups can affect its biological activity [3].

Rhomotoxin was acetylated to obtain monoacetate (Japanese Rhododendron II). The melting points of rhomotoxin and Japanese Rhododendron III are identical (284–286 °C). IR and Rf of silica gel thin layer chromatography are identical. Therefore, rhomotoxin is Japanese Rhododendron III [4].

Pharmacology

The pharmacological effects of rhomotoxin are mainly displayed in three aspects: (1) analgesic effect: the suspension, infusion, and alcohol made of rhomotoxin fruit have an analgesic effect in mice [2]; (2) effects on cardiovascular system: the resting heart rate is a bad prognostic indicator of hypertension and other cardiovascular diseases. Rhomotoxin can reduce the resting heart rate; (3) kidney protection: rhomotoxin promotes good blood pressure and a slow heart rate and provides kidney protection. Its mechanism may be related to a reduction in the content of angiotensin II (Ang II), increasing endothelial nitric oxide synthase (eNOS) content, and slowing heart rate [5, 6].

The plasma concentration of mice decreased rapidly over time following intravenous injection of rhomotoxin; there was no residue after 6 h. Rhomotoxin is rapidly distributed to various organs through the blood circulation and reaches a peak within 5–15 min. The gallbladder (including bile) contained the highest amount, followed by the liver and kidneys. It affects, in order, the gallbladder (including bile), liver, kidney, thyroid, stomach, adrenal gland, heart, lungs, and brain. Rhomotoxin is excreted mainly by the kidney and digestive tract. More prototype drugs are in the urine than in the stool. Rhomotoxin has a stimulating effect on the stomach. A certain amount of rhomotoxin also entered the thyroid and adrenal glands. A very small amount was found in brain tissue, demonstratinge that rhomotoxin does not easily pass through the blood-brain barrier. Plasma dialysis showed that the plasma protein binding rate of rhomotoxin can reach as high as 60% in 30 min [7].

Clinical Application

Rhomotoxin acts as an antihypertensive drug, and it is able to lower the blood pressure in patients with severe hypertension. Its pharmacological effects are related to parasympathetic function. Large-dosage-induced low blood pressure can be restored with ephedrine. Its antihypertensive effect lasts for 0.5–1 h. If blood pressure has not dropped to a satisfactory level, rhomotoxin must not have been added because an excessive drop in blood pressure would lead to shock. Oral administration with aluminum hydroxide can reduce rhomotoxin-induced stomach discomfort. Intramuscular injection can also cause local pain. In addition, procaine reduces its antihypertensive effect. Secondary hypertension and malignant hypertension patients should use it with caution, critical and dying patients should be disabled (including, for example, severe heart failure, cardiomyopathy, atrioventricular block, and severe ventricular arrhythmias).

Discussion

Rhomotoxin exerts antihypertensive effects and also has a protective effect on the kidney. It is quite stable. However, further research on its exact mechanism is needed.

Fig. 1 Rhododendron molle (Bl.) G. Don

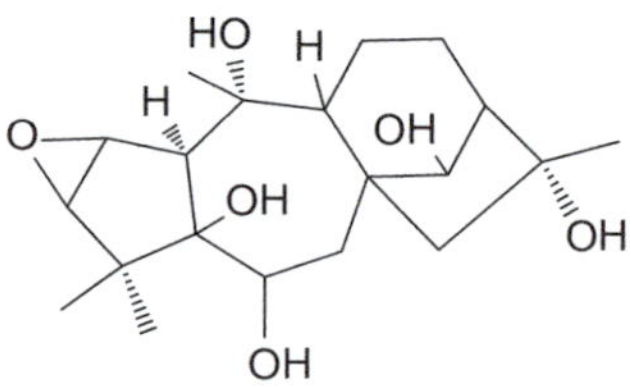

Fig. 2 Chemical structure

References

1. Deng DJ, et al. A preliminary study on the antihypertensive and slow heart rate of Chinese herbal medicine. Wuhan New Med. 1973;3(2):1.
2. Qin YN, et al. Study on analgesic active ingredients of sheep wander. J Xuzhou Med Coll. 1980;2
3. Yi WP, Qiu HY, et al. Extraction, remodeling and pharmacological effects of rhomotoxin. J Chin New Drugs. 2006;15(19):1655–8.
4. Deng DJ, et al. Refining and structural identification of rhomotoxin. J Hosp Pharm. 1981;1(3):7.
5. Cheng HZ, Ding BP, Huang ZH, et al. Antihypertensive and kidney protection of rhomotoxin. J Clin Exp Med. 2011;1(10):2.
6. Yi WP, Qiu HY, Zhang GH, et al. Effects of structural changes of rhomotoxin to apoptosis-related proteins in rat cardiomyocytes. J Pract Med. 2006;22(8):877.
7. Liu YJ, Wang YZ, Mao HY, et al. The distribution and excretion of 3H-Rhomotoxin. J Wuhan Med Coll. 1985;2(1):132.

Rutin

Bi-Yu Hou, Li Zhang, Jin-Hua Wang, and Guan-Hua Du

Abstract Rutin is widely found in nature and is almost contained in all of the Rutaceae and Sectaceae plants. As a flavonoid substance, rutin has a significant protective effect on the cardiovascular system. Rutin and its derivatives have a strong free radical scavenging effect, of which rutin has the strongest antioxidant activity. Clinically, rutin is mainly used for the adjuvant treatment of hypertension and treatment for the prevention of other bleedings due to lack of rutin, such as cerebral hemorrhage. Structural modification of rutin can generate a series of derivatives with good activities, such as troxerutin, which is well water-soluble and conducive to human absorption. Troxerutin has a more significant efficacy than rutin and is used in the treatment of varicose veins/venous disorders. Since rutin has a mild effect with low cost and less adverse reactions, especially its remarkable effect on acute cerebral infarction, it is of great worth on the promotion and application of rutin.

Keywords Rutin · Flavonoids · Cardiovascular protection · Antihemorrhagic

Alias: Rutin, vitamin P, purple quercetin, purple skin
Origin: Remote lemongrass herb (Fig. 1)
Chemical name (Fig. 2)

2-(3,4-Dihydroxyphenyl)-5,7-dihydroxy-4-oxo-4H-chromen-3-yl-6-*O*-(6-deoxy-alpha-L-mannopyranosyl)-D-glucopyranoside

Molecular formula, $C_{27}H_{30}O_{16}$; **MW**, 610.51; **CAS**, 153-18-4
Rutin Derivatives (Fig. 3)

B.-Y. Hou · L. Zhang · J.-H. Wang · G.-H. Du (✉)
Beijing Key Laboratory of Drug Targets Identification and Drug Screening, Centre for Pharmaceutical Screening, Institute of Materia Medica, Chinese Academy of Medical Sciences and Peking Union Medical College, Beijing, China
e-mail: dugh@imm.ac.cn

G.-H. Du, *Natural Small Molecule Drugs from Plants*,
https://doi.org/10.1007/978-981-10-8022-7_23

Properties

Appearance: light yellow or yellow-green needle crystal or crystalline powder, tastes slightly bitter, usually contains three crystal water, melting point at 176–178 °C. *Solubility*: Rutin is soluble in methanol, pyridine, alkaline solution, and boiling water and hardly soluble in cold water, chloroform, carbon disulfide, ether, benzene, and petroleum ether.

Dosage Forms and Indications

The product is included in the department of chemicals and preparations.

At present, the main development of the dosage form is the increase of its solubility to improve bioavailability. The main dosage forms are rutin cyclodextrin saturation, HPMC controlled release tablets, solid dispersion tablets, coprecipitate, and rutin effervescent particles. In addition, compound prescription such as compound rutin tablets and compound rutin capsules has been prescribed mainly for the prevention and treatment of cerebral hemorrhage, hypertension, retinal hemorrhage, hemorrhagic purpura, acute hemorrhagic nephritis, and chronic bronchitis [1].

Literature

Rutin is widely found in nature and is almost contained in all of the Rutaceae and Sectaceae plants, especially abundant in Rutaceae, Rutaceae, Epacridaceae of leguminous, buckwheat of Polygonaceae, Eulali of *Hypericum*, *Berchemia polyphylla var. leioclada* of Tetranychidae, and wild wutong leaves of thistle, which are also used as raw materials in rutin extraction. In addition, it also exists in the *Ilex pubescens* of Aquifoliaceae, *Forsythia* of Oleaceae, *pagoda tree* pod of Leguminosae, tobacco, jujube, apricots, flavedo, tomatoes, and other plants.

At present, rutin in China mainly extracted from the *Sophora japonica* Linn of *leguminous*, which is listed in top grade of Shen Nong's *Classic Materia Medica.* In addition, tartary buckwheat, which is rich in rutin and flavonoids, is native to India and now produced in China's northwest, southwest, north, south, and other places. According to Xinnong's book and "Qi Min surgery miscellaneous," rutin is called a health-care factor. *Eucalyptus youmanii* is native to Australia. It is abundant in rutin and was introduced into China's Sichuan Province by Maoyu Chen et al. since 1997. It has been used as the raw material for rutin, bringing great industrial development value [2].

History of R&D

In the mid-1930s, Hungarian scientist Szent Gyorgy firstly separated the flavonoid mixture. After the German pharmacy firstly made it into ranosine in 1942, the concept of vitamin P has been established worldwide. Further study proved that rutin was the most important flavonoids of vitamin P. These compounds were certified to have effects on many diseases in medical.

Recently, the research of rutin mainly focuses on the extraction process improvement, pharmacological effects, and pharmacodynamics research, aiming at improvement of its bioavailability through the development of different dosage forms. As for the extraction process, new extraction and purification methods have been developed since the original alkali extraction acid precipitation method. These methods greatly improve its extraction efficiency and reduce cost, including hot water precipitation, hot water extraction with macroporous resin adsorbing purification, ultrasonic radiation, hot water extraction with alcohol precipitation, cold alkali percolation extraction with acid precipitation, continuous reflux extraction, ethanol extraction, supercritical CO_2 extraction, and enzymatic hydrolysis [3].

In recent years, advanced rutin dosage forms, such as rutin cyclodextrin saturation, HPMC controlled release tablets, solid dispersion tablets, coprecipitate, and rutin effervescent particles, greatly improve the rutin dissolution rate and its bioavailability.

Structural modification of rutin generated a series of derivatives with good activities, such as diethylamine ethyl rutin, diethylamine methyl rutin, and rutin methyl ether lysine salts. 7, 3′, 4′-Tris [*O*- (2-hydroxyethyl)] rutin is a semisynthetic product of rutin by hydroxyethylation and is a main component of Venoruton [1]. Troxerutin is well water-soluble, which is conducive to human absorption, and has a more significant efficacy than rutin.

Pharmacology

As a flavonoid substance, rutin has a significant protective effect on the cardiovascular system, including the endothelium-dependent vasodilation through NO-guanylate cyclase pathway, antagonization on platelet-activating factor (PAF), inhibition on subsequent reactions induced by PAF binding to its specific membrane receptor, and protection of myocardial cells [4].

Rutin also has good free radical scavenging effects. Studies showed that rutin and its derivatives had a strong free radical scavenging effect, of which rutin possessed the strongest antioxidant activity. Rutin removed superoxide anion and hydroxyl radicals, exerted a strong anti-lipid peroxidation, protected mitochondria, and enhanced the activity of superoxide dismutase (SOD).

Animal experiments showed that rutin could inhibit freezing-restraint stress and acid-induced gastric mucosal injury, protect the gastric mucosa, promote gastric mucosal cell synthesis, and release endogenous NO; it also could promote female adolescent rats' and lactating rats' immune organ thymus and spleen development, enhance immunity, activate neuro-U2 receptor, enhance insulin sensitivity, and lower blood sugar. Besides, rutin also has effects of antivirus, antibacterial, and anti-inflammatory, analgesic, and so on [5].

Clinical Application

Rutin is mainly used for the adjuvant treatment of hypertension and treatment for the prevention of other bleedings due to lack of rutin, such as cerebral hemorrhage, retinal hemorrhage, purpura, acute hemorrhagic nephritis, chronic bronchitis, and abnormal blood osmolality, restoration of capillary elastic embolism, and also for the prevention and treatment of diabetes and hyperlipidemia [6]. Troxerutin, the most important active ingredient in hydroxy rutin, is used in the treatment of varicose veins/venous disorders, hemorrhoids, lymphedema, and postoperative edema, treatment of thrombosis and cerebrovascular disease, and also in the treatment of diabetes and liver disease. Since rutin has a mild effect with low cost and less adverse reactions, especially its remarkable effect on acute cerebral infarction, it is of great worth on promotion and application of rutin [7].

Discussion

Flavonoids are widely distributed natural products in plants. In plants, most of them are contained in sugar and glycosides; some of them are in the free form of aglycones. Many studies have shown that flavonoids have a significant protective effect on the cardiovascular system and the liver, in addition to anti-inflammatory, improving the body immunity, and so on. As a flavonoid substance, rutin has a variety of pharmaceutical properties including reducing capillary permeability, anti-inflammatory, anti-allergic, antitumor, antibacterial, antivirus, and inhibition of aldose reductase activity. However, rutin in the current application for treatment is still relatively weak; it is necessary to find high-activity application of rutin.

Fig. 1 Remote lemongrass herb

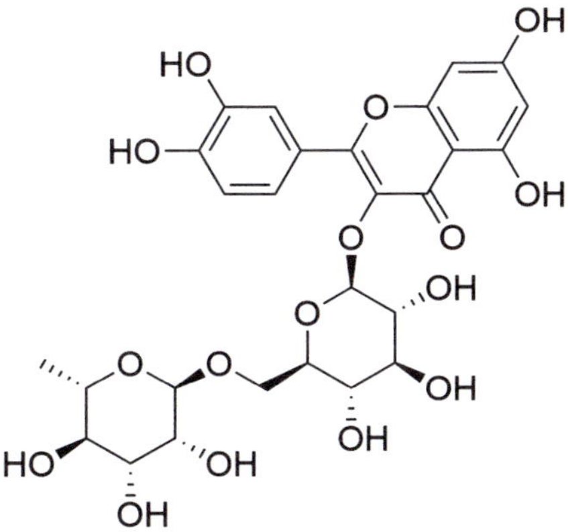

Fig. 2 Rutin chemical structure

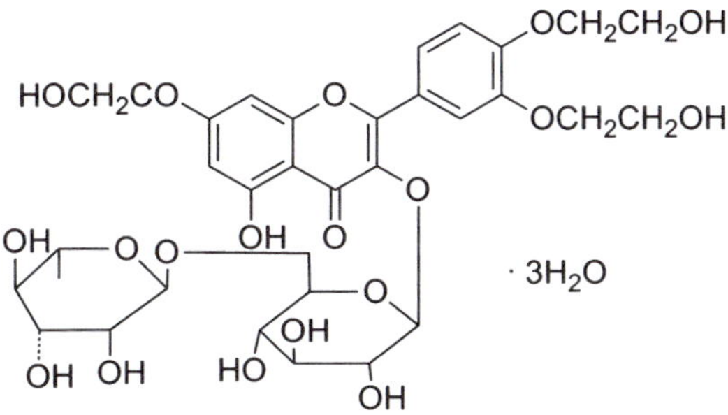

Fig. 3 Rutin derivative-troxerutin

References

1. Jiankun T, Yonghe Z, Wenzhong S. Research advancement in application of rutin and its derivatives. J Tianzhong. 2007;5:18–21.
2. Yushan L. Progress on resources, pharmacological effect and major forms of rutin. Amino Acids Biotiv Resour. 2013;03:13–6.
3. Xiangying M, Guo L, Yuxin L, et al. Resources, application and extraction and purification method of rutin. Acad Period Changchun Coll Tradit Chin Med. 2003;02:61–4.
4. Zhihe Z, Liping C, Ling Z. Progress on pharmacological action and preparation of rutin. Her Med. 2007;7:758–60.
5. Quanjiang L, Yang T. Research and prospect of rutin. J Chin Med Inf. 2002;04:39–42.
6. Jing L. Clinical pharmaceutical special of rutin. Chin J Clin Pharmacol. 2009;03:256.
7. Yuanqing F. Pharmacological properties and clinical application of rutin. Guid China Med. 2012;07:59–60.

Salvia miltiorrhiza Bunge (Danshen)

Li Zhang and Guan-Hua Du

Abstract Danshen is the dry root and rhizome of *Salvia miltiorrhiza* Bunge and is a traditional Chinese herbal medicine. The study proved that Danshen extracts and their components have many pharmacological actions. The lipophylic and hydrosoluble compounds from Danshen have been developed into many preparations for clinical application. Danshen preparations are mainly used in treating cardiovascular and cerebrovascular diseases, liver diseases, renal failure, cancer, and so on. Danshen and its preparations are the important drugs commonly used in clinical practice.

Keywords *Salvia miltiorrhiza* Bunge · Components · Pharmacological effects

Origin: *Salvia miltiorrhiza* Bunge (Fig. 1)

Dosage Forms

Danshen 丹参 (*Salvia miltiorrhiza* Bunge) is a traditional Chinese herbal medicine and is recorded in the *Pharmacopoeia of the People's Republic of China* (2015) and *United States Pharmacopeia* (40).

Danshen preparation forms include injections, tablets, capsules, dripping pills, etc., such as Danshen injectable powder, Danshen glucose injection, sodium tanshinone IIA sulfonate injection, Danshen injection, compound Danshen injection, Danshen polyphenolic acid injection, Danshen polyphenolic salt injection, tanshinone tablet, Danshen tablet, compound Danshen tablet, compound Danshen buccal tablet, Danshen heart-comforting capsules, tanshinone capsules, tanshinone

L. Zhang · G.-H. Du (✉)
Beijing Key Laboratory of Drug Targets Identification and Drug Screening, Centre for Pharmaceutical Screening, Institute of Materia Medica, Chinese Academy of Medical Sciences and Peking Union Medical College, Beijing, China
e-mail: dugh@imm.ac.cn

G.-H. Du, *Natural Small Molecule Drugs from Plants*,
https://doi.org/10.1007/978-981-10-8022-7_24

dripping pills, compound Danshen preparations soft capsules, coronary Danshen dripping pills, etc.

Indications

Danshen preparations are mainly used in treating cardiovascular and cerebrovascular diseases, liver diseases, renal failure, cancer, and so on.

Literature

Danshen is the dry root and rhizome of *Salvia miltiorrhiza* Bunge, which belong to the family of *Lamiaceae*, genus of *Salvia*. Danshen was firstly recorded in Shen Nong's Classic of Materia Medica. Later, Danshen was recorded in the other ancient monographs, for example, Wu Pu's Materia Medica, Miscellaneous Records of Famous Physicians, and Grand Compendium of Materia Medica. The functions of Danshen have "eliminating the pathogenic factors in heart and abdomen", "replenishing the heart, resuscitation, and relieving mental stress", "quieting vital fetus and removing the dead fetus, arresting hemorrhage, mitigating leucorrhea and regulating the irregular channels of women", "promoting blood circulation, regulating pericardium, invigorating qi and nourishing the blood"[1].

Danshen is widely distributed throughout China and is mainly distributed in Jiangsu, Anhui, Hebei, Sichuan, Henan, Hubei, Fujian, and Shanxi provinces.

History of R&D

Research on the chemical components of Danshen was initiated in the 1930s. Much research on the lipophylic components have been carried out. More than 40 lipophylic compounds have been identified. The research on the hydrosoluble components of Danshen was conducted later. Since the discovery of tanshinol, Chinese scholars isolated a series of hydrosoluble compounds from Danshen. These compounds have the structure of phenolic acid. They are named as salvianolic acids in alphabetic order, salvianolic acids A, B, C, D, E, G, H, tetramethyl salvianolic acid F, isosalvianolic acid C, rosmarinic acid, lithospermic acid, etc. The lipophylic and hydrosoluble compounds have been developed into many preparations for clinical application (Fig. 2).

Pharmacology

The study proved that Danshen extracts and their components have many pharmacological actions [2]. It was demonstrated that salvianolic acids have strong antioxidant actions. It has been shown by experiments that salvianolic acids can scavenge oxygen-free radicals and inhibit the peroxide reactions of lipids. Salvianolic acids also have significant inhibiting effects on the aggregation of platelets. Many experiments showed that Danshen can significantly protect the heart and brain against ischemia-reperfusion injury. Danshen also showed obvious protective effects on liver injuries and renal function impairment caused by multiple factors.

Clinical Application

Danshen preparations have been used to treat many diseases, such as cardiovascular disease, stroke, refractory hepatic diseases, pulmonary diseases, hypertension, infectious diseases, cancer, adhesive ileus, etc. Although Danshen has little toxicity, some adverse reactions have been reported. Danshen preparations may lead to adverse cardiovascular effects and should be used with caution.

Discussion

Danshen is a traditional herbal drug with long history of clinic application in China. Danshen and its components have various functions. It can be used for the treatment of many diseases and is an important drug commonly used in clinical practice.

Fig. 1 *Salvia miltiorrhiza* Bunge

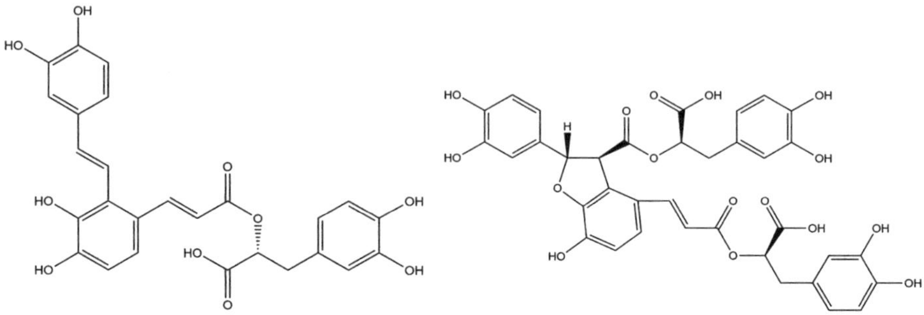

Salvianolic acid A Salvianolic acid B

Fig. 2 The structures of salvianolic acid A and salvianolic acid B

References

1. Yan XJ. Cyclopedia of Danshen. Beijing: People's Medical Publishing House; 2013.
2. Du GH, Zhang JT. The general situation and progress of the modern research of red sage root (Radix Salviae Miltiorrhizae). Her Med. 2004;23(7):435–40.

Tanshinone IIA

Yu-Cai Chen, Lian-Hua Fang, and Guan-Hua Du

Abstract Tanshinone IIA is extracted from the roots of *Salvia miltiorrhiza* Bge., which exhibits a variety of pharmacological effects, such as promoting blood circulation and opening the energy channels. In the long history of morphine R&D, the efficacy of tanshinone IIA is mainly concentrated in interfering with the pathological process of cardiovascular diseases. It can reduce intracellular calcium concentration, attenuate the damages from the reactive oxygen species to vascular endothelial cells, and lower the risk of atherosclerosis. For better water solubility, the sodium sulfonate group is introduced for clinical application. The research on tanshinone IIA continues. Current researches show that tanshinone may play a positive role in the treatment of a variety of diseases and has good clinical application prospects.

Keywords Tanshinone IIA · Cardiovascular disease · *Salvia miltiorrhiza* Bge.

Origin: *Salvia miltiorrhiza* (Fig. 1)
Chemical name (Fig. 2)

Phenanthro[1, 2-b]furan-10,11-dione, 6, 7, 8, 9-tetrahydro-1, 6, 6-trimethyl

Molecular formula, $C_{19}H_{18}O_3$; **MW**, 294.33; **CAS**, 568-72-9

Y.-C. Chen · L.-H. Fang · G.-H. Du (✉)
Beijing Key Laboratory of Drug Targets Identification and Drug Screening, Centre for Pharmaceutical Screening, Institute of Materia Medica, Chinese Academy of Medical Sciences and Peking Union Medical College, Beijing, China
e-mail: dugh@imm.ac.cn

G.-H. Du, *Natural Small Molecule Drugs from Plants*,
https://doi.org/10.1007/978-981-10-8022-7_25

Properties

Appearance: brick-red crystalline powder. *Melting point*: 209–210 °C. *Solubility*: absorbs moisture easily, very soluble in hot water, slightly soluble in methanol or ethanol, and practically insoluble in chloroform

Dosage Forms and Indications

This product is included in the tenth volume of national standards for chemical drugs.

At present, tanshinone II is clinically used in the form of sodium sulfonate, a derivative of it. And the dosage form of it is injection. It is mainly applied to the treatment of ischemic cardiovascular and cerebrovascular diseases such as coronary heart disease, premature ventricular contraction, cerebral insufficiency, cerebral thrombosis and cerebral embolism, and peripheral circulatory disorders such as various arterial occlusive diseases, vasculitis, and microcirculatory disorders caused by diabetes. It can also be used for the adjuvant therapy for hypertension, hyperlipidemia, and other diseases. Tanshinone capsules are made of total flavonoids extracted from *Salvia miltiorrhiza* and can be used for the treatment of adolescent acne.

Literature

Tanshinone IIA mainly exists in the roots of *Salvia miltiorrhiza* Bge. Traditional Chinese medicine Danshen, slightly cold and bitter, is often used for activating blood circulation and eliminating stasis in Chinese medicine. It can also promote blood circulation, open the energy channels, calm the heart, cool the blood, relieve swelling, remove twinge in the heart and stomach, and remove carbuncle and erysipelas. It is classified as high grade in *Shen Nong's Herbal Classic*. It has been used as a traditional Chinese medicine in clinical practice for thousands of years. And it is not only recorded in the leading medical works through the ages but also in CP and USP.

History of R&D

The study of the chemical composition of *Salvia miltiorrhiza* began in the 1930s. Japanese scholars first extracted three kinds of liposoluble components from Danshen. They are tanshinones I, II, and III. After that, Chinese scholars carried out an in-depth systematic exploration in the chemical research and found that tanshinone II is a mixture of two components, named tanshinone IIA and tanshinone IIB, respectively, and elucidated their chemical structures [1].

Since the discovery of tanshinone, many researchers studied in depth the liposoluble components of Danshen and found that the variety of compounds has reached up to 40, such as tanshinone I, tanshinone IIA, tanshinone IIB, tanshinone III, isotanshinone I, isotanshinone IIA, tanshinone, isocryptotanshinone, hydroxy tanshinone IIA, dihydro tanshinone I, L-dihydrogen tanshinone I, *Salvia* new quinone, *Salvia* new quinone B, and so on.

After decades of efforts, we have a comprehensive understanding of the liposoluble components of *Salvia miltiorrhiza*. With the application of new technology and the discovery of the bioactivity of liposoluble components, the study on the chemical composition of liposoluble components of Danshen is further concerned by researchers. And some new compounds were reported. Some tanshinones with diterpene-quinone structure were discovered, such as neocryptotanshinone II and oleoyl neocryptotanshinone [2].

Pharmacology

Long-term researches show that tanshinone exhibits a variety of pharmacological effects, such as the protection of the cardiovascular system, anti-infective effects, and antioxidant effects.

As for the domestic pharmacological study of tanshinone IIA, it was first started in the Shanghai Institute of Chinese Medicine. Professor Ding Guangsheng discussed the cardiovascular effects of tanshinone IIA. He found that intraperitoneal injection of tanshinone IIA sodium sulfonate (200 mg/kg) significantly prolonged the survival time of mice under hypoxia atmosphere, and an increase in cardiac output was observed in a dog under anesthesia with a one-time intravenous injection of 20 mg/kg [3]. It is worth mentioning that Tang Jixue, professor of Chinese Academy of Medical Sciences, simultaneously studied the anti-infective effect of total tanshinone, including tanshinone IIA. She found that tanshinone has a significant inhibitory effect on *Staphylococcus aureus*, especially resistant strains. Its minimum inhibitory concentration to *Mycobacterium tuberculosis* H37Rv is lower than 1.5 μg/mL, at which concentration of the antibacterial effect is stronger than that of berberine [4].

In recent years, the mechanism of tanshinone IIA is increasingly brought to further research. Tanshinone IIA can increase the activity of superoxide dismutase (SOD) and interfere with the pathological process of many diseases, especially in cardiovascular diseases. It can attenuate the damages from the reactive oxygen species to vascular endothelial cells, lower the risk of atherosclerosis, and reduce the formation of atheromatous plaque. Intraperitoneal injection of tanshinone 40 mg/kg can significantly improve myocardial energy metabolism and alleviate myocardial ischemic injury [5].

Tanshinone IIA can reduce intracellular calcium concentration through a variety of ways [6]. It can not only protect cardiac muscles and reduce the incidence of arrhythmia by preventing myocardial calcium overload and reducing myocardial hypoxia injury but also inhibit the development of cardiac hypertrophy via reducing the activity of calcineurin and the expression of its protein.

With the progress of cell biology and molecular biology, new therapy pathways and targets have been found, which provide new ideas and methods for the study of the antitumor effect of traditional Chinese medicine. Recent studies indicate that tanshinone IIA, from the traditional Chinese medicine, has cytotoxic effects on many kinds of tumor cells and can play an antitumor effect through the mechanism of killing, inducing differentiation and apoptosis, inhibiting the invasion and migration of tumor cells, and especially inducing the differentiation of tumor cells. Recent studies show that tanshinone, such as diphosphate ketone, can reverse the drug resistance of tumor cells, which is highly thought by researchers. The antitumor mechanism of tanshinone may be related to the inhibition of DNA synthesis in tumor cells and the expression of various genes connected with the proliferation, differentiation, and apoptosis of tumor cells. It may also have something to do with the inhibition of telomerase activity in tumor cells, changes in antigen expression on tumor cell surface, etc. [7].

Clinical Application

Tanshinone IIA is a diterpenoid quinone liposoluble ingredient of high content, and its chemical structure is the most representative in Danshen. But its monomer is of low water solubility and bioavailability. The introduction of sodium sulfonate group can not only greatly enhance its water solubility but also ensure its higher efficacy. However, there are reports that, because of it strong polarity, it is so difficult for tanshinone IIA sodium sulfonate to pass through the biofilm with a double structure that it can hardly pass through the blood-brain barrier, thus making its concentration in the central nervous system too low to display therapeutic effects. Due to its fast metabolic rate, we cannot give full play to its efficacy [8]. In addition to tanshinone IIA sodium sulfonate, there are a variety of preparations of *Salvia* common ketone used clinically, such as tanshinone tablets, tanshinone capsules, tanshinone injection, Danshen Shuxin capsule, compound Danshen soft capsules, and compound Danshen particles.

Discussion

Salvia is the main medical material for the treatment of cardiovascular diseases in traditional Chinese medicine since ancient times. The pharmacological effects of tanshinone IIA have been discovered in recent decades, prompting researchers to comprehensively evaluate the effects of the compounds. Comprehensive analysis of the pharmacological action and mechanism of tanshinone IIA show that tanshinone IIA is a multi-target drug and is involved in redox reaction, energy metabolism, ion channel status, and other physiological processes. Current researches show that tanshinone may play a positive role in the treatment of a variety of diseases and has good clinical application prospects.

Fig. 1 *Salvia miltiorrhiza*

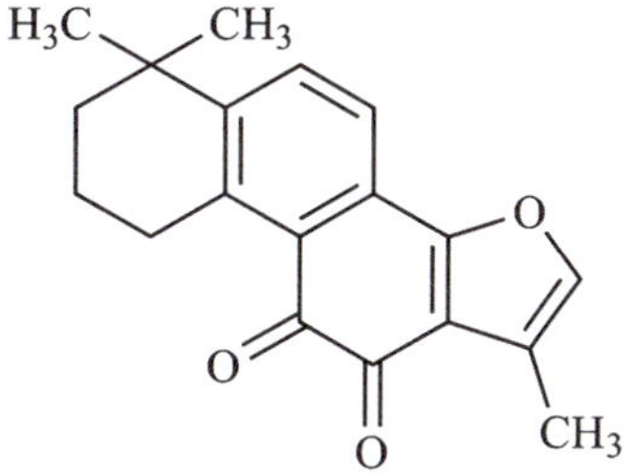

Fig. 2 Chemical structure of tanshinone IIA

References

1. Chen WZ. Pharmacological action of Salvia miltiorrhiza. Acta Pharm Sin. 1984;19(11):876–80.
2. Du GH, Zhang JT. The general situation and progress of the modern research of red sage root. Her Med. 2004;23(6):355–60.
3. Chen WZ, Dong YL, Wang CG, Ding GS. Pharmacological study of tanshinone IIA sodium sulfonate. Acta Pharm Sin. 1979;05:277–83.
4. Gao YG, Song YM, Yang YY, Tang JX, Liu WF. Pharmacological study of tanshinone. Acta Pharm Sin. 1979;14(02):75–82.
5. Hu H, Zhai C, Qian G, Gu A, Liu J, Ying F, et al. Protective effects of tanshinone IIA on myocardial ischemia reperfusion injury by reducing oxidative stress, HMGB1 expression, and inflammatory reaction. Pharm Biol. 2015;53(12):1752–8.
6. Wang J, Jiang Q, Wan L, Yang K, Zhang Y, Chen Y, et al. Sodium tanshinone IIA sulfonate inhibits canonical transient receptor potential expression in pulmonary arterial smooth muscle from pulmonary hypertensive rats. Am J Respir Cell Mol Biol. 2013;48(1):125–34.
7. Zhang MT, Qian YH, Tang AQ. Advances in pharmacological effects of tanshinone IIA. Med Recapitulate. 2010;16(17):2661–4.
8. Bai FM. Clinical application of tanshinone IIA sodium sulfonate. China Pharm. 2012;23(31):2971–3.

Tetrandrine

Huan Yang, Yue-Hua Wang, and Guan-Hua Du

Abstract Tetrandrine (Tet) is a dibenzyl isoquinoline alkaloid, which is extracted from the root of *Stephania tetrandra* S. Moore (Fĕn Fáng Jĭ) of Menispermaceae. Tetrandrine has strong biological effects on the cardiovascular diseases, fibrosis, tumors, portal vein and pulmonary hypertension, liver diseases, and immunological diseases. Current researches showed that tetrandrine is superior to conventional therapies in clinical treatment of silicosis. Other pharmacological effects of tetrandrine are still under research. As a traditional Chinese calcium antagonist, tetrandrine has significant effects on cardiovascular and inflammatory diseases with relatively fewer side effects, which grants it great potential for further explorations and clinical applications.

Keywords Tetrandrine · Alkaloid · Cardiovascular diseases · Clinical applications

Origin: *Stephania tetrandra* S. Moore (Fig. 1)
Chemical name (Fig. 2)

(1β)-6,6′,7,12-Tetramethoxy-2,2′-dimethylberbaman (S,S)-(+)-Tetrandrin

Molecular formula, $C_{38}H_{42}O_6N_2$; **MW**, 622.75; **CAS**, 518-34-3

Properties

Appearance: Needle-like crystals (ether). *Solubility*: Hardly soluble in water and petroleum ether; soluble in ether and some organic solvents. *Melting point*: 219–222 °C. *Specific optical rotation*: 285° ($c = 1$, $CHCl_3$); sensitive to light.

H. Yang · Y.-H. Wang · G.-H. Du (✉)
Beijing Key Laboratory of Drug Targets Identification and Drug Screening, Centre for Pharmaceutical Screening, Institute of Materia Medica, Chinese Academy of Medical Sciences and Peking Union Medical College, Beijing, China
e-mail: dugh@imm.ac.cn

G.-H. Du, *Natural Small Molecule Drugs from Plants*,
https://doi.org/10.1007/978-981-10-8022-7_26

Dosage Forms and Indications

This product is included in national standards for chemical drugs (Volume 14), *British Pharmacopoeia* (2017), and *European Pharmacopoeia* (9.0th ed.).

Tetrandrine is used for the treatment of mild to moderate hypertension and hypertensive crisis, rheumatism, silicosis, etc.

Literature

Tetrandrine (Tet) is the dibenzyl isoquinoline alkaloid extracted from the root of *Stephania tetrandra* S. Moore (Fěn Fáng Jǐ) of Menispermaceae. It is the main bioactive component of the plant.

As "Shennong Bencao Jing" records: *Aristolochia fangchi* (Fáng Jǐ), the dry roots of *Stephania tetrandra* S. Moore (Fěn Fáng Jǐ) of Menispermaceae. Indications: cold, fever, epilepsy, urinary tract diseases. The roots of this herb should be dug in autumn. After removing the rough skin, they are half-dried and cut into sections or longitudinal pieces. Further dried, then they can be used as drug [1, 2].

Since *Stephania tetrandra* S. Moore (Fěn Fáng Jǐ) and Radix Aristolochiae Fangchi (Guǎng Fáng Jǐ) have similar medicine name and functions, they are often used confusedly. Radix Aristolochiae Fangchi is the plant in Aristolochiaceae of Aristolochia L. It is similar to *Stephania tetrandra* S. Moore, but it contains aristolochic acid which can cause kidney damage. The State Food and Drug Administration stipulated that it cannot be used as a drug since September 30, 2004 [2, 3].

History of R&D

Recent studies have shown that tetrandrine has a variety of biological effects and very good applicational prospects in the treatment of fibrosis and portal vein and pulmonary hypertension, the regulation of immunologic function, as well as the prevention and treatment of tumor [2, 4].

As early as 1988, tetrandrine has been found having the effect on blocking the Ca^{2+} channel and was quickly applied into the pharmacological research in the field of cardiovascular and inflammatory diseases [5]. Results from a large number of studies have shown that tetrandrine has good effects on antihypertension, arrhythmia, myocardial ischemia, inflammation, and so on [6]. As a traditional Chinese calcium antagonist, tetrandrine has a broad prospects in clinical applications of cardiovascular and inflammatory diseases [7].

In the early 1990s, the application of tetrandrine was extended. During that time, researchers conducted many studies about its protective effects on liver, lung, and mitochondria, which opened a new field for the treatment of liver disease [8]. In 2002, it was found that tetrandrine can inhibit the synthesis of DNA and RNA in tumor cells, which provided a new method for the treatment of cancer [9].

At present, the prevention and treatment effects of hypertension, fibrosis, digestive diseases, tumors, rheumatoid arthritis, and other autoimmune diseases of tetrandrine have been confirmed, as well as the function of reducing portal hypertension and pulmonary hypertension, while its other pharmacological effects are to be explored in further study [10].

Pharmacology

Tetrandrine has analgesic, anti-inflammatory, and anti-allergic effects and has a wide range of usage on the cardiovascular system owing to its antihypertensive, anti-myocardial ischemia/reperfusion injury and antiarrhythmic effects. It can inhibit the platelet aggregation induced by ADP, collagen, and arachidonic acid in vitro and can also restrain the platelet adhesion and thrombosis (in rabbits).

Tetrandrine also has anticancer effects. Studies have shown that tetrandrine has a strong inhibitory effect on the DNA and RNA synthesis in L7712 and S180 (cancer cells), which can significantly suppress the growth of Wacker sarcoma W256. Besides that, tetrandrine has the ability to relax the striated muscle, and its methyl iodide or methyl bromide derivatives can also affect the muscles. Notably, tetrandrine can prevent silicosis and has a preferable outcome on the clinical treatment of such disease. In addition, tetrandrine also owns antipyretic, diuretic, and anti-allergic shock effects [2, 3, 7].

Clinical Application

Tetrandrine is used for the treatment of hypertension, angina, termination of paroxysmal supraventricular tachycardia, pulmonary fibrosis, and other diseases in clinical application, and it also has strong antitumor effects. Tetrandrine was also approved for lowering blood glucose and free radical damage; its treatment effect on silicosis is significant and superior to conventional immunosuppressive and cytotoxic drugs [3, 7].

Discussion

Researchers have carried out a large quantity of researches on the pharmacological effects of tetrandrine and found out that tetrandrine has a wide range of pharmacological effects on the cardiovascular system. The current research of tetrandrine on the treatment of silicosis has achieved satisfactory results, which showed that tetrandrine is superior to conventional immunosuppression and cytotoxic drugs in clinical treatment of silicosis. Tetrandrine is used as broad-spectrum anti-inflammatory drug and has a certain degree of immunosuppressive effect. To sum up, tetrandrine can be applied for treating several kinds of diseases and has fewer side effects, which grants it a bright future for further explorations and clinical applications.

Fig. 1 *Stephania tetrandra*

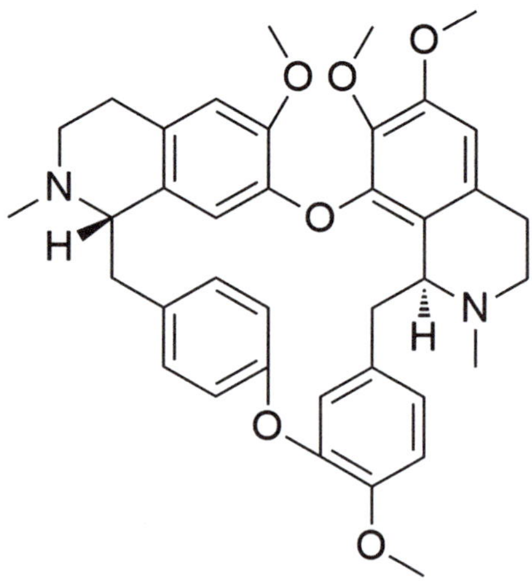

Fig. 2 The structure formula of tetrandrine

References

1. National Pharmacopoeia Commission. Pharmacopoeia of the People's Republic of China (volume 1). Beijing: China Medical Science and Technology Press; 2015. p. 148–9.
2. Zhen P. Research progress of tetrandrine. J Med Res. 2004;33(8):79–82.
3. Zhang XH. Study on the identification and clinical application of Stephania tetrandra S. Moore and Radix Aristolochiae Fangchi. Chin Mod Med. 2015;22(24):111–3.
4. Wang H, Luo SD, Cai HS. Pharmacological research progress of tetrandrine. Chin J Pharm. 2000;35(12):800–2.
5. King VF, Garcia ML, Himmel D. Interaction of tetrandrine with slowly inactivating calcium channels. Characterization of calcium channel modulation by an alkaloid of Chinese medicinal herb origin. J Biol Chem. 1988;263:2238–44.
6. Liu YQ, Karpinski E, Rao MR. Tetrandrine: a novel calcium channel antagonist inhibits type I calcium channels in neuroblastoma cells. Neuropharmacology. 1991;30(12A):1325–31.
7. Wang ZR. Research progress on pharmacological action of tetrandrine. Chin Pharmacol Bull. 2000;16(5):488–92.
8. Li DG, Liu YL, Lu HM, Jiang ZM, Xu QF. Effects of tetrandrine on mitochondria in rats with hepatic fibrosis. Chin J Dig. 1994;14(6):339–42.
9. Chen YJ. Potenital role of tetrandrine in cancer therapy. Acta Pharmacol Sin. 2002;23(12):1102–6.
10. Kwan CY. Traditional herbal medicine research with special reference to tetrandirne and related bis-benzylisoquinoline alkaloids: a preface to this special issue on herbal drugs. Acta Pharmacol Sin. 2002;23(12):I.

Yohimbine

Wei-Qi Fu, Wan Li, Jin-Hua Wang, and Guan-Hua Du

Abstract Yohimbine is a natural alkaloid extracted from *yohimbe* barks and has an aphrodisiac effect. So far, yohimbine is the preferred natural drug product to treat impotence and improve male sexual function. However, a long-term medication of yohimbine may bring some side effects to patients. Therefore, in order to promote the clinical application of the drug, it is necessary to make further studies to diminish side effects and determine its applicable populations. In general, yohimbine demonstrates attractive pharmacological effects in clinics, and it is the ultimate goal to innovate new dosage forms for yohimbine.

Keywords Yohimbine · Natural alkaloid · Impotence · Aphrodisiac effect

Origin: *Corynanthe johimbe* Schum (Fig. 1)
Alias: Corynine, menolysin, yohimbine hydrochloride
Chemical name (Fig. 2)

17-α-Hydroxyyohimban-16-α-carboxylic acid methyl ester

Molecular formula, $C_{21}H_{26}N_2O_3$; **MW**, 354.45; **CAS**, 146-48-5

W.-Q. Fu · W. Li · J.-H. Wang · G.-H. Du (✉)
Beijing Key Laboratory of Drug Targets Identification and Drug Screening, Centre for Pharmaceutical Screening, Institute of Materia Medica, Chinese Academy of Medical Sciences and Peking Union Medical College, Beijing, China
e-mail: dugh@imm.ac.cn

G.-H. Du, *Natural Small Molecule Drugs from Plants*,
https://doi.org/10.1007/978-981-10-8022-7_27

Properties

Appearance: white powder. *Solubility*: soluble in ethanol, chloroform, and hot benzene; slightly soluble in water and ether, usually salified by hydrochloric acid to increase its solubility in water. *Specific rotatory power* (°): [α]D22 +105° (in water). *Melting point*: 241–246 °C.

Dosage Forms and Indications

This product is listed in the 2017 edition of the *British Pharmacopoeia*, 40 editions of the *American Pharmacopoeia*, and 9.0 edition of the *European Pharmacopoeia*.

The main clinical application of yohimbine includes tablets and injections. It is mainly used to treat various types of impotence and sexual dysfunction in men.

Literature

Yohimbine is a natural alkaloid. It was first extracted from the barks of *Corynanthe yohimbe*, a species of Rubiaceae trees in West Africa. It was reported that, in the dried bark of *Pausinystalia johimbe*, the content of mixed alkaloids is higher than 6.1%, in which the main component is yohimbine, indicating a great prospect for development [1]. *Yohimbe* bark has been used as an aphrodisiac in Africa since ancient times. In 1900, it was applied by Kowit and Muller to patients with impotence and paralytic insensitivity caused by neurasthenia and obtained curative effect. From then on, clinical application of yohimbine began. Currently, yohimbine is a pure plant preparation in the treatment of erectile dysfunction with more affirmation and more applications.

History of R&D

Yohimbine has been used as an aphrodisiac for many years. At first, pharmacologists attributed its aphrodisiac effects to psychological effects similar to placebo or increasement of peripheral vascular congestion, rather than real sexual stimulation. Physiologists at the Stanford University first conducted a study on the pharmacological effects of yohimbine and found that yohimbine could increase the mating ability of rats [2], which was then published on *Science* in 1984 [3]. In addition, researchers in the Queensland University in Canada conducted experiments on 23 patients with sexual dysfunction. Six of them recovered after taking the drug for 10 weeks. In 1987, Canadian scientists confirmed that yohimbine treatment in psychogenic impotence was safe and effective and this drug could restore the patient's sexual ability [4]. Besides, they proved that this medicine showed good curative effects on organic impotence.

Pharmacology

Studies have shown that yohimbine has extensive pharmacological effects and has been developed for the clinical treatment of arteriosclerosis, rheumatism, and other diseases. The most obvious pharmacological action is in the treatment of male sexual dysfunction. Yohimbine tablets have been approved by the FDA and circulate in international markets. Yohimbine can selectively block the presynaptic alpha 2 receptors and promote the release of norepinephrine [5]. It stimulates more norepinephrine released by cavernous nerve endings and reduces reflux of phallic vein, which is conducive to congestive erection. A small amount of application can make the perineum swell and stimulate the erection center at the spinal cord, leading to sexual hyperfunction [6]. Yohimbine hydrochloride also has a psychological stimulant effect and increases libido. Like other types of adrenergic blocking drugs, yohimbine's resistance to adrenergic mediator in blood circulation is much stronger than to sympathetic nerve impulse. Again, like tolazoline, yohimbine shows slight effect in resisting adrenergic response in ocular smooth muscle. This drug does not block the frequency and inotropic effects of epinephrine on mammalian hearts. Yohimbine has minor direct effects on smooth muscle, and its effect on the central nervous system is far less than that of ergot alkaloids, because yohimbine performs an excited-to-paralyzed action. This drug produces diuretic effect, probably due to the stimulation of the hypothalamus, resulting in release of posterior pituitary hormone. In addition, yohimbine has a significant local anesthetic effect [7].

Clinical Application

The clinical single agent is used for the treatment of male impotence and sexual dysfunction. According to clinical observations, yohimbine therapy is not related to the degree of erectile dysfunction, and it can be effective in mild, moderate, and severe abnormalities [8].

Discussion

As early as 70 years ago, it has been recognized that *yohimbe* extracts have unique healthcare effects. Yohimbine is a natural product that can make the human body stronger. By taking yohimbine together with intensive training, athletes can achieve extremely high standard of physical strength. At present, yohimbine is the first choice of natural drugs to treat impotence and improve sexual function in male, and it is effective and convenient to use. However, there may be some side effects with long-term medication. So it is necessary to make further studies to eliminate side effects and determine the applicable population of the drug. In summary, although the evaluation work of yohimbine is still in progress, the attractive pharmacological effects of its clinical manifestations should not be ignored. In the future, it will be imperative to innovate controlled release preparations of yohimbine. The ultimate goals of its dosage form should be constant speed, long-term effect, and low toxicity.

Fig. 1 *Corynanthe johimbe* Schum

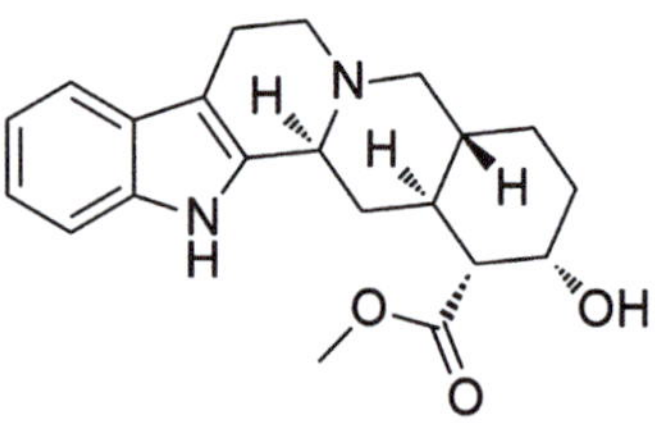

Fig. 2 Structural formula of yohimbine

References

1. Clark E, Davidson SJ. Abstract, 9th annual meeting of the International Academy of Sex Research, Harriman, N.Y., 22 to 26 November 1983; J. Clark, Monoaminergic modulation of copulation in male rats, thesis, Stanford University. 1983.
2. Huang X, Lin H. Clinical application of yohimbine. Strait Pharm J. 1999;02:4–5.
3. Clark JT, Smith ER, Davidson JM. Enhancement of sexual motivation in male rats by yohimbine. Science. 1984;225(4664):847–9.
4. Reid K, Surridge DH, Morales A, et al. Double-blind trial of yohimbine in treatment of psychogenic impotence. Lancet. 1987;2(8556):421–3.
5. Marwaha J, Aghajanian GK. Relative potencies of alpha-1 and alpha-2 antagonists in the locus coeruleus, dorsal raphe and dorsal lateral geniculate nuclei: an electrophysiological study. J Pharmacol Exp Ther. 1982;222(2):287–93.
6. Zheng P. The pharmacological mechanism and clinical application of yohimbine for impotence [J]. Chin Pharm Bull. 1993;06:418–20.
7. Morals A, et al. Yohimbine effection in treatment of impotence. Jurol. 1982;123(1):45.
8. Susset JG, Tessier CD, Wincze J, et al. Effect of yohimbine hydrochloride on erectile impotence: a double-blind study [J]. J Urol. 1989;141(6):1360–3.

Part II
Natural Small Molecule Drugs for the Prevention and Treatment of Neuropsychiatric Diseases

Overview

In the long history of human understanding natural drugs, the discovery of drugs that act on the nervous system is a unique process. Most of the drugs were originally used as poisons, which was the main basis for people recognizing the close relationship between drugs and toxicity. In the long process of human struggling against nature, especially for the hunting race, in order to better capture the ferocious wild animals, they found the substances with neurotoxicity. The toxicities of these substances were very significant and strong, therefore attracting people's attention. After a series of studies, a number of natural small molecule drugs with selective toxicity were discovered, such as tubocurarine, cissampelosine methiodide, pilocarpine, physostigmine, and so on.

In order to demonstrate the toxic characteristics and rules of these toxic small molecule drugs, scientists have carried out systematic and in-depth studies of their action mechanisms; discovered the roles of neural signal transduction, agonism, and antagonism on receptors; and greatly promoted the development of neuroscience. We are now able to understand the mechanism of neural signal transduction at the molecular level, which is closely related to the action mechanisms of these drugs. For these drugs, their toxic, pharmacological, and physiological effects are the quantitative change process of the same action. Their actions increase with the increase of dosage, and they appear toxic when their doses reach a certain level. Therefore, a suitable dose is a good medicine for disease, and the overdose is a fatal poison.

Many natural small molecule drugs have been discovered, which act on the nervous system; some of them act on the cholinergic nervous system, such as cholinergic receptor agonists, antagonists, and cholinesterase inhibitors. Among these natural small molecule drugs, some of which mainly show peripheral action and some of which mainly show central action, they have played active roles in the treatment of many diseases, such as galantamine, huperzine A, and so on. In addition, there are several drugs that act on other nervous systems, such as ephedrine, which acts on the adrenergic nervous system, and caffeine, a central stimulant.

In the study of anticholinergic drugs, the study of the natural cholinergic M receptor antagonists also led to favorable effects in the clinical treatment of acute diseases, such as shock. It is worth mentioning that the study of scopolamine and its derivatives promoted the study of microcirculation and therefore effectively improved the treatment effects of the acute diseases caused by microcirculation disorders. Based on the studies of the natural products scopolamine and so on, the development of the drug "654-2" has rescued a large number of critical patients, and now it is still an important drug for clinical emergency treatment.

The pharmacological study of ephedrine is a major discovery by Chinese scholars in the study of traditional medicines. In the 1920s, the pharmacologists demonstrated that ephedrine induced smooth muscle relaxation after the experiments. This discovery not only provides a new drug for the treatment of diseases, but the mechanism study also made progress, which better confirms the effectiveness and material basis of Chinese medicine in treating diseases.

In the natural drugs that act on the nervous system, there is a particular drug, opium. Opium is a natural drug with strong analgesic effect, used for the treatment of many diseases. Because of its addiction, it has brought great disaster to mankind in the process of its long-term use. So far, opium, as a starting point, has not only promoted the human understanding of central analgesic mechanism, and the discovery of the opioid receptor, but also promoted the discovery of various analgesic medicines, cough medicines and so on. At the same time, due to the use of opioid, a number of addictive drugs appeared. Opium, as a drug, fully performs the dual natures of drugs with therapeutic and toxic effects. Its effect is remarkable in relieving human suffering, and its evil is also extremely serious in misuse.

Among the small molecule drugs discovered and used for the treatment of neuropsychiatric system, some of which have been clinically proven to have very limited therapeutic efficacy, and their alternative drugs with better efficacy have been developed. So this kind of drugs should be evaluated objectively and withdrawn from the clinic according to specific conditions. There are also many medicines with the features of being effective and safe. Although these drugs were ever used in clinical, other chemical drugs developed later with some special characteristics, especially with more clear mechanism, gradually replaced these drugs. These early discovered drugs, such as rotundine, strychnine, chelidonine, and so on, need more in-depth studies to comprehensively understand their characteristics and rules. We hope these natural medicines will play greater roles in the prevention and treatment of nervous system diseases.

Anisodine

Shou-Bao Wang, Xiu-Ying Yang, and Guan-Hua Du

Abstract Anisodine was separated from the herbal medicine *Anisodus tanguticus* from Qinghai province, and its hydrobromide form was an originally created new drug in China. It competes with acetylcholine in the M cholinergic receptor to prevent acetylcholine from binding M cholinergic receptor, thus blocking the nerve impulse transmission and interfering with the physiological function based on the cholinergic neurotransmission. Anisodine has weaker intensity of action than atropine, as well as toxicity. Oral anisodine is absorbed rapidly and completely. At first, anisodine was mainly used in the treatment of various diseases of the central nervous system. The efficacy of anisodine for the treatment of cerebrovascular disease was validated from 1970 to 1975. Recently, compound anisodine injection based on a combination of procaine and anisodine hydrobromide showed a good effect for a variety of ischemic eye diseases induced by a variety of causes and other related eye diseases, especially superior to traditional therapies for ischemic lesions of the optic nerve, retina, and choroid. This drug has been widely used in China to promote the treatment for various eye diseases which achieved exciting results.

Keywords Anisodine · M cholinergic receptor · Compound anisodine injection · Ischemic eye diseases

Alias: AT-3, 703, daturamine, α-hydroxyscopolamine
Origin: *Anisodus tanguticus* (Fig. 1)
Chemical name (Fig. 2)

2,3-Dihydroxy-2-phenylpropanoic acid 9-methyl-3-oxa-9-azatricyclo[3.3.1.$0^{2,4}$] nonane-7alpha-yl ester;

S.-B. Wang · X.-Y. Yang · G.-H. Du (✉)
Beijing Key Laboratory of Drug Targets Identification and Drug Screening, National Centre for Pharmaceutical Screening, Institute of Materia Medica, Chinese Academy of Medical Sciences and Peking Union Medical College, Beijing, China
e-mail: dugh@imm.ac.cn

G.-H. Du, *Natural Small Molecule Drugs from Plants*,
https://doi.org/10.1007/978-981-10-8022-7_28

9-methyl-3-oxa-9-azatricyclo[3.2.1.$0^{2,4}$]non-7-yl α-hydroxy-α-(hydroxymethyl) benzeneacetate

Molecular formula, $C_{17}H_{21}NO_5$; **MW**, 319.36; **CAS**, 52646-92-1

Properties

Appearance: white crystals or crystalline powder. *Flash point*: 253.2 °C. *Melting point*: 126–128 °C (acetone-H_2O), 190–192 °C (95% ethanol), and 197–200 °C (absolute ethanol). MS m/e (%): 319(51), 154(14), 138(100), 96(5), 95(5), 94(36), 97(15), 96(5), 42(15), 137(25), and 119(23).

Dosage Forms and Indications

Anisodine hydrobromide is recorded in the second volume of national standards for chemical drugs of the People's Republic of China. The formulation consists of injection and tablet.

It is used clinically for treatment of migraine, retinal vascular spasm, ischemic lesions, shrinking of the optic nerve, retina, and choroid, etc. It could improve the functional recovery after acute paralysis induced by inflammation of the nervous system and cerebrovascular disease, paralysis agitants, and carbon monoxide-induced toxic encephalopathy. It was also used for intravenous combined anesthesia, organophosphorus poisoning, bronchitis, asthma, and prevention and treatment of seasickness.

Literature

Anisodine is isolated from the roots of the plant Solanaceae *Anisodus tanguticus* (Maxim.) Pasch. (*Scopolia tangutica* Maxim.), which has a common name of Zhang-Liu-Shen, growing at the plateau with an altitude of 1700–4300 m and richly distributed in Tibet, Sichuan, Qinghai, Gansu, and other provinces in China.

History of R&D

Anisodine hydrobromide was an originally created new drug in China. It was separated from the herbal medicine *Anisodus tanguticus* from Qinghai province. In the early 1960s, *Anisodus tanguticus* was given the common name of Zhang-Liu-Shen in Qinghai province, similar to the common name Zhang Liu of *Radix Phytolaccae*. As a result, the root was misused as *Radix Phytolaccae* until the occurrence of atropine poisoning symptoms. Histological identification found that it is *Anisodus tanguticus* (Maxim.) Pasch, not *Radix Phytolaccae*. In order to make full use of the wild plant's resources and develop the products of henbane drugs in China, the Chinese Academy of Medical Sciences carried out the systemic research on the plant [1].

In the process of separating chemical components from the stem and leaves, a researcher mistakenly knocked the sample marked as At-2 into his eye, and the pupil enlarged immediately and recovered the next day. This accidental phenomenon aroused the great interest of pharmacological researchers. They soon carried out a comprehensive and systematic study of At-2 and compared it with atropine. Experimental results showed that the effect of At-2 on the central nervous system was weaker than that of atropine, but the peripheral effects of both were similar. Chemistry studies have shown that the chemical structure of At-2 is 6S-hydroxyl scopolamine. Therefore, it was named anisodamine according to the botanical name of its origin. At that time, the pediatric department in Beijing Friendship Hospital is using high-dose atropine to rescue some children with shock caused by toxic dysentery. However, adverse side effects of atropine due to strong inhibition of gland secretion diminished the therapeutic benefits of atropine on shock in some cases, so the urgent need for drugs with atropine-like therapeutic effect but not the side effects, i.e., inhibition of gland secretion, remained. The pharmacological effects of anisodamine are in line with this requirement, which inhibits glandular secretion by several times as much as atropine. Anisodamine is applied into the clinical trial for the first time in April 1965, which is the origin of anisodamine trade name "654" (synthetic products known as 654-2; see "anisodamine") [2]. Because anisodamine had a significant effect on different types of shock, and its main mechanism is to improve microcirculation disorders, it could receive the similar effect and greatly reduce the mortality rate on different diseases such as toxic dysentery and fulminant meningitis and of critically sick children with large leaf pneumonia. The invention and application of anisodamine promoted the studies of chemical constituents from *Anisodus tanguticus* and their pharmacological action [3].

In 1965, the Institute of Materia Medica, Chinese Academy of Medical Sciences, found a new alkaloid named as anisodine from *Anisodus tanguticus*. In 1967, the First Pharmaceutical Factory of Chengdu isolated anisodine from the extraction liquor of *Anisodus tanguticus* for atropine production, and then massive production came true. At that time, the drug played a pivotal role in the clinic, which once reached an annual output containing 33 kg of hydrobromide citrate powder, 1.588 million vial of injection, and 3.36 million tablets. Excessive collection of plant resources resulted in a shortage of sources and decline in production. At present, there is only a small-scale production in Yunnan Kunming Pharmaceutical Factory. The department of ophthalmology from People's Liberation Army General Hospital complexes anisodine with other drugs to form *compound anisodine hydrobromide injection* approved as one of the class II new drugs by SFDA. The full chemical synthesis of anisodine was successfully completed in 1975 [4].

Pharmacology

Central anticholinergic effects: anisodine competes with acetylcholine in the M cholinergic receptor to prevent acetylcholine binding M cholinergic receptor, thus blocking the nerve impulse transmission and interfering with the physiological function based on the cholinergic neurotransmission. It has similar or slightly weaker effects than that of atropine on electrical activity of the brain, on pain caused by tremorine, and on tremor caused by arecoline in mice. Its effect is 19 times weaker than those of scopolamine [5].

Peripheral anticholinergic effects: anisodine possesses antispasmodic and anti-asthma effects and inhibits saliva secretion and mydriasis, which are weaker than those of atropine.

It acts against organic phosphate pesticide poisoning.

Anti-shock effect: anisodine prolonged significantly the survival time of animals or reduced mortality in patients with clinical shock through directly relieving vasospasm of vascular smooth muscle, antagonizing adrenaline-induced vasoconstriction, and improving microcirculation [6].

Anti-cerebral ischemic disease: the effect of anisodine on cerebral blood circulation is mainly to regulate vasomotor, decrease cerebral vascular resistance, and increase cerebral blood flow, thereby improving the symptoms of cerebral ischemia.

The effects of anisodine on the eye: a clinical study found that anisodine exerted a significant outcome to treat retinal vascular spasm, ischemic optic neuropathy, and other ocular diseases through adjusting the function of blood vessels in the eye, alleviating vascular spasm in the eye, and increasing blood supply to the eye.

Treatment of motion sickness: anisodine performed superior to scopolamine and other commonly used drugs at domestic and abroad in the anti-seasick test using waves stronger than grade 5.

Oral anisodine is absorbed rapidly and completely. The blood concentration was close to zero at 1 h after intravenous injection. It can be distributed in all tissues and organs in the body and finally excreted through the kidney. Anisodine has weaker intensity of action than atropine, as well as toxicity.

Clinical Applications

In the past, anisodine was mainly used in the treatment of various diseases of the central nervous system. The efficacy of anisodine for the treatment of cerebrovascular disease was validated from 1970 to 1975. The combination of procaine and anisodine hydrobromide was regarded as a good choice for the recovery of visual function in patients with glaucoma whose intraocular pressure has been controlled. It had a certain therapeutic effect on patients with very low vision and deserves clinical application [7]. When used for ischemic optic neuropathy treatment, anisodine significantly promoted the recovery of the visual field, with the effective rate as high as 82.14% compared with conventional therapeutic regimen using corticosteroids combined with vasodilators, thrombolytic agents, vitamins, and antibiotics. Compound anisodine based on a combination of procaine and anisodine hydrobromide showed a good effect for a variety of ischemic eye diseases induced by a variety of causes and other related eye diseases, especially superior to traditional therapies for ischemic lesions of the optic nerve, retina, and choroid [8].

Oral administration: 1–4 mg every time, three to four times/day.
Intramuscular injection: 5–10 mg every time, one to two times/day;
Intravenous injection: for infection toxic shock, the dose was modulated in accordance with the condition, for adult 10–40 mg every time and for children 0.3–2.0 mg per kilogram of body weight, and administration was repeated every 10–30 min. This drug of 5–10 mg can be added to 200 ml 5% glucose solution for intravenous infusion. The interval can be extended with the improved condition until withdrawal. Increase the dosage at appropriate discretion when the situation deteriorates. The amount of anisodine depends on the condition when used for organophosphate poisoning.

Discussion

The systematic study on *Anisodus tanguticus* is an important achievement and paradigm of drug research in China. It has developed anisodamine and anisodine, which were already successfully used for clinical application and saved many lives. Scopolamine with similar structure and similar pharmacological effects has also been found and is often used as a tool for research because of its strong central action. At present, the clinical application of anisodine has exceeded the original application and awareness. Anisodine has obvious advantages over conventional therapeutic regime using corticosteroids combined with vasodilators, thrombolytic agents, vitamins, and antibiotics for the treatment of ischemic optic neuropathy. Compound anisodine injection developed by China People's Liberation Army General Hospital consisting of procaine and hydrobromide anisodine has been clinically applied for many years, which showed convinced curative effects for a variety of ischemic eye diseases and other related eye disease, especially superior to traditional therapy for ischemic lesions of the optic nerve, retina, and choroid. The drug has saved a large number of patients with primary and secondary trauma from blindness in recent years. The drug has been widely used in the country to promote the use for various eye diseases which achieved exciting results. With the understanding of the mechanisms and characteristics, these drugs will give access to greater achievement in the clinic.

Fig. 1 *Anisodus tanguticus*

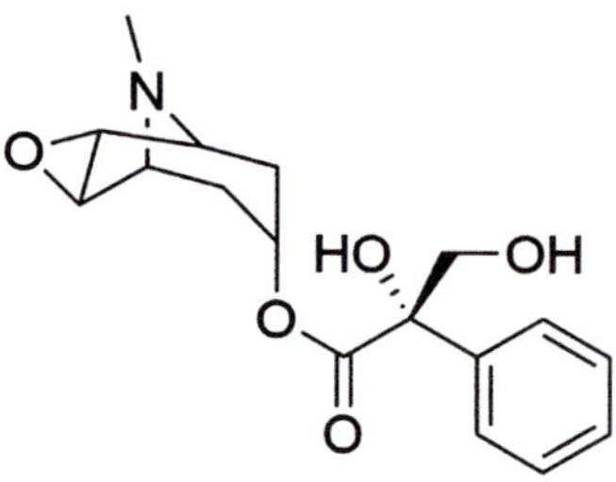

Fig. 2 Chemical structure of anisodine

References

1. Xie J, Wang L, Yongqi L, et al. Chemical structure of anisodine. Chin Sci Bull. 1975;10:52–3.
2. The First Pharmaceutical Factory of Chengdu. Production process of anisodine hydrobromide. Pharm Ind. 1975;5:12–5.
3. Weijian C. Pharmacological and clinical application of anisodine. Jiangsu Med J. 1976;2:50–2.
4. Xie J, Jin Z, Chunzhen Z, et al. New progress in the synthesis of anisodine. J Chin Acad Med Sci. 1982;4(2):92–6.
5. Varma DR, Yue TL. Adrenoceptor blocking properties of atropine-like agents anisodamine and anisodine on brain and cardiovascular tissues of rats. Br J Pharmacol. 1986;87(3):587–94.
6. Xiu RJ, Hammerschmidt DE, Coppo PA, et al. Anisodamine inhibits thromboxane synthesis, granulocyte aggregation, and platelet aggregation. A possible mechanism for its efficacy in bacteremic shock. J Am Med Assn. 1982;247:1458–60.
7. Chao L, Lu N. Compound hydrobromide on the recovery of visual function in the control of glaucoma with intraocular pressure. J Clin Ophthalmol. 2002;10(3):246–7.
8. Juan P, Hu X, Gao D, et al. Clinical application of compound anisodine injection in ophthalmology. Int J Ophthalmol. 2007;7(4):1124–7.

Atropine

Lian-Hua Fang, Jin-Hua Wang, and Guan-Hua Du

Abstract Atropine is a poisonous white crystalline alkaloid extracted from roots and leaves of *dianqiecao* (*Belladonnae Herba*) and *mantuoluo* (*Mandala*). Atropine sulphate is commonly used in clinics. Atropine is a blocker of typical M-choline receptor. Therefore, atropine sulphate was mainly used to treat toxic shock and organic phosphorus pesticide poisoning, to relieve visceral colic, as preanaesthetic medication and to reduce bronchial mucus secretion. The indications of atropine sulphate eye gel are iridocyclitis, fundus examination and mydriasis. However, clinical application of atropine sulphate was limited, because of its toxicity and small dosage window. After a prolonged endeavour, some progress has been made in respects such as fast action, longer duration of action, reduction of side effects and convenient use.

Keywords Atropine · Alkaloid · M-choline receptor · Toxic shock · Organic phosphorus pesticide poisoning

Alias: Atropine, atropine sulphate, atropinol, borotropin
Origin: *Mandala* (Fig. 1)
Chemical name (Fig. 2)

α-(Hydroxymethyl)benzeneacetic acid (3-endo)-8-methyl-8-azabicyclo [3.2.1] octyl easter

Molecular formula, $C_{17}H_{23}NO_3$; **MW**, 289.37; **CAS**, 51-55-8

L.-H. Fang · J.-H. Wang · G.-H. Du (✉)
Beijing Key Laboratory of Drug Targets Identification and Drug Screening, Centre for Pharmaceutical Screening, Institute of Materia Medica, Chinese Academy of Medical Sciences and Peking Union Medical College, Beijing, China
e-mail: dugh@imm.ac.cn

G.-H. Du, *Natural Small Molecule Drugs from Plants*,
https://doi.org/10.1007/978-981-10-8022-7_29

Derivatives

Sulphate monohydrate, $(C_{17}H_{23}NO_3)_2 \cdot H_2SO_4 \cdot H_2O$; MW, 694.84; CAS, 5908–99-6

Properties

Appearance: atropine appears as colourless, odourless crystals or a white crystalline powder. *Solubility*: very soluble in water and soluble in ethanol. *Melting point*: melting point of atropine isn't higher than 189 °C (melting time decomposition) (Chinese Pharmacopoeia), 114–118 °C (United States Pharmacopeia) and 115–119 °C (British Pharmacopoeia). The chemical structure of atropine is made up of amino alcohol esters. It is easy for atropine to be hydrolysed into tropine and despun tropic acid under alkaline condition. Atropine is stable in faintly acid and neutral aqueous solution, most stable at pH 3.5–4.0.

Dosage Forms and Indications

This product was recorded in the Pharmacopoeia of the People's Republic of China (2015), the British Pharmacopoeia (2017), the United States Pharmacopeia (40), the Japanese Pharmacopoeia (17th ed.), the Indian Pharmacopoeia (2010), the European Pharmacopoeia (9.0th ed.), the International Pharmacopoeia (5th ed.) and the Korean Pharmacopoeia (10th ed.).

Atropine sulphate is commonly used in clinics. Dosage forms are injection, tablet and eye ointment; atropine sulphate was mainly used to treat toxic shock and organic phosphorus pesticide poisoning, to relieve visceral colic, as preanaesthetic medication and to reduce bronchial mucus secretion. The indications of atropine sulphate eye gel are iridocyclitis, fundus examination and mydriasis.

Literature

Atropine is a poisonous white crystalline alkaloid extracted from roots and leaves of *dianqiecao*颠茄草 (*Belladonnae Herba*) and *mantuoluo*曼陀罗(*Mandala*).

Atropa belladonna L., which belongs to the *Solanaceae* belladonna genera, is a perennial herb. Originating in middle and south areas of Europe and Asia Minor, *Atropa belladonna* L. was recorded in the British Pharmacopoeia in 1809 and was widely planted in the USA, Europe and India. *Atropa belladonna* L. was introduced in China in the 1930s and is cultivated in Shandong, Zhejiang, Beijing and Shanghai.

Atropa belladonna L. tastes bitter and spicy and is poisonous. The effective components are alkaloids, which are distributed in roots (0.6%), leaves (0.4%), stem (0.15%), flower (0.19%), fruit (0.21%) and seed (0.31%). The major effective component is hyoscyamine. Left-handed scopolamine existed in plants is instability in storage, process and preparation and is gradually transformed into racanisodamine (atropine).

Varieties of belladonna alkaloids widely exist in seeds of *Hyoscyamus niger* L., roots of *Scopolia japonica* Maxim, leaves of *Atropa belladonna* L., flowers of *Datura metel* L. and roots and seeds of *Anisodus tanguticus* Pascher. Atropine, scopolamine, anisodamine and anisodine are the four major scopolia alkaloids [1].

History of R&D

Mandragora (mandrake) was described for treatment of wounds, gout and sleeplessness and as a love potion in the fourth century BC by Theophrastus. Atropine extracted from the Egyptian henbane was used by Cleopatra in the last century BC to dilate her pupils in the hope that she would appear more alluring. In the Renaissance, women used the juice of the berries of *Atropa belladonna* to enlarge the pupils of their eyes for cosmetic reasons.

It isn't until the first century AD that Dioscorides found that wine containing mandrake can be used as an anaesthetic treatment for pain or sleeplessness in surgery or cautery. The combination of extracts containing tropane alkaloids and opium was used to treat diseases, which was popular in the Roman and Islamic Empires and Europe. The combination was replaced by the use of ether, chloroform and other modern anaesthetics about 100 years ago.

The mydriatic effects of atropine were studied by the German chemist Friedlieb Ferdinand Runge (1795–1867). In 1831, the German pharmacist Heinrich F. G. Mein (1799–1864) succeeded in separating pure atropine from plants. The substance was first synthesized by German chemist Richard Willstätter in 1901. Total yield was only 0.75%, for processes of synthesis were very complex [2].

Tropic ketone from pharmaceutical industry can't meet the needs during World War I. In 1917, English chemist Robert Robinson discovered the simple synthesis of tropic ketone, which became the classic course in organic synthesis, and total yield increased to 17% [2].

In 1889, Richard Willstätter first confirmed the chemical structure of atropine. Atropine was first synthesized by A. Ladenburg. Homatropine, a kind of tropic alkaline ester, is used in the diagnosis and treatment in ophthalmology, and it has a shorter acting time than atropine. Quaternary ammonium compounds of atropine obtained by alkylation of nitrogen atoms have anticonvulsant function, which does not affect the central nervous system, due to their polarity. In 1970, atropine sulphate was synthesized in Hangzhou, the location of the first pharmaceutical factory in China, which increased the yield, reduced the cost and met the requirements of clinics [3].

Pharmacology

Atropine is a blocker of typical M-choline receptor. In addition to terminating the gastrointestinal smooth muscle spasm, inhibiting glands, dilating pupils, increasing intraocular tension, adjusting vision through paralysis, accelerating heart rate and dilating bronchi, large doses of atropine dilate blood vessels, terminating the spasmodic contraction and improving minicirculation. Atropine can excite or inhibit the central nervous system in a dose-dependent manner. Atropine exerts longer and stronger effect on heart, intestine and bronchial smooth muscle than other belladonna alkaloids. Atropine also relaxes the pupillary sphincter and the ciliary muscle and dilates the pupils by blocking M-choline receptor in ocular tissue.

Blockers of M-choline receptor included atropine, scopolamine, anisodamine and anisodine. Belladonnas not only block M-choline receptor in internal organ cells but also in the central nervous system. Compared with atropine, scopolamine has an oxygen bridge, which increases central nervous system function. The oxygen bridge of scopolamine is partially broken and then becomes anisodamine [4], which is difficult to pass through the blood-brain barrier, and symptoms caused by atropine in the central nervous system were less than that caused by atropine.

Peak concentration of plasma can be reached at 15–20 min after intramuscular injection of atropine and at 1–2 h after oral administration and can last for 4–6 h. Most of the atropine can be absorbed by the gastrointestinal tract and other mucous membranes, and a little of the atropine can be absorbed by the eyes and skin. The $t_{1/2}$ is 3.7–4.3 h. Binding rate of plasma protein is 14–22%. Volume of distribution is 1.7 L/kg after oral administration. Atropine can rapidly distribute to different organ systems and pass the blood-brain barrier and the placenta.

After absorption by the eye's conjunctiva, 30% of the products are excreted unchanged via the kidneys; the others become metabolites by hydrolysis and glucuronidation or glucosidation. After 1% gel eye drop, enlarged pupil function lasts for 7–10 days, and regulatory paralysis lasts for 7–12 days.

Clinical Application

Atropine has been widely used in various ways: as relief to internal colic, as preanaesthetic medication, as antishock and antiarrhythmic drug, as treatment to organic phosphorus pesticide poisoning and so on. Clinical use of atropine was increased including myopia prophylaxis, prevention of carotid sinus syndrome in internal carotid artery stenosis patients with carotid artery stenting and treatment of fever caused by transfusion reaction, and when combined with other drugs, it is used for treatment of pulmonary tuberculosis haemoptysis, infantile asthmatoid bronchitis, chilblain and migraine.

In 1971, Bedrossian first reported that in patients with myopia, monocular long-term usage of atropine (0.5% or 1%, qn) could make the development of myopia to stop or slow down [5]. The mechanism of myopia prophylaxis is to block M-choline receptor in retina tissue, including choroid, retina pigment epithelium and sclera. The therapeutic effect is positive, but long-term efficacy, optimum concentration and usage, side effects and, in relation to the eyes, a tenfold increase in risk need to be further discussed [6].

Clinical studies showed that in internal carotid artery stenosis patients with carotid artery stenting, preoperative prophylactic intravenous injection of atropine (1 mg) reduced the incidence of carotid sinus syndrome intra- and postoperatively, which demonstrated that atropine exerted a preventive effect on the syndrome in these patients [7].

Common adverse reactions are thirst, dizziness, mydriasis, erubescence, accelerated heart rate, excitement, dysphoria, delirious speech and convulsions. Patients with glaucoma, prostatic hyperplasia and hyperpyrexia are disabled. Ophthalmic administration may lead to dryness of skin and mucous membrane, flushing and tachycardia. Allergic reactions, such as itchy and swelling eyelids and conjunctival injection, occur in a few patients; thus they should immediately stop taking it. It was reported that therapeutic dose of atropine may lead to mental disorders [8].

Discussion

Atropine sulphate is a kind of M-choline receptor blocker, whose clinical application was limited, because of its toxicity and small dosage window. After a prolonged endeavour, some progress has been made in respects such as fast action, longer duration of action, reduction of side effects and convenient use. To meet the needs of clinical treatment, atropine sulphate can be developed into other dosage forms [9], such as sublingual tablet, as well as new dosage forms and clinical applications.

Fig. 1 *Mandala*

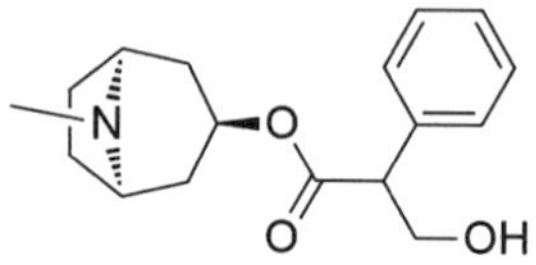

Fig. 2 Chemical structure of atropine

References

1. Chinese Pharmacopoeia Commission. Chinese pharmacopoeia. Beijing: The Medical Science and Technology Press of China; 2015. p. 1335–7.
2. Brief history of drug discovery – atropine. http://news.anychem.com/detail/YRGzaa.html.
3. Hangzhou first pharmaceutical factory. Synthesis process of atropine sulfate. Chin J Pharm. 1972;8:15–7.
4. Shi BX. Reasonable application of belladonna drugs to rescue organophosphorus pesticide poisoning. New Drugs Clin Remedies. 1986;5(5):288–91.
5. Bedrossian RH. The effect of atropine on myopia. Ann Ophthalmol. 1971;3(8):891–7.
6. Chen M, Wang LC. Research progress of atropine prevention and treatment of myopia. J Med Forum. 2010;31(3):125–7.
7. Liu JG, Wang Z, Liu ZZ, et al. Preventive application of atropine to patients receiving carotid arteries stenting. Neural Inj Funct Reconstr. 2013;8(3):192–3.
8. Liu FR, Yi CY. Therapeutic dose of atropine to psychiatric disorders in 1 case. Contemp Med. 2013;19(25):48.
9. Qin ZG, Chen ZH, Li RD. The research progress of dosage forms of atropine sulfate. Chin Foreign Med Treat. 2009;33:168–9.

Borneol

Li-Song Sheng, Li-Da Du, Gui-Fen Qiang, and Guan-Hua Du

Abstract Borneol is a bicyclic organic compound and a terpene derivative. It's derived from fresh branches and leaves of *Cinnamomum camphora* (L.) Presl. The borneol is one of the earliest natural medicines of single organic ingredient used in the history. Because of the peculiar aromatic odor, it has been widely used in cosmetic ingredients and drugs. Many researches have taken blood-brain barrier permeability as an indication of inducing resuscitation in multiple preparations. Lots of structure modifications in borneol are performed to find effective derivatives.

Keywords Borneol · Blood-brain barrier · Camphol

Alias: Bingpian Borneolum
Origin: Cinnamomum longepaniculatum (Fig. 1)
Chemical name (Fig. 2)

endo-1,7,7-Trimethyl-bicyclo[2.2.1]heptan-2-ol

Molecular formula, $C_{10}H_{18}O$; **MW**, 154.25; **CAS**, 464-45-9
Borneol derivatives (Fig. 3)

L.-S. Sheng · G.-F. Qiang · G.-H. Du (✉)
Beijing Key Laboratory of Drug Targets Identification and Drug Screening, Centre for Pharmaceutical Screening, Institute of Materia Medica, Chinese Academy of Medical Sciences and Peking Union Medical College, Beijing, China
e-mail: dugh@imm.ac.cn

L.-D. Du
School of Biomedical Sciences, Chinese University of Hong Kong, Hong Kong, China

G.-H. Du, *Natural Small Molecule Drugs from Plants*,
https://doi.org/10.1007/978-981-10-8022-7_30

Properties

Appearance: colorless to white lumps. *Odor*: pungent, camphor-like. *Density*: 1.011 g/cm^3 (20 °C). *Melting point*: 208 °C (406 °F; 481 K). *Boiling point*: 213 °C (415 °F; 486 K). *Solubility*: slightly soluble in water (D-form), soluble in chloroform, ethanol, acetone, ether, benzene, toluene, decalin, and tetralin. *Flash point*: 65 °C (149 °F; 338 K). It's stable under sealed condition while volatile in the air.

Dosage Forms

The Borneol is included in the *Pharmacopoeia of the People's Republic of China* (2015), *British Pharmacopoeia* (2017), and *European Pharmacopoeia* (9.0th ed.).

It is usually used in the compatibility with other Chinese herbs, especially in the Chinese traditional patent medicine, seldom in the decoction. Borneol is also used for the external application.

Indications

The main efficacy of borneol is to induce resuscitation (with aromatic stimulation), clear stagnated fire (fever feeling), remove nebula for improving eyesight, and relieve swelling and pain. The indications of borneol are sore throat, aphthous, red eyes, purulent ear discharge, convulsions, febrile delirium, sudden faint due to qi depression, stroke, and coma. In Chinese traditional medicine, the borneol is often used as an envoy drug and combined with other drugs but is not used as a single medicine with the inexact efficacy.

Literature

Borneol is an analgetic, antibacterial, and resuscitation-inducing norborneol derived from fresh branches and leaves of *Cinnamomum camphora* (L.) Presl. Far more than 2000 years ago, it has been introduced to China [1]. In China, it has been firstly recorded in *Ming Yi Bie Lu* and then included in *Tang Ben Cao*. It was recorded in history that borneol was derived from *Dryobalanops camphora gaertner* and then precipitated from the resin to form the natural crystal compound or distilled from the trunk and cooled down to form the crystal compound, which is certified from Indonesia. In China, the natural borneol mainly relied on imports. In recent years, it was extracted from the Lauraceae plants, including *Cinnamomum camphora*, *Cinnamomum longepaniculatum*, and *Cinnamomum burmannii*, which greatly increases the resources of natural borneol for China. *Cinnamomum camphora* is mainly distributed in Jiangxi and Fujian provinces with 81.78% of borneol. *Cinnamomum longepaniculatum* is mainly distributed in Hunan and Sichuan provinces with 77.57% of borneol. *Cinnamomum burmannii* is mainly distributed in Yunnan and Guangxi provinces with 70.81% of borneol. Among them, *Cinnamomum camphora* contains more borneol than the other two types [2].

History of R&D

Borneol has been widely used worldwide. It has been systematically studied since 1803 in Dutch literature. This might be because borneol was originated from Indonesia which had been the colony of the Netherlands since the seventeenth century. Stockman [3] reviewed the borneol systematically and conducted the preliminary pharmacological experiments. He used frogs as model animals to study the general pharmacological effects of borneol on nerve stimulation, muscles, and all kinds of tissues and organs. The pharmacology experiments in guinea pigs, rabbits, cats, and dogs showed that low dose (0.5 g) in rabbits had no significant pharmacological effects while high dose (2 g) could produce a strong effect on reducing fever. After he compared borneol with camphor and menthol, he found that borneol has a simple chemical structure and wide pharmacological effects but lacks specificity; thus, it was used as envoy drugs in traditional Chinese medicine and combined with other drugs. Moreover, borneol has some toxicity, so it was believed that it might not be used to specific diseases. The current pharmacological studies of borneol focus on crossing blood-brain barrier and its mechanism, as well as promoting the penetration of blood-brain barrier after compatibility with other drugs, which has been started by Qizhong Mo in Shanghai Institute of Materia Medica, Chinese Academy of Sciences since 1982 [4].

About the synthesis and modification of borneol, Jackson first reported that borneol can be synthesized from camphor by adding reaction, which is the first time for the research of borneol aiming at the chemical structure [5]. The structure of borneol contains hydroxyl group, which is the main reaction site. The common modification reaction is mainly ester addition reaction. Qide Wu et al. synthesized a series of ester derivatives of natural borneol and studied its biological properties. It was found that (+) – 4-methoxybenzoic acid borneol ester had a significant effect on the opening of the blood-brain barrier and was less toxic than borneol [6]. Because of the unique chemical structure of borneol and relatively low molecular weight, borneol is often modified to observe whether the drug has such pharmacological effects of antitumor, increasing the penetration of blood-brain barrier, antibacterial, antioxidant, and others. Up to date, there is no druggability report based on borneol modification.

Pharmacological Effects

The main pharmacological effects of borneol include anti-inflammatory, antibacterial, central nervous system, and antifertility effects [7]. Guangchi Jiang found that intraperitoneal injection of borneol at 3.5 mL/kg can significantly inhibit foot swelling caused by egg white in rats. Borneol can inhibit and kill *Staphylococcus aureus*, B-type *Streptococcus*, and other five common cells with the minimum inhibitory concentration (MIC) of 1.0–2.0% and the lowest bactericidal concentration (MFC) of 1.5–2.0%. There was significant odinopoeia effect on the late pregnant mice after given 112 mg/kg borneol. Qide Liu et al. found that 10% borneol paraffin oil at the dose of 1 mg/kg by oral gavage can significantly increase the concentration of gentamicin in rat brain tissue, suggesting that borneol can change the blood-brain barrier permeability. The current mechanisms of anti-inflammatory effects include inhibition of inflammatory factors of interleukin-1β, tumor necrosis factor-α, and cell adhesion molecule-1 expression. The mechanisms of central nervous system effects are involved in inhibiting p-glycoprotein, opening the intercellular tight junction, increasing the number of pinocytotic vesicles, and improving the phospholipid molecule arrangement of epithelial cell membrane. In addition, borneol also affects the level of nitric oxide and inhibits the elevation of Ca^{2+} concentration.

Borneol has a simple structure and a wide range of plant sources. Moreover, a series of synthetic methods of borneol were developed, so many analogues are used in clinical practice, such as camphor, isoborneol, and l-borneolum (Ai Pian). As a dehydrogenation compound of borneol, camphor has strong cardiotonic, anti-inflammatory, and antibacterial pharmacological effects, but it is generally not as medicinal because of its obvious toxicity. The heterogeneous compounds of borneol, isoborneol, and l-borneolum (Ai Pian) have similar pharmacological effects as borneol. Guangchi Jiang found that both borneol and isoborneol have analgesic effects and can prolong the sleep time induced by pentobarbital [8]. Isoborneol can significantly prolong the survival time in anti-hypoxia tolerance tests. However, the effect of borneol is not significant. Borneol and isoborneol have small local irritation. In mucosal administration, the irritation of isoborneol is slightly stronger than borneol, while it is opposite at intramuscular injection. The pharmacokinetics of borneol in human are carried out in the form of compatibility. Jun Guo et al. determined the concentration of borneol in Su Xiao Jiu Xin Pill and found that the absorption of borneol reached the peak at 16 min [9]. Shanshan Hu et al. determined the pharmacokinetics of borneol, and Danshen determined the borneol compatibility in rabbits by LC-MS method and found that metabolites of borneol were borneol acetate with the rapid distribution and time to peak at 30 min [10].

Clinical Application

As a traditional Chinese medicine, borneol is commonly used as envoy drugs in the compatibility of traditional Chinese medicine. On behalf of combination drugs such as Danshen dripping pills, Niu Huang Jie Du pills, and watermelon cream, its effect is significant because of its special aromatic smell. Borneol with a certain irritation, oral administration may cause the gastrointestinal discomforts, severely causes vomiting and other adverse reactions.

Overview

The borneol is one of the earliest natural medicines of single organic ingredient used in the history. Since the Southern dynasty, there were records of overseas tributes. Because of its special aromatic smell, wide pharmacological effects, and relatively low toxicity, it has been widely used since ancient times. With in-depth pharmacological research of borneol, the continuous expansion of plant resources, and the continuous improvement of chemical synthesis process, the borneol still has good application prospect and clinical needs.

Fig. 1 Cinnamomum longepaniculatum

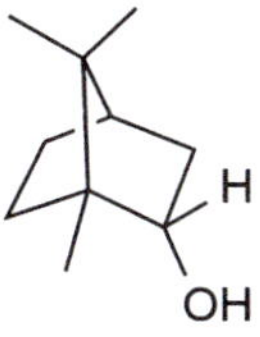

Fig. 2 The structure of borneol

Fig. 3 Structure of borneol derivatives

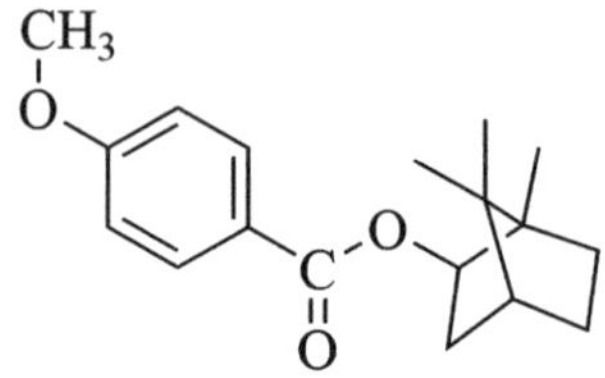

(+) -4-borneol-methoxybenzoic acid ester

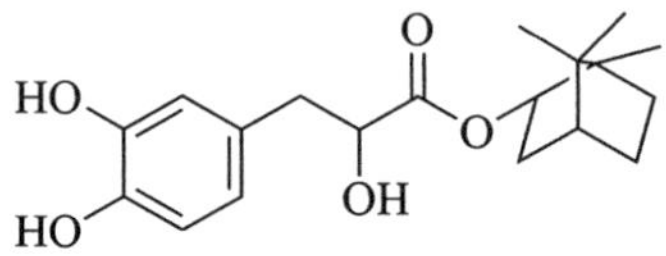

Salvianic Acid ester of borneol

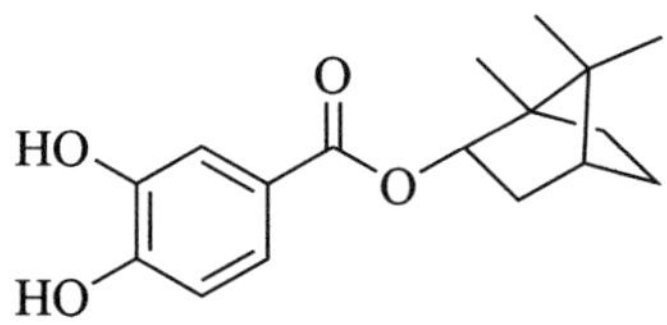

Protocatechuic acid ester of borneol

References

1. Zhao SX. Borneol-the earliest used natural organic small molecular drugs in the world. Asian Pac Tradit Med. 2006;2:24–5.
2. Chen JN, Zeng HF, Li G, et al. Analysis of borneol and camphora constituents of essential oils. J Chin Med Mater. 2005;28(9):781.
3. Stockman R. The physiological action of borneol. A contribution to the pharmacology of the camphor group. J Physiol. 1888;9(2–3):65–91.
4. Mo QZ, Gong B, Qian XP. Study on the dynamics of 3H – Borneol: a preliminary research on the mechanism of aromatic inducing resuscitation. Zhongchengyao Yanjiu. 1982;8:5–7.
5. Jackson CL, Menke AE. A new method of preparing borneol from camphor. Proc Am Acad Arts Sci. 1882:93–5.
6. Wu QD. Synthesis of ester derivatives of natural borneol and its pharmacokinetic study in rats [D] Guangzhou University of Chinese Medicine; 2011.
7. Wu SR, Cheng G, Feng Y. Advances in pharmacological effects of borneol. Chin Tradit Herb Drug. 2001;32(12):1143–5.
8. Jiang GC, Feng XJ, Huang L, et al. Pharmacological effects of borneol and isolated brain on mice and rabbits. West China J Pharm Sci. 1989;4(1):23–5.
9. Guo J, Huang X, Wang LL, et al. Simultaneous determination of borneol and ligustrazine in blood by GC-FID method. Chin Tradit Herb Drug. 2003;34(8):730–2.
10. Hu SS. The study of monarch drug of Salvia miltiorrhiza's impact on the envoy drug borneol in pharmacokinetics [D] Northwest University; 2008.

Breviscapine

Dan-Shu Wang, Yu Yan, Lian-Hua Fang, and Guan-Hua Du

Abstract Breviscapine is extracted from *Erigeron breviscapus*, the main component of which is scutellarin. Breviscapine has a good pharmacological activity in the treatment of cardiovascular and cerebrovascular diseases. The oral bioavailability of ordinary tablets and granules is very low; at the same time, the half-life of ordinary injection and injection powder is short, eliminating quickly in the body. Therefore, the new preparations for breviscapine such as liposomal injections, microcapsules, sublingual tablets, orally disintegrating tablets, self-emulsifying soft capsules, and other quick-release preparations are in the stage of development.

Keywords Breviscapine · Scutellarin · Cardiovascular diseases · Cerebrovascular diseases

Alias: Scutellarin
Origin: *Erigeron breviscapus* (Fig. 1)
Chemical name (Fig. 2)

7-(β-D-Glucopyranuronosyloxy)-5,6-dihydroxy-2-(4-hydroxyphenyl)-4H-1-benzopyran-4-one

Molecular formula, $C_{21}H_{18}O_{12}$; **MW**, 462.36; **CAS**, 27740-01-8

D.-S. Wang · Y. Yan · L.-H. Fang · G.-H. Du (✉)
Beijing Key Laboratory of Drug Targets Identification and Drug Screening, Centre for Pharmaceutical Screening, Institute of Materia Medica, Chinese Academy of Medical Sciences and Peking Union Medical College, Beijing, China
e-mail: dugh@imm.ac.cn

G.-H. Du, *Natural Small Molecule Drugs from Plants*,
https://doi.org/10.1007/978-981-10-8022-7_31

Properties

Appearance: yellow powder, with a certain hygroscopicity, odorless, tasteless, or slightly salty. *Solubility*: soluble in methanol and pyridine and diluted alkali solution, slightly soluble in hot water (ethanol and ethyl acetate), almost insoluble in water (ether, chloroform, benzene, and acetone). It has a maximum absorption at 284 ± 2 nm and 335 ± 2 nm wavelengths.

Dosage Forms and Indications

Breviscapine is recorded in the *Pharmacopoeia of the People's Republic of China* (2015).

Tablets: 20 mg/tablet, 40 mg/tablet; Injection: 5 mg/ml. Ischemic cerebrovascular disease, such as cerebral thrombosis, cerebral embolism, and so on.

Literature

Breviscapine is extracted from *Erigeron breviscapus*, the main component of which is scutellarin. It is the herb of *Erigeron breviscapus (Vant.) Hand.-Mazz*, distributed in Yunnan, Guangxi province, and other places, especially in Yunnan.

Erigeron breviscapus was first contained in *Southern Yunnan Materia Medica*: "*Erigeron breviscapus* is mainly used for bruises, with the effects of dispelling cold and antagonizing inflammation." *Erigeron breviscapus* is also widely distributed in ethnic minority areas, so it becomes a folk herb in many nations.

History of R&D

The research on the chemical composition of *Erigeron breviscapus* began in the late 1970s; Yunnan Institute of Materia Medica was the first to carry on the work [1]. In the 1970s, Yunnan Institute of Materia Medica sent researchers to collect folk remedies. A 90-year old senior herbalist doctor from Qiubei County of Wenshan Zhuang and Miao Autonomous Prefecture said that *Erigeron breviscapus* could be used for the treatment of cerebral hemiplegia. The clinical trial demonstrates that the extract of *Erigeron breviscapus* has a significant effect on cerebral hemiplegia, coronary heart disease, cerebral thrombosis, cerebral hemorrhage, cerebral embolism, microcirculation, and other diseases [2]. Since then, *Erigeron breviscapus* gradually gets into the vision of researchers.

G. Zemplén and L. Farkas et al. synthesized 5,6,7,4′-tetraacetoxy-baicalein in 1974; it turned into 7-hydroxy-5,6,4′-triacetoxyflavone after reaction, combined with bromine acetyl glucoside generating into breviscapine, which achieved the total synthesis of breviscapine, making it easy to proceed production of industrial scale [3].

Currently ordinary tablets, granules, injections, and injection powder are listed in the market. The oral bioavailability of ordinary tablets and granules is very low; at the same time, the half-life of ordinary injection and injection powder is short, eliminating quickly in the body. It is inconvenient for patients to use. The new preparations for breviscapine such as liposomal injections, microcapsules, sublingual tablets, orally disintegrating tablets, self-emulsifying soft capsules, and other quick-release preparations are in the stage of development.

It is difficult to improve the absorption of scutellarin and other flavonoids. At present the preferred method to improve solubility at abroad is modifying chemical structure of the compound, such as esterification, benzyl esterification, and hydroxyacetamide, so as to improve the pharmacokinetic properties, esterifying the carboxyl group with polyethylene glycol (PEG) to improve the water solubility of scutellarin and hydrolyzing glucuronic acid to improve the fat-soluble scutellarin and its oral absorption [4].

Pharmacology

Since 1979, breviscapine has been made into injection and tablets used in clinical, but the further research on pharmacological effects and large-scale development of breviscapine began until the late 1980s. The study found that breviscapine has a good pharmacological activity in the treatment of cardiovascular and cerebrovascular diseases.

Breviscapine can be used for patients with coronary heart disease angina pectoris by inhibiting the aggregation of platelet and erythrocyte, increasing blood flow, dilating blood vessels, and improving microcirculation. It can antagonize heart failure by blocking the β-receptor, inhibiting Ca^{2+} influx and K^+ efflux. The mechanism of breviscapine is not clear; on one hand, it may be due to the inhibition of intracellular calcium release and, on the other hand, may be stimulating the baroreceptors of blood vessel walls, so as to increase the incoming of parasympathetic impulses, directly activate the central parasympathetic nucleus and acetylcholine receptors, and then increase the activity of parasympathetic nerve. At the same time, the sympathetic impulse reduces because of the sensitization of baroreceptor, so that the sympathetic activity and blood pressure decreases. In addition, a large number of studies have shown that breviscapine can inhibit cardiomyocyte hypertrophy, hyperplasia, and reverse vascular remodeling [5].

In addition to cardiovascular system, breviscapine also plays a critical role in the dilation of cerebral blood vessels. It can reduce cerebral vascular resistance and has a protective effect on the brain with ischemia reperfusion injury, which is mainly through upregulating the Nrf2/HO-1 pathway [6]. Breviscapine also has a certain role in diabetic neuropathy [7].

Clinical Application

Breviscapine is a cerebrovascular dilator, with effects of increasing cerebral blood flow, reducing cerebrovascular resistance, and improving microcirculation. It is used clinically for the treatment of ischemic cerebrovascular disease such as cerebral thrombosis. Breviscapine has less adverse reactions, which occasionally can cause high fever, chills, and other infusion reactions. Individual patients may have rash, dry mouth, fatigue, and other symptoms, disappearing after withdrawal. Breviscapine injection added with low molecular weight dextran had caused acute gastrointestinal bleeding; therefore, the elderly or patients with a history of upper gastrointestinal bleeding should use cautiously [8].

Discussion

Breviscapine shows a good pharmacological activity in the treatment of cardiovascular and cerebrovascular diseases, but the study on its pharmacological action is not enough, especially at the cellular and molecular level. In addition, the low bioavailability of breviscapine also affects its use in clinical practice, so it is the focus of future research to develop breviscapine with high bioavailability and good pharmacological activity.

Fig. 1 *Erigeron breviscapus*

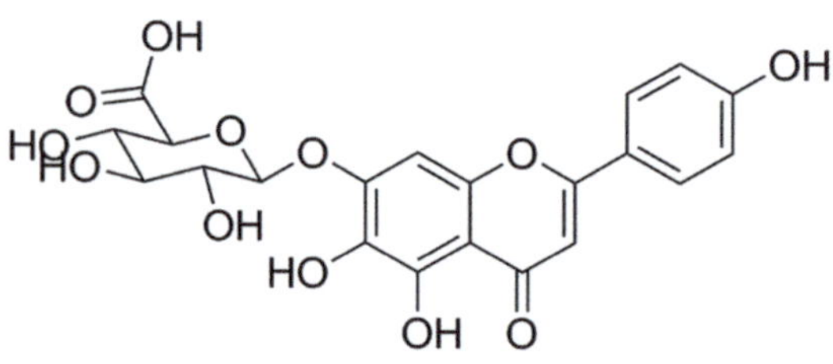

Fig. 2 The structure of breviscapine

References

1. Yunnan Institute of Materia Medica. Study on the chemical components of Erigeron breviscapus. Chin Herb Med Commun. 1976;11:11–4.
2. Yang WY, Zhang Y. The folk application status of Erigeron breviscapus. Chin Med Newsp. 2005;5(31):A07.
3. Cui JM, Wu S. Advances in research on breviscapine. Nat Prod Res Dev. 2003;15(3):255–8.
4. Yang LM, Gu J, Lin MJ, et al. Advances in research on breviscapine. Tianjin Pharm. 2010;22(1):56–60.
5. Zhang XD, Liu J, Zhang WB, et al. Research advance on cardiovascular pharmacological and clinical application of breviscapine. China Pharm. 2007;16(21):3–5.
6. Guo C, Zhu YR, Weng Y, et al. Therapeutic time window and underlying therapeutic mechanism of breviscapine injection against cerebral ischemia/reperfusion injury in rats. J Ethnopharmacol. 2014;151(1):660–6.
7. Zheng CJ, Ou WL, Shen HY, et al. Combined therapy of diabetic peripheral neuropathy with breviscapine and mecobalamin a systematic review and a meta-analysis of Chinese studies. Biomed Res Int. 2015;2015:680756.
8. Wei N. Advances of breviscapine in clinical applications and its side-effects. China Pharm. 2011;4(2):144–6.

Caffeine

Lin Wang, Li-Da Du, and Guan-Hua Du

Abstract Caffeine is a kind of xanthine alkaloid and mainly found in coffee, tea, Paraguay Holly (Mate), and the fruit or leaf taken from guarana. Caffeine shows the function of dispelling fatigue and exciting nerve. The commonly used clinical preparations include caffeine sodium benzoate and ergotamine caffeine. The adverse reactions include nausea, vomiting, abdominal pain, and fatigue. Overdose of caffeine will cause damage to the human body. Due to addiction at high dose, caffeine is included in the scope of controlled psychotropic drugs. Caffeine is mainly metabolized in the liver, and only 1–5% of it is excreted as metabolites through urine. Theobromine is the main active metabolite of caffeine and reported to cause activities in the cardiovascular system.

Keywords Coffee · Alkaloid · Addiction

Alias: Caffeinum
Origin: Coffee berry (Fig. 1)
Chemical name (Fig. 2)

1,3,7-Trimethylxanthine

Molecular formula, $C_8H_{10}N_4O_2$; **MW**, 194.2; **CAS**, 58-08-2

L. Wang · G.-H. Du (✉)
Beijing Key Laboratory of Drug Targets Identification and Drug Screening, Centre for Pharmaceutical Screening, Institute of Materia Medica, Chinese Academy of Medical Sciences and Peking Union Medical College, Beijing, China
e-mail: dugh@imm.ac.cn

L.-D. Du
School of Biomedical Sciences, Chinese University of Hong Kong, Hong Kong, China

G.-H. Du, *Natural Small Molecule Drugs from Plants*,
https://doi.org/10.1007/978-981-10-8022-7_32

Properties

Appearance: odorless silky needle-like crystal or crystal powder with the color of white or a little yellowish green. *Solubility*: weathering, easily dissolved in water or chloroform and slightly soluble in water, ethanol, or acetone, very slightly dissolved in ether. *Melting point*: 235–238 °C.

Dosage Forms

This product is included in the Pharmacopoeia of the People's Republic of China (2015), the British Pharmacopoeia (2017), the United States Pharmacopeia (40), the Japanese Pharmacopoeia (17th ed.), the European Pharmacopoeia (9.0th ed.), the Indian Pharmacopoeia (2010), and the International Pharmacopoeia (5th ed.).

Commonly used dosage forms of caffeine include tablet, powder, and injection. Mainly used dosage forms in the market include caffeine citrate tablets, amidopyrine caffeine tablets, amidopyrine caffeine, children acetaminophen aspirin caffeine tablets, ergotamine caffeine tablets, caffeine sodium benzoate injection, cafe bromine agent, etc.

Indications

1. To antagonize acute infection, central circulation failure, and respiratory failure caused by the toxicity of ethanol, hypnotics, narcotic, and narcotic analgesics
2. To prevent and treat premature infant apnea or paroxysmal dyspnea
3. To treat migraine in combination with ergotamine and general headaches by making compound preparations with acetylsalicylic acid (aspirin, acetaminophen)
4. To treat psychoneurosis in combination with bromide, adjust the excitement and inhibition process of the cerebral cortex, and restore the balance between excitement and inhibition

Literature

Caffeine is a kind of xanthine alkaloid. It is mainly found in coffee, tea, Paraguay Holly (Mate), and the fruit or leaf taken from guarana. There is also little caffeine in cocoa tree, kola nut, and generation of tea Holly. Up to now, caffeine was found in more than 60 kinds of fruit, leaf, and seeds in the world [1]. The main source of caffeine is the coffee beans. The original field of coffee is the plateau section of Kaffa province in southwest of Ethiopia. However, Brazil now becomes the world's largest coffee producer. Tea, native to China, is another important source of caffeine. In addition, chocolate made from cocoa powder also contains a small amount of caffeine.

There was no clear historical record for the discovery of coffee. As early as the Stone Age, human beings have begun to use caffeine. The history of tea can be traced back to the tea tree found in the Yangshao culture. As Lu Yu referred in *The Book of Tea* in the Tang dynasty, the history of tea as a drink started from the "the mythical emperor Shennong" [2]. Like coffee berries and tea leaves, kola nut also has an ancient origin. The earliest evidence for using cocoa was the residue found in the jar from the period of ancient Mayan civilization in the eighth century BC. Many West African civilizations reported to recover energy and alleviate hunger from chewing kola nut individually or in group. The earliest written records about coffee may be the "Bunchum" by al-Razi, a Persian physician in the ninth century. During the late sixteenth century, European residents in Egypt recorded the use of coffee. Probably at this time, coffee has been used widely in the near east.

History of R&D

In 1819, Friedrich Ferdinand Runge, a German chemist, firstly isolated pure caffeine from cocoa beans and realized its extraction from tea later. The chemical structure was determined by Stenhouse. In 1899, Fisehe firstly succeeded the synthesis of caffeine. China isolated caffeine from tea in 1950 and started to artificially synthesize it in 1958 [3].

Caffeine is the stimulant of the central nervous system. The main pharmacological effects of caffeine include the following:

1. As the stimulant of the central nervous system. Small doses (<3 mg/kg of body mass, ~200 mg) can excite the cerebral cortex and improve the thinking activity and inductivity to the outside. High doses (5–13 mg/kg of body mass) can excite the respiratory center of the medulla oblongata and vascular motorium and increase the rate and depth of breath.
2. As a cardiovascular drug with the positive function of promoting the synthesis and release of catecholamine. A small dose (<3 mg/kg of body mass, ~200 mg) can increase the heart rate and blood pressure and other positive heart functions.
3. To stimulate the pituitary-adrenocortical axis, causing the increase of the synthesis of adrenocorticotropic hormone and cortisol. Caffeine (674 mg/mL) stimulates gastric branch of the vagus nerve by exciting the vagus center of the central nervous system, which causes the increase of gastric acid secretion and gastric hypersecretion. Caffeine also stimulated the release of histamine from gastric mast cells, causing the increase of acid secretion from stomach cells.
4. To promote Ca^{2+} release from the sarcoplasmic reticulum and strengthen the contractility of external skeletal muscle [4–6].

After oral administration, caffeine is quickly and completely absorbed in the gastrointestinal tract and reaches peak concentration at about 15–60 min, and the peak concentration sustains until 120 min. The first-pass effect of caffeine is weak. After absorption, it completely enters into systemic tissues and passes into them with freedom. The half-life of caffeine varies greatly among different individuals. Caffeine is mainly metabolized in the liver, and only 1–5% of it is excreted as metabolites through urine. Metabolites of caffeine include vice xanthine (1,7-dimethyl xanthine, 84%), theobromine (12%), and theophylline (4%). These compounds are further metabolized and finally excreted through the urine [7, 8]. Theobromine is the main active metabolite of caffeine and reported to have activities of causing diuresis, stimulating the cardiovascular system, causing the relaxation of smooth muscle, increasing the activity of glands, etc.

Clinical Application

The commonly used clinical preparations include caffeine sodium benzoate and ergotamine caffeine. The preparation of caffeine sodium benzoate (injection) is constituted of 0.12 g/ml of caffeine, 0.13 g/ml of sodium benzoate, and cafe bromine mixture (oral liquid). In 200 ml of caffeine sodium benzoate (injection), there are 0.05–2 g of caffeine sodium benzoate and 1–10 g of sodium bromide. The indications of caffeine sodium benzoate include (1) treating respiratory and circulation failure caused by poisoning with anesthetics and analgesics or poisoning with acute infection and hypnotics, (2) treating neurosis in combination with bromide and restoring the balance of excitement and inhibition of the cerebral cortex, (3) treating general headaches as compound preparations with acetylsalicylic acid, and (4) treating migraine when it was made into compound preparations with ergotamine. Each piece contains 1 mg of tartaric acid ergotamine and 100 mg of caffeine. Excessive dosage or subcutaneous injection of this preparation commonly induces nausea, vomiting, abdominal discomfort, diarrhea, muscle weakness, and even chest pain.

Ergotamine caffeine can correct the excessive expansion of cerebral arteries and abnormal heart beat back to normal. Clinically, it can be used for migraine headaches, cerebral artery dilated headache, or headache caused by histamine. However, it is invalid in the prevention of headaches. The adverse reactions include nausea, vomiting, abdominal pain, and fatigue. Other common symptoms include numbness and tingling of the hands, toes, and face and swelling of the foot and lower limb. Overdose causes severe poisoning, mental disorder, ataxia, convulsions, gray chills of the hand and foot, sensory disturbance, and even death due to coma and respiratory paralysis [9]. Caffeine citrate preparation, including injection and oral solution, is the only internationally approved drug for the treatment of premature infant apnea.

Discussion

Caffeine shows the function of dispelling fatigue and exciting nerve. Caffeine is clinically used in the treatment of neurasthenia, for coma recovery, and in the treatment of migraine and neurosis as compound preparations. However, large doses or long-time use of caffeine will cause damage to the human body. Especially one needs to pay attention to caffeine addiction. When the dose was increased due to drug resistance, caffeine will not only act on the brain cortex but will also directly excite the medulla oblongata, inducing clonic seizures and bone tremor, causing the damage to important organs such as the liver, stomach, and kidney. It also induces respiratory inflammation and breast tumor in women. It even leads to low intelligence and body deformation in the next generation of addicts [10]. Therefore, caffeine was also included in the scope of controlled psychotropic drugs.

Fig. 1 Coffee berry

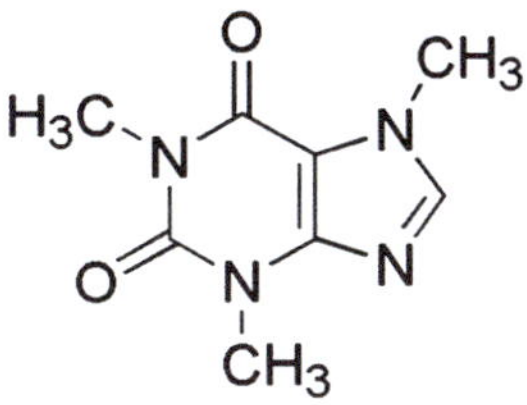

Fig. 2 Structure of caffeine

References

1. Yu-guan Z, Le S, Li X, et al. Caffeine in acute or chronic pain control and its possible mechanism. J Chin Acad Med Sci. 2014;6:697–700.
2. Chao-ran Y, Zhong-qing W. The pharmacological effects and application of caffeine. J Med Grad Stud. 2005;18(3):270–2.

3. Jia-bin H, Jing C, Wang J-k, et al. Progress in caffeine preparation methods. Chem Ind Vertical Horizontal. 2003;17(3):13–6.
4. Spriet LL. Exercise and sport performance with low doses of caffeine. Sports Med. 2014;44(2):S175–84.
5. Hai-xia L, Rong C, Dan Z. Progress in the synthesis and pharmacological effects of caffeine. Pharm J West Chin. 2011;2(2):182–7.
6. Pohanka M. The perspective of caffeine and caffeine derived compounds in therapy. Bratisl Lek Listy. 2015;116(9):520–30.
7. Ying C. Reevaluation of caffeine as a central stimulant. Foreign Med Physiol Pathol Sci Clin Pathol. 1997;4:375–8.
8. Yao C, Hong-hao Z. The research progress of caffeine metabolism and its application. Prog Physiol Sci. 2010;41(4):256–60.
9. Jian-liang F, Yu L. Comparison between imitrex injection and ergotamine caffeine tablets in the treatment of migraine. Chin J New Drugs Clin Remedies. 1999;2:82–4.
10. D'Ambrosio SM. Evaluation of the genotoxicity data on caffeine. Regul Toxicol Pharmacol. 1994;19(3):243–81.

Camphor

Jun-Ke Song, Li-Da Du, Gui-Fen Qiang, and Guan-Hua Du

Abstract Camphor was recorded in many ancient books of traditional Chinese medicine. There has been a long history for traditional Chinese medicine to use camphor. The research and development process of camphor has gone through from the natural product extraction to the modern chemical drug synthesis. Camphor is mainly used for the treatment of pruritic skin diseases, fibrous tissue inflammation, neuralgia, and influenza. It is toxic to human. Overdose of camphor can cause irritability, drowsiness, muscle spasms, vomiting, convulsions, epilepsy, and other symptoms. Camphor has wide pharmacological activities, but its toxicity limits the clinical application.

Keywords Camphor · Pruritic skin diseases · Pharmacological activities · Toxicity

Alias: DL-Camphor; 2-camphanone
Origin: Root bark of bodinier cinnamon (Fig. 1)
Chemical name (Fig. 2)

4,7,7-trimethylbicyclo[2.2.1]heptan-3-one

Molecular formula, $C_{10}H_{16}O$; **MW**, 152.24; **CAS**, 76-22-2

J.-K. Song · G.-F. Qiang · G.-H. Du (✉)
Beijing Key Laboratory of Drug Targets Identification and Drug Screening, Centre for Pharmaceutical Screening, Institute of Materia Medica, Chinese Academy of Medical Sciences and Peking Union Medical College, Beijing, China
e-mail: dugh@imm.ac.cn

L.-D. Du
School of Biomedical Sciences, Chinese University of Hong Kong, Hong Kong, China

G.-H. Du, *Natural Small Molecule Drugs from Plants*,
https://doi.org/10.1007/978-981-10-8022-7_33

Properties

Appearance: Camphor is a colorless or white-colored crystalline powder with a strong mothball-like odor. It is volatile at room temperature. *Melting point*: 176–181 °C for natural camphor and 174–179 °C for synthetic camphor. *Boiling point*: 204 °C. *Specific optical rotation*: The specific rotation of natural camphor is +41 to +44°, and synthetic camphor is −1.5 to +1.5°. *Solubility*: Camphor is slightly soluble in water with a water solubility of 0.12 g/100 mL (25 °C) and soluble in ethanol, ether, chloroform, carbon disulfide, and volatile or nonvolatile oils [1, 2].

Dosage Forms and Indications

This product has been recorded in the *Pharmacopoeia of the People's Republic of China* (2015), the *British Pharmacopoeia* (2017), the *United States Pharmacopeia* (40), the *Japanese Pharmacopoeia* (17th ed.), and the *European Pharmacopoeia* (9.0th ed.). There are lots of camphor-containing prescriptions with indications including herpes zoster and neurodermatitis.

Literature

Camphor was recorded in the ancient books of traditional Chinese medicine, such as *Pin Hui Jing Yao, Ben Cao Gang Mu, and Sheng Lian Fang.* There has been a long history for traditional Chinese medicine to use camphor.

History of R&D

The research and development process of camphor has gone through from the natural product extraction to the modern chemical drug synthesis [3, 4]. For a long time, the Chinese extracted camphor mainly from camphor tree (*Cinnamomum camphora*), root bark of bodinier cinnamon, and Yunnan camphor tree. With the development of chemical industry, human beings started to use chemical synthesis methods to obtain a large amount of camphor. At present, the chemical synthesis process of camphor in China has been well developed. The synthetic camphor is divided into industrial and pharmaceutical grades. The industrial grade camphor has a content of up to 96% or higher, and the pharmaceutical grade camphor with high purity can meet the standard of pharmacopeia.

Pharmacology

Camphor is toxic to human. Overdose of camphor can cause irritability, drowsiness, muscle spasms, vomiting, convulsions, epilepsy, and other symptoms. The lethal dose of camphor is 50–500 mg/kg (oral administration). In general, 2 g of camphor can cause serious toxicity, and 4 g of camphor will produce fatal toxicity.

Camphor can cause a cold sensation similar to mint when applied to human skin. In addition, it has a slightly local anesthetic effect. It can act on the gastrointestinal mucosa to produce a certain degree of stimulating effect. An appropriate dose of camphor can make people feel warm and comfortable in the stomach, but high dose of camphor will cause nausea and vomiting reaction. The effect of camphor on the central nervous system is obvious. It can act on the motor area of cerebral cortex and brain stem to produce the epilepsy-like seizures.

It is generally believed that camphor may have some therapeutic effects on patients with acute heart failure or recurrent collapse. A low dose (50 mg) may treat mild heart fatigue and other illnesses. The oxidized camphor metabolite has a more obvious cardiotonic, hypertensive, and respiratory excitement effects.

Camphor can be absorbed by the body after oral administration easily through the mucous, subcutaneous, and muscle. The *in vivo* metabolism of camphor occurs mainly in the liver. It is firstly oxidized into camphorol and then goes through phase II metabolism to produce glucuronide conjugate with glucuronic acid. Finally, the vast majority of glucuronide conjugate is excreted from the urine.

Clinical Application

Camphor is mainly used for the treatment of pruritic skin diseases, fibrous tissue inflammation, neuralgia, and influenza.

Discussion

It has been a long time for human beings to use camphor, which has the pharmacological effects of excitement, cardiotonic, anti-inflammation, analgesic, antibacterial, antitussive, penetration enhancer, and acaricidal. It is noteworthy that camphor is toxic to various human organs and tissues with more toxicity in the heart, liver, ovary, testis, and nervous system [5, 6], especially for the pregnant women and infants. In addition, the investigation of the American Association of Poison Control Centers (AAPCC) suggested that more than 10,000 of people have been poisoned due to direct or indirect contact with camphor [7]. Taken together, camphor has wide pharmacological activities, but its toxicity limits the clinical application.

Fig. 1 Root bark of bodinier cinnamon

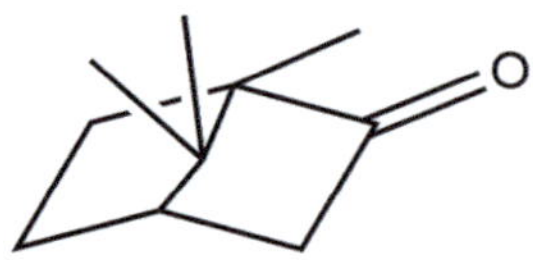

Fig. 2 The chemical structure of camphor

References

1. Mann J, Davidson RS, Hobbs JB, et al. Natural products: their chemistry and biological significance. London: Longman Scientific & Technical; 1994.
2. Xiong Y, Wu XR, Tu XM. Advances in the pharmacy of camphor. Lab Med Clin. 2009;6(12):999–1001.
3. Chen F, Chen SH. Brief introduction to the development of synthetic camphor. J Chem Ind For Prod. 1997;31(2):20–3.
4. Tang QY, Li JB, Lei CX. Extracting camphor and camphor oil from camphor trees. J Yiyang Teach Coll. 2000;17(5):49–50.
5. Ding YG, Ma HM, Zhang BL. Research on pharmacology, toxicology and safety of camphor: review and prospect. Chin J Pharmacovigilance. 2012;9(1):38–42.
6. Smith AG, Margolis G. Camphor poisoning: anatomical and pharmacologic study; report of a fatal case; experimental investigation of protective action of barbiturate. Am J Pathol. 1954;30(5):857–69.
7. Manoguerra AS, Erdman AR, Wax PM, et al. Camphor poisoning: an evidence-based practice guideline for out-of-hospital management. Clin Toxicol. 2006;44(4):357–70.

Chelidonine

Jia-Lin Sun, Li-Da Du, and Guan-Hua Du

Abstract Chelidonine, also named ranunculine, is a benzophenanthridine alkaloid isolated from the whole plant of *Chelidonium majus.* L. It has been used as a pain-killer in clinical treatment based on its morphine-like analgesic effect. Pharmacological researches in recent years have demonstrated that chelidonine played a role in antitumor, antibacterial, spasmolysis activities, and so on. And the antitumor activity is drawing more and more attention. However, its mechanisms of function remain unclear, which needs to be clarified in the future.

Keywords *Chelidonium majus.* L · Chelidonine · Analgesia · Antitumor

Alias: chelidonine; chelidonin; (+)-chelidonine; stylophorin; stylophorine; stylophoron; chelidoniny; helidonine; khelidonin; 11β-Hydroxychelidonane
Origin: *Chelidonium majus.* L (Fig. 1)
Chemical name (Fig. 2)

[(56-*S*-(5*β*a,6*β*,12*β*a)]-5b,6,7,12b,13,14-hexahydro-13-methyl[1,3] benxodioxolo[5,6-C]-1,3-dioxolo[4,5-I] phenanthridin-6-ol

Molecular formula, $C_{20}H_{19}NO_5$; **MW**, 353.37; **CAS**, 476-32-4

J.-L. Sun
Department of Pharmacy, The Affiliated Hospital of Qingdao University, Qingdao, China

L.-D. Du
School of Biomedical Sciences, Chinese University of Hong Kong, Hong Kong, China

G.-H. Du (✉)
Beijing Key Laboratory of Drug Targets Identification and Drug Screening, Centre for Pharmaceutical Screening, Institute of Materia Medica, Chinese Academy of Medical Sciences and Peking Union Medical College, Beijing, China
e-mail: dugh@imm.ac.cn

G.-H. Du, *Natural Small Molecule Drugs from Plants*,
https://doi.org/10.1007/978-981-10-8022-7_34

Properties

Appearance: white crystal or crystalline powder, monoclinic prism crystal. *Solubility*: insoluble in water; soluble in ethanol, chloroform, and amyl alcohol. *Melting point*: 135–136 °C. *Boiling point*: 220 °C. *Specific optical rotation*: +115°.

Dosage Forms and Indications

Current clinical dosage form is chelidonine injection which is mainly used for relieving the pain of gastrointestinal colic, gastric and duodenal ulcers, renal colic, dysmenorrhea, and biliary tract ascariasis. This injection can also be used for women's sterilization surgery when combined with local anesthetics. In addition, it is often used for the treatment of rheumatoid arthritis, trauma, and cancerous pain due to its good effect on analgesia.

Literature

Chelidonine, also named ranunculine, is a kind of alkaloid isolated from the whole plant of *Chelidonium majus.* L [1]. As a kind of traditional Chinese medicine, *Chelidonium majus.* L was firstly recorded in Chiu Huang Pen Ts'ao of Ming dynasty. In the seventeenth century, it has been used for the treatment of jaundice, biliary colic, and cholelithiasis. In 1941, a researcher from Soviet Union reported that it showed good effect for the treatment of cutaneous tuberculosis. This plant has multiple effects on different diseases, including anti-inflammatory, analgesia, relieving cough, inducing diuresis, anticancer, antifungal, relieving asthma, and detoxification. It was recorded in the Pharmacopoeia of the People's Republic of China (1977) [1].

The material basis for *Chelidonium majus.* L is alkaloid. Among which, chelidonine, a benzophenanthridine alkaloid, is the main effective ingredient [2]. Although great progresses have been made in the chemical synthesis of chelidonine recently, the biosynthetic pathways of chelidonine remain unclear, which still need in-depth study and exploration in the future.

History of R&D

Chelidonine was isolated from *Chelidonium majus.* L in 1824 for the first time. Pharmacological studies found that it exerted antitumor effect via different mechanisms. However, its poor bioavailability limited its application. Some researchers developed a chelidonine/polylactic acid copolymer nanoparticle using nanotechnology. The nano-chelidonine showed good tissue distribution without causing any toxicity in mice and could enter into the brain tissues, showing great application potential. In addition, nano-chelidonine showed protective effect on liver injury induced by cadmium in mice [3].

In 1981, an Austrian researcher isolated a novel compound ukrain (phosphorothioate derivative of chelidonine) from *Chelidonium majus.* L. Ukrain could inhibit tumor cell proliferation through a variety of mechanisms. Currently, ukrain is used as an antitumor drug for the clinical treatment of lung cancer, breast cancer, prostate cancer, and pancreatic cancer [4]. In 2004, quarter amine chelidonine phosphorothioate derivatives have been applied for an invention patent as anticancer drugs in China.

Chelidonine has morphine-like analgesic effects suggesting that its analgesic effect is mainly peripheral and cannot be antagonized by the morphine receptor antagonist naloxone [5]. The 6-alkoxy and 6-acyloxy derivatives of chelidonine can inhibit the central nervous system, especially for nerve terminal, and show sedative and hypnotic effects [6].

Pharmacology

Chelidonine has a variety of pharmacological effects, mainly manifested in the following aspects:

1. The effect on nervous system (analgesia, sedation)

Chelidonine, as a protopine, has analgesia effect similar to that of morphine. After intragastric administration (5–20 mg/kg) to mice, it showed a dose-dependent analgesic effect, which could sustain 4–48 h [5]. Chelidonine derivatives also have sedative and hypnotic effects [6].

2. The effect on cardiovascular system

Chelidonine has many effects on cardiovascular system, including exciting heart, expanding coronary blood vessels, and increasing blood pressure. Chelidonine (0.01–0.02 mg) leads to the excitement of frog heart in vitro, as well as slowdown of heart beat. Higher dosage (0.05 mg) of chelidonine could cause arrhythmia and diastolic cardiac arrest [7].

3. The antitumor effect

Chelidonine is a toxic substance that can influence mitosis. In vitro studies have shown that chelidonine had significant inhibitory effects for gastric cancer, leukemia, nasopharyngeal carcinoma, and hepatoma carcinoma cells. It also could delay the growth of malignant tumors. Its antitumor effects were mediated by different mechanisms, which still need further studies [8].

4. The effect on smooth muscle

Chelidonine has spasmolytic and diastolic effects and can inhibit a variety of smooth muscle spasm. Obvious spasmolytic effects for gastrointestinal tract and bronchial and urinary system have been reported [9].

5. The antibacterial effect

Chelidonine has antibacterial effect. It inhibits *Mycobacterium tuberculosis* in vivo and inhibits alpha *Streptococcus*, *Diplococcus pneumoniae*, and other gram-positive bacteria in vitro. In addition, chelidonine also shows inhibitory effect on Kauffman-Wolf *Trichophyta* and *Epidermophyton floccosum* [10].

6. Other pharmacological effects

Chelidonine showed protective effects on cadmium chloride-induced liver and kidney toxicity in rats. Chelidonine can inhibit the growth of human keratinocytes. Guinea pig test confirmed that chelidonine (4–10 mg/kg) could prevent or delay the anaphylactic shock. In addition, chelidonine has anti-mite effect and good synergistic effect with trichlorfon or omethoate to prevent cotton bollworm.

Clinical Application

Chelidonine has been used as a painkiller in clinical application due to its significant analgesic effect. Chelidonine phosphate can be used for treatment of gastrointestinal and ulcer pain, as a substitute of morphine preparations. Wei Tong Shu capsule is used to treat pain caused by gastric convulsion, chronic gastritis, gastric ulcer, and duodenal ulcer. Its main effective compositions are chelidonine and protopine. Fu Fang Zhong Yao Tong An injection has collateral dredging and analgesic effects and can be used for the treatment of moderate pain caused by chemoradiotherapy or non-chemoradiotherapy of gastric cancer, lung cancer, and liver cancer. Its two main components are chelidonine and sinomenine. Ukrain (NSC-631570), a phosphorothioate derivative of chelidonine, has been used in clinical treatment as an effective antitumor drug for the treatment of lung cancer, breast cancer, prostate cancer, and pancreatic cancer.

Discussion

Chelidonine has various pharmacological effects, including antitumor, analgesiam, and spasmolysis, and is receiving more and more attention. However, its mechanisms of function remain to be clarified. With the development of science and technology, chelidonine will have broader prospects in new drug research and development.

Fig. 1 *Chelidonium majus.* L

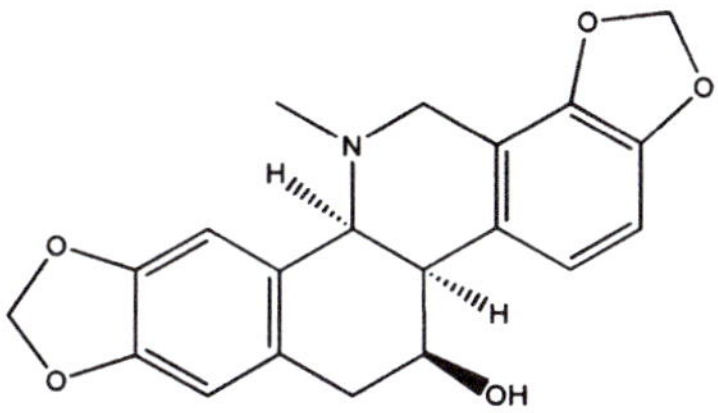

Fig. 2 The structure of chelidonine

References

1. State Administration of Traditional Chinese Medicine. Chinese materia medica. Shanghai: Shanghai Science and Technology Press; 1999.
2. Lei QF, Zhao XL, Xu LJ, et al. Chemical constituents of plants from tribe chelidonieae and their bioactivities. Chin Herb Med. 2014;6(1):1–21.
3. Paul A, Das S, Das J, et al. Cytotoxicity and apoptotic signalling cascade induced by chelidonine-loaded PLGA nanoparticles in HepG2 cells in vitro and bioavailability of nano-chelidonine in mice in vivo. Toxicol Lett. 2013;222(1):10–22.
4. Habermehl D, Kammerer B, Handrick R, et al. Proapoptotic activity of Ukrain is based on Chelidonium majus L. alkaloids and mediated via a mitochondrial death pathway. BMC Cancer. 2006;6:14.
5. He ZM, Tong JM, Gong FC. Study on the analgesic effect of chelidonine. Chin Herb Med. 2003;34(9):837–8.
6. Jagiello-Wójtowicz E, Chodkowska A, Gadzikowska M, et al. Preliminary evaluation of CNS effects of 6-O-substituted chelidonine derivatives. Acta Pol Pharm. 2003;60(2):133–8.
7. Ji YB. Pharmacology and application of active ingredients of natural medicine. Beijing: Science Press; 2007. p. 119–20.
8. Zou X, Wang YM, Wang JQ, et al. Advances in pharmacological effects of chelidonine. Drugs Clin. 2014;29(11):1326–30.
9. Zhang BH, Ma L, Ma JJ, et al. Research on pharmacological effects of chelidonine. Tradit Chin Med Bull. 1985;10(1):41–4.
10. Niu CQ, He LY. Review on the research of *Chelidonium majus* L. Chin Pharm J. 1994;29(3):138–40.

Cissampelini Methiodidum

Jie Yu, Li-Da Du, Xiu-Ping Chen, Shou-Bao Wang, and Guan-Hua Du

Abstract Cissampelini Methiodidum is a methiodide of the Yahonglong alkaloid, extracted from Yahonglong. Cissampelini Methiodidum is a muscle relaxant with exact curative effect successfully developed in China. However, Cissampelini Methiodidum has poor water solubility. Therefore, it is necessary to explore the analogue of Cissampelini Methiodidum with water solubility and strong effect of muscle relaxation. In the long history of morphine R&D, there are many analogues of Cissampelini Methiodidum found and observed to be satisfactory muscle relaxation effects in a clinical trial. Due to its side effects such as respiratory depression, affecting the mitosis of mammalian cell and chromosomal division, and the discovery of more effective and safer alternative drugs, the clinical use of Cissampelini Methiodidum gradually decreased. However, Cissampelini Methiodidum is one of the important landmarks in the muscular relaxant research area in China, which provides a good example for exploring drugs from traditional herbs.

Keywords Cissampelini Methiodidum · Alkaloid · Muscle relaxant

Origin: Yahunu (Fig. 1)

Alias: Xishengteng alkaloid II, Yahunu, Yahong Long, Hayatin Mee
Chemical name (Fig. 2)

(±)-Curine dimethiodide

J. Yu · X.-P. Chen (✉)
State Key Laboratory of Quality Research in Chinese Medicine, Institute of Chinese Medical Sciences, University of Macau, Macao, China
e-mail: xpchen@umac.mo

L.-D. Du
School of Biomedical Sciences, Chinese University of Hong Kong, Hong Kong, China

S.-B. Wang · G.-H. Du
Beijing Key Laboratory of Drug Targets Identification and Drug Screening, Centre for Pharmaceutical Screening, Institute of Materia Medica, Chinese Academy of Medical Sciences and Peking Union Medical College, Beijing, China

G.-H. Du, *Natural Small Molecule Drugs from Plants*,
https://doi.org/10.1007/978-981-10-8022-7_35

Molecular formula, $C_{38}H_{44}O_6N_2I_2$; **MW**, 878.13; **CAS**, 22260-42-0

Properties

Appearance: light yellow, odorless, and crystalline substance with slightly bitter taste. *Solubility*: slightly soluble in hot water; very slightly soluble in cold water or ethanol; insoluble in ether and chloroform. *Melting point*: 277–281 °C. Cissampelini Methiodidum in the clinical use is the racemic methiodide of curarine [1].

Dosage Forms

Cissampelini Methiodidum is recorded in the *Pharmacopoeia of the People's Republic of China* (1977).

Cissampelini Methiodidum injection

Indications

Cissampelini Methiodidum injection is mainly used as muscle relaxant for shallow general anesthesia surgery, cooperatively used with diethyl ether, fluothane, methoxyflurane, and other inhalation anesthesia as well as intravenous administration of procaine, herbal anesthesia, and anesthetization by acupuncture.

Literature

Cissampelini Methiodidum is a methiodide of the Yahonglong alkaloid (Xishengteng alkaloid II), extracted from the roots and stems of Yahonglong (*Cissampelos pareira* L.). "Daijisong" is the trade name [1, 2]. Yahonglong is the whole dry plant of Xishengteng (*Cissampelos pareira* L.) from Menispermaceae family and is traditionally used in Dai nationality [3]. It is found in the northwest of Guangxi, southwest of Guizhou, and southern of Yunnan. It grows in tropical riverside, sandy beaches, wasteland, slopes, or wetlands in bushes. It has been excavated in spring and summer, and then the sediment was removed and dried. Cissampelini Methiodidum is a muscle relaxant with exact curative effect successfully developed in China.

History of R&D

Tubocurarine is a natural bisbenzylisoquinoline alkaloid isolated from Menispermaceae plants. It is one of the major toxic components of tubocurare. In the early 1950s, Pradhan reported a new muscle relaxant, Hayatin, extracted from Xishengteng of India [4–6]. The racemic d, l-tubocurarine extracted from Yahonglong (Dai drug) by scientists in China was methylated to make iodomethane salt which was named as Cissampelini Methiodidum (trade name is Daijisong) [1].

Another powerful non-depolarizing muscle relaxant, dimethyl racemic curarine chloromethane salt, was developed in China in November 1980. It is an analogue of Cissampelini Methiodidum and was made from dimethyl Yahonglong alkaloid through chloromethylation. Its water solubility is ten times higher that of Cissampelini Methiodidum, which not only solves the problem of poor water solubility of Cissampelini Methiodidum but also intensifies the effect of muscle relaxation. Pharmacological studies have shown that its effect on muscle relaxation in rabbits, cats, and dogs is 1–2.5 times stronger than that of tubocurarine and 0.5–3 times than Cissampelini Methiodidum. It has similar efficacy to that of tubocurarine in terms of mouse neuromuscular blocking. Its muscle relaxation effect also can be antagonized by the neostigmine. When combined with a variety of anesthesia, such as intravenous anesthesia, diethyl ether, fluothane, and herbal anesthesia, satisfactory muscle relaxation effects were observed in a clinical trial [7].

Pharmacology

Cissampelini Methiodidum is a non-depolarizing muscle relaxant that blocks the striated muscle at the neuromuscular junction. It can make all the striated muscle relax until the respiratory muscle paralysis. During its respiratory depression, effective artificial respiration can gradually restore breathing. The time of onset is less than 1 min and has a duration of action of 20 min in intravenous administration. Due to the accumulation effect, repeated administration in 30 min should reduce the dosage. Synergistic effect was observed in combination with diethyl ether. It barely affects the circulatory system, which was applicable to use as muscle relaxant in shallow general anesthesia surgery [3].

Clinical study found that Cissampelini Methiodidum may cause respiratory paralysis and even respiratory cessation due to relaxation of respiratory muscles. This could be managed by artificial respiration such as mouth-to-mouth resuscitation. Artificial breathing equipment and related personnel are needed to ensure medical safety when using this drug. Its respiratory depression effect could be enhanced in old patients and those with electrolyte disturbances and shock, as well as patients administrated with inhibitory central nervous system agents such as morphine, pethidine, thiopental sodium, etc. Under such circumstances, dosage should be reduced. The muscle relaxation effect of this drug can be antagonized by administration of neostigmine (l–2 mg) [3].

Clinical Application

Cissampelini Methiodidum can relax the muscles in the shallow general anesthesia, and thereby, it could avoid deep anesthesia and reduce the dosage of anesthetic, which was beneficial for operation. It can be used in various surgeries in thoracic, abdomen, urology, obstetrics, gynecology, and orthopedics. It can be used in combination with diethyl ether, fluothane, methoxyflurane, as well as intravenous administration of procaine, herbal anesthesia, and anesthetization by acupuncture.

Discussion

For a long time, all muscle relaxants used in China were imported from other countries. Cissampelini Methiodidum, a derivative of tubocurarine isolated from Dai medicine, has been widely used as a muscle relaxant in shallow general anesthesia surgery in China [7]. The Cissampelini Methiodidum injection has been included in *Pharmacopoeia of the People's Republic of China* (1977). The research and clinical use of Cissampelini Methiodidum has promoted the development of more than 20 botanical muscle relaxants in China, which promoted the research in this area [7, 8]

Recently, the clinical use of Cissampelini Methiodidum was gradually decreased due to its side effects such as respiratory depression, affecting the mitosis of mammalian cell and chromosomal division, and was replaced by more effective and safer alternative drugs [9]. However, Cissampelini Methiodidum is one of the important landmarks in the muscular relaxant research area in China, which provides a good example for exploring drugs from traditional herbs.

Fig. 1 Yahunu

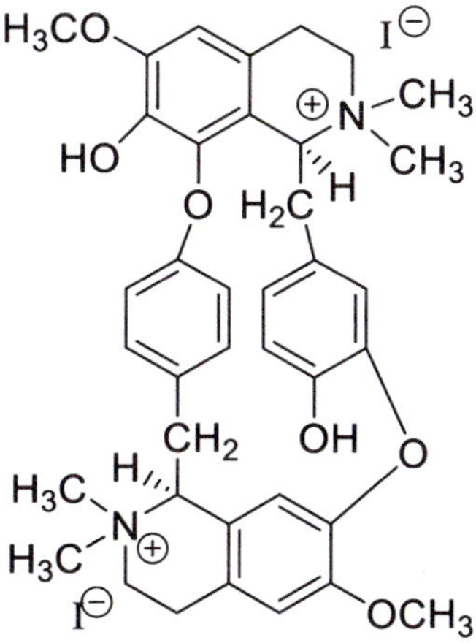

Fig. 2 The structure of Cissampelini Methiodidum

References

1. Zhong LJ. Regard of the structure of Daijisong. J Chemother. 1980;38(6):567–72.
2. Jean-Michel L, Jean G, Stephane M, et al. Crystal structure of Daijisong. Anal Sci. 2004;20:105–6.
3. Ming Y. The injection of Daijisong. J Chin Pharm Ind. 1976;4.
4. Pradhan SN, De NN. Hayatin methiodide: a new curariform drug. Br J Pharmacol Chemother. 1953;8(4):399–405.
5. Pradhan SN, Pandey K, Badola RP. A clinical trial of Hayatin methiodide as a relaxant in 100 cases. Br J Anaesth. 1964;36(36):604–11.
6. Santoro D, Passantino A, Ricciardi CA, et al. Diuretics in the "Istituzioni di Materia Medica e Terapeutica" (1862–1864) by Giovanni Pagano of Naples. J Nephrol. 2013;26(Suppl. 22):139–42.
7. Yinghe Y. A review of domestic new drugs in 1980~1981. Chin Pharm J. 1982;17(7):419–25.
8. Zhiqing C. A research overview of muscular relaxant from Chinese herbal medicine. Henan Tradit Chin Med. 1977;4:75–9.
9. Zhijun H, Xu W. Effects of d-tubocurarine on the mitosis of V79 cells and mouse bone marrow cells. Cancerous Distortion: Mutat. 2000;12(1):8–11.

Cocaine

Ying Zhao, Jin-Hua Wang, Xiu-Ying Yang, and Guan-Hua Du

Abstract Cocaine, is also called cocain, was mainly separated from the leaves of coca. Cocaine is white crystal powder and odorless. It is insoluble in water, soluble in chloroform, ethyl ether, ethanol, and other organic solvents. In 1855, cocaine was first extracted from the leaves of herbal ingredients and was used as a local anesthetic in 1840. At present, the clinical dosage form of cocaine is injection, and it is mainly used for local anesthesia in various operations. Cocaine is the first natural alkaloid with local anesthesia effect and is replaced by other anesthetics for its toxic and addictive properties now.

Keywords Cocaine · Coca · Anesthesia · Addiction

Alias: Cocain
Origin: Coca (Fig. 1)
Chemical name (Fig. 2)

Methyl (1R,2R,3S,5S)-3- (benzoyloxy)-8-methyl-8-azabicyclo[3.2.1] octane-2-carboxylate

Molecular formula, $C_{17}H_{21}NO_4$; **MW**, 303; **CAS**, 50-36-2

Properties

Appearance: pure white crystal, odorless. *Solubility*: insoluble in water (1:600); soluble in chloroform (1:0.7), ethyl ether (1:3), ethanol (1:7), and other organic solvents; soluble in acetone, benzene, ethyl acetate, two carbonyl sulfide, and

Y. Zhao · J.-H. Wang · X.-Y. Yang · G.-H. Du (✉)
Beijing Key Laboratory of Drug Targets Identification and Drug Screening, National Centre for Pharmaceutical Screening, Institute of Materia Medica, Chinese Academy of Medical Sciences and Peking Union Medical College, Beijing, China
e-mail: dugh@imm.ac.cn

G.-H. Du, *Natural Small Molecule Drugs from Plants*,
https://doi.org/10.1007/978-981-10-8022-7_36

petroleum ether; show monoclinic flaky crystal in ethanol. *Melting point*: 98° (it can be slowly evaporated over 90°), the specific rotation of −16°.

Its Hydrochloride Form

Appearance: is crystalline or crystalline powder. *Solubility*: soluble in water and soluble in ethanol and chloroform but insoluble in ether.

Dosage Forms

This product is recorded in the *Pharmacopoeia of the People's Republic of China* (2015), *Japanese Pharmacopoeia* (17th ed.), *British Pharmacopoeia* (2017), *United States Pharmacopeia* (40), and *European Pharmacopoeia* (9.0th ed.).

At present, the clinical dosage form is injection.

Indications

It is mainly used for local anesthesia in various operations. It is also suitable for nasal, pharyngeal, ear, urethra, vagina, and other operations (with 5–10% solution). 2% to 3% solution is used for eye surgery.

Literature

Cocaine was mainly separated in the leaves of coca (古柯). Coca is a tropical montane evergreen shrub that grows in the Andes of South America. In the thirteenth century, the coca plant was admired by the Inca Empire of Peru. As early as the sixteenth century, the Spanish explorers noticed that Native Americans refreshed them by chewing coca plants. Also it was even recorded that cocaine could alleviate the feeling of hunger and thirst, so it is called "holy grass." In 1750, the first samples of the ancient tree were shipped to Europe by the French botanist Suyi Yu. In 1781, revolution of indigenous Bolivia La Paz broke out. The soldiers were supported by coca leaf in the case of lack of rations for a period of time, which makes a lot of Europeans to understand the effect of coca leaf. In modern medicine, cocaine is used as a local anesthetic or vasoconstrictor. Because of its good anesthetic effect and strong penetrating power, it is mainly used for surface anesthesia [1, 2].

History of R&D

In 1855, cocaine was first extracted from the leaves of herbal ingredients and named *Erythroxylon* by the German chemist Friedrich (G. Friedrich) (1828–1890) [3]. In 1859, the German chemist Niemann (Albert Niemann) (1834–1861) defined a higher purity of material, named cocaine [4]. In 1880, cocaine was used as a local anesthetic by the "father of modern surgery," William Stewart Halsted. In 1884, cocaine was firstly recommended to be used as a local anesthetic, sexual stimulant, and antidepressant by Sigmund Freud, a famous psychologist in Austria (1856–1939). He called it "a magical substance" [5, 6].

Pharmacology

Cocaine can block the generation and transmission of nerve impulses in the epidermis and mucous membranes, which can last up to 20–40 min. Once it is absorbed in local or scattered to other parts of the body, the local anesthetic action disappears. It was generally used as a local anesthetic at concentration of 5–10%. Cocaine is used for topical mucosal anesthesia in clinic. Cocaine can cause changes of neurotransmitter in the central nervous system, which can lead to seizures or insanity [7]. Cocaine stimulated the cerebral cortex, medulla oblongata, and spinal cord of the central nervous system by strengthening chemicals in the body activity and excited mood. In addition, cocaine can cause systemic vasoconstriction, which results in a number of cardiovascular complications.

Recent studies have shown that cocaine regulates gene expression associated with certain systems in the nucleus accumbens (NAc). Repeated cocaine effects lead to the change of gene expression in NAc and neuronal morphology in rodent animals. Cocaine directly induced the modification of NAc chromatin, leading to changes in histone acetylation and phosphorylation. The nucleus accumbens is a key center in the brain's reward circuit. Cocaine addiction can cause persistent changes in gene expression in the loop which may be the basis of certain behavioral effects of cocaine addiction [8]. The study also found that G9a overexpression can block repeated cocaine-induced enhancement of gene expression. The inhibition of G9a and dimethylated H3K9 under repeated cocaine promotes the hobby of cocaine, partly due to the transcriptional activation of the neuronal dendritic plasticity abnormality and the downregulation of G9a expression in NAc increasing the density of neuronal dendritic spines [9, 10]. Cocaine can pass through the blood-brain barrier and accumulate in the central nervous system. Cocaine can also pass through the placental barrier.

Clinical Application

Cocaine is the first natural alkaloid with local anesthesia effect, long-term local anesthetics, high lipid solubility and strong penetrating power, and good affinity for nerve tissue, resulting in a good surface anesthetic effect. It was commonly used in clinic for local anesthesia, such as in the mouth, nose, pharynx, ear, urethra, and vagina, and other surgical anesthesia. The main side effect of cocaine is that it can excite the cerebral cortex at a small dose of application and result in euphoria, which has a strong potential for drug abuse and dependence. High-dose application excites respiratory, cardiovascular, and vomiting center, and serious eclampsia can occur and finally from excitement to inhibition, respiratory depression, heart failure, and even death. Severe cardiovascular disease, hypertension, hyperthyroidism, and glaucoma patients are banned. Cocaine had side effects on a damaging effect on the digestive system, immune system, cardiovascular system, and urogenital system.

Discussion

Cocaine is the first natural alkaloid found in human beings. Cocaine is a long-acting ester local anesthetic and has high fat solubility, strong penetrability, good affinity for nerve tissue, and good local surface anesthesia. Contraction of blood vessels caused by cocaine may be associated with blocking the reuptake of norepinephrine at nerve endings. Recently, it has been replaced by other anesthetics for its toxic and addictive properties.

Fig. 1 Coca

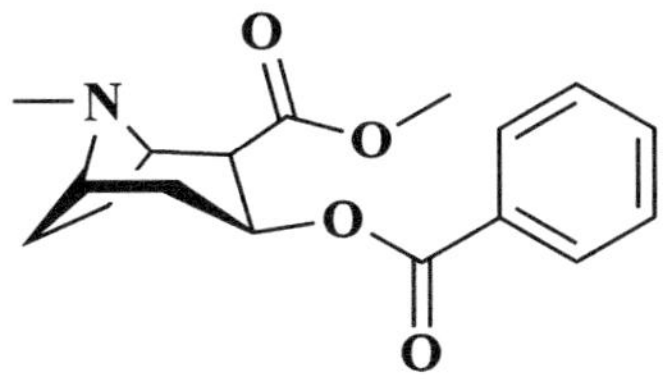

Fig. 2 Chemical structure of cocaine

References

1. Halsted W. Practical comments on the use and abuse of cocaine. NY Med J. 1885;42:294–5.
2. Corning JL. An experimental study. NY Med J. 1885;42:483.
3. Gaedcke F. Ueber das Erythroxylin, dargestellt aus den Blättern des in Südamerika cultivirten Strauches Erythroxylon Coca Lam. Arch Pharm. 1855;132(2):141–50.
4. Gootenberg P, editor. Cocaine: global histories. London: Routledge; 1999. ISBN 0-203-02646-2
5. Madge T. White mischief: a cultural history of cocaine. Edinburgh: Mainstream Publishing Company; 2001. ISBN 978-1-84018-405-1
6. Spillane JF. Cocaine: from medical marvel to modern menace in the United States, 1884–1920. Baltimore/London: The Johns Hopkins University Press; 2000. ISBN 0-8018-6230-2
7. Bocklisch C, Pascoli V, Wong JC, et al. Cocaine disinhibits dopamine neurons by potentiation of GABA transmission in the ventral tegmental area. Science. 2013;341(6153):1521–5.
8. Sartor GC, Powell SK, Brothers SP, et al. Epigenetic readers of lysine acetylation regulate cocaine-induced plasticity. J Neurosci Off J Soc Neurosci. 2015;35(45):15062–72.
9. Maze I, Covington HE, Dietz DM, et al. Essential role of the histone methyltransferase G9a in cocaine-induced plasticity. Science. 2010;327(5962):213–6.
10. Nyssen L, Brabant C, Didone V, et al. Response to novelty and cocaine stimulant effects: lack of stability across environments in female Swiss mice. Psychopharmacology. 2015;233:1–10.

Cyclandelate

Yin-Zhong Ma, Gui-Fen Qiang, and Guan-Hua Du

Abstract Cyclandelate is extracted from almond, which is a kind of nut from Xinjiang, China. Traditional Chinese medicine believes that almond can promote blood circulation to dispel blood stasis, lubricate bowels to relieve constipation, and relieve cough and asthma. The chemical structure and effect of cyclandelate are similar to papaverine. It can directly relax vascular smooth muscle and relieve the spasm of ileum and uterus smooth muscle induced by acetylcholine, histamine, and barium chloride in guinea pig. This effect is three to five times stronger than papaverine. It was reported that cyclandelate can promote the collateral circulation but has little effect on respiration, blood pressure, cardiac output, and myocardial oxygen consumption. It is safe for long-term administration. Cyclandelate can be used for clinical treatment of cerebral arteriosclerosis, cerebral vascular accident and its sequelae, post-traumatic brain syndrome, coronary arteriosclerosis, hypertensive heart disease, Raynaud's disease, thromboangiitis obliterans, acrocyanosis, and Meniere's disease. However, cyclandelate has not yet been approved by the Food and Drug Administration in the United States, Canada, and other countries because it easily causes white blood cell deficiency. It has been withdrawn from the market after drug approval in Japan, France, and other countries in the 1970s.

Keywords Cyclandelate · Papaverine · Cerebral arteriosclerosis · White blood cell deficiency

Origin: Almond (Fig. 1)
Chemical name (Fig. 2)

3,3,5-Trimethylcyclohexyl 2-hydroxy-2-phenylacetate

Y.-Z. Ma · G.-F. Qiang · G.-H. Du (✉)
Beijing Key Laboratory of Drug Targets Identification and Drug Screening, Centre for Pharmaceutical Screening, Institute of Materia Medica, Chinese Academy of Medical Sciences and Peking Union Medical College, Beijing, China
e-mail: dugh@imm.ac.cn

G.-H. Du, *Natural Small Molecule Drugs from Plants*,
https://doi.org/10.1007/978-981-10-8022-7_37

Molecular formula, $C_{17}H_{24}O_3$; **MW**, 276.37; **CAS**, 456-59-7

Properties

Appearance: white or almost white amorphous powder, special smell, and bitter taste. *Solubility*: very soluble in ethanol or acetone and almost insoluble in water. *Melting point*: 50–62 °C.

Dosage Forms

Cyclandelate was included in the Pharmacopoeia of the People's Republic of China (2015) and in the United States Pharmacopeia (40). The dose form is capsule, 100 mg/granule.

Indications

The indications of cyclandelate are arteriosclerosis obliterans, acrocyanosis, cerebral arteriosclerosis, cerebral insufficiency, cerebrovascular disease, brain trauma, and post-traumatic brain syndrome.

Literature

Almond (*Amygdalus communis* L., 扁桃) is a kind of nut from Xinjiang, China. Traditional Chinese medicine believes that almond can promote blood circulation to dispel blood stasis, lubricate bowels to relieve constipation, and relieve cough and asthma. The indications include amenorrhea, fever, wind arthralgia, malaria, blood stasis and pain, injuries, and blood stasis with constipation.

History of R&D

The chemical synthesis of cyclandelate was first synthesized by Funcke et al. from Elan Corporation in Ireland using a-hydroxyphenylacetic acid and cis-3,3,5-cycloalkyl cyclohexanol. The raw material of mandelic acid was obtained by hydrolysis after benzaldehyde reacted to sodium cyanide. However, sodium cyanide is hypertoxic, and because of its unique structure of a-hydroxy acid, it is easy to decompose under the acid condition, which leads to more by-products and low yield. A series of improved methods have been developed, which can reduce the environmental pollution under the premise of ensuring the yield of Zn/HCOONH4/C2H5OH system [1]. The method was used to synthesize cyclandelate since then. However, this drug has not yet been approved by the Food and Drug Administration in the United States, Canada, and other countries because it easily causes white blood cell deficiency. It has been withdrawn from the market after drug approval in Japan, France, and other countries in the 1970s.

Pharmacology

The chemical structure and effect of cyclandelate are similar to papaverine. It can directly relax vascular smooth muscle and relieve the spasm of ileum and uterus smooth muscle induced by acetylcholine, histamine, and barium chloride in guinea pig. This effect is three to five times stronger than papaverine. Cyclandelate can also expand the cardiovascular, cerebrovascular, and renal blood vessels and limb peripheral vascular and coronary artery, increase blood flow, and promote blood circulation [2]. It can also increase the tolerance to hypoxia, but the effect on human cerebral blood flow has not been confirmed. It was reported that cyclandelate can promote collateral circulation but has little effect on respiration, blood pressure, cardiac output, and myocardial oxygen consumption. It is safe for long-term administration [3].

Clinical Application

Cyclandelate can be used for clinical treatment of cerebral arteriosclerosis, cerebral vascular accident and its sequelae, post-traumatic brain syndrome, coronary arteriosclerosis, hypertensive heart disease, Raynaud's disease, thromboangiitis obliterans, acrocyanosis, and Meniere's disease [4, 5].

Discussion

Similar to papaverine, cyclandelate can directly relax vascular smooth muscle and especially alleviate uterine smooth muscle spasm. The effect is three to five times stronger than papaverine. In addition, cyclandelate can expand coronary blood flow and promote blood circulation but can hardly affect respiration, blood pressure, cardiac output, and myocardial oxygen consumption. The long-term administration of cyclandelate is safer than papaverine [6].

Fig. 1 Almond

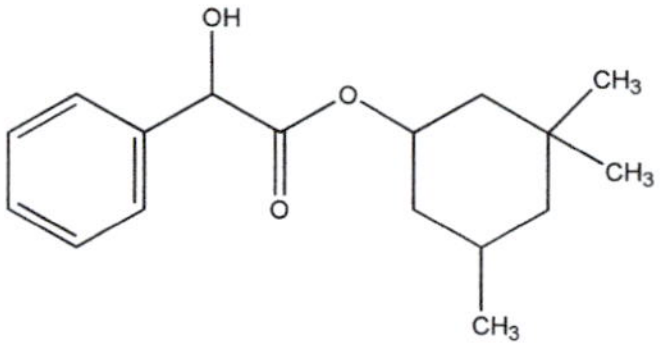

Fig. 2 The chemical structure of cyclandelate

References

1. Funcke AB, Ernsting MJ, Rekker RF, Nauta WT. Studies of spasmolytics. I. Mandelic acid esters. Arzneimittelforschung. 1953;3(10):503–6.
2. Diener HC, Krupp P, Schmitt T, Steitz G, Milde K, Freytag S. Cyclandelate in the prophylaxis of migraine: a placebo-controlled study. Cephalalgia. 2001;21(1):66–70.
3. CameronBD,etal.Themetabolicfateofthecoronaryvasodilator4-(3,4,5-Trimethoxycinnamoyl)-1-(N-pyrrolidinocarbonylmethyl)piperazine (cinepazide) in the rat, dog and man[J]. Xenobiotica. 1976;6(7):441–55.
4. Laporte JR, Capellà D, Juan J. Agranulocytosis induced by cinepazide. Eur J Clin Pharmacol. 1990;38(4):387–8.
5. Takeda M, Tanaka T, Okochi M. New drugs for Alzheimer's disease in Japan. Psychiatry Clin Neurosci. 2011;65(5):399–404.
6. Sidney M. Wolfe, M.D. Differences in the number of drug safety withdrawals: United States, United Kingdom, Germany and France 1970–1992. Drug Safety Withdrawals 1995.

Ephedrine

Jin-Hua Wang, Xiu-Ying Yang, and Guan-Hua Du

Abstract Ephedrine, also called ephedrine ingot, L-ephedrine, is from dry grass stems of ephedra plants such as ephedra, Zhong ephedra, and Equisetum ephedra. Ephedrine is a white needle-like crystal or crystalline powder and odorless. It is soluble in water and can be dissolved in ethanol but is insoluble in chloroform or ether. Dosage forms of ephedrine are mainly tablets, injections, and eye drops, and they were used widely in clinical practice. Ephedrine was first isolated from the herb "Shuanghui Ephedra" in 1885. Ephedrine was successfully synthesized in 1920. In 1926, ephedrine was approved in clinical practice by the FDA of the United States. Structure of ephedrine was defined in 1926. Ephedrine hydrochloride was used for the treatment of bronchial asthma, hypotension, and central excitation. The side effects of ephedrine hydrochloride are anxiety, insomnia, palpitations, sweating, and other symptoms. Ephedrine is rarely used for asthma treatment now.

Keywords Ephedrine · Asthma · Drug tolerance · Addiction

Alias: Ephedrine, ephedrine ingot, levapedin
Origin: *Ephedra sinica* Stapf (Fig. 1)
Chemical name (Fig. 2)

(R,S)-2-(methylamino)-1-phenylpropan-1-ol

Molecular formula, $C_{10}H_{15}NO$; **MW**, 165; **CAS**, 299-42-3
Ephedrine derivative (Fig. 3)

Molecular formula, $C_{10}H_{15}NO{\cdot}HCl$; **MW**, 201.7; **CAS**, 134-71-4

J.-H. Wang · X.-Y. Yang · G.-H. Du (✉)
Beijing Key Laboratory of Drug Targets Identification and Drug Screening, National Centre for Pharmaceutical Screening, Institute of Materia Medica, Chinese Academy of Medical Sciences and Peking Union Medical College, Beijing, China
e-mail: dugh@imm.ac.cn

G.-H. Du, *Natural Small Molecule Drugs from Plants*,
https://doi.org/10.1007/978-981-10-8022-7_38

Properties

Appearance: a white needle-like crystal or crystalline powder, odorless, and tastes bitter. *Solubility*: soluble in water and can be dissolved in ethanol but insoluble in chloroform or ether. *Melting point*: 217–220 °C.

Dosage Form and Indications

This product was recorded in the *Pharmacopoeia of the People's Republic of China* (2015), the *British Pharmacopoeia* (2017), the *United States Pharmacopeia* (40), the *Indian Pharmacopoeia* (2010), the *European Pharmacopoeia* (9th ed.), and *The International Pharmacopoeia* (5th ed.).

Dosage forms are mainly tablets, injections, and eye drops.

Ephedrine was clinically used in three aspects: (1) prevention for bronchial asthma and relief of mild asthma, (2) treatment for subarachnoid anesthesia or epidural anesthesia caused by low blood pressure and chronic hypotension, and (3) treatment of nasal congestions and swelling caused by nasal congestion.

Literature

Ephedrine is an alkaloid, which is from dry grass stems of ephedra plants such as ephedra, zhong ephedra, or ephedra (麻黄、中麻黄或木贼麻黄). Ephedra is an herb recorded in "Chinese Pharmacopoeia," which can be used for sweating and cold, adjustment of lung function, asthma, and swelling. Ephedrine can treat cold, chest tightness, bronchial asthma, and so on. Ephedra was used as a drug in China for more than 2000 years, which was recorded in history books of materia medica. Ephedra is also known as Longsa recorded in the book *Shen Nong's Herbal Classic of Materia Medica*, Bixiang recorded in the book *Another record by famous record*, and Gougu recorded in the book *Guang Ya*. Ephedra was firstly recorded in the book *Shen Nong's Classic of Materia Medica*. Ephedra tastes bitter and its color is yellow. In the book *Shen Nong's Materia Medica of 100 kinds of herbs*, ephedra was described as no tasteless, and ephedra reveals the pores of the skin, but also deep into the sputum and stop the blood, where the drug can't reach. Ephedra was used as Zuozhi medicine and widely used for the treatment of cold, skin diseases, ulcers, swelling and pain, and injury bruises during the Song, Jin, Yuan, Ming, and Qing dynasty. Ephedra was recorded in many books, such as *Compendium of Materia Medica*, *Synopsis of Prescriptions of the Golden Chamber*, *Treatise on Febrile Diseases Caused by Cold*, *Prescriptions Worth a Thousand in Gold for Every Emergency*, *Hua Zi all herbal medicine*, and *Theory on Drugs*.

History of R&D

In traditional Chinese medicine (TCM), ephedrine and pseudoephedrine are main active ingredients in "herbal ephedra" [1]. Organic chemist, Changi Changyi (1844–1929), from Japan first isolated ephedrine from the herb "Shuanghui Ephedra" in 1885 [2, 3]. Ephedrine was successfully synthesized by chemist Ernst Späth from Austria (1886–1946) in 1920 [4]. Kehui Chen (1898–1988) and its co-colleague Carl F. Schmidt extracted ephedrine from ephedra in 1923 [5] and then clarified the pharmacological effects of ephedrine, which attracted the attention from Europe and the United States. Ephedrine was approved in clinical practice by the FDA of the United States. Structure of ephedrine was analyzed by Changi Changyi in 1926.

Pharmacology

Ephedrine has a similar effect with adrenaline. Ephedrine can bind α, β receptors and activate the receptor as adrenaline. Ephedrine makes the adrenergic nerve endings release medium and indirectly play the role of adrenaline. Ephedrine is stable and its effect is weaker than adrenaline, and the central excitatory effect is more obvious. Ephedrine can relax bronchial smooth muscle but less durable than adrenaline. Ephedrine can also relieve bronchial spasm, excite the heart, enhance myocardial contractility, and speed up the heart rate. However, its role in increasing the heart rate may be weakened by the vagus nerve excitement caused by increased blood pressure [6]. Ephedrine accelerates heart rate and increases cardiac contractility by stimulating the beta receptor. Ephedrine stimulates the alpha receptor and contracts the arterial effect, but the contraction of the rabbit aorta is dominated by direct action. In recent years, ephedra has been found to have a role inhibiting the formation of acute blood clotting, promote fat synthesis in fat, and scavenge oxygen-free radicals.

Clinical Application

In clinical practice, ephedrine hydrochloride was used for the treatment of bronchial asthma, hypotension, central excitation [7], and toxicity by morphine and barbiturate. Ephedrine hydrochloride was also used to eliminate nasal mucosal congestion. Drug tolerance can occur quickly when ephedrine hydrochloride was used often in a short time. Ephedrine hydrochloride can also be used for hypotension and chronic hypotension caused by subarachnoid anesthesia or epidural anesthesia [8, 9]. The side effects of ephedrine hydrochloride are anxiety, insomnia, palpitations, sweating, and other symptoms by central excitement caused by the long-term use of large doses which can be cured by hypnotic sedative drugs. Acute drug tolerance can be induced by repeated use in short time.

Discussion

Efficacy of ephedrine lasts longer than adrenaline but the effect is poor. Ephedrine may be effective when the symptoms are mild. It is best to choose adrenaline when the symptoms are serious. Tremor, anxiety, insomnia, palpitations, and other symptoms can be caused by long-term repeated use of ephedrine, which can be reduced in combination with phenobarbital. Ephedrine is contraindicated in patients with hyperthyroidism, hypertension, atherosclerosis, and angina. Addiction and tolerance can be induced by long-term repeated use of ephedrine. Bogey, paclitaxel, and other monoamine oxidase inhibitors were forbidden to use with to prevent high blood pressure. As new drugs continue to emerge, ephedrine is rarely used for asthma treatment.

Fig. 1 *Ephedra sinica* Stapf

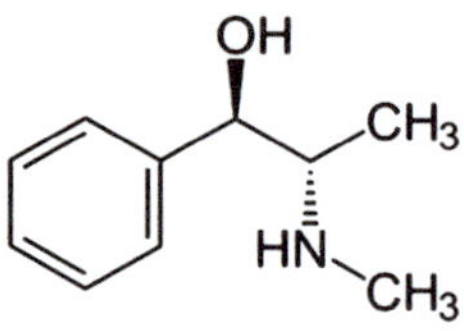

Fig. 2 Chemical structure of ephedrine

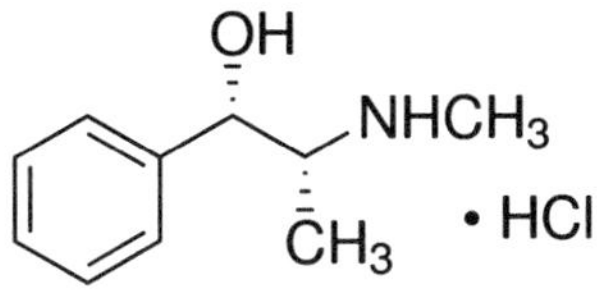

Fig. 3 Chemical structure of ephedrine hydrochloride

References

1. Botanical nomenclature, Shanghai, 1917;1004.
2. Lock M. East Asian medicine in urban Japan: varieties of medical experience. University of California Press; Reprint edition 1984. ISBN 0-520-05231-5.
3. Nagai N, Pharm Z 1887;xxxii: 700.
4. Späth E, Gökring R, Monatsh F. Chemistry. 1920;xli:319.
5. Chen K, Schmidt CF. J Pharmacol Exp Ther. 1924;xxiv:339.
6. Ruben JE, Kamsler PM, Howell WL Jr. The spinal anaesthetic effects of ephedrine sulfate: a preliminary report. Science. 1948;107(2774):223.
7. von Mutius E, Drazen JM. A patient with asthma seeks medical advice in 1828, 1928, and 2012. N Engl J Med. 2012;366(9):827–34.
8. Cannon C, Cowan F, Koppanyi T, et al. Explanation of cocaine desensitization of blood pressure responses to ephedrine. Science. 1961;134(3485):1075–7.
9. Eldaba AA, Amr YM. Intravenous granisetron attenuates hypotension during spinal anesthesia in cesarean delivery: a double-blind, prospective randomized controlled study. J Anaesthesiol Clin Pharmacol. 2015;31(3):329–32.

Ergometrine and Ergotamine

Yin-Zhong Ma, Gui-Fen Qiang, and Guan-Hua Du

Abstract Ergotamine and ergometrine belong to two kinds of ergot alkaloids. Ergot is the sclerotium of ergot fungus that grows in the rye and related cereal and forage plants. Traditional Chinese medicine believes that ergot is tasteless, tepid, toxic, liver meridian, and kidney meridian. Ergot can be used for uterine contraction, uterine bleeding, and postpartum hemorrhage after delivery, incomplete recovery of uterus, retrogression, etc. According to modern research, ergotamine or ergot can block the a-adrenergic receptor, so that the pressor effect of adrenaline was reversed. They do not prevent the release of sympathetic mediators. Ergonovine maleate is mainly used for the prevention and treatment of the weak uterine contraction or poor contraction caused by uterine bleeding after delivery or abortion. It is also used for incomplete recovery of postpartum uterus to accelerate the recovery of the uterus. Ergotamine tartrate is mainly used for the treatment of migraine by relieving the symptoms. Because ergonovine has been discontinued, oxytocin and prostaglandin preparation are the mainstream at present. The combination of ergotamine tartrate and caffeine can increase the absorption of ergotamine and enhance the contraction of blood vessels. Therefore, ergotamine-caffeine tablets have better effect on the early stage of migraine, but it is not suitable for long-term administration due to its side effects.

Keywords Ergotamine · Ergometrine · a-Adrenergic receptor · Uterine contraction

Origin: *Claviceps purpurea* (Fig. 1)
Chemical name (Fig. 2)

(6aR,9R)-N-((S)-1-Hydroxypropan-2-yl)-7-methyl-4,6,6a,7,8,9-hexahydroindolo[4,3-fg]quinoline-9-carboxamide

Y.-Z. Ma · G.-F. Qiang · G.-H. Du (✉)
Beijing Key Laboratory of Drug Targets Identification and Drug Screening, Centre for Pharmaceutical Screening, Institute of Materia Medica, Chinese Academy of Medical Sciences and Peking Union Medical College, Beijing, China
e-mail: dugh@imm.ac.cn

G.-H. Du, *Natural Small Molecule Drugs from Plants*,
https://doi.org/10.1007/978-981-10-8022-7_39

Molecular formula, $C_{23}H_{27}N_3O_6$; **MW**, 441.48; **CAS**, 60-79-7

Ergotamine
Chemical name (Fig. 3)

Ergotaman-3′,6′,18-trione, 12′-hydroxy-2′-metyl-(5′α)-phenylmethyl

Molecular formula, $C_{70}H_{76}N_{10}O_{16}$; **MW**, 1313.41; **CAS**, 379-79-3

Properties

Appearance: White or slightly yellow fine crystalline powder with slight hygroscopicity. *Solubility*: Slightly soluble in water and ethanol, insoluble in ether and chloroform.

Dosage Forms

Ergometrine maleate and ergotamine tartrate were both included in *Pharmacopoeia of the People's Republic of China (2015)*, *British Pharmacopoeia* (2017), *United States Pharmacopeia* (40), *The Japanese Pharmacopoeia* (17th ed.), *European Pharmacopoeia* (9.0th ed.), and *The International Pharmacopoeia* (5th ed.).

Indications

Ergometrine maleate is used in the form of extract tumor injection. As a uterotonic, ergometrine maleate is mainly used for the prevention and treatment of uterine bleeding caused by weak or poor uterine contraction after delivery or abortion, as well as the incomplete recovery of postpartum uterus to accelerate the recovery of uterus.

Ergotamine tartrate is used for the treatment of migraine in the form of tablet and parenteral solution. It also can be used for other neurogenic headache as well.

Literature

Ergotamine and ergometrine belong to two kinds of ergot alkaloids. Ergot (*Claviceps purpurea, 麦角*) is the sclerotium of ergot fungus that grows in the rye and related cereal and forage plants. The ergot is grown in the northeast and north China and Xinjiang, Jiangsu, Zhejiang, Sichuan, and other places. At present, the method of growth of ergot is inoculation cultivation.

Traditional Chinese medicine believes that ergot is tasteless, tepid, toxic, liver meridian, and kidney meridian. The traditional usage of ergot is oral administration of extractum, 2–4 ml each time, daily amount of 6–8 ml. Its dosage forms also contain infusion, pills, injection, etc. *Pharmacology of Chinese Medicine* records ergot can be used for uterine contraction [1], uterine bleeding, and postpartum hemorrhage after delivery, incomplete recovery of uterus, retrogression, etc. The *Plateau Chinese herbal medicine treatment manual* records that ergot can cure migraine [2].

History of R&D

As a kind of traditional Chinese medicine, ergot was originally used for hemostasis and involution of uterus by stimulating the uterine muscle powerfully and persistently. After the modernization the traditional Chinese medicine was putting forward, by making several kinds of wild ergot into extractum, we found that all of them can promote the contraction of the rabbit uterus in vivo and in vitro. In isolated rabbit uterus, the effect of ergot on postpartum uterus and pregnant uterus was stronger than nonpregnant uterus. Among a variety of active ingredients in the ergot, ergonovine has the strongest effect on uterine, followed by ergotamine. Seventeen of the 131 strains of ergot derived from the host plants of our country are able to produce ergot alkaloids in wheat culture medium. Among the strains, *Claviceps microcephala* Ce-3 produces the most ergonovine. Therefore, it became the main bacteria used in the process of industrialization [3]. The study suggested that ergotamine or ergot can block the a-adrenergic receptor, so that the pressor effect of adrenaline was reversed. They do not prevent the release of sympathetic mediators. It showed that the effect of ergotamine is similar to norepinephrine on the hypogastric nerve of guinea pig in vitro. Low dose of ergot can stimulate the brain (slow heart rate, breathing, convulsions, etc.) whereas high dose can cause deaths due to the delayed paralysis. In addition, it can enhance the sedative and hypnotic effects of barbiturates, morphine, and methadone.

Pharmacology

1. The effect of uterus excitation

   The effect of ergot on the uterine muscle is similar to the posterior pituitary preparation, which directly acts on the uterine muscle with more powerful and lasting effects. The pregnant uterus is more sensitive to ergot. Low dose of ergot has significant effects on the uterus of labor and new postpartum with severely tetanic contraction. The extracts from several kinds of wild ergot can promote the contraction and increase the tension of uterus in rabbit in vivo and in vitro [4]. Among a variety of active ingredients in the ergot, ergonovine has the strongest effect on uterus, followed by ergotamine.

2. The effects on cardiovascular

   In the isolated vascular experiment, ergotamine can cause the contraction of peripheral vascular smooth muscle and hypertension due to vasoconstriction and boost and eventually produce compensatory bradycardia. Ergot has less effect on the heart than blood vessels, but when the heart rate drops to a certain extent, it can enhance the cardiac contractility. High dose of ergotamine can damage vascular endothelial cells (the mechanism is unknown), leading to gangrene. Ergotine has the strongest effect on blood vessels, and ergometrine hardly affects blood pressure and damages to vascular endothelial cells [5]. Ergotine can cause vasoconstriction and decrease the arterial pulse amplitude, which may be related to the therapeutic effect on migraine.
3. The effects on the nervous system

   Ergotamine or ergot can block the a-adrenergic receptor and reverse the pressor effect of adrenaline without preventing the release of sympathetic neurotransmitters. Low dose of ergot stimulates the brain (slow down heart rate, breathing, convulsions, etc.), whereas high dose causes deaths due to the delayed paralysis. Ergotamine enhances the sedative and hypnotic effects of barbiturates, morphine, and methadone [6].

Clinical Application

Ergonovine maleate is mainly used for the prevention and treatment of the weak uterine contraction or poor contraction caused by uterine bleeding after delivery or abortion. It is also used for incomplete recovery of postpartum uterus to accelerate the recovery of the uterus.

Ergotamine tartrate is mainly used for the treatment of migraine by relieving the symptoms. However, it is only suitable for short-term usage of headache attacks, normally does not prevent migraine. The combination of ergotamine tartrate and caffeine is better than single oral administration with the mild side effects. It also can be used for other neurogenic headaches.

Discussion

Because ergonovine has been discontinued, oxytocin and prostaglandin preparation are the mainstream at present [7]. The combination of ergotamine tartrate and caffeine can increase the absorption of ergotamine and enhance the contraction of blood vessels. Therefore, ergotamine-caffeine tablets have better effect on the early stage of migraine, but it is not suitable for long-term administration due to its side effects.

Fig. 1 *Claviceps purpurea*

Fig. 2 The chemical structure of ergometrine

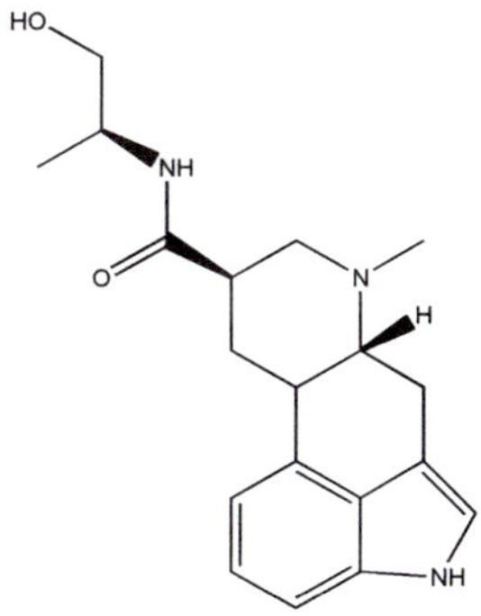

Fig. 3 The chemical structure of Ergotamine tartrate

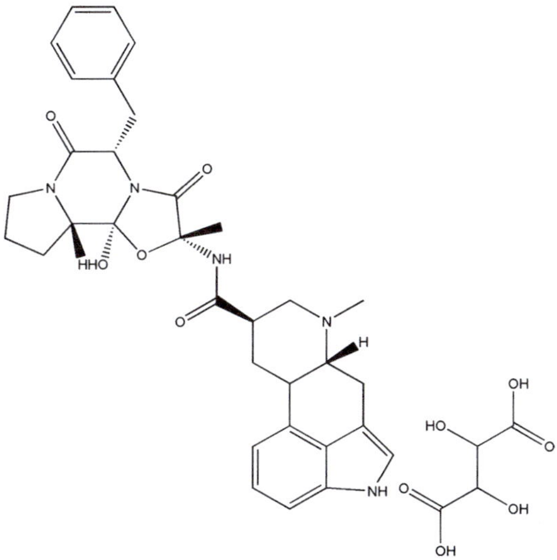

References

1. Mou HY. Pharmacology of traditional Chinese medicine. 1952;150–3.
2. Ruoergai County. Treatment manual of Chinese herb from plateau. 1971;95–6.
3. Fang QC. A new method of isolating ergometrine. Acta Pharm Sin. 1963;10(12):712–9.
4. He HX, Zhu P, Yue DC. Clinical application of ergot alkaloids and their pharmacological activities. Chin Pharm J. 1992;27(4):198–201.
5. Chinese Pharmacopoeia Commission. Pharmacopoeia of People's Republic of China (2). Beijing: China Medical Science and Technology Press; 2010. p. 44–5.
6. Gao JJ. Evaluation of postpartum hemorrhage drug oxytocin and ergometrine curative effect. Chin Foreign Med Treat. 2009;22:81.
7. Li ZY, Ren JH. The reason and thinking of the national basic drug injection production. Chin Pharmeuticals. 2013;22(24):8–9.

Erycibe Alkaloid II

Zhi-Hong Yang, Li-Da Du, Shou-Bao Wang, and Guan-Hua Du

Abstract Erycibe alkaloid II belongs to scopolamine alkaloids, whose chemical structure is different from other known cholinomimetic drugs. Erycibe alkaloid II has the remarkable therapeutic effect on glaucoma by contracting pupil, reducing intraocular pressure, and improving coefficient of aqueous outflow. Its pharmacological potency is similar to that of pilocarpine, while its miotic effect is stronger than that of pilocarpine. Erycibe alkaloid II is suitable for long-term use, because its long-term efficacy was not decreased, without obvious adverse reactions and side effects. So it is an ideal drug for the treatment of glaucoma. Although the content of erycibe alkaloid II is only about one-hundred-thousandth of the original plant, the scientists have done plenty of research on its synthesis and structure-activity relationship. In the foreseeable future, it is expected that erycibe alkaloid II and its new generation of derivatives will have a wide application prospect in the field of glaucoma treatment.

Keywords Erycibe alkaloid II · Cholinergic drug · Contracting pupil · Glaucoma

Alias: Baogongteng A
Origin: Obtuseleaf Erycibe Stem (Ding Gong Teng) (Fig. 1)
Chemical name (Fig. 2)

2β-Hydroxyl-6β-acetyloxy-8-azabicyalo[3,2,1]octane

Molecular formula, $C_9H_{15}NO_3$; **MW**, 185.22; **CAS**, 74239-84-2

Z.-H. Yang
Institute of Medicinal Plant Development, Chinese Academy of Medical Sciences and Peking Union Medical College, Beijing, China

L.-D. Du
School of Biomedical Sciences, Chinese University of Hong Kong, Hong Kong, China

S.-B. Wang · G.-H. Du (✉)
Beijing Key Laboratory of Drug Targets Identification and Drug Screening, Centre for Pharmaceutical Screening, Institute of Materia Medica, Chinese Academy of Medical Sciences and Peking Union Medical College, Beijing, China
e-mail: dugh@imm.ac.cn

G.-H. Du, *Natural Small Molecule Drugs from Plants*,
https://doi.org/10.1007/978-981-10-8022-7_40

Properties

Appearance: white sticky substance. Its benzoate is a white fine acicular crystal. $[\alpha]_D^{20} = -5.56°(C = 0.90, CHCl_3)$. *Melting point*: 159–160 °C [1].

Dosage Forms

Eye drops: 0.05%, 1% (W/V) aqueous solution.

Erycibe alkaloid II can contract the pupil and decrease the intraocular pressure with lasting effect. It is applicable to treat the multiple types of glaucoma, including open-angle glaucoma, acute angle-closure glaucoma, and chronic angle-closure glaucoma, and poor control of intraocular pressure after glaucoma surgery.

Attention: asthma patients should take caution.

Literature

Obtuseleaf Erycibe Stem (Ding Gong Teng) is one of the Chinese herbal medicines recorded by the *Pharmacopoeia of the People's Republic of China* (2015). It is a dried rattan from the plants of Convolvulaceae *Erycibe obtusifolia Benth.* or *Erycibe schmidtii Craib.* and can be harvested all year. *Erycibe obtusifolia Benth.* distributes in the midland and coastal islands of Guangdong province, and *Erycibe schmidtii Craib.* distributes in the southeast of Yunnan province, the southwest to east of Guangxi province, Guangdong province, and Hainan province. The indications of Ding Gong Teng included rheumatoid arthritis, hemiplegia, swelling, and pain. So Ding Gong Teng is used as a folk medicine to relieve the symptoms of rheumatism in Guangdong and Guangxi provinces.

Erycibe alkaloid II belongs to scopolamine alkaloids, whose chemical structure is different from other known cholinomimetic drugs. It has remarkable effects in the treatment of glaucoma by contracting the pupils, reducing the intraocular pressure, and improving the aqueous humor coefficient. It has rare adverse reactions and mild side effects. So erycibe alkaloid II is an ideal drug for the treatment of glaucoma.

The content of erycibe alkaloid II is only about one-hundred-thousandth of the original plant. Therefore, efficient synthesis of erycibe alkaloid II is very important for the further research.

History of R&D

Ding Gong Teng, as a folk herbal medicine for sweating and defervescence, is commonly known as "Zhu Mu Jiao" in Lufeng County of Guangdong province. Because of its serious side effects, Ding Gong Teng was once only applied to those patients with strong physical constitution. However, in the 1970s, pharmacological studies indicated that the water decoction of Ding Gong Teng showed a strong pupil contraction that was rare for other Chinese medicinal herbs. In that period of time, the source of pilocarpine and physostigmine used for pupil contraction was lacking in China, so a plenty of chemical and pharmacological research has been done in order to develop a new miotic for the treatment of glaucoma.

There were few studies reported on the chemical composition of Obtuseleaf Erycibe Stem. In 1967, Sasorith, Souvan K. did some work about the chemical composition of *Erycibe elliptilimba* originated from Laos. That study reported that there were flavonoids in its leaves and tannin in its xylem, respectively, and its stems are not poisonous. Chinese scientists extracted two monomer compounds from the stem of *Erycibe obtusifolia Benth.* collected from Lufeng County, named as Baogongteng A (erycibe alkaloid II) and Baogongteng B. Erycibe alkaloid II as the active ingredient for pupil contraction is a colorless jelly with alkalescence. It is unstable, but its benzoate and 3,5-dinitrobenzoic acid salt are stable. The content of erycibe alkaloid II is only about one-hundred-thousandth of the original plant. The miotic effects of 0.025% eye drops consisting of benzoate of erycibe alkaloid II are equal to or even more than that of 2% pilocarpine nitrate [2].

In the 1980s, the pharmacological studies of glaucoma treatment and the synthesis studies of erycibe alkaloid II were carried out promptly. Pharmacological studies indicate that the tropane alkaloids, known as atropine and anisodamine, show anticholinergic activity, while erycibe alkaloid II is opposite with cholinergic activity. Cholinergic drugs, taking pilocarpine as a representative, have strong antiglaucoma activity, all of which are the agonist of muscarinic receptor (M receptor). Erycibe alkaloid II, a full agonist of M receptor, has distinct chemical structure from other known cholinergic drugs. It is competitive with M receptor blocker such as atropine and so on. Erycibe alkaloid II has no significant inhibitory effect on the activity of cholinesterase but has a strong role in the pupil contraction and the excitation of intestinal smooth muscle. Its efficacy is 80 times higher than that of pilocarpine. Significant depressurization and bradycardia were caused after intravenous injection of 0.5 mL 0.025% erycibe alkaloid II in rabbit. Natural erycibe alkaloid II is the levo isomer with secondary amine. Compared with erycibe alkaloid II, N-methylation derivative and acetylation derivative have obviously decreased cholinergic activity, even no efficacy for diacetylation derivative. It indicates that different substitution patterns (secondary amine, tertiary amine, or amide) of nitrogen atom in the chemical structure of erycibe alkaloid II cause obvious pharmacological differences [3]. The $C_2\beta$-OH group of erycibe alkaloid II may be one of the key groups to maintain its miotic activity [4].

Pharmacology

Erycibe alkaloid II can reduce the intraocular pressure in normal rabbit and the experimental high intraocular pressure induced by water overload. Its effect is stronger than that of pilocarpine, which is not suitable for long-term use. Local infusion of erycibe alkaloid II has no effect on systemic blood pressure in rabbit, and intravenous administration can lower blood pressure without affecting intraocular pressure.

Erycibe alkaloid II is a cholinergic drug while has no inhibitory effect on cholinesterase. It shows muscarinic effect via direct interaction with M-cholinergic receptor [5]. Through the M_3 receptor on cell membrane, erycibe alkaloid II can trigger to increase the Ca^{2+} concentration in the human eye ciliary muscle cells. Then it takes on the pupil contraction effect and the intraocular pressure reduction effect mediated by the M_3 receptor subtype [6]. The mechanism of erycibe alkaloid II is coupled with the cyclic nucleotide system [7, 8].

Moreover, the animal experiments indicate that erycibe alkaloid II may improve cardiac function via slowing heart rate, enhancing cardiac contractility, reducing oxygen consumption, strengthening full oxidation of acidic metabolites and sodium pump activity, etc. [9].

Clinical Application

The miotic effect, intraocular pressure reduction, and improvement of the coefficient of aqueous outflow are similar between erycibe alkaloid II eye drop and pilocarpine. Furthermore, it has strong myotic effect than pilocarpine without obvious side effects. So it is suitable for long-term application.

Erycibe alkaloid II has been shown a good dose-dependent response in both the miosis and the intraocular pressure reduction. The intraocular pressure reduction effects of 0.01% and 0.05% erycibe alkaloid II are similar to that of 1% pilocarpine, and 0.25% erycibe alkaloid II has stronger activity than 1% pilocarpine. The miotic effects of 0.01%, 0.05%, and 0.25% erycibe alkaloid II are similar to that of 1% pilocarpine. There is no significant difference in the intraocular pressure reduction response between 0.05% erycibe alkaloid α and 1% pilocarpine after chronic administration (1 month). Both erycibe alkaloid II and pilocarpine are able to improve significantly the coefficient of aqueous outflow without the obvious side effects [7].

Overdose of oral administration or injection of erycibe alkaloid II will cause poisoning, mainly shown as sweating, salivation, asthma, abdominal pain, diarrhea, limb numbness, miosis, blood pressure reduction, bradycardia, etc. It should be paid attention. Local application of erycibe alkaloid II may cause different degrees of blurred vision, conjunctival congestion, and foreign body sensation [10].

Discussion

Erycibe alkaloid II has the remarkable therapeutic effect on glaucoma by contracting pupil, reducing intraocular pressure, and improving coefficient of aqueous outflow. Its pharmacological potency is similar to that of pilocarpine, while its miotic effect is stronger than that of pilocarpine. Erycibe alkaloid II is suitable for long-term use, because its long-term efficacy was not decreased, without obvious adverse reactions and side effects. So it is an ideal drug for the treatment of glaucoma.

Although the content of erycibe alkaloid II is only about one-hundred-thousandth of the original plant, the scientists have done plenty of research on its synthesis and structure-activity relationship. In the foreseeable future, it is expected that erycibe alkaloid II and its new generation of derivatives will have a wide application prospect in the field of glaucoma treatment.

Fig. 1 Obtuseleaf Erycibe Stem (Ding Gong Teng)

Fig. 2 Chemical structure of erycibe alkaloid II

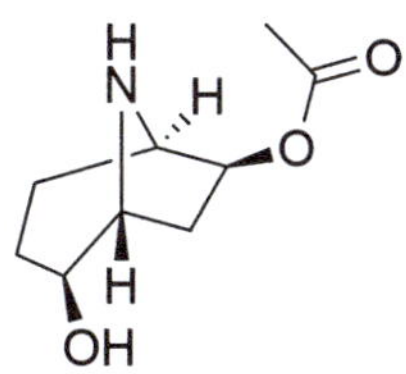

References

1. Fang YW, Zhao JJ, Bian ZL. Structure determination of memantine Dinggongteng EOB. Chemistry. 1981;4:17–8.
2. Yao TR, Chen ZN. Chemical investigation of Chinese medicinal herb, Baogongteng I: the isolation and preliminary study on a new myotic constituent, Bao Gong Teng A. Acta Pharm Sin. 1979;14(12):731–5.
3. Yao TR, Chen ZN, Yi DN, et al. Chemical investigation of Chinese medicinal herb, Baogongteng. Acta Pharm Sin. 1981;16(8):582–8.
4. He HM, Shen JX, Ma XY, et al. The synthesis and pharmacological activities of analogs of erycibe alkaloid II. Acta Pharm Sin. 1989;24(5):335–40.
5. Wang XT, Huang MY, Liang SW. The miotic pharmacological research of Ding gongteng. J New Med. 1978;6:279–80.
6. Huang WY, Peng DY, Zeng SJ, et al. The effect of erycibele erycibe alkaloid on intracellular Ca^{2+} mobilization of cultured human ciliary muscle cells. Eyescience. 1999;15(4):212–4.
7. Zhou WB, Peng DY, Chen XQ, et al. Studies on Ding gong teng in lowering of intraocular pressure and miosis. Eyescience. 1986;2(3):150–3.
8. Zeng SJ, Zhang YB, Peng DW, et al. Studies of the mechanism of lowering intraocular pressure on Erycibele alkaloid. Chin J Ophthalmol. 1999;35(3):171–3.
9. Cheng BH, Tang HM, Li L, et al. Dynamic observation on the effect of Baogongteng-A on K^+, H^+ and pO_2 and inotropism and chronotropisms of rat heart. J Shanghai Second Med Univ. 1986;2:130–2. 170, 189
10. Huang HH, Yen DH, Wu ML, et al. Acute Erycibe henryi Prain ("Ting Kung Teng") poisoning. Clin Toxicol (Phila). 2006;44(1):71–5.

Eserine

Yi-Huang Lin, Lian-Hua Fang, and Guan-Hua Du

Abstract Eserine is an alkaloid extracted from *Physostigma venenosum*, which is a reversible acetylcholinesterase (AChE) inhibitor. Eserine is the same as acetylcholine, which can combine with cholinesterase, while AChE inhibitor will combine more tightly with cholinesterase, which leads to slow hydrolysis, inactive enzyme, cumulative ACh, and emergent biological activities. Eserine is employed clinically to treat glaucoma. Compared with pilocarpine, it is provided with shorter effective time and stronger effect; meanwhile it is more provocative and poorly tolerated, so it is most often used to locally treat acute glaucoma. For eserine is toxic, unstable, and unsafe, it is only used early in ophthalmology. With the expansion of its applications and the development of analytic techniques, researchers got over some barriers, and the studies about eserine were taken more seriously in recent years.

Keywords Eserine · Alkaloid · AChE inhibitor · Glaucoma

Origin: *Physostigma venenosum* (Fig. 1)
Chemical name (Fig. 2)

(3aR,8aS)-1,3a,8-Trimethyl-1H,2H,3H,3aH,8H,8aH-pyrrolo[2,3-b]indol-5-yl N-methylcarbamate

Molecular formula, $C_{15}H_{21}N_3O_2$; **MW**, 275.35; **CAS**, 57-47-6

Y.-H. Lin · L.-H. Fang · G.-H. Du (✉)
Beijing Key Laboratory of Drug Targets Identification and Drug Screening, Centre for Pharmaceutical Screening, Institute of Materia Medica, Chinese Academy of Medical Sciences and Peking Union Medical College, Beijing, China
e-mail: dugh@imm.ac.cn

G.-H. Du, *Natural Small Molecule Drugs from Plants*,
https://doi.org/10.1007/978-981-10-8022-7_41

Properties

Appearance: flaky crystal. *Solubility*: slightly soluble in water; soluble in ethanol, benzene, and fatty oil. *Melting point*: 102–104 °C. *Specific optical rotation*: −120° in benzene and −76° in chloroform, respectively.

Dosage Forms and Indications

Eserine is recorded in the Pharmacopoeia of the People's Republic of China (1963). In the present, as one of cholinesterase inhibitor drugs, injection, eye drops, and oculentum of eserine salicylate were used to treat glaucoma or analepsis.

Literature

Eserine is an alkaloid derived from *dubiandou* 毒扁豆 (*Physostigma venenosum*). Early, *Physostigma venenosum* was called as éséré (the West African name), which means "Trial bean." In the middle of the nineteenth century, a European visited Old Calabar, an eastern province of Nigeria, and was aware of a native belief: Individuals were fed up with "Trial bean" to determine whether they were innocent or guilty of some serious misdemeanors. Later, missionaries brought "Trial bean" to Great Britain, and until 1864, chemists extracted eserine from *Physostigma venenosum* or Calabar bean for the first time [1–3].

History of R&D

Eserine was first discovered as a reversible AChE inhibitor, and it is also a tertiary amine and easily crosses the blood-brain barrier. In 1846, Robert Christison observed that the extract from Calabar bean caused cardiac arrest and death; he personally ate a certain amount of the extract and felt extremely feeble but luckily survived. In 1855, Christison reported that some kind of substances in the Calabar bean possessed strong biological activity.

In 1864, chemists afforded crystal pure extract which was named as eserine. After that, Thomas Richard Fraser and Douglas Argyll Robertson cooperated to employ eserine in experimental ophthalmology, and the results showed that the antagonistic effect of eserine on mydriasis is induced by atropine. In 1875, Ludwig Laqueur declared that eserine could also be employed to depress intraocular pressure and first used as a treatment for glaucoma. In 1925, Edgar Stedman and George Barger determined the structure of eserine, which belongs to a natural product whose structure is characterized with hexahydropyrrolo[2, 3]indole. In 1935, Percy Lavon Julian completed the chemical synthesis of its racemate for the first time [1, 2, 4].

Pharmacology

Eserine was first discovered as one of AChE inhibitors. AChE inhibitor is the same as ACh, which can combine with cholinesterase, while AChE inhibitor will combine more tightly with cholinesterase, which leads to slow hydrolysis, inactive enzyme, cumulative ACh, and emergent biological activities [4]. Although eserine does not directly activate M and N receptor, it can cross the central nervous system and strongly militate the central and peripheral nervous systems [5].

When locally using eserine in the eyes, the effect is similar to pilocarpine but more powerful and durable. It can activate AChR of iridis sphincter, representing that the pupil is narrowed and the intraocular pressure is depressed, which is more obvious when being used to treat glaucoma patients. When being absorbed, the effect of eserine is similar to neostigmine, which is called as M- and N-like effects, representing that smooth muscle is activated strongly. After crossing the central nervous system, eserine can inhibit the activities induced by AChE, and it is presented as "activate previously, inhibit later." It is noted that the effect of eserine is dependent on the status of the central nervous system [5].

Clinical Application

Though eserine as AChE inhibitor was found to have a long history, now it is still employed clinically to treat glaucoma locally. Compared with pilocarpine, it is provided with shorter effective time and stronger effect; meanwhile it is more provocative and poorly tolerated, so it is most often used to locally treat acute glaucoma. All symptoms, including asthenic bulbar paralysis, AD, atropinism, and so on, induced by AChE or other cholinergic drugs can be treated with eserine; however, in most cases, it is not a good choice of a drug, because of its low specific detoxication and considerable risk [4, 5].

Discussion

For eserine is toxic, unstable, and unsafe, it is only used early in ophthalmology. With the expansion of its applications and the development of analytic techniques, researchers got over some barriers, for example, low drug dose, low concentration in blood and brain, and inconvenient determination [6, 7]. In recent years, the studies about eserine were taken more seriously; for example, eserine was widely employed in the study of physiological and pathological mechanisms as a cholinergic tool [8, 9].

Fig. 1 *Physostigma venenosum*

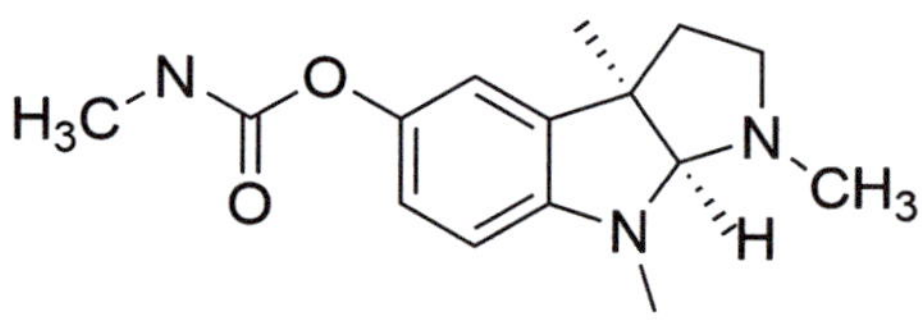

Fig. 2 The structure of eserine

References

1. QE JI. Progress on mechanism and supersession of physostigmine. Foreign Med Sci Sect Pharm. 1993;20(4):201–3.
2. Zhou YY. Review of the synthesis of physostigmine. J Yunnan National Univ. 2015;24(3):165–75.
3. Proudfoot A. The early toxicology of physostigmine. Toxicol Rev. 2006;25(2):99–138.
4. Run GY. The development history of natural medicine transform to chemical medicine. Heilongjiang: Heilongjiang University of Chinese Medicine; 2013.
5. Yang SJ. Pharmacology. Beijing: People's Medical Publishing House; 2012.
6. Boyer AW, Somani SM. Pharmacokinetics of biliary excretion of physostigmine in rat. Arch Int Pharmacodyn Ther. 1986;278(2):180–92.
7. Walter K, Muller M, Barkworth MF, et al. Pharmacokinetics of physostigmine in man following a single application of a transdermal system. Br J Clin Pharmacol. 1995;39(1):59–63.
8. Meuret P, Backman SB, Bonhomme V, et al. Physostigmine reverses propofol-induced unconsciousness and attenuation of the auditory steady state response and bispectral index in human volunteers. Anesthesiology. 2000;93(3):708–17.
9. Kirrane RM, Mitropoulou V, Nunn M, et al. Physostigmine and cognition in schizotypal personality disorder. Schizophr Res. 2001;48(1):1–5.

Galantamine

Tian-Tian Lei, Jin-Hua Wang, and Guan-Hua Du

Abstract Galantamine is an Amaryllidaceae alkaloid which is first obtained from the plant of snowdrop (*Galanthus nivalis*). Although the plants of *Galanthus* are native species in many areas in Europe, its earliest pharmaceutical applications are seldom known. To be used as an antidote may be the plants of *Galanthus*' oldest medicinal records. In the early 1950s, Soviet Union's scientists started modern medicine research of galantamine. Initially, galantamine hydrobromide was used in anesthesia to antagonize the effects of non-depolarizing muscle relaxants. Galantamine is a reversible acetylcholinesterase (AChE) inhibitor, which is highly selective and competitive. It blocks the degradation of acetylcholine (Ach) by AChE and thus achieves efficacy. Galantamine hydrobromide launched into market firstly in 1995 and has been marketed in more than 20 countries for the indication of mild to moderate Alzheimer's disease (AD). Although galantamine hydrobromide has good tolerance and safety, it only temporarily alleviates the symptoms of AD patients.

Keywords Galantamine · Antidote · AChE inhibitor · AD

Alias: Galanthamine
Origin: Snowdrop (Fig. 1)
Chemical name (Fig. 2)

(4aS,6R,8aS)-4a,5,9,10,11,12-Hexahydro-3-methoxy-11-methyl-6H-benzofurol[3a,3,2,-ef][2]benzazepin-6-ol

Molecular formula, $C_{17}H_{21}NO_3$; **MW**, 287.36; **CAS**, 357-70-0

T.-T. Lei · J.-H. Wang · G.-H. Du (✉)
Beijing Key Laboratory of Drug Targets Identification and Drug Screening, Centre for Pharmaceutical Screening, Institute of Materia Medica, Chinese Academy of Medical Sciences and Peking Union Medical College, Beijing, China
e-mail: dugh@imm.ac.cn

G.-H. Du, *Natural Small Molecule Drugs from Plants*,
https://doi.org/10.1007/978-981-10-8022-7_42

Properties

Hydrobromide of galantamine can be used as drug although galantamine can't be used as drugs.

Appearance: an almost white powder. *Solubility*: soluble in water; slightly soluble in alcohol; and insoluble in acetone, trichloromethane, and diethyl ether [1]. *Melting point*: 269–270 °C. *Specific optical rotation*: −90 to −100°.

Dosage Forms

Galantamine hydrobromide is recorded in the *Pharmacopoeia of the People's Republic of China* (2015), the *British Pharmacopoeia 2017*, the *United States Pharmacopeia* (40), and the *European Pharmacopoeia* (9.0th ed.).

Indications

Its formulation includes capsule, tablet, and oral solution. It is used in the symptomatic treatment of mild to moderately severe dementia in Alzheimer's disease.

Literature

Galantamine is an Amaryllidaceae alkaloid which is first obtained from the plant of snowdrop (*Galanthus nivalis*), and nowadays, it's extracted from the plants of *Narcissus* and *Galanthus species* or obtained from chemosynthesis [2]. In many areas in Europe such as Bulgaria, eastern Turkey, and the Caucasus, the plants of *Galanthus* are native species. But its earliest pharmaceutical applications are seldom known. Plaitakis and Duvoisin hypothesized that "moly" in ancient Homer's epic might be snowdrops. In Homer's epic Odyssey, "moly" was used as an antidote by Odysseus against Circe's poisonous drugs. To be used as an antidote may be the plants of *Galanthus*' oldest medicinal records. But there is not much evidence for that [3]. There is little evidence of the traditional application of the plants of *Galanthus*, and it is certain that until the Second World War, the plants of *Galanthus* and other genera of Amaryllidaceae were not frequently used in European drugs [4].

Italian scholar Iannello studied *Pancratium illyricum* L., an Amaryllidaceae species endemic to Sardinia, and isolated a new galantamine-type alkaloid in its leaves. This new galantamine-type compound exhibited a pronounced in vitro AChE inhibitory activity (IC50 = 3.5 ± 1.1 μM) in comparison with the reference standard galantamine hydrobromide (IC50 = 1.5 ± 0.2 μM) [5].

History of R&D

In the early 1950s, Soviet Union's scientists started modern medicine research of galantamine. In 1951, the Soviet Union's pharmacologist Mashkovsky and Kruglikova-Lvova firstly proved its AChE-inhibiting properties, which was published in the paper. In 1957, the Bulgarian pharmacologist Paskov et al. found that the plants of *Galanthus* were the richest sources of galantamine, which opened a way for its commercial development by the company Sopharma in Bulgaria. Galantamine hydrobromide was launched into market under the trade name *Nivalin*. Initially, Nivalin was used in anesthesia to antagonize the effects of non-depolarizing muscle relaxants, and since then, it was rapidly introduced in other areas of medicine.

In 1976, the former Soviet Union's scientist found that galantamine combined with electrical stimulation could significantly antagonize and improve scopolamine-induced experimental memory loss in rats. In 1988, Bulgarian scientists found that oral administration of galantamine hydrobromide, combined with B vitamins, could strengthen galantamine's action on the central nervous system's excitement, promote the formation of conditioned reflex, and improve memory. In 1990, American researchers reported that galantamine could improve memory impairment in mice and further hypothesized that galantamine might be effective in the treatment of central cholinergic disorders in human Alzheimer's disease.

Galantamine hydrobromide was launched into market for the indication of mild to moderate Alzheimer's disease firstly in 1995, developed by Sanochemia Pharmaceuticals. In the United States, galantamine hydrobromide was developed by Janssen Pharmaceuticals, a subsidiary of Johnson & Johnson, and was listed under the trade name Razadyne in 2001. So far, galantamine hydrobromide has been marketed in more than 20 countries and regions, such as the United States, Europe, and Japan.

Pharmacology

Acetylcholine (Ach) is associated with brain learning and memory, which is mainly hydrolyzed and metabolized by AChE. When AD occurs, neurons in the basal forebrain region of the brain are lost, resulting in the reduction of synthesis, storage and release of ACh, which leads to a variety of clinical manifestations especially memory and recognition dysfunction. Galantamine is a reversible AChE inhibitor, which is highly selective and competitive. It competitively binds to AChE in a synaptic gap with Ach, blocks the degradation of Ach by AChE, and thus achieves efficacy. Besides, it has the activity as an allosteric modulator of nicotinic acetylcholine receptor (nAChR). nAChR is mainly located in the muscarinic choline receptor's presynaptic structure. It can promote the release of Ach, which is critical to maintain the survival and function of cholinergic neurons. The study found that in cerebral cortex of the patients of AD, nAChR was significantly reduced. Galantamine binds to AChR's allosteric site to increase the transmission of acetylcholine signaling [6].

Galantamine has a 53-fold greater inhibitory activity in vitro for AChE than for butyrylcholinesterase (BChE). The low affinity of galantamine for BChE may be responsible for a tolerability advantage. BChE, which is primarily found in plasma, is not thought to be directly involved in the cholinergic disruption seen in early AD, and its inhibition may contribute to peripheral side effects [7].

It was recorded in *Martindale: The Complete Drug Reference* that an initial oral dose of 4 mg is given twice daily with food for 4 weeks and then increased to 8 mg twice daily. This dose should be maintained for at least 4 weeks; thereafter, the dose may be further increased to 12 mg twice daily according to response and tolerance. The clinical benefit of galantamine should be reassessed, preferably within the first 3 months and thereafter on a regular basis.

Galantamine is metabolized by hepatic cytochrome P450 enzymes CYP3A4 and CYP2D6, glucuronidated, and excreted unchanged in the urine. After intravenous injection or oral administration, about 20% of the dose was excreted as unchanged galantamine in the urine, and the other was metabolized by the hepar.

Clinical Application

Before the 1990s, galantamine was mainly used to treat poliomyelitis sequelae, muscle atrophy, postoperative intestinal muscle paralysis, urinary retention, and myasthenia gravis. In the 1990s, it was found that galantamine had improved memory impairment in mice, suggesting that it might be effective for central cholinergic disorders in Alzheimer's disease.

In a 3–6 months' well-designed clinical trial, recipients of galantamine achieved significant improvements in cognitive symptoms compared to placebo recipients. Galantamine also improved activities of daily living in these patients and significantly reduced the requirement for caregiver assistance with activities of daily living [8]. Clinical development of new indications of galanthamine is currently underway, such as smoking cessation and improving cognitive impairment in schizophrenia and mania.

Discussion

As a natural source of galantamine, snowdrops' information on the initial use as drugs is difficult to obtain, which made it mysterious for us. But due to lack of information, it's much harder to explore the history of galantamine as drug. The drug development's history of galantamine shows the importance of accessible information about traditional uses of nature products and sharing of information.

Galantamine hydrobromide has good tolerance and safety, which is suitable for elderly AD patients to use, but it only temporarily alleviates the symptoms of AD patients. It's expected scientists have more in-depth study of the pathogenesis of AD to develop more effective drugs. Will the next candidate also come from natural products? We are expecting a lot.

Fig. 1 Snowdrop

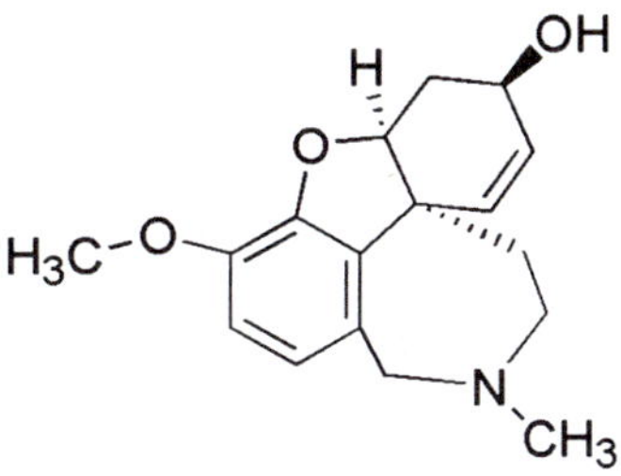

Fig. 2 Chemical structure of galantamine

References

1. National Pharmacopoeia Commission. Pharmacopoeia of the People's Republic of China (2015) part 2. Beijing: Chemical Industry Press; 2015.
2. Heinrich M, Teoh HL. Galanthamine from snowdrop—the development of a modern drug against Alzheimer's disease from local Caucasian knowledge. J Ethnopharmacol. 2004;92(2–3):147–62.
3. Plaitakis A, Duvoisin RC. Homer's moly identified as Galanthus nivalis L.: physiologic antidote to stramonium poisoning. Clin Neuropharmacol. 1983;6(1):1–5.
4. Heinrich M. Snowdrops: the heralds of spring and a modern drug for Alzheimer's disease. Pharm J. 2004;273(7330):905–6.
5. Iannello C, Pigni NB. A potent acetylcholinesterase inhibitor from Pancratium illyricum L. Fitoterapia. 2014;92:163–7.

6. Cong W, Xinhong S. Research progress of galantamine in the treatment of Alzheimer's disease. Pharm Clin Res. 2008;16(1):45–9. 6
7. Lilienfeld S. Galantamine—a novel cholinergic drug with a unique dual mode of action for the treatment of patients with Alzheimer's disease. CNS Drug Rev. 2006;8(2):159–76.
8. Scott LJ, Goa KL. Galantamine: a review of its use in Alzheimer's disease. Drugs. 2000;60(5):1095–122.

Gastrodin

Xiao-Cong Pang, Ai-Lin Liu, and Guan-Hua Du

Abstract Gastrodin is extracted from *Gastrodia elata*, which is widely used for the treatment of vertigo, headache, neuralgia, neurasthenia syndrome, and brain traumatic syndrome. Studies showed that gastrodin has many pharmacological effects, including hypnosis, sedation, and anticonvulsive. At present, to reduce the frequency of administration, the fluctuations of blood drug concentration, as well as side effects, some new dosage forms have been developed in recent years, such as *Gastrodia* dispersible tablets, *Gastrodia* sustained-release tablet, sustained-release pellets and osmotic pump tablets, etc. Although gastrodin has been applied clinically, a lot of studies still need to be done to confirm the efficacy and safety of gastrodin, and the other compositions of *Gastrodia elata* are subjected to further investigation.

Keywords Gastrodin · *Gastrodia elata* · Acetagastrodin · Efficacy

Origin: *Gastrodia elata* (Fig. 1)
Chemical name (Fig. 2)

(2R,3S,4S,5R,6S)-2-(hydroxymethyl)-6-[4-(hydroxymethyl)phenoxy]oxane-3,4,5-triol

Molecular formula, $C_{13}H_{18}O_7$; **MW**, 286.28; **CAS**, 62499-27-8

Derivative name (Fig. 3): (2R,3S,4S,5R,6S)-2-(hydroxymethyl)-6-[4-(hydroxymethyl)phenoxy]oxane-3,4,5-triol

X.-C. Pang · A.-L. Liu · G.-H. Du (✉)
Beijing Key Laboratory of Drug Targets Identification and Drug Screening, Centre for Pharmaceutical Screening, Institute of Materia Medica, Chinese Academy of Medical Sciences and Peking Union Medical College, Beijing, China
e-mail: dugh@imm.ac.cn

G.-H. Du, *Natural Small Molecule Drugs from Plants*,
https://doi.org/10.1007/978-981-10-8022-7_43

Properties

Appearance: white prismatic crystal, odorless, and bitter; little hygroscopicity. *Solubility*: soluble in water, methanol, ethanol, acetone, and acetonyl acetate, insoluble in chloroform, and poorly soluble in ethyl ether. *Melting point*: −153 to 156 °C. *Specific optical rotation*: −68 to −72° [1].

Dosage Forms and Indications

Gastrodin is included in the *International Pharmacopoeia* (5th ed.).

The dosage forms include injections, tablets, and capsules. They are widely used for the treatment of vertigo (Meniere's disease, medicinal dizziness, traumatic vertigo, sudden deafness, vestibular neuronitis, vertebrobasilar insufficiency, etc.), headache (vascular headache, migraine, etc.), neuralgia (trigeminal neuralgia, sciatica, etc.), neurasthenia syndrome, and brain traumatic syndrome [1].

Literature

Gastrodin is extracted from the dry root of the orchid plant, *Gastrodia elata*. The source of raw materials is Yunnan Province. Traditional Chinese medicine *Gastrodia elata* (*Gastrodia elata* Bl.) is also known as Chi Jian, Ding Feng Cao, Shui Yang Yu, and so on, belonging to special symbiosis perennial herb of Orchidaceae and *Armillaria mellea*. The medicinal part is its tubers. It has sweet flavor and liver channel tropism and is neutral in nature. Traditional Chinese medicine believes that its main functions are to expel wind and relieve convulsion, calm the liver and suppress liver-yang, and expel wind evil and channel. There are 23 chemical ingredients extracted from *Gastrodia elata*, including gastrodin, gastrodia glycosides, vanilla alcohol, vanilla aldehyde, beta steroid valley alcohol, p-hydroxybenzoic acid, carrot glycosides, and so on, among which the highest content of single active ingredient is *Gastrodia* glycosides, namely, gastrodin [2].

History of R&D

It has been more than 30 years since the chemical composition of *Gastrodia elata* was analyzed. It includes phenols, organic acids, sterols, and so on. At present, the studies of the composition of *Gastrodia elata* focus on the fingerprint, but the studies of gastrodin were centered on the activity. The studies showed that gastrodin has many pharmacological effects with safety and few side effects; moreover, it is not easy to accumulate in the body. Based on traditional medicinal experience of *Gastrodia elata*, gastrodin has been developed as a medicine for the treatment of insomnia, headaches, and other symptoms related to the functions and indications of *Gastrodia elata*. At present, many pharmaceutical factories in China have produced active pharmaceutical ingredients (API) and general preparation of gastrodin, such as *Gastrodia* tablet, *Gastrodia* capsule, and *Gastrodia* injection. To reduce the frequency of administration, the fluctuations of blood drug concentration, as well as side effects, some new dosage forms have been developed in recent years, such as *Gastrodia* dispersible tablets, *Gastrodia* sustained-release tablet, sustained-release pellets and osmotic pump tablets, etc. [1]. Although gastrodin has been applied clinically, a lot of studies still need to be done to confirm the pharmacological effects of gastrodin. Recent studies found that other compositions of *Gastrodia elata* also showed rather excellent biological activity, suggesting that the study of *Gastrodia elata* and gastrodin needs further investigation.

Pharmacology

Gastrodin has the pharmacological effects of hypnosis, sedation, and anticonvulsive. The pentobarbital sodium-induced sleep time test and sodium barbital-induced sleep onset latency test demonstrated that gastrodin can prolong the sleeping time and shorten the sleep onset latency in mouse, suggesting that gastrodin has the sedative and hypnotic effects [3].

Gastrodin also has antiepileptic effect. It suggested that gastrodin may inhibit the hippocampal excitatory neurotransmitter receptors of amino acid glutamate, activate hippocampal inhibitory neurotransmitter receptor of GABA, reduce the excitability of the cerebral cortex, and thus attenuate the development of epilepsy to play a role of antiepileptic effect [4].

Additionally, research has shown that gastrodin has certain pharmacological effect on scavenging free radicals and inhibiting the central nerve cell apoptosis, etc., which may be related to antiaging and neural protection [5]. According to the available references, the researches of pharmacology and clinical efficacy of gastrodin are very preliminary; more research evidences still need to confirm whether gastrodin is the active constituent of rhizoma gastrodiae.

Clinical Application

Gastrodin has sedative and hypnotic, analgesia, and immune enhancement function and is used for the treatment of tension headache to relieve pain and intense sentiment and thus further relieve the related symptoms [6]. Clinical usage of gastrodin acupoint injection to treat insomnia also achieves certain effect [7]. Moreover, gastrodin agents have been used for the treatment of cardiovascular diseases and microcirculation system, such as hypertension and coronary heart disease without obvious side effects [8, 9].

It is also reported that gastrodin can treat epilepsy, vascular dementia, deafness, tinnitus, herpes zoster, osteoarthritis, cognitive impairment, and so on but still needs further investigation.

Discussion

Rhizoma gastrodiae is a traditional Chinese medicine used for a long history without obvious adverse effects. Gastrodin has certain sedative, analgesic, antihypertensive effects. It can protect the neurons and vascular endothelial cells, inhibit ischemia-reperfusion injury, also dilate blood vessels, improve myocardial microcirculation, increase myocardial nutritional blood flow, and so on. But due to the lack of deep and normative research, the effectiveness of the study remains to be confirmed. Currently, the main dosage forms of gastrodin are injection, tablets, and capsules as well. In order to improve the targeting of gastrodin and improve the patient's compliance, the study on new dosage forms of gastrodin needs to be done in the future.

Fig. 1 *Gastrodia elata* Bl.

Fig. 2 Chemical structure of gastrodin

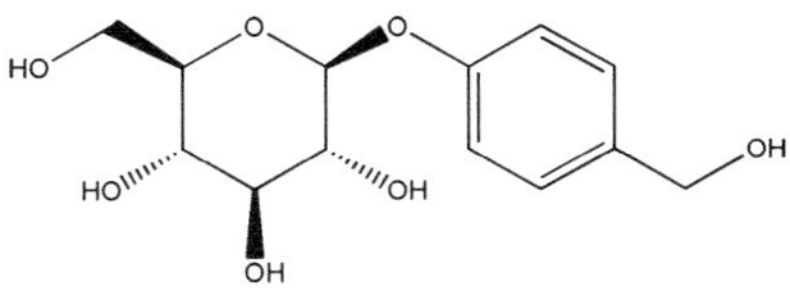

Fig. 3 Chemical structure of acetagastrodin

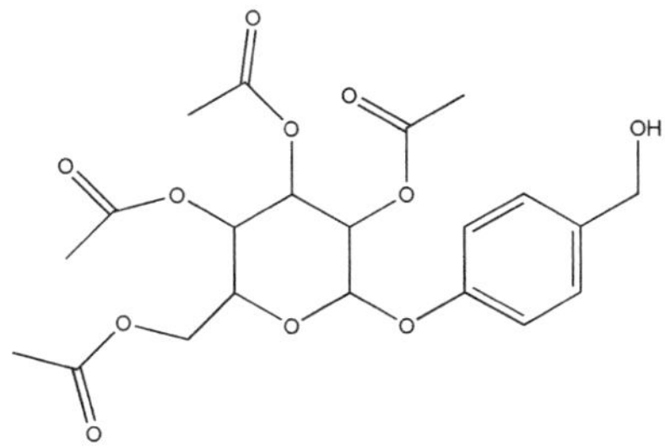

References

1. Chen C, Li XS, Zhou JJ, et al. Advances in extraction and purification of Gastrodin. J Shaanxi Univ Technol (Nat Sci Ed). 2013;29(3):4–5.
2. Yang SL, Lan J, Xu JT. Research progress in Rhizoma gastrodiae. Chin Tradit Herb Drug. 2000;31(1):66–9.
3. Yu P, Yu DZ. Experimental research on the improvement of sleep function in certain Rhizoma gastrodiae products. Mod Prev Med. 2000;34(16):3056–7.
4. Zou N, Lu JT, et al. The sedative and hypnotic effect of gastrodin in mice. Lishizhen Med Mater Med Res. 2011;22(9):807–9.
5. Wang Q, Chen GS, Zeng S. Distribution and metabolism of gastrodin in rat brain. J Pharm Biomed Anal. 2008;46(2):399–4.
6. Qiu CL, Yi WJ. Gastrodin treatment of hypertonic headache in liver Yang. Mod J Integr Tradit Chin West Med. 2011;20(24):3045–6.
7. Shi L, Chen L, Zhang JL. 40 cases of gastrodin acupoint injection in the treatment of insomnia. J External Ther Tradit Chin Med. 2010;19(1):44.
8. Zhang Q, Yang YM, Yu GY. Effects of gastrodin injection on blood pressure and vasoactive substances in treatment of old patients with refractory hypertension: a randomized controlled trial. J China Int Med. 2008;6(7):695–9.
9. Miao K, Wang M, Xie Z. An observation of the clinical effect of gastrodin in the treatment of coronary heart angina. Tianjin Pharm. 2000;21(6):26–7.

Ginkgolide B

Wen Zhang, Qi-Meng Zhou, and Guan-Hua Du

Abstract Ginkgolide B belongs to families of ginkgo diterpene lipids, which are terpenes extracted from *Ginkgo biloba*. Ginkgolide B is both the platelet-activating factor (PAF) receptor antagonist and a nonspecific central nervous glycine receptor antagonist which can upregulate BDNF and inhibits the release of intracellular calcium ion and NO. Due to the complexity of its structure, its solubility and bioavailability are the most important factors affecting druggability. With the further study, this compound will be used in clinical better.

Keywords Ginkgolide B · *Ginkgo biloba* · PAFR antagonist · Glycine receptor antagonist

Alias: BN52021, gingko lactone, ginkgolide b, ginkgolides
Origin: *Ginkgo biloba* (Fig. 1)
Chemical name (Fig. 2)

5H-Dicyclopenta[b,c]furan-3,5a(6H)-diacetic acid, 6-tert-butyl-3a-carboxy-hexahydro-a5a,1,2,3,5,8-hexahydroxy-a3-methyl-, tri-g-lactone (8CI)

Molecular formula, $C_{20}H_{24}O_{10}$; **MW**, 424.40; **CAS**, 15291-77-7

W. Zhang · Q.-M. Zhou · G.-H. Du (✉)
Beijing Key Laboratory of Drug Targets Identification and Drug Screening, Centre for Pharmaceutical Screening, Institute of Materia Medica, Chinese Academy of Medical Sciences and Peking Union Medical College, Beijing, China
e-mail: dugh@imm.ac.cn

G.-H. Du, *Natural Small Molecule Drugs from Plants*,
https://doi.org/10.1007/978-981-10-8022-7_44

Properties

Appearance: ginkgolide B is a white powder. *Solubility*: weighed 0.98 g and dissolves in 1 L water (calculation). *Melting point*: the compound melts at 280 °C (calculation).

Derivatives (Fig. 3)

From 1932, people have found there are ten kinds of ginkgo diterpene lipids (A, B, C, J, K, L, M, N, P, Q) (Fig. 3). The bioavailability of ginkgolides, especially A and B, is better than that of ginkgetin extracted from ginkgo. So ginkgolide is the main component among various kinds of *Ginkgo biloba* preparations in the market.

The hydroxyl of ginkgolide B is the main target to be modified. Park et al. [4] found that introducing large or aromatic group with strong electron withdrawing group at C-10 accounts for enhanced antagonism against PAF. There is another finding that tert-butyl group in C/D/F ring is essential for pharmacological action of ginkgolides [5]. The study found that C1-OH is extremely important for the antagonism against glycine receptor of ginkgolides, and a variety of modifications will lead to decreased activity, which are lower than the activity of natural hydroxyl [6]. Its multiple rings in the structure lead to dissolve difficultly. The solubility can be improved by the modification of C1, such as C-10 hydroxyl is added to salt, amino acid, and glycosylation.

Dosage Forms

Ginkgolide infection and diterpene ginkgolides meglumine injection

Indications

The injection consists of multiple components, and there are three kinds of substance mainly in diterpene ginkgolides meglumine injection. It promotes blood circulation by removing blood stasis and activates the channels and collaterals for cerebral ischemia treatment.

Literature

Ginkgo biloba is a perennial tree belonging to Ginkgoaceae (*Ginkgo biloba* L., Yin Xing). Its seeds, also known as ginkgo, have a long history of application in traditional medicine for its medicinal value. With promoting blood circulation by removing blood stasis, *Ginkgo biloba* has a long history of its application in traditional Chinese medicine.

In recent years, it has been widely used in the treatment of cerebral ischemic diseases. After further study, it was found that the effective component is ginkgolide compound, which laid the foundation for the development of small molecule drugs.

History of R&D

The leaf of *Ginkgo biloba* is sweet, bitter, astringent, and mild. With activating blood circulation to remove blood stasis, dredging collaterals, relieving pain, and asthma, it is applied for treating blood stasis, chest pain, stroke, lung deficiency cough, and hyperlipidemia.

As early as in the 1930s, scientists began to study ginkgo and harvest a variety of ginkgolide [1]. Ginkgolide B was isolated and its structure illustrated by Wan Shan et al. in 1967[2]. Subsequently, it was synthesized by Corey et al. in 1988[3]. The preparation of effective components of Ginkgo biloba has been listed at home and abroad with development.

The *Ginkgo biloba* extract consisting of ticlopidine is widely used in overseas market for the treatment of cerebral infarction. Intravenous injection of ginkgolide B is applied for the treatment of inflammation, but the project has been in the state, and there is no progress in it for many years. In addition, whether other extract of *Ginkgo biloba* can be used for the treatment of stable angina, cognitive impairment is still in preclinical research status. Approval of ginkgo preparations is more in domestic than abroad. Among of these, there are three kinds of chemical drugs: ginkgo biloba injection, ginkgo mihuan oral liquid, and ginkgo leaf extract and dipyridamole injection. Others belong to Chinese traditional drugs, which are made into various preparations, such as injection, decoction, syrup, granule, pill, tablet, capsule, tincture, pill, and drop.

Pharmacology

Ginkgolide B has a variety of pharmacological effects, mainly in cardiovascular and cerebrovascular protection and neuroprotection. On the one hand, with the strongest activity in nature, the platelet-activating factor (PAF) receptor antagonist is recognized by the public [7]. PAF can inhibit the downstream inflammatory cascade for anti-inflammatory. It also can alleviate cerebral infraction volume and cerebral edema induced by cerebral ischemia among the animal model of cerebral ischemia reperfusion through unknown mechanism. By inhibiting NF-κB and TNF-α, the apoptosis of nerve cell can be alleviated. The compound protects the abdominal cavity and intestinal tract by inhibiting release of neutrophil chemotactic protein by macrophage. There are other studies demonstrated that its similar compound XQ-1H affects platelet aggregation by activating MMP-9 to reduce blood clot with 15.6 mg/kg administrated into gaster [8]. Also, there are studies indicating that 1.4 mol/L ginkgolide B preincubated for 1 h is almost completely inhibiting migration of monocytes induced by oxidized LDL, protecting human umbilical vein endothelial cells and reducing atherosclerosis [9]. Antioxidant is performed by inhibiting ethanol-induced oxidase. On the other hand, it acts as a nonspecific antagonist against central nervous glycine receptor to upregulate BDNF and inhibit the release of intracellular calcium ion and NO. So it has certain preventive or therapeutic effect on CNS diseases, such as Alzheimer's disease. It also plays a role in antagonism against GABA type-A receptor (IC_{50} is 7.3×10^{-5} mol/L) to protect CNS from damage [10].

Clinical Application

Oral and injection preparations are applied widely in clinical therapy. Made in oral preparation, flavonoids are the main component of raw material. There are little chemical compounds in injection, and ginkgolides are the main component whose content is different from each other. Listed ginkgolide injection is applied for ischemia stroke formed by blood stasis with promoting blood circulation by removing blood stasis and clearing and activating the channels and collaterals. Compared with Shuxuening injection in clinical trials, ginkgolides injection is consistent in improving indicators effect, such as protection in nerve injury and others. The other ginkgo preparations are consistent with pharmacological effects, which are mainly used for anticoagulation, prevention and recovery of stroke, nerve protection, and cognitive function.

Discussion

As one of the most important natural herbal medicines applied in the prevention and treatment of cardiovascular and cerebrovascular diseases worldwide, ginkgolide has become the focus of the study because of its unique mechanism, relatively clear target, and low toxicity. Due to the complexity of its structure, its solubility and bioavailability are the most important factors affecting the patent medicine, and the basis of pharmacological effect is also demonstrated by the structural modification of ginkgolides. On the other hand, the detailed mechanism of its pharmacological effect is not very clear, and it is also the focus of the next phase. Further study will make this compound used better in clinical applications.

Fig. 1 *Ginkgo biloba*

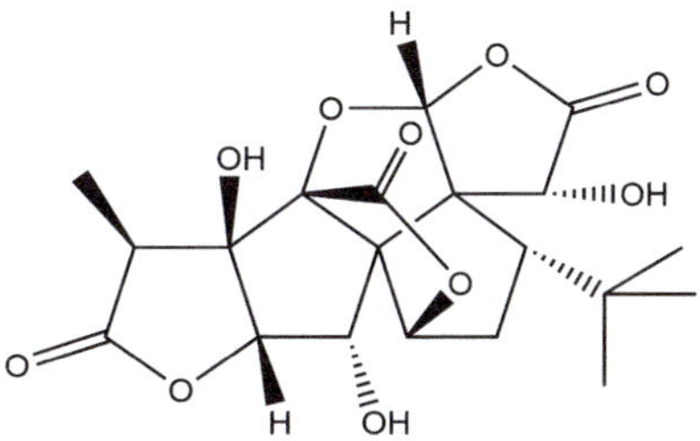

Fig. 2 Chemical structure of ginkgolide B

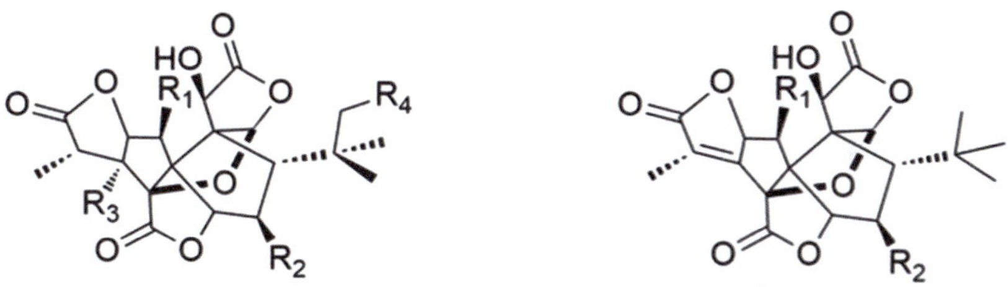

1 Ginkgolide A (GA)	R_1 = H	R_2 = H	R_3 = OH	R_4 = H
2 Ginkgolide B (GB)	R_1 = OH	R_2 = H	R_3 = OH	R_4 = H
3 Ginkgolide C (GC)	R_1 = OH	R_2 = OH	R_3 = OH	R_4 = H
4 Ginkgolide J (GJ)	R_1 = H	R_2 = OH	R_3 = OH	R_4 = H
7 Ginkgolide M (GM)	R_1 = OH	R_2 = OH	R_3 = H	R_4 = H
9 Ginkgolide P (GP)	R_1 = H	R_2 = H	R_3 = OH	R_4 = OH
10 Ginkgolide Q (GQ)	R_1 = OH	R_2 = H	R_3 = OH	R_4 = OH

5 Ginkgolide K (GK)	R_1 = OH	R_2 = H
6 Ginkgolide L (GL)	R_1 = H	R_2 = H
8 Ginkgolide N (GN)	R_1 = OH	R_2 = OH

Fig. 3 Chemical structure of ten kinds of ginkgo diterpenoids

References

1. Furukawa S. Constituents of Ginkgo biloba leaves. Sci Pap Inst Phys Chem Res. 1932;19:27–38.
2. Woods MC, Miura I, Nakadaira Y, Ginkgolides V. Some aspects of their N.M.R. spectra. Tetrahedron Lett. 1967;(4):321–6.
3. Corey E, Kang M, Desai AK, Ghosh, Houpis IN. J Am Chem Soc. 1988;110:649.
4. Park PU, Pyo S, Lee SK. Ginkgolide derivatives. 1996.
5. Corey E, Gavai AV. Simple analogs of ginkgolide B which are highly active antagonists of platelet activating factor. Tetrahedron Lett. 1989;30(50):6959–62.
6. Kondratskaya E, Krishtal O. Effects of Ginkgo biloba extract constituents on glycine-activated strychnine-sensitive receptors in hippocampal pyramidal neurons of the rat. Neurophysiology. 2002;34(2–3):155–7.
7. Braquet P. Proofs of involvement of PAF-acether in various immune disorders using BN 52021 (ginkgolide B): a powerful PAF-acether antagonist isolated from Ginkgo biloba L. Adv Prostaglandin Thromboxane Leukot Res. 1985;16:179–98.
8. Fang W, Sha L, Kodithuwakku ND. Attenuated blood-brain barrier dysfunction by XQ-1H following ischemic stroke in hyperlipidemic rats. Mol Neurobiol. 2015;52(1):162–75.
9. Liu X, Sun W, Zhao Y. Ginkgolide B inhibits JAMA, Cx43, and VE-cadherin expression and reduces monocyte transmigration in oxidized LDL-stimulated human umbilical vein endothelial cells. Oxidative Med Cell Longev. 2015;2015:907926.
10. Ivic L, Sands TT, Fishkin N. Terpene trilactones from Ginkgo biloba are antagonists of cortical glycine and GABAA receptors. J Biol Chem. 2003;278(49):49279–85.

Huperzine A

Wen-Wen Lian, Ai-Lin Liu, and Guan-Hua Du

Abstract Huperzine A is obtained from *Huperzia serrata*, which is a highly effective and selective reversible inhibitor of acetylcholinesterase. It shows ability to improve learning and memory and was approved for treatment in Alzheimer's disease (AD) in China in 1996. The low efficiency (only about six ten thousandths) of extraction of huperzine A from *Huperzia serrata* and the complexity of the full synthesis and structure modification of huperzine A limit its large-scale development and utilization. ZT-1, a derivative of huperzine A, was found to be with similar AChE inhibitory activity and lower butyrylcholinesterase inhibitory activity and toxicity compared with huperzine A. It can go through the blood-brain barrier, and its oral bioavailability and duration of action are comparable with huperzine A. ZT-1 is a promising anti-AD drug and has approved patents in China, America, Japan, and Europe.

Keywords Huperzine A · *Huperzia serrata* · Alzheimer's disease · ZT-1

Alias: Fordine, Huperzine, Li Min
Origin: *Huperzia serrata (Thunb.) Trevis* (Fig. 1)
Chemical name (Fig. 2)

(5*R*,9*R*,11*E*)-5-amino-11-ethylidene-5,8,9,10-tetrahydro-7-methyl-5,9-methanocycloocta[b]pyridin-2(1H)-one

Molecular formula, $C_{15}H_{18}N_2O$; **MW,** 242.3; **CAS,** 102518-79-6

W.-W. Lian · A.-L. Liu · G.-H. Du (✉)
Beijing Key Laboratory of Drug Targets Identification and Drug Screening, Centre for Pharmaceutical Screening, Institute of Materia Medica, Chinese Academy of Medical Sciences and Peking Union Medical College, Beijing, China
e-mail: dugh@imm.ac.cn

G.-H. Du, *Natural Small Molecule Drugs from Plants*,
https://doi.org/10.1007/978-981-10-8022-7_45

Properties

Appearance: white crystalline powder. Bitter with hygroscopicity. *Solubility*: easily soluble in chloroform, soluble in methanol and ethanol, and slightly soluble in water. *Melting point*: 211–216 °C.

Dosage Forms

It is reported in *Pharmacopoeia of the People's Republic of China (2015)*. Currently huperzine A has three kinds of formulations, including tablets, injections, and capsules.

Indications

It was mainly applied to age-related memory dysfunction and Alzheimer and has a good effect on improving memory function. It can be used to treat various types of Alzheimer's disease and also myasthenia gravis.

Literature

Huperzine A is obtained from *Huperzia serrata*, which is the perennial fern. It shows activities in antipyretic, hemostasis, and dehumidification and is used for the treatment in folk of pneumonia, lung abscess, hematemesis, hematochezia, traumatic injury, etc. [1].

History of R&D

In the 1980s, Chinese scholars isolated huperzine A from *Lycopodiales*, *Huperziaceae*, *Phlegmariurus fordii*, and *Huperzia serrata (Thunb.) Trevis* [2, 3]. At present, about 120 chemical components have been isolated and identified from the plant, including 90 lycopodium alkaloids and 32 lycopodium triterpenes. Huperzine A has the most potent inhibition on acetylcholinesterase activity, followed by huperzine B and 6β-hydroxy huperzine A. These three compounds belong to lycodine-type lycopodium alkaloids. Apart from *Huperzia serrata*, *Phlegmariurus fordii* is in the southern China, and *Lycopodiastrum casuauinoides* is in the Central China. *Lycopodium selago* close to *Huperzia serrata* genetically can be considered as the source of huperzine A. The low efficiency (only about six ten thousandths) of extraction of huperzine A from *Huperzia serrata*, coupled with slow growth of plants, results in expensive raw materials of huperzine A. The full synthesis of huperzine A is complex and costly. Therefore, it is a focus to develop biotransformation or semisynthesis with other alkaloids as lead compounds on the basis of the intrinsic relationship among different kinds of alkaloids.

After the determination of chemical structure of huperzine A in 1986, it was found to be the same alkaloid as selagine separated from *Lycopodium selago* by Valenta in the 1960s, so it was classified as lycodine-type alkaloid. Huperzine A is a potent reversible inhibitor of AChE, and its ability to improve learning and memory has been validated in animal models. It was approved for treatment in Alzheimer's disease (AD) in China in 1996 [4].

Due to the high cost of extraction of huperzine A, research on its chemical synthesis has been the focus at home and abroad since 1986. It has been found that the chiral structure of huperzine A is essential for its biological activity; the inhibitory activity on AChE of natural products (-) - huperzine A is twice as its raceme and 38–50 times as its enantiomer (+) - huperzine A which is not a natural product, so the chemical synthesis of natural product (-) - huperzine A has received extensive attention. The chemical preparation method of (-) - huperzine A can be divided into asymmetric synthesis and raceme separation and is limited to a small amount preparation in laboratory for the high cost.

Considering the difficulty of realizing unique bridge ring and amino structure of (-) - huperzine A and achieving its full synthesis and structural modification, scientists are trying to synthesize analogues of huperzine A with simple structure and AchE inhibitory activity. It was found that the activity of spearmint huperzine A was similar to that of huperzine A, with improved selectivity and poor chemical stability. It was modified structurally to obtain ZT-1 (Fig. 3) [5]. The AChE inhibitory activity of ZT-1 is comparable with that of huperzine A, but the inhibitory activity on butyrylcholinesterase and the toxicity are lower than huperzine A. It can go through the blood-brain barrier, and its oral bioavailability and duration of action are comparable with huperzine A. It is a promising anti-AD drug and has approved patents in China, America, Japan, and Europe.

Pharmacology

Huperzine A has the ability to enhance learning and memory, improve spatial memory, and can be used for age-related dementia, vascular dementia, and other neurodegenerative diseases. Compared with the current anti-AD drugs, huperzine A can go through the blood-brain barrier, with a high oral bioavailability and longer time inhibition on AChE [6].

As a highly selective AChE reversible inhibitor, huperzine A can inhibit AChE, reduce acetylcholine hydrolysis, and improve the level of acetylcholine in the synaptic gap. This inhibition is reversible, lasts for a long time, shows no drug dependence if repeated administration, and does not induce significant liver toxicity. X-ray diffraction results show that the direct binding of huperzine A to AChE active sites inhibits the binding of AChE to its substrate [7].

In addition to the potent inhibition on AChE, huperzine A only shows a weak inhibitory effect on the butyrylcholinesterase; also protects neurons by inhibiting oxidative stress, reducing somatostatin, reducing the content of glutamate, decreasing the increased intracellular calcium, and inhibiting neuronal apoptosis; further improves AD-related cognitive function and reduces the symptoms of AD patients.

Clinical Application

Since 1994, huperzine A has been approved for clinical use in improving memory and cognitive impairment in old patients with memory loss and dementia. A large number of domestic clinical studies have found that huperzine A shows therapeutic effect on learning and cognitive dysfunction of vascular dementia, mental retardation, and schizophrenia patients with mild adverse reactions.

Discussion

Alzheimer's disease is a common degenerative disease in old people with cognitive impairment and progressive damage in memory. Huperzine A is a highly effective and selective reversible inhibitor of acetylcholinesterase, which is easy to go through the blood-brain barrier, with high oral bioavailability, a longer duration, and less adverse reactions. Huperzine A shows significant effect in improving the memory and cognitive impairment. It not only inhibits the acetylcholinesterase but also plays a role in neuronal protection by various mechanisms. However, the low content of huperzine A in plants and the complexity synthesis limit its large-scale development and utilization. Research on synthesis of huperzine A deserves more efforts [8].

Fig. 1 *Huperzia serrata (Thunb.) Trevis*

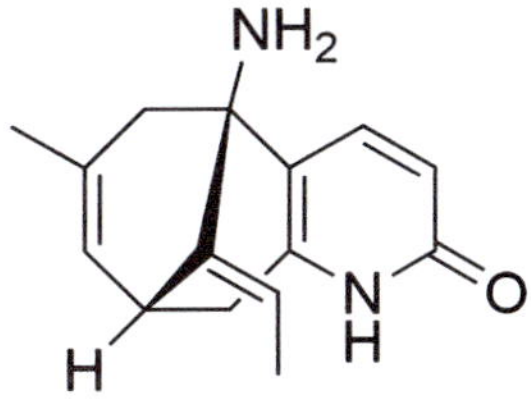

Fig. 2 Chemical structure of huperzine A

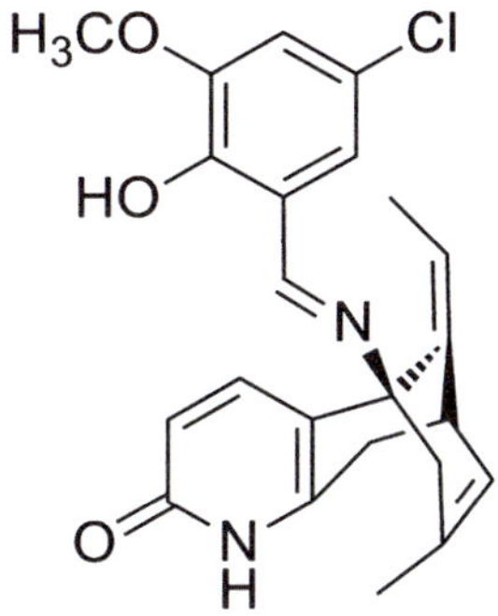

Fig. 3 The chemical structure of ZT-1

References

1. Jiangsu New Medical College. Great dictionary of Chinese medicine. Shang Hai: Shanghai Science and Technology Press; 1986.
2. Liu J, Yu C, Zhou Y, et al. Chemical studies on Huperzine A and Huperzine B. J Chem. 1986;44(10):1035–40.
3. Xu Z, Chu B, Luan X, et al. Structural determination of Fuding. Med J Chin People's Liberation Army. 1985;10(4):263–4.4.
4. Tang XC. Huperzine A shuangyipian: a promising drug for Alzheimer's disease. Zhongguo Yao Li Xue Bao. 1996;17(6):481–4.
5. Ma X, Gang DR. The lycopodium alkaloids. Nat Prod Rep. 2004;21(6):752–72.
6. Wang R, Yan H, Tang XC. Progress in studies of huperzine A, a nature cholinesterase inhibitor from Chinese herbal medicine. Acta Pharmacol Sin. 2006;27(1):1–26.
7. Raves ML, Harel M, Pang YP, et al. Structure of acetylcholinesterase complexed with the nootropic alkaloid, (−)-huperzine A. Nat Struct Biol. 1997;4(1):57–63.
8. Ma X, Tan C, Zhu D, et al. Huperzine A from Huperzia species–an ethnopharmacolgical review. J Ethnopharmacol. 2007;113(1):15–34.

l-Borneolum

Li-Song Sheng, Gui-Fen Qiang, and Guan-Hua Du

Abstract The l-borneolum is similar to the commonly used medicine, borneol, which has an polar opposite optical rotation. It was originally from the Compositae Aenean *Blumea balsamifera* (L.) DC (Ai Na Xiang). The plant resources of borneol are known as the "woody incense," and the plant resources of l-borneolum are known as "herbal fragrance." It is considered that woody incense is better than herbal fragrance, so l-borneolum's usage is not as common as borneol. Because of large demand and short supply of borneol, l-borneolum will also keep a high demand in the clinical application.

Keywords L-borneolum · Ai pian · Camphor · Herbal fragrance

Origin: *Blumea balsamifera* (L.) DC (Fig. 1)
Alias: Ai Pian, Ainaoxiang
Chemical name (Fig. 2)

(1R,2R,4S)-1,7,7-Trimethylbicyclo[2.2.1]heptan-3-ol

Molecular formula, $C_{10}H_{18}O$; **MW**, 154.25; **CAS**, 507-70-0

Properties

Appearance: white translucent flakes, bulk or granular crystals. With a clear aroma, spicy, cool, volatile, black smoke, and yellow flame when lit without residue left. *Melting point*: 201–205 °C. Soluble in ethanol, chloroform, or ether, almost

L.-S. Sheng · G.-F. Qiang · G.-H. Du (✉)
Beijing Key Laboratory of Drug Targets Identification and Drug Screening, Centre for Pharmaceutical Screening, Institute of Materia Medica, Chinese Academy of Medical Sciences and Peking Union Medical College, Beijing, China
e-mail: dugh@imm.ac.cn

G.-H. Du, *Natural Small Molecule Drugs from Plants*,
https://doi.org/10.1007/978-981-10-8022-7_46

insoluble in water. *Specific optical rotation*: −36.5° to −38.5° in 5% (g/mL) ethanol solution, stable under sealed condition at room temperature.

Pungent, bitter, slightly cold; heart, spleen, and lung meridians entered. (In Chinese traditional medicine's opinion)

Dosage Forms

l-Borneolum was included in the *Pharmacopoeia of the People's Republic of China* (2015) and *British Pharmacopoeia* (2017).

It is used in the dosage form of pills or grinded powder for external application. Modern preparation includes soft capsule.

Indications

Sore throat, aphthous, red eyes, purulent ear discharge, convulsions, febrile delirium, sudden faint due to qi depression, stroke, and coma [1]

Literature

l-Borneolum was originally from the Compositae Aenean *Blumea balsamifera* (L.) DC (Ai Na Xiang). Its crystals were made after being extracted from the fresh leaves of Ai Na Xiang. It is an optical isomer with natural borneol. It's usually used for making spice the same as natural borneol, synthesized borneol, and so on [1].

Ai Na Xiang is a traditional Chinese medicine alias as Da Feng Ai, Niu Er Ai, etc. It has been recorded firstly in *Guang Zhi*, "Ai Na Xiang came from the west country and looks like the artemisia." It has been recorded in many Chinese materia medica books, such as *Hai Yao Ben Cao*, *Ling Nan Cai Yaolu*, etc. In *Zhong hua Ben cao* (Chinese Materia Medica), the literature described it as "Ai Na Xiang (l-borneolum) has the fragrance because it is like the artemisia. It can also be prepared to the borneol, so it is called "Bing Pian Ai." It is believed that this is the earliest records that Ai Na Xiang was made into borneol [2]. In China, the authentic product area of l-borneolum is in Luodian, Guizhou Province [3]. There are no other plant resources reported for producing l-borneolum in China.

History of R&D

l-Borneolum is one of the few drugs used from ancient times which is a single organic small molecule. With the progress of technology in modern times, it was found that there are optical differences by measuring the polarimetry, and accordingly the borneol is divided into three drugs: natural borneol (d-borneol), ai pian (l-borneol), and artificial borneol (synthetic borneol or dl-borneol).

Because of the special flavor, the modification process of l-borneolum was to focus on the medicinal pathways, as well as spices and other areas, such as the formation of camphor by dehydrogenation (Fig. 2) or the formation of isoborneol by altering the three-dimensional structure (Fig. 2). The chemical structure of l-borneol is relatively simple, containing ten carbon atoms and spirocyclic alcohol, so most of the modern modification process (Fig. 2) is to modify by esterification, such as producing the bornyl acetate by reacting with acetic acid. Due to its special three-dimensional structure and chiral position, the general modification process is to modify the other structures as a part of the structure. Accordingly, it produces the compounds with special three-dimensional structure and novel pharmacological effects (Fig. 3).

The researches of l-borneolum are inextricably linked to borneol, but compared with the borneol, l-borneolum still has significant disadvantages. For example, borneol is used in the wider region than 1-borneolum which is mainly used in China. The pharmacological studies of borneol in modern times are much more than l-borneolum as well. The plant resources of borneol are the *Cinnamomum camphora*, known as the "woody incense," and the plant resources of l-borneolum are Ai Na Xiang, known as "herbal fragrance." From the Chinese traditional habits, it is thought that woody incense is better than herbal fragrance, so l-borneolum's usage is limited. The purification process of borneol was done earlier than l-borneolum, and the process is excellent, with less camphor, isosorbide, and other toxic components. l-Borneolum has less purification process due to the small distribution range. Until the 1970s, l-borneolum was produced using the traditional manual process, leading to the low rate of borneol and high toxic components, finally affecting its clinical usage. In the 1980s and 1990s of the twentieth century, with the reduction of plant resources of natural borneol, the increasing demands of borneol and the poor synthesis process of borneol, the plant cultivation and purification studies of l-borneolum were developed for a while. However, with the continuous discovery of plant resources of natural borneol and the improvement of synthesis process, the research of l-borneolum was back to normal.

Pharmacology

The main effects of l-borneolum are to induce resuscitation (with aromatic stimulation), clear stagnated fire (fever feeling), remove nebula to improve eyesight, and relieve swelling and pain. In traditional Chinese medicine, borneol is often used as an envoy drug and combined with other drugs.

The modern pharmacological researches showed that l-borneolum can cross the blood-brain barrier (BBB) through increasing cell membrane fluidity, Na^+ -K^+-ATPase activity, decreasing membrane potential, and regulating intracellular calcium concentration, which is involved in the effect of resuscitation [4].

The key brain-protective mechanism of l-borneolum and synthetic borneol is closely related to the regulation of P-glycoprotein pathway, lipid peroxidation, and nitric oxide pathway. In addition, it can regulate the calcium pathway, which is the main biological mechanism of "Xin-floating-heart" medicinal property of borneol and resuscitation. Based on the statistical results of strength integral law, it demonstrated that the brain-protective effect of l-borneolum is stronger than synthetic borneol, suggesting that we should prefer l-borneolum for the treatment of cerebrovascular disease [5].

Clinical Application

l-Borneolum is used as a kind of borneol in clinical usage all along, but its clinical usage is fewer than borneol considering its clinical efficacy is weaker than borneol. The main reasons may be the following aspects: the discovery of l-borneolum is later than borneol, and the processing technology of l-borneolum is far behind borneol which leads to high impurities. l-Borneolum has less adverse reactions the same as borneol, including gastrointestinal irritation, reproductive toxicity, and allergic reactions.

Discussion

l-Borneolum is still widely used in the clinical application currently. With the continuous development of borneol researches, the research trends of l-Borneolum will still be hot. Because of large demand and short supply of borneol, l-borneolum will also keep a high demand in the clinical application.

Fig. 1 *Blumea balsamifera* (L.) DC

Fig. 2 Structure of l-borneolum

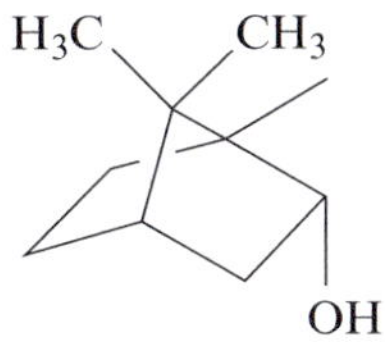

Fig. 3 Structure of l-borneolum's derivatives

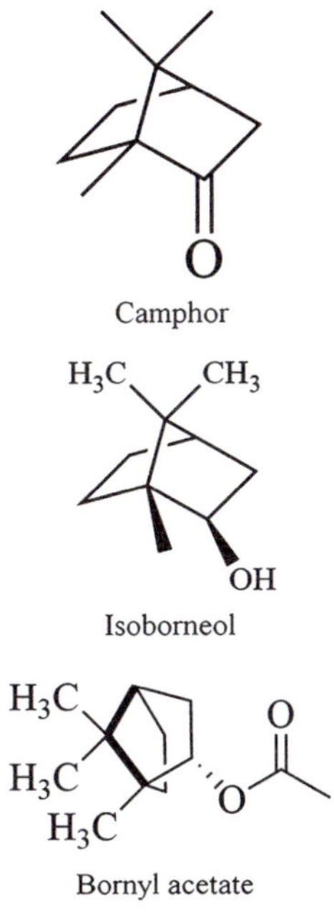

References

1. Chinese Pharmacopoeia. 2015:88.
2. Zhong Hua Ben Cao (Chinese Materia Medica). 1998:7–6765.
3. Ye Q. Cultivating and processing of l-borneolum. Shanghai J Tradit Chin Med. 1957;7:013.
4. Chen ZZ, Du SY, Lu Y, et al. Research of l-Borneolum and muscone mechanism for resuscitation-inducing aromatic herbs. J Chin Med Mater. 2014;3:460–4.
5. Tian W, Wang J, Gao T, et al. Impact of l-borneolum andsynthesised borneol on cerebral ischemia and hypoxia model mice. Pharmacol Clin Chin Mater Med. 2013;29(2):53–6.

Lobeline

Shi-Feng Chu, Ling-Lei Kong, and Nai-Hong Chen

Abstract Lobeline is mainly present in *Lobelia chinensis*, *Lobelia inflata*, *Campanula medium*, *Lobelia hassleri*, and *Lobelia nicotianaefolia* (Wang et al. Changchun Univ Tradit Chin Med. 2008;9–14). It is a selective activator for the peripheral chemical receptors, then reflexively excites the respiratory center, and deepens breathing. Further study showed lobeline has the effect of inhibiting drug addiction by inhibiting the dopamine release and relieving the psychiatric symptoms due to the nicotine withdrawal. Moreover, it also showed anticancer effect by inhibiting the uptake of oxygen in mice. In clinical practice, it is mainly used for the treatment of neonatal asphyxia, carbon monoxide cause suffocation, and so on.

Keywords Lobeline · Respiratory · Drug addiction

Origin: *Lobelia chinensis* (Fig. 1)
Chemical name (Fig. 2)

2-((2R,6S)-6-((S)-2-Hydroxy-2-phenylethyl)-1-methylpiperidin-2-yl)-1-phenylethanone

Molecular formula, $C_{22}H_{27}NO_2$; **MW**, 337.45; **CAS**, 90-69-7

S.-F. Chu · N.-H. Chen (✉)
State Key Laboratory of Bioactive Substances and Functions of Natural Medicines, Institute of Materia Medica & Neuroscience Center, Institute of Materia Medica, Chinese Academy of Medical Sciences and Peking Union Medical College, Beijing, China
e-mail: chennh@imm.ac.cn

L.-L. Kong
Beijing Key Laboratory of Drug Targets Identification and Drug Screening, Centre for Pharmaceutical Screening, Chinese Academy of Medical Sciences and Peking Union Medical College, Beijing, China

G.-H. Du, *Natural Small Molecule Drugs from Plants*,
https://doi.org/10.1007/978-981-10-8022-7_47

Properties

Appearance: White crystalline or granular powder, odorless, and bitter. *Solubility*: Soluble in ethanol or chloroform, slightly soluble in water. *Melting point*: 130–131 °C.

Dosage Forms and Indications

Lobeline was recorded in chemical drug and preparations as lobeline hydrochloride, which is prepared as injection for the treatment of central respiratory inhibition induced by a variety of reasons. In addition, it is also used for the treatment of neonatal stasis, carbon monoxide, opioid poisoning, and so on.

Literature

Lobeline is mainly present in *Lobelia chinensis*, *Lobelia inflata*, *Campanula medium*, *Lobelia hassleri*, and *Lobelia nicotianaefolia* [1]. *Lobelia inflata* has also been considered as an Indian tobacco and has been used for the treatment of respiratory diseases for a long history. It was also a treatment of asthma by American Aborigines. In the United States during the nineteenth century, doctors use *Lobelia inflata* as a vomiting agent, to remove the poison from the body. It is also called "vomit grass." Now *Lobelia inflata* is still used to clear throat, bronchial, lung, and other respiratory mucus [2].

History of R&D

In the 1930s, the chemical synthesis process of lobeline was completed, and its artificial synthesis was realized, which was found in *Lobelia inflata* from the North American Campanulaceae. It was commonly used as a lobeline hydrochloride. Because of its structure similar to nicotine, it was initially used for the treatment of respiratory diseases. Further study found that lobeline can selectively excite the carotid sinus peripheral chemoreceptors, then reflect the excitement of the medullary breathing center, and enhance respiratory function. Therefore, it is widely used as a respiratory stimulant [3].

Although lobeline showed similar biological activity with nicotine, its potency is just only 1/5~1/20 of nicotine. Hence, lobeline was used as a substitute for nicotine in many smoking cessation products. However, in 1993, the Food and Drug Administration (FDA) banned the sale of smoking cessation products containing lobeline because it was ineffective in helping people quit smoking. However, the research of lobeline in drug addiction still continues.

Pharmacology

The pharmacological effects of lobeline are extensive, mainly manifested as nicotine-like effect. On the one hand, lobeline can selectively excite the carotid sinus and aortic body chemoreceptors and induce reflective excitement of the respiratory center and vagus center. Lobeline showed better excitatory effect for the respiratory inhibition caused by morphine. There is bronchiectasis effect directly when lobeline is inhaled, hence, against pilocarpine and acetylcholine-induced tracheal contraction. When the dose increased, lobeline can directly stimulate the respiratory center and excite vagal center (reducing heart rate) and vomiting center in medulla oblongata. In addition, lobeline showed dual role in the regulation of ganglion, as manifested excitement and inhibition in chronological order. Lobeline has a zebra-like effect on the striated muscle. In addition, it also has a certain anticancer effect, as manifested by significantly inhibiting the uptake of oxygen in mouse ascites cancer cells [4].

Clinical Application

It is mainly used in the treatment of (1) neonatal asphyxia, (2) suffocation caused by carbon monoxide, (3) poisoning induced by inhalation of anesthetics and other central inhibitors (such as opioids and barbiturates), and (4) respiratory failure caused by pneumonia, diphtheria, and other infectious diseases. Lobeline also showed side effects, such as nausea, vomiting, cough, headache, palpitations, and other adverse reactions.

Discussion

Respiratory failure is a common clinical critical symptom, although the mechanical ventilation was fully affirmed in the rescue of patients with acute respiratory failure life, prolong survival of patients, but also cause serious complications, reduce the quality of life of patients. Respiratory stimulants are a class of drugs that can directly or indirectly excite the medullary breathing center and enhance breathing; these drugs can enhance ventilation function and improve ventilation quality. Compared with mechanical ventilation, it is easy to use, has less complications, and is suitable for long-term use. In accordance with the classification of respiratory stimulants, lobeline is a selective activator for peripheral chemical receptors and then a reflexive excitement of respiratory center, and it deepens breathing. In addition, due to the similarity of nicotine, it has also been used in the study of drug addiction. It has been shown that the lobeline may have the effect of inhibiting drug addiction, and its mechanism may be related to its inhibition of dopamine release. Lobeline also relieves psychiatric symptoms due to nicotine withdrawal, such as depression, anxiety, and so on. In addition, there are studies found that lobeline improved neurodegenerative diseases, which may be related to the binding of choline receptors. As a drug, it has been used for nearly a century and has made a greater contribution in protecting human health.

Fig. 1 *Lobelia chinensis*

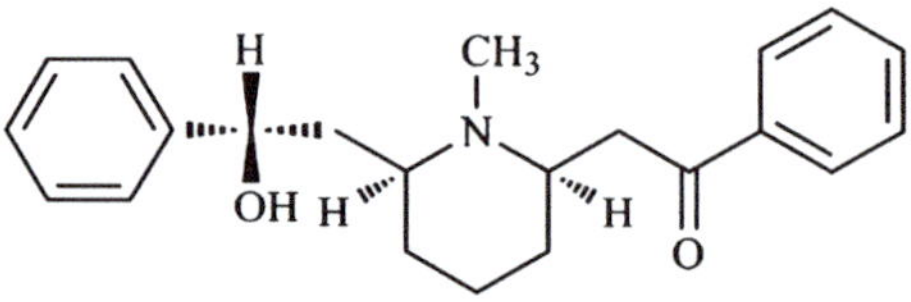

Fig. 2 Chemical structure of Lobeline

References

1. Wang et al. Chemical study of Lobelia chinensis. Changchun Univ Tradit Chin Med. 2008;9–14.
2. Kowalczyk-Bronisz SH. Studies on vomitory effect of some acridine compounds. Arch Immunol Ther Exp. 1980;28(5):777–82.
3. Benwell ME, Balfour DJ. The influence of lobeline on nucleus accumbens dopamine and locomotor responses to nicotine in nicotine-pretreated rats. Br J Pharmacol. 1998;125(6):1115–9.
4. Chen MW, Chen WR, Zhang JM, et al. Lobelia chinensis: chemical constituents and anticancer activity perspective. Chin J Nat Med. 2014;12(2):103–7.

Menthol

Ying Zhao, Li-Da Du, and Guan-Hua Du

Abstract Menthol is a covalent organic compound made synthetically or obtained from peppermint or other mint oils with flavoring and local anesthetic properties. The main form of menthol occurring in nature is (−)-menthol, which is assigned to the (1R, 2S, 5R) configuration. Menthol has a large amount of clinical application and a wide range of pharmacological effects, especially its odor. Menthol has local anesthetic and counterirritant qualities, and it is widely used to relieve minor throat irritation. When added to pharmaceuticals and foods, menthol functions as a fortifier for peppermint flavors. It also has a counterirritant effect on skin and mucous membranes, thereby producing a local analgesic or anesthetic effect. It is an important external drug excipient and active ingredient and has good application prospects.

Keywords Menthol · Mint · Analgesic · Counterirritant

Alias: BoHe ice, Menthol, BoHe frost
Origin: Mint (Fig. 1)
Chemical name (Fig. 2)

(1α,2β,5α)-5-methyl-2-(1-methylethyl) cyclohexanol

Molecular formula, $C_{10}H_{20}O$; **MW**, 156.27; **CAS**, 89-78-1
Derivatives (Fig. 3)

Y. Zhao · G.-H. Du (✉)
Beijing Key Laboratory of Drug Targets Identification and Drug Screening, Centre for Pharmaceutical Screening, Institute of Materia Medica, Chinese Academy of Medical Sciences and Peking Union Medical College, Beijing, China
e-mail: dugh@imm.ac.cn

L.-D. Du
School of Biomedical Sciences, Chinese University of Hong Kong, Hong Kong, China

G.-H. Du, *Natural Small Molecule Drugs from Plants*,
https://doi.org/10.1007/978-981-10-8022-7_48

Properties

Appearance: colorless or white acicular prismatic crystals or white crystalline powder. *Solubility*: easily dissolved in ethanol, chloroform, ether, liquid paraffin, or volatile oil and soluble in water. *Odor*: a cool, refreshing, and pleasant mint aroma, sweet odor, and taste cool early after burning. *Boiling point*: 212 °C. *Melting point*: 41–43 °C. *Specific optical rotation*: −49 to −50°.

Dosage Forms

This product is recorded in the *Pharmacopoeia of the People's Republic of China* (2015), the *British Pharmacopoeia* (2017), the *United States Pharmacopeia* (40), the *Japanese Pharmacopoeia* (17th ed.), and the *European Pharmacopoeia* (9th ed.).

The current clinical dosage includes tablets, ointment, and spirits.

Indications

For external application, it is applied locally and often used to relieve pain and itching. A nasal drip is used in the head cold. Inhalation or spray is used for sore throat. Oral administration can promote digestion.

Literature

Mentha is a kind of saturated cyclic alcohol, obtained by steam distillation, freezing, and recrystallization of the fresh stem and leaves of the mint *Mentha haplocalyx* Briq. It is the main component of peppermint and spearmint oil, with the evacuation of wind heat, heat relieving, dispelling swelling, and relieving sore throat pain effects [1].

Mint is widely distributed in the temperate regions of the northern hemisphere and a few species in the southern hemisphere [2]. It is a widely used Chinese herbal medicine. It has a long history of medicinal use, and the earliest record of mint was found in Tang materia medica. As early as 2000 years ago, the ancients had been known to collect mint for food and medicine. Medical scientist Li Shizhen Ming recorded medicinal value of mint in the "Compendium of Materia Medica" that "Mint symplectic can diverge, dedicated to eliminating wind heat, throat mouth diseases, scrofula, sores, itchy rash hidden wind. The portion of "Dan Daquan" book which talks about the nasal part contains many references to mint medicine treatment. In the records of flora of China, there are 12 species of mint plants, which are mainly distributed in the northeast, East China, and Xinjiang.

History of R&D

In the world of aromatic chemicals, menthol is often described as a unique source of skin and mucous membranes. Menthol is mainly extracted from natural plants. The vast majority of natural menthol in the global market is extracted from numerous Asian mint. In 2006, the world's natural menthol production is 12,800 tons or more. However, due to various factors, the production of natural menthol is becoming less and less. Since the 1960s, Japan, Germany, and other countries had developed the synthetic menthol products. In 1974 the German company Symrise and Japan Takasago Fluidic Systems company launched chemical synthesis of menthol. At present, attention has been paid to the study of menthol analogy substances and the effect of the changes of molecular structure on the aroma and cool sensation. Some studies have shown that hydroxyl groups located in the branched or 1, 4 chain cannot show a sense of cool. The researchers believed that menthol analogy substances with optical activity should be able to show different, surprising cooling effect, and the structure of hydroxyl alkyl groups is the key point to have a cooling effect [3–6]. The Japanese chemist Ryoji Noyo and his colleagues found that in BINAP and rhodium complexes as catalysts for asymmetric hydrogenation reaction, the synthesis of menthol is very effective. Because of the contribution of asymmetric organic synthesis, they won the 2001 Nobel Prize in Chemistry.

Pharmacology

Menthol has a wide range of pharmacological effects. Menthol can stimulate the nerve endings of skin and slowly penetrate into the skin, resulting in prolonged congestion and reflex caused by deep vascular changes, adjusting the vascular function, and achieving therapeutic effects in topical application. Topical application of compound has anti-inflammatory, analgesic, and anesthetic effects [7–10].

In respiratory system: For treatment of bronchitis, it could reduce the respiratory tract foam sputum and increase the effective ventilation cavity road. For treatment of rhinitis and laryngitis, it exerts mitigation by promoting secretion and diluting viscous mucus.

In digestive system: Menthol showed the powerful effect on bile.

In central nervous system: A small amount of peppermint can stimulate the central nervous system through the peripheral nerve to expand the skin capillaries, promote the secretion of sweat glands, increase heat dissipation, and show sweating antipyretic effect.

In addition, menthol has penetration-enhancing effect on many kinds of drugs; the mechanism is related to the changes of the ultrastructure of the skin, which is expected to be widely used in transdermal drug delivery agents. In vitro experiments, it had strong antibacterial activity against *Staphylococcus aureus and Bacillus proteus*. Recently, British scientists found that mint leaves can prevent cancer lesions in the blood vessel growth to decrease blood supply for, showing a certain anti-tumor effect [11, 12].

Clinical Application

It can be clinically used to assist anesthesia, postoperative analgesia, intercostal nerve block, trigeminal nerve block, occipital nerve block, and so on. When applied locally, it can promote blood circulation, diminish inflammation, relieve itching, relieve pain, relieve edema, and so on.

Discussion

Peppermint has strong physiological and pharmacological activities. However, up to now, only a few studies have been done on the chemical constituents of the plant. Natural menthol has a large amount of clinical application and a wide range of pharmacological effects, especially its special odor. It is an important external drug excipient and active ingredient and has good application prospects.

Fig. 1 Mint

Fig. 2 The structure of menthol

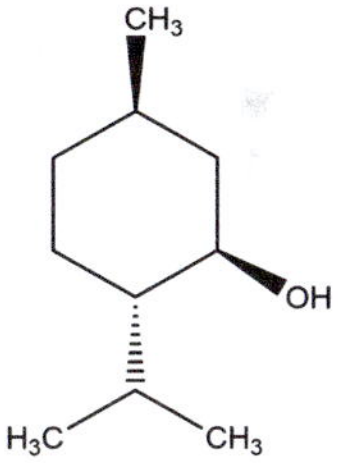

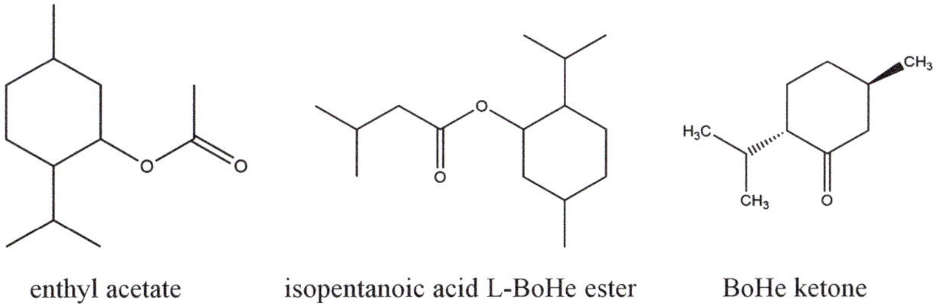

Fig. 3 The structure of menthol derivatives

References

1. National Pharmacopoeia Committee. Pharmacopoeia of the People's Republic of China (2015). Beijing: Beijing chemical industry; 2015. p. 420–1.
2. Hou ZH. Mint, peppermint oil and menthol. Chin J Pharm. 1955;3(7):321–2.
3. Min M. Aromatic raw material features: menthol (menthol)-history, consumption, synthesis, products, substitute products and derivatives. Toothpaste Ind. 2008;4:47–51.
4. Maodao G. Synthetic menthol. Chem Bull. 1964;12:4–5.
5. Shan H, Lu T, Biao J. Present situation and development of synthesis of levo menthol. Org Chem. 2009;24(6):884–90.
6. Kumobayash H, Sayo N, Akutagawa S, et al. Industrial asymmetric synthesis by use of metal-BINAP catalysts. Nippon Kagaku Zassi. 1997;12:835–46.
7. Xinsheng Y, Wu M. Natural medicine chemistry. Beijing: People's Medical Publishing House; 2003.
8. Lu Y. The medicinal value and function of mint. Capital Med. 2007;14(8):44–5.
9. Wang H, Wu Z. Study on pharmacological action of mint and its active components. Chin Herb Med. 1998;29(6):422–4.
10. Shen M, Xiaomeng L, Qiyuan S. Advances in studies on chemical constituents and pharmacological effects of mint. Chin Arch Tradit Chin Med. 2012;30(7):1484–7.
11. Sharma V, Hussain S, Gupta M, et al. In vitro anticancer activity of extracts of Mentha Spp. against human cancer cells.[J]. Indian J Biochem Biophys. 2014;51(5):416–9.
12. Sun Z, Wang H, Wang J, et al. Chemical composition and anti-inflammatory, cytotoxic and antioxidant activities of essential oil from leaves of Mentha piperita grown in China[J]. Plos One. 2014;9(12):e114767.

Morphine

Ling-Lei Kong, Jin-Hua Wang, and Guan-Hua Du

Abstract Morphine is extracted from poppy, which is the most important alkaloid in opium. Morphine has a strong analgesic effect, especially for moderate and severe cancer pain. However, morphine has serious side effects, such as addiction, respiratory depression, acute poisoning, and death. Therefore, in regulating the clinical application of morphine, it is necessary to explore the alternative to morphine with strong analgesic role and minor side effects. In the long history of morphine R&D, there are many opioid drugs to be found and they make great contributions for the treatment of pain and human disease. Therefore, it is the ultimate goal to develop new drugs by applying advanced technology and protect human health, as far as possible to reduce possible harm.

Keywords Morphine · Alkaloid · Analgesic · Addiction

Origin: Opium poppy (Fig. 1)
Chemical name (Fig. 2)

7,8-Didehydro-4,5α-epoxy-17-methylmorphinan-3,6α-diol

Molecular formula, C17H19NO3; **MW**, 285.34; **CAS**, 57-27-2
Morphine derivatives (Fig. 3)

L.-L. Kong · J.-H. Wang · G.-H. Du (✉)
Beijing Key Laboratory of Drug Targets Identification and Drug Screening, Centre for Pharmaceutical Screening, Institute of Materia Medica, Chinese Academy of Medical Sciences and Peking Union Medical College, Beijing, China
e-mail: dugh@imm.ac.cn

G.-H. Du, *Natural Small Molecule Drugs from Plants*,
https://doi.org/10.1007/978-981-10-8022-7_49

Properties

Appearance: white or almost white crystalline powder or colorless, silky needles or cubical masses, efflorescent in a dry atmosphere. *Solubility*: soluble in water, slightly soluble in ethanol, practically insoluble in chloroform or ether. *Specific optical rotation*: −110.0° to 115.0°

Dosage Forms

Morphine is recorded in the *Pharmacopoeia of the People's Republic of China* (2015), *British Pharmacopoeia* (2017), *United States Pharmacopeia* (40), *European Pharmacopoeia* (9.0th ed.), *The Japanese Pharmacopoeia (17th ed.)*, and *The International Pharmacopoeia* (5th ed.).

The dosage forms of current clinical use are tablets, injection, and sustained-release tablet.

Indications

Morphine is mainly used to treat both acute and chronic severe pain. It is also used for pain caused by myocardial infarction and for labor pains. Morphine relieves pulmonary edema symptoms; anesthesia and preoperative administration can make the patient quiet and drowsy; compound formula of morphine was used for acute and chronic diarrhea.

Literature

Morphine is extracted from the plant poppy. Poppy is native in the West Asia region, with the scientific name "somniferum," meaning "hypnosis," reflecting its narcotic. As opium poppy has the effect of euphoria, Sumerians thought that opium poppy was a "happy plant" [1]. The ancient Babylonians spread the poppy seeds and techniques of their planting to the ancient Egyptians, known as the "god flower." In the third century BC, the poppy was gradually recorded in the ancient Greek and ancient Roman books; the ancient Greeks called it "the flower of the dream," while the Roman poet Virgil called it the "hypnotic flower." The great poet Homer described that "poppy was given by Helen, the daughter of Zeus, to Telemachus and his friends to help them forget their grief over Odysseus' absence." In the seventh century BC, Arab traders brought opium to India and China. During the tenth to thirteenth centuries, opium was spread to whole Europe from Asia.

Opium was separated from desiccated juice of poppy fruit. The magical effect of opium was ascribed to the prosperity of opium, but many people have died of excessive use of opium. From the sixteenth century, manuscripts can be found describing drug abuse and tolerance in Turkey, Egypt, Germany, and England. Nowhere was the problem of addiction greater than in China where the practice of smoking opium began in the mid-seventeenth century after tobacco smoking was banned. Efforts to suppress the sale and use of opium failed for the British, later joined by the French. Chinese government was forced to permit opium trade and consumption.

All morphine in current clinical application is from opium poppy, which is cultivated. All methods of synthesis are not suitable for large-scale industrial production. But as a complex molecule, the total synthesis of morphine alkaloids will bring the challenges inspiration to the scientist. In 1952, Gates reported that the total synthesis of morphine was completed by the 23-step reaction at a total yield of 0.01% [2]. In recent years, the Heck reaction, as the representative of the organic palladium reagent, is widely used in the organic synthesis, making morphine total synthesis more ingenious and economical [3].

History of R&D

Morphine is the most important alkaloid in opium (about 10–15%). It was found that smoking opium can give people the feeling and experience of euphoria for a long time. The ancient Greek doctor Hippocrates, who was revered as the "father of medicine," had found that opium had an analgesic effect on visceral diseases, gynecology, and infectious diseases. But until the time of ancient Rome, Galen defined opium as an effective analgesic applied to clinical.

In December 1804, morphine was discovered as the first active alkaloid extracted from the opium poppy plant by Friedrich Sertürner in Paderborn, Germany [4]. In 1817, in Sertürner 'experiments, he and dogs were administered with morphine, and both of them almost died of morphine. Sertürner originally named the substance as morphium after the Greek god of dreams, Morpheus, as it has a tendency to cause sleep [5]. This discovery also brought a new era of human analgesic drug research.

Due to complex structure and specific physical and chemical properties, morphine attracted many chemists' interest after separation and purification. However, there was no modern spectral analysis and single crystal diffraction technique. It was very difficult to determine the structure of morphine by elemental analysis and chemical degradation alone. Thanks to the long-term accumulation and lasting interest, British chemist Robinson in 1925 put forward the correct molecular structure of morphine. After that, the chemist Eisleb had studied the structure of morphine in detail. It is believed that the circular system in the molecule was a necessary group for morphine analgesia, which laid the foundation for the study of opioid analgesic drugs.

Morphine has a strong analgesic effect, especially for moderate and severe cancer pain. However, morphine has serious side effects, such as addiction, respiratory depression, acute poisoning, and death. Therefore, in regulating the clinical application of morphine at the same time, it is necessary to explore the alternative to morphine with strong analgesic role and minor side effects.

In 1917, thebaine derivative oxycodone was used clinically as a strong opioid analgesic drug for the treatment of severe pain. In 1939, drawing on the chemical structure of morphine and atropine, the chemist Eisleb developed the analgesic pethidine [6]. Methadone was developed in Germany around 1937–1939 by Gustav Ehrhart and Max Bockmühl at the Farbwerke Hoechst who were looking for a synthetic opioid that could be created with readily available precursors, to solve Germany's opium shortage problem [7]. It was approved for use in the United States in 1947.

Pentazocine was developed by the Sterling Drug Company, Sterling-Winthrop Research Institute, of Rensselaer, New York, which can relieve severe acute pain and chronic pain, with nonaddictive effect. It was approved by the Food and Drug Administration in 1967 after being favorably reviewed following testing on 12,000 patients in the United States. Pentazocine was already being sold in Mexico, England, and Argentina in different trade names. Pentazocine is a breakthrough in the field of analgesic research. Subsequently, the powerful opioid analgesic drugs were gradually emerging, such as buprenorphine and fentanyl, 50 and 100 times stronger than morphine, and analgesic effect is extremely fast.

In order to treat severe pain and ensure the life quality of patients, opioid drugs with strong analgesic effect were developed. But at the same time, taking into account strong addiction of the strong opioid analgesic drug, the development of weak opioid drugs for moderate pain was also carried out. Just like morphine, codeine was also extracted from the poppy fruit, which is the first emergence of weak opioid analgesic drugs, and used to treat pain, cough, and diarrhea in clinic. On the basis of the chemical structure of codeine, dihydrocodeine, as a weak opioid drug, was discovered, which analgesic effect was stronger and addiction was weaker than codeine.

In the 1970s, tramadol was launched and marketed as "Tramal" by the German pharmaceutical company Grünenthal GmbH in West Germany [8], and 20 years later, it was launched in countries such as the United Kingdom, the United States, and Australia. It is marketed in many brand names worldwide. After R&D of more than 100 years, opioid drugs were increased and have been widely used in clinic as analgesic drugs.

Pharmacology

Morphine is subject to extensive first-pass metabolism; only 40–50% of the dose reaches the central nervous system when taken orally. Morphine is metabolized primarily in the liver, and approximately 87% of a dose is excreted in the urine after 72 h of administration. Morphine is metabolized primarily into morphine-3-glucuronide and morphine-6-glucuronide via glucuronidation by phase II metabolism enzyme UDP-glucuronosyl transferase-2B7. The half-life is about 2.5–3.5 h [9].

Morphine, as an agonist of opioid receptor, has many pharmacological effects:

1. Analgesic: morphine can activate the opioid receptor and produce strong analgesic effect by simulating the role of endogenous enkephalin. It is better for persistent dull pain than intermittent sharp pain and visceral cramps.
2. Sedation: in the analgesic at the same time, there is a clear sedative effect, sometimes produce euphoria, which can improve the tension of the patient.
3. Respiratory depression: morphine can inhibit the respiratory center and reduce the sensitivity of the respiratory center to carbon dioxide. The inhibition degree of breathing is parallel to the dose of morphine, and excessive doses can cause respiratory failure or death.
4. Antitussive: morphine can inhibit the cough center, resulting in antitussive effect.
5. Excited smooth muscle: morphine can excite the digestive tract smooth muscle, leading to constipation, and increase tension of bile duct, ureter, and bronchial smooth muscle.
6. The cardiovascular system: morphine can promote the release of endogenous histamine and lead to peripheral vascular dilatation, blood pressure decrease, cerebrovascular expansion, and intracranial pressure increase.
7. Antiemetic and miotic effect: morphine can play antiemetic effect through the central nervous system, and the most important feature with morphine application is the pinhole-like pupil.

Clinical Application

1. Used for acute short-term pain, such as surgery, trauma, burn-associated severe pain, and cancer pain.
2. Used for patients with myocardial infarction and plays a role in sedation, reduces cardiac load, and eases fear.
3. Used in patients with acute pulmonary edema for its cardiovascular effects and to allay anxiety. Forbid using in the treatment of pulmonary edema resulting from a chemical respiratory irritant.
4. Used for anesthesia and preoperative administration to make the patient quiet and drowsy.
5. Used parenterally for analgesia during labor.
6. Opium tincture or compound camphor tincture can be used for acute and chronic diarrhea.

Side effects: Potentially serious side effects include respiratory depression and low blood pressure. Morphine has a high potential for addiction and abuse. If the dose is reduced after long-term use, withdrawal may occur. Common side effects are drowsiness, vomiting, headache, tired feeling, anxiety, mild itching, and constipation. Caution is advised when used during pregnancy or breastfeeding, as morphine will affect the baby.

Discussion

In the long history of morphine R&D, there are many opioid drugs to be found and they make great contributions for the treatment of pain and human disease. But at the same time, morphine, as a representative drug, also brings catastrophe to humans. Therefore, it is the ultimate goal to develop new drugs by applying advanced technology and protect human health, as far as possible to reduce possible harm.

Fig. 1 Opium poppy

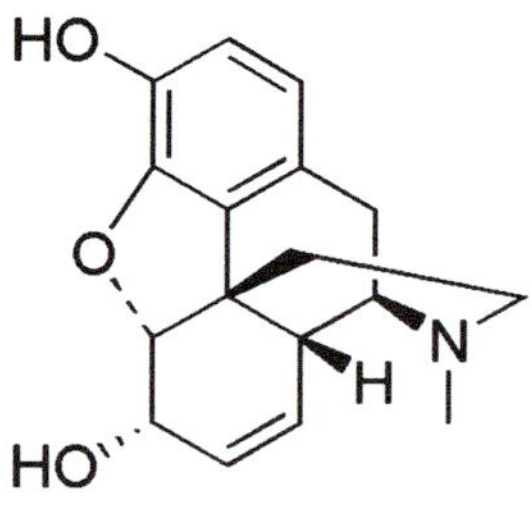

Fig. 2 The structure of morphine

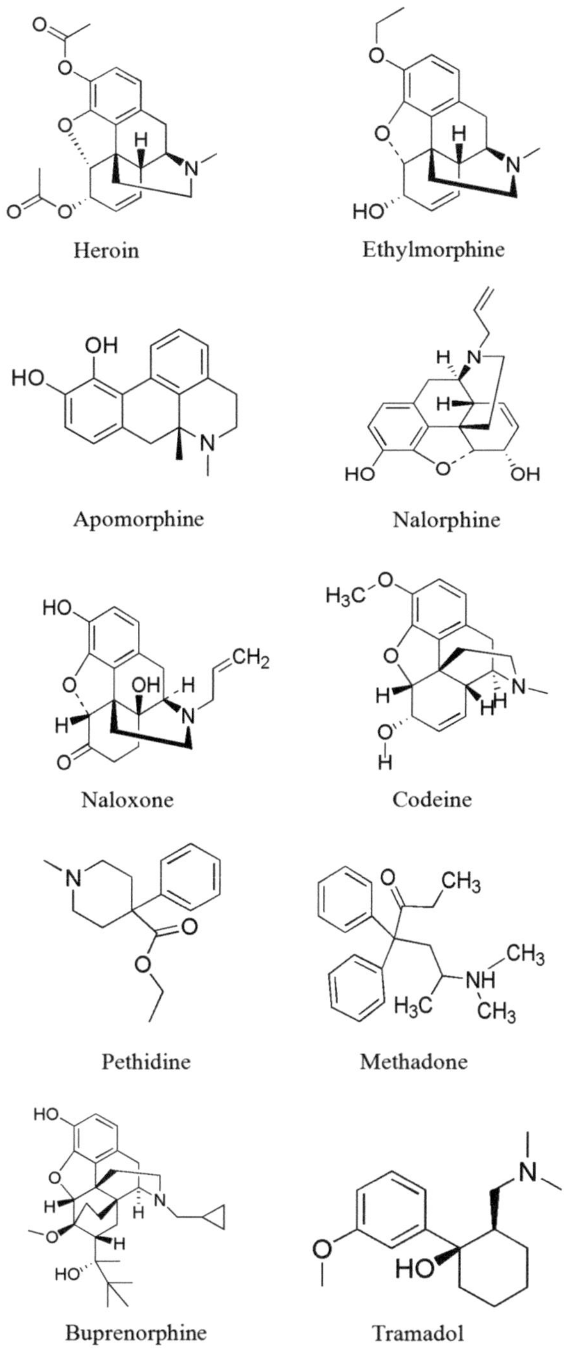

Fig. 3 The structure of morphine derivatives

References

1. Brownstein MJ. A brief history of opiates, opioid peptides, and opioid receptor. Proc Natl Acad Sci U S A. 1993;90:5391–3.
2. Gates M, Tschudi G. The synthesis of morphine. J Am Chem Soc. 1952;74:1109–0.
3. Novak BH, Hudlicky T, Reed JW, et al. Morphine synthesis and biosynthesis – an update. Curr Org Chem. 2000;4:343–62.
4. Serturner FW. Trommsdorff's Journal der Pharmazie fur Aerzte. Apotheker und Chemisten. 1806;14:47–93.
5. Serturner FW. Ueber das Morphium, eine neue salzfahige Grundlage, und die Mekonsaure, als Hauptbestandtheile des Opiums. Gilbert's Annalen der Physik. 1817;55:56–89.
6. Eisleb O, Schaumann O. Dolantain, a new antispasmodic and analgesic. Deut Med Wochschr. 1939;65:967–8.
7. Scott CC, Chen KK. The action of 1,1-diphenyl-1(dimethylaminoisopropyl) butanone-2, a potent analgesic agent. J Pharmacol Exp Ther. 1946;87:63–71.
8. Preston KL, Jasinski DR, Testa M. Abuse potential and pharmacological comparison of tramadol and morphine. Drug Alcohol Depend. 1991;27:7–17.
9. Trescot AM, Datta S, Lee M, Hansen H. Opioid pharmacology. Pain Physician. 2008;11(2 Suppl):S133–53.

Picrotoxin

Huan-Li Xu, Li-Da Du, and Guan-Hua Du

Abstract Picrotoxin is extracted from the seeds of Radix Cocculi Trilobi. The pharmacological effects of picrotoxin are focused on the central nervous system and GABA receptor. Picrotoxin also showed direct effects on the heart, including inhibiting cardiac-specific conduction system and decreasing heart rate and cardiac contractility, as well as cardiac action potential. As a GABA receptor antagonist, picrotoxin is extensively studied for its strong effects on the mesencephalon and oblongata. Its application in serious barbital poisoning patients achieves satisfactory results. However, picrotoxin overdose may cause cramp, respiratory center inhibition, and vomiting. Thus, the clinical symptoms of the patients should be watched closely, and their respiration, pulse, and blood pressure should be monitored when using picrotoxin.

Keywords Picrotoxin · Sesquiterpene · GABA receptor · Overdose

Alias: Picrotin; picrotoxinin; sesquiterpene; cocculin
Origin: Radix Cocculi Trilobi (Fig. 1)
Chemical name (Fig. 2)

3,6-Methano-8h-1,5,7-trioxacyclopenta(ij)cycloprop(a)azulene-4,8(3h)-dione

Molecular formula, $C_{30}H_{34}O_{13}$; **MW**, 602.58; **CAS**, 124-87-8

H.-L. Xu
Department of Pharmacology, School of Basic Medical Sciences, Capital Medical University, Beijing, China

L.-D. Du
School of Biomedical Sciences, Chinese University of Hong Kong, Hong Kong, China

G.-H. Du (✉)
Beijing Key Laboratory of Drug Targets Identification and Drug Screening, Centre for Pharmaceutical Screening, Institute of Materia Medica, Chinese Academy of Medical Sciences and Peking Union Medical College, Beijing, China
e-mail: dugh@imm.ac.cn

G.-H. Du, *Natural Small Molecule Drugs from Plants*,
https://doi.org/10.1007/978-981-10-8022-7_50

Properties

Appearance: colorless, soft, and glossy rhombic pillar-shaped crystal or crystalline fine powder. It is odorless and tastes very bitter. It is steady in the air, but it metamorphoses easily when exposed to light. *Solubility*: its solubility in water is 0.3%; soluble in diluted acid; slightly soluble in ether or chloroform.

Dosage Forms

It is recorded in the *Pharmacopoeia of the People's Republic of China* (1953).

Its current dosage form is injection and is clinically used for rescuing barbital poisoning.

Literature

Menispermaceae contains 71 genera with 450 species, which are mainly distributed in tropical and subtropical areas. *Anamirta cocculus* is a Southeast Asian and Indian climbing plant. The plant has large-stem (up to 10 cm in diameter) and corky gray bark with white wood. The flowers are small, yellowish-white, and sweet-scented, about 6–10 cm across. The fruit is a drupe, about 1 cm in diameter when dried [1].

Menispermaceae contains many components and quaternary alkaloids, such as berberine, palmatine, magnoflorine, and columbamine, which can be extracted from its stem and roots [2]. Picrotoxin, a sesquiterpene, can be found in the seeds. The tertiary alkaloids menispermine and paramenispermine can be found in its seed shells [3].

History of R&D

The seeds of Menispermaceae are traditionally used as fishing bait or pesticide, as well as pediculicide [4]. Once it was used for brewing beer to impart a more intoxicating quality to beer in the nineteenth century [5]. In 1812, cocculin was firstly isolated by Pierre Boullay and named picrotoxin, a combination of the Greek words "picros" and "toxicon." Picrotoxin showed strong physiological action and was once extensively studied.

The action mechanism of picrotoxin is a controversial issue. Newland et al. considered that picrotoxin enhanced the occurrence of a desensitized state or an allosterically blocked state of $GABA_A$ receptor [6], while Jong et al. showed that picrotoxin acts as a noncompetitive channel blocker for the $GABA_A$ receptor chloride channels, rather than a receptor antagonist [7].

Pharmacology

The pharmacological effects of picrotoxin are focused on the central nervous system and GABA receptor. Han et al. showed that picrotoxin significantly inhibited the nociceptive discharges of parafascicular neurons of the thalamus in rabbit [8]. It was also showed that picrotoxin, as a GABA receptor antagonist, could enhance the effects of electroanalgesia. These results provided theoretical basis for algogenesis and pain management [9]. Xu et al. showed that GABA receptor might be involved in decreased blood pressure induced by hypoxia in rabbits and picrotoxin could block the decreased blood pressure induced by hypoxia [10]. This result provided strong evidence for the treatment of decreased blood pressure induced by hypoxia using picrotoxin. Picrotoxin also showed direct effects on the heart. It inhibits cardiac-specific conduction system and decreases heart rate and cardiac contractility, as well as cardiac action potential [11]. Picrotoxin mainly activates the mesencephalon and oblongata as a GABA receptor antagonist, and high dosage of picrotoxin may also activate the spinal cord and cerebrum.

Clinical Application

Acute barbital poisoning is very common and may be life-threatening. Many studies reported the superior effects of picrotoxin in rescuing barbital poisoning [12, 13]. Picrotoxin can be used in serious barbital poisoning patients when conventional central nervous system stimulants are ineffective. The dosage and route of administration depend on the severity and state of the illness. It was usually administrated intramuscularly or intravenously (3–9 mg), once every 15–30 min, until corneal and deglutition reflex or slight tremble in the face and four limbs appears.

Discussion

As a GABA receptor antagonist, picrotoxin is extensively studied for its strong effects on the mesencephalon and oblongata. Its application in serious barbital poisoning patients achieves satisfactory results. However, picrotoxin overdose may cause cramp, respiratory center inhibition, and vomiting. Thus, the clinical symptoms of the patients should be watched closely, and their respiration, pulse, and blood pressure should be monitored when using picrotoxin.

Fig. 1 Radix Cocculi Trilobi

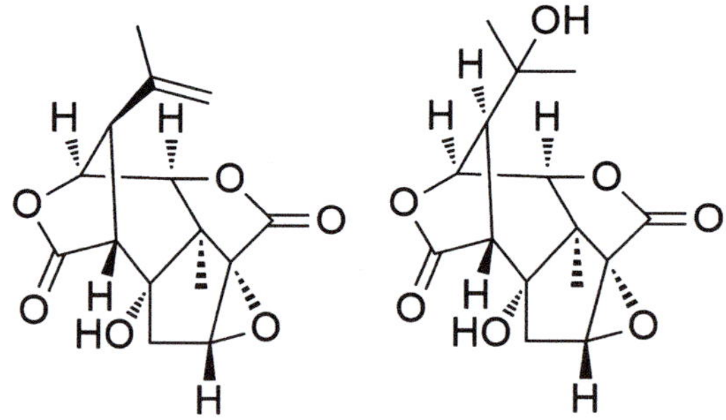

Fig. 2 Chemical structure of picrotoxin

References

1. Dichoso WC. Research information series on ecosystems–volume 12 no.2. Useful plant species with toxic substance. DENR, 2000.
2. Verpoorte R, Siwon J, Tieken MEM, Baerheim SA. Studies on Indonesian medicinal plants. V. The alkaloids of Anamirta cocculus. J Nat Prod. 1981;44(2):221–4.
3. Allen AH. Bitters of cocculus indicus. Commercial organic analysis -Vol. III part III. 2nd ed. Philadelphia: P. Blakiston, Son &Co; 1896.
4. Duke JA. Ethnobotanical uses. Germplasm resources information network: Dr. Duke's phytochemical and ethnobotanical databases. USDA, 1995.
5. Porter N. Webster's revised unabridged dictionary. Springfield: G. & C. Merriam Co; 1913.

6. Newland CF, Cull-Candy SG. On the mechanism of action of picrotoxin on GABA receptor channels in dissociated sympathetic neurones of the rat. J Physiol. 1992;447(1):191–213.
7. Rho JM, Donevan SD, Rogawski MA. Direct activation of GABA-A receptors by barbiturates in cultured rat hippocampal neurons. J Physiol. 1996;497(2):509–22.
8. Han D, Zhang G, Li X. Effect of picrotoxin on nociceptive discharges of parafascicular neurons of thalamus. J Hubei Med Coll. 1990;11(1):21–3.
9. Zhang Z. Effect of intracerebroventricular injection of picrotoxin on electroacupuncture in rats. J Wuhan Med Coll. 1985;4:292–3.
10. Xu C, He J, Meng X, et al. Blocking effect of picrotoxin on decreased blood pressure induced by hypoxia in rabbits. Chin J Appl Physiol. 1998;14(2):123–4.
11. Yuan X, Xu J, Shi H, et al. Effect of picrotoxin on rate, contractility, and action potential of isolated frog heart. J Third Mil Med Univ. 1986;8(Suppl):37–41.
12. Mao Y, Wang Y. Cases report of three barbital poisoning patients rescued using picrotoxin. Jiangxi Med J. 1965;5(8):955.
13. Sierra-Paredes G, Galán-Valiente J, Vazquez-Illanes MD, et al. Extracellular amino acids in the rat hippocampus during picrotoxin threshold seizures in chronic microdialysis experiments. Neurosci Lett. 1998;248(1):53–6.

Pilocarpine

Yi-Huang Lin, Lian-Hua Fang, and Guan-Hua Du

Abstract Pilocarpine is an alkaloid extracted from *Pilocarpus jaborandi*, which is one of cholinomimetic drugs. Pilocarpine activates cholinergic M-receptor and has an obvious effect on eyes and salivary glands. Pilocarpine nitrate is mainly used to treat glaucoma clinically. However, the interaction time is extremely short due to its low bioavailability. Therefore, many researchers still make a contribution to improve dosage forms of pilocarpine, including increasing intraocular permeability and bioavailability, cutting times of medication, reducing toxicity, and improving efficacy.

Keywords Pilocarpine · Alkaloid · Cholinergic M-receptor · Glaucoma

Origin: *Pilocarpus jaborandi* (Fig. 1)
Chemical name (Fig. 2)

(3S,4R)-3-Ethyldihydro-4-((1-methyl-1H-imidazol-5-yl)methyl)-2(3H)-furanone mononitrate

Molecular formula, $C_{11}H_{16}N_2O_2$; **MW,** 208.26; **CAS**, 92-13-7

Properties

Appearance: colorless crystal or white crystalline powder. *Solubility*: freely soluble in water; slightly soluble in ethanol; insoluble in chloroform or diethyl ether. *Melting point*: 174–178 °C.

Y.-H. Lin · L.-H. Fang · G.-H. Du (✉)
Beijing Key Laboratory of Drug Targets Identification and Drug Screening, Centre for Pharmaceutical Screening, Institute of Materia Medica, Chinese Academy of Medical Sciences and Peking Union Medical College, Beijing, China
e-mail: dugh@imm.ac.cn

G.-H. Du, *Natural Small Molecule Drugs from Plants*,
https://doi.org/10.1007/978-981-10-8022-7_51

Dosage Forms and Indications

Pilocarpine is recorded in *Pharmacopoeia of the People's Republic of China* (2015), *British Pharmacopoeia* (2017), *United States Pharmacopeia* (40), and the *International Pharmacopoeia* (5th ed.). Pilocarpine nitrate eye drops now are used to treat glaucoma clinically.

Literature

Pilocarpine is an alkaloid derived from *maoguoyunxiang* 毛果芸香 (*Pilocarpus jaborandi*) in South America. The alkaloid content is 0.16–0.84% approximately. As one of cholinomimetic drugs, it is used to treat glaucoma with the actions of myosis and the depression of intraocular pressure. In the 1930s, the chemical synthesis of pilocarpine was completed successfully; however, they didn't achieve commercial production, and pilocarpine was extracted from plants only at that time [1, 2].

History of R&D

It has a history of hundreds of years since pilocarpine was used to treat glaucoma [3]. In 1874, a Brazilian, Coutinhou, authenticated that the people who crews the leaves of *Pilocarpus jaborandi* would increase salivary secretion, and he extracted alkaloid from the leaves in 1875. Soon, Weber observed the effect of the alkaloid on the pupil, sweat gland, and salivary gland. Since then, the alkaloid became a drug used for glaucoma clinically by depressing intraocular pressure, as a myotic. Now, its main dosage form is still aqueous solution [4].

In 1933, the chemical synthesis of pilocarpine was firstly reported. However, pilocarpine couldn't be used for treatments, because its synthetic route is so long and focuses on isopilocarpine, of which the pharmacological activities are 1/20–1/50 of pilocarpine. In 1972, DeGraw successfully synthesized the cis-homopilopic acid employing catalytic hydrogenation of precious metals and obtained pilocarpine as the main part product. Therefore, the study on the production of pilocarpine by chemical synthesis has made new progress and has been artificially synthesized [5].

Pharmacology

Pilocarpine activates cholinergic M-receptor and has an obvious effect on eyes and salivary glands. Pilocarpine nitrate eye drops take part in the actions of myosis, depressing intraocular pressure and alleviating cyclospasm. It increases glandular secretions at 10–20 mg, i.h., including the sweat gland, salivary gland, lacrimal gland, gastric gland, pancreas, intestinal gland, respiratory mucosa, and so on. Pilocarpine activates intestinal smooth muscle and promotes its tension and peristalsis. It induces asthma by activating bronchial smooth muscle and activates smooth muscles of uterus, bladder, gallbladder, and biliary passage as well [4].

Clinical Application

Pilocarpine nitrate is mainly used to treat glaucoma clinically. Characterized with the progressive cupping of the optic disk, hypopsia, and elevated intraocular pressure, the severe patients will go blind. Patients with angle-closure glaucoma (congestive glaucoma) generally have the narrow anterior chamber angle, the obstruction of aqueous humor outflow, and the elevation of intraocular pressure, and these can be reversed by a low-concentration pilocarpine. But it is noted that a high-concentration pilocarpine will promote the progress of glaucoma. Pilocarpine is also used to treat open-angle glaucoma. The mechanism of the action is not entirely clear. Using atropine and pilocarpine alternately prevents posterior synechiae. In addition, pilocarpine is orally used to treat Zagari's disease after neck radiotherapy, increasing salivary secretion and sweat secretion [4].

Discussion

As an agonist of cholinergic M-receptor, pilocarpine, used to treat glaucoma historically and still availably [6], is able to narrow the pupil and depress the intraocular pressure. However, after using the eye drops, it will be diluted immediately by the tear in the conjunctival sac and eliminated soon through lacrimal passage, so the interaction time is extremely short due to its low bioavailability. Many researchers still make a contribution to improve dosage forms of pilocarpine, including increasing intraocular permeability and bioavailability, cutting times of medication, reducing toxicity, and improving efficacy [7–10]. Besides, pilocarpine plays an important role in studies of cholinergic nerve function and receptor theory.

Fig. 1 *Pilocarpus jaborandi*

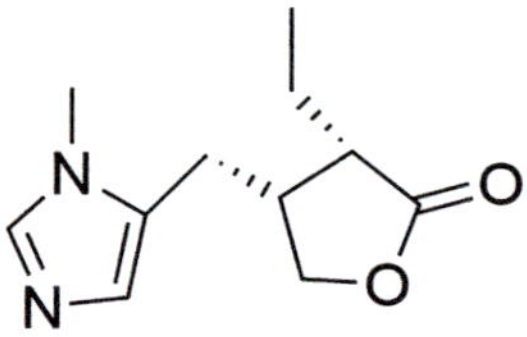

Fig. 2 The structure of pilocarpine

References

1. Wang XT. The chemical synthesis DL-pilocarpine. Yi Yao Gong Ye. 1978;12:5–8.
2. Wang XT. The chemical synthesis pilocarpine. J New Med. 1978;6:275–6.
3. Lekha Nair K, Vidyanand S, James J, et al. Pilocarpine-loaded poly (DL-lactic-co-glycolic acid) nanoparticles as potential candidates for controlled drug delivery with enhanced ocular pharmacological response. J Appl Polym Sci. 2012;124(3):2030–6.
4. Yang SJ. Pharmacology. Beijing: People's Medical Publishing House; 2012.
5. Al-Badr AA. Pilocarpine. Anal Profiles Drug Subst. 1983;12(8):385–432.
6. Tang XL. Advancement of drug treatment in glaucoma. Chin Pharm J. 2005;40(18):1369–72.
7. Li N, Zhuang C, Wang M, et al. Liposome coated with low molecular weight chitosan and its potential use in ocular drug delivery. Int J Pharm. 2009;379(1):131–8.
8. Desai SD, Blanchard J. Pluronic F127-based ocular delivery system containing biodegradable polyisobutylcyanoacrylate nanocapsules of pilocarpine. Drug Deliv. 2000;7(4):201–7.
9. Gavini E, Chetoni P, Cossu M, et al. PLGA microspheres for the ocular delivery of a peptide drug, vancomycin using emulsification/spray-drying as the preparation method: in vitro/in vivo studies. Eur J Pharm Biopharm. 2004;57(2):207–12.
10. Vandamme TF. Microemulsions as ocular drug delivery systems: recent developments and future challenges. Prog Retin Eye Res. 2002;21(1):15–34.

Rotundine

Tian-Yi Yuan, Li-Da Du, and Guan-Hua Du

Abstract Rotundine is a kind of organic alkaloid with analgesic activity extracted from traditional Chinese medicinal plant named *Corydalis rhizoma*. It has been used in a lot of countries as an alternative to anxiolytic and sedative drugs of the benzodiazepine group and analgesics such as opiates. Rotundine has potent effects on the central nervous system but has nothing to do with opioid receptors. Therefore, it has little addiction effects. Recent studies also found other pharmacology effects of Rotundine, such as cardiovascular protective effects and antitumor effects. Because of its low toxicity and less adverse reactions, Rotundine has attracted a lot of attention in pain management. It has great market prospects and deserves further development and utilization.

Keywords Rotundine · Organic alkaloid · Analgesia · Addiction

Alias:

(−)-Gindarine; (13aS)-2,3,9,10-Tetramethoxy-5,6,13,13a-tetrahydro-8H-dibenzo[a,g]quinolizine; (S)-Tetrahydropalmitine; [13aS,(−)]-5,8,13,13a-Tetrahydro-2,3,9,10-tetramethoxy-6H-dibenzo[a,g]quinolizine; Gindarine; (13aS)-2,3,9,10-tetramethoxy-6,8,13,13a-tetrahydro-5H-isoquinolino[2,1-b]isoquinoline; TETRAHYDROPALMATINE; L-(P); L-Tetrahydropalmatine; L-THP

Origin: *Curcuma longa* (Fig. 1)
Chemical name (Fig. 2)

(S)-2,3,9,10-tetramethoxy-5,8,13,13a-tetrahydro-6H-dibenzo[a,g]quinolizine

Molecular formula, $C_{21}H_{25}NO_4$; **MW**, 355.428; **CAS**, 10097-84-4

T.-Y. Yuan · G.-H. Du (✉)
Beijing Key Laboratory of Drug Targets Identification and Drug Screening, Centre for Pharmaceutical Screening, Institute of Materia Medica, Chinese Academy of Medical Sciences and Peking Union Medical College, Beijing, China
e-mail: dugh@imm.ac.cn

L.-D. Du
School of Biomedical Sciences, Chinese University of Hong Kong, Hong Kong, China

G.-H. Du, *Natural Small Molecule Drugs from Plants*,
https://doi.org/10.1007/978-981-10-8022-7_52

Properties

Appearance: off-white solid, odorless, tasteless, and turns yellow under light and heat. It should be protected from light and air. *Solubility*: this product is usually dissolved in chloroform, slightly soluble in ethanol or ether, insoluble in water, and soluble in dilute sulfuric acid. *Melting point*: 141–144 °C. *Specific optical rotation*: −290 to −300°.

Dosage Forms and Indications

This product is found in the *Pharmacopoeia of the People's Republic of China (2015)*.

1. Rotundine sulfate injection is used for pain induced by gastric ulcer and duodenal ulcer, dysmenorrhea, uterine contraction pain, tension insomnia, spasmodic cough, and so on.
2. Rotundine tablets is used for headache and dysmenorrhea and helps with sleep.
3. Rotundine hydrochloride tablets: analgesic.
4. Tetrahydropalmatine sulfate is a racemic tetrahydropalmatine, and its effective component is the levo isomer, Rotundine. It is used for moderate pain, making it especially suitable for patients who cannot sleep because of pain and tension.

Literature

Rotundine is a kind of organic alkaloid with analgesic activity extracted from a traditional Chinese medicinal plant named *Corydalis rhizoma* (Yan Hu, Xuan Hu, or Yuan Hu). Yan Hu Suo is the dry tuber of the perennial herbaceous plant *Corydalis yanhusuo* W. T. Wang, which belongs to the *Papaveraceae* family. According to Chinese theories of medicine, Yan Hu Suo effectively unclogs blood vessels, dissolves blood stasis, and alleviates pain. Yan Hu Suo was firstly recorded in *Lei Gong's Treatise on the Preparation*. According to *Ben Cao Gang Mu*, Princess Jingmu of the Ming Dynasty (more than 500 years ago) developed epigastric pain when eating soba noodles while angry. A famous Chinese doctor, Li Shizhen, was consulted, who recalled a statement from *Lei Gong's Treatise on the Preparation* recommending "Yan Hu as quickly as possible in the event of cardiodynia." Thus, the princess was given Yan Hu Suo soaked in warm wine, and then she stopped vomiting and was relieved of her stomachache.

Yan Hu Suo grows in Anhui, Jiangsu, Zhejiang, Hubei, Henan, and other provinces. At present, Shaanxi, Gansu, and other places are also introduced to its cultivation but it is mainly in Zhejiang. Yan Hu Suo grows in the mountain forest, mainly by artificial cultivation. *Chinese Pharmacopoeia* prescribed medicinal varieties for Yan Hu Suo, which is the dry root tuber of *Corydalis yanhusuo* W.T. Wang. In early summer, the plants are dug when the stems and leaves become wilted. The fibrous roots are removed, washed, and put into boiling water to cook. At last, the roots are dried in the sun ready for use. *Chinese Pharmacopoeia* has the same plants with the above mentioned *Corydalis*, including the Northeast *Corydalis*, Tu Yan Hu, and North Korea *Corydalis*. These medicinal herbs contain similar chemical composition to the abovementioned Yan Hu Suo. At present, Yan Hu Suo preparations can be used clinically as analgesic and topical narcotic drugs.

History of R&D

From the 1920s, many researchers have focused on the alkaloid compounds contained in the Yan Hu Suo and extracted more than 30 different kinds of alkaloids. Among them Rotundine is one of the most important active compounds and has been successfully applied in clinical practice.

Tetrahydropalmatine (THP) was firstly discovered by Zhao Chenggu, a famous medicinal phytochemist in China [1]. He isolated 13 kinds of alkaloids and identified the structure of 6 of them. In 1936, Huang Minglong, a famous organic chemist, proved THP is a racemate. From 1933, Wang Jingxi and Lu Zihui began to study the pharmacological effects of THP and found that THP could induce catalepsy in animals. Rotundine is a levo-THP and obtained from the resolution of DL-THP during 1959–1964. From 1956 to 1965, the famous Chinese neuropharmacologist, Jin Guozhang, identified Rotundine as the main effective component in Yan Hu Suo and further studied its underlying mechanisms. He also found that *Stephania* contains abundant L-THP, which could become the source of clinical medication.

In 1965, Rotundine was recorded in the textbooks of pharmacology. After over 20 years of clinical practices, people have proved that Rotundine has reliable efficacy and low side effect. In 1977, Rotundine was recorded in the *Chinese Pharmacopoeia*. In 1979, THP achieved a full synthesis, and currently Rotundine is mainly produced by artificial synthesis. From 1980, the study on the mechanisms of Rotundine made great progress. The analgesic effect of Rotundine has nothing to do with opioid receptors or prostaglandins. Rotundine does not belong to narcotic analgesics and antipyretic analgesics. Further, Rotundine has been found to have no affinity with M-cholinergic receptor and GABA system in the brain. But it has affinity to dopamine receptors [2]. The exact mechanisms and pharmacology effects of Rotundine still need more research.

Pharmacology

1. Effects on the central nervous system: Rotundine has significant analgesic, sedative, and hypnotic effects, and its mechanism has nothing to do with opioid receptors. It has no obvious addiction effect. Using Rotundine combined with pethidine can enhance the analgesic effect, reduce the amount of pethidine, and further reduce the occurrence of drug dependence. The present study suggests that Rotundine is a central dopamine receptor blocker that inhibits the transmission of peripheral pain information by blocking the D2 receptor of the striatum and nucleus accumbens and the PAG-RgpI-spinal dorsal horn nerve pathway [3]. In addition, Rotundine also has a suppression of the strengthening of the motor response induced by drug abuse, that is, inhibition of behavioral sensitization.
2. Effects on the cardio-cerebrovascular system: Rotundine has inhibitory effect on the injuries caused by cerebral ischemia and reperfusion [4] and myocardial ischemia and reperfusion [5]. It also could lower the blood pressure [6]. The pharmacological mechanism of Rotundine on the cardiovascular and cerebrovascular aspects is quite complex and may relate to the regulation of inflammation and inhibition of apoptosis. It is also found that Rotundine acts on α-receptors and has a noncompetitive antagonistic effect on calcium [7].
3. Rotundine can reverse the drug resistance of tumor cells; the mechanism may be through the downregulating of the expression of P-glycoprotein (P-gp) and the upregulating of the expression of topo II in tumor cells [8].
4. Rotundine has an excitatory effect on the endocrine system, such as the pituitary-adrenal system, and also has a protective effect on the liver [9].
5. Drug metabolisms: The absorption of Rotundine by oral administration is complete. At 15 min it can be absorbed by 40–50%. It begins to work at 10–30 min, and the effect can last 2–5 h. In vivo distribution is through fat and the lung, liver, and kidney. If administered by subcutaneous injection, after 12 h, Rotundine will be discharged by 80% with the urine [10].

Clinical Application

1. Pain induced by gastrointestinal and hepatobiliary disorders, menstrual pain, and pain after childbirth.
2. Mild trauma and postoperative pain.
3. Headache insomnia and spasmodic cough.
4. There is a report about the usage of Rotundine in arrhythmia caused by I–III hypertension and other diseases.

5. The hypnotic effect of Rotundine happens in 15 min after administration and then appears after 2 h. Because the analgesic effect happens at the same time, Rotundine is particularly suitable for insomnia patients with pain.
6. Side effects: drowsiness, dizziness, fatigue, and nausea. Large dose of Rotundine can inhibit the respiratory center and may cause extrapyramidal symptoms.

Discussion

Rotundine, as a natural small molecule drug from plants, has analgesic, antitussive, and hypnotic effects. Rotundine has lower toxicity and less adverse reactions, and its application is quite safe. When the analgesic effect is achieved, mild hypnotic effect occurs at the same time. It is particularly suitable for patients who cannot sleep because of pain. On chronic persistent pain and visceral dull pain, Rotundine shows better effects. But it has worse effects on acute pain or terminal cancer pain. Because of the lack of addiction and dependence, Rotundine has now attracted worldwide attention. It has great market prospects and deserves further development and utilization.

Fig. 1 *Corydalis rhizoma* (Yan Hu Suo, 延胡索)

Fig. 2 Chemical structure of Rotundine

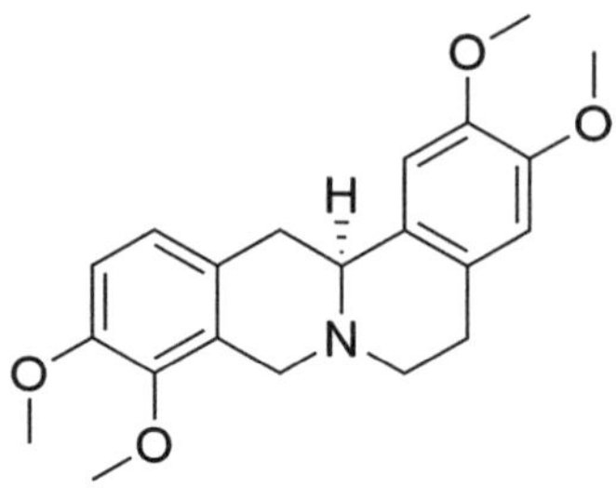

References

1. Zhao CG. The studies on several traditional Chinese medicines. Chin Sci Bull. 1953;08:59–62. (Chinese)
2. Jin GZ. Advances in pharmacological studies of L-tetrahydropalmatine and its second-generation new drug. Acta Pharm Sin. 1987;06:472–80. (Chinese)
3. Chu H, Jin G, Friedman E, et al. Recent development in studies of tetrahydroprotoberberines: mechanism in antinociception and drug addiction. Cell Mol Neurobiol. 2008;28(4):491–9.
4. Mao XW, Pan CS, Huang P, et al. Levo-tetrahydropalmatine attenuates mouse blood-brain barrier injury induced by focal cerebral ischemia and reperfusion: involvement of Src kinase. Sci Rep. 2015;5:11155.
5. Han Y, Zhang W, Tang Y, et al. l-Tetrahydropalmatine, an active component of Corydalis yanhusuo W.T. Wang, protects against myocardial ischaemia-reperfusion injury in rats. PLoS One. 2012;7(6):e38627.
6. Chueh FY, Hsieh MT, Chen CF, et al. Hypotensive and bradycardic effects of dl-tetrahydropalmatine mediated by decrease in hypothalamic serotonin release in the rat. Jpn J Pharmacol. 1995;69(2):177–80.
7. Meng HX, Wang B, Liu JX. Effect of salvianolic acid B and tetrahydropalmatine on the L-type calcium channel of rat ventricular myocytes. Zhongguo Zhong Xi Yi Jie He Za Zhi. 2011;31(11):1514–7.
8. Sun S, Chen Z, Li L, et al. The two enantiomers of tetrahydropalmatine are inhibitors of P-gp, but not inhibitors of MRP1 or BCRP. Xenobiotica. 2012;42(12):1197–205.
9. Du YD ZYZ. Research progress of pharmacology and clinical application of rotundine. Northwest Pharm J. 2012;01:91–4. (Chinese)
10. Liu P, Li W, Li ZH, et al. Comparisons of pharmacokinetic and tissue distribution profile of four major bioactive components after oral administration of Xiang-Fu-Si-Wu decoction effective fraction in normal and dysmenorrheal symptom rats. J Ethnopharmacol. 2014;154(3):696–703.

Scopolamine

Lv-Jie Xu, Ai-Lin Liu, and Guan-Hua Du

Abstract Scopolamine is a type of alkaloid from a variety of Solanaceae plants, which is the main active ingredient of these plants. It was first isolated from the *Scopolia japonica*, so named as scopolamine. Due to its obvious deficiencies in solubility and usage, its structure was modified as scopolamine butylbromide, which is widely used in clinical for the treatment of gastrointestinal motility hyperthyroidism or cramps, biliary colic, renal colic, and so on. With the deeper understanding of this drug, clinical application will be given more scientific and reasonable guidance.

Keywords Scopolamine · Scopolamine butylbromide · Anti-choline drug

Alias: Hyoscine; Transderm Scop; Scopine (-)-tropate; Hyosol; Scopine tropate
Origin: *Scopolia japonica* (Fig. 1)
Chemical name (Fig. 2)

9-Methyl-3-oxa-9-azatricyclo[3.3.1.02,4]non-7-yl ester-alpha-(hydroxymethyl) phenylacetate (Fig. 2)

Molecular formula, $C_{17}H_{21}NO_4$; **MW**, 303.35; **CAS**, 51-34-3

Properties

Appearance: a kind of viscous liquid, brown color. *Solubility*: soluble in ethanol, ethyl ether, chloroform, acetone, and water, very soluble in hot water, slightly soluble in benzene and petroleum ether, and also soluble in cold water. It can generate various crystals with multiple inorganic or organic acids. *Melting point*: 59 ± 1.0 °C.

L.-J. Xu · A.-L. Liu (✉) · G.-H. Du (✉)
Beijing Key Laboratory of Drug Targets, Research and Drug Screening, National Centre for Pharmaceutical Screening, Institute of Materia Medica, Chinese Academy of Medical Sciences and Peking Union Medical College, Beijing, China
e-mail: liuailin@imm.ac.cn; dugh@imm.ac.cn

G.-H. Du, *Natural Small Molecule Drugs from Plants*,
https://doi.org/10.1007/978-981-10-8022-7_53

Boiling point: 460.3 ± 1.6 °C. *Specific optical rotation*: [α] −28° (in water), −18° (in ethanol) [1].

Derivatives (Fig. 3)

Scopolamine butylbromide, which is one of the derivatives of scopolamine, is a synthesized drug from scopolamine and used in clinic. Scopolamine and its derivatives hyoscyamine, anisodamine, and anisodine also have some pharmaceutical functions. The structures of the derivatives are introduced in Fig. 3.

Dosage Forms and Indications

Scopolamine butylbromide is reported in *Pharmacopoeia of the People's Republic of China* (2015), United States Pharmacopeia (40), the *Japanese Pharmacopoeia* (17th ed.), and *European Pharmacopoeia* (9.0th ed.).

Scopolamine butylbromide has been developed into injections, capsules, and tablets for clinical use [2]. Its indications are as follows: (1) used for administration before anesthesia, motion sickness, and Parkinson's disease; (2) used for mydriasis and relieving smooth muscle spasm, especially the gastrointestinal tract and dilation; (3) used for rescuing organophosphate poisoning; (4) mainly used for atropine allergy patients, but also for mild iridocyclitis; and (5) used for bronchial asthma and asthma-like bronchitis [3].

Literature

Scopolamine is a type of alkaloid that exists in a variety of Solanaceae plants such as *Scopolia japonica*, *Datura metel* L., and so on. It is the main active ingredient in these plants.

Chinese medicine *Scopolia japonica* has effects on diet stagnation and choke with nausea. It is documented in *Bencao Shiyi*, *Yao Xing Kao*, *Gangmu Shiyi*, *Kangxi Jixia Gewu Bian*, and other works, with a thousand years of application history. It grows in the dunes and sand and is distributed in the northeast, north, northwest, and other regions in China.

Chinese medicine *Datura metel* L. is documented in *Bencao Gangmu*, *Shengcao Yaoxing Beiyao*, *Bencao Biandu*, and *Luchuan Bencao*, with a long application history of more than 2000 years. It is distributed in tropical and subtropical regions and generally cultivated in temperate regions. There are wild plants in Taiwan, Fujian, Guangdong, Guangxi, Yunnan, Guizhou, Liaoning, Hebei, Henan, and China's other provinces. And more cultivated plants are distributed in Jiangsu, Zhejiang, Hubei, Sichuan, Shanghai, Nanjing, and other regions in the southern and northern of China. In addition, scopolamine can also be originated from *Atropa belladonna*, *Hyoscyamus niger*, *Datura metel*, *Datura innoxia*, *Scopolia carniolica*, and so on.

Apart from scopolamine, several other chemical ingredients also exist in *Scopolia japonica*, including hyoscyamine, anisodamine, anisodine, and so on. Hyoscyamine is an inhibitor of parasympathetic nerve, with the analgesic and antispasmodic functions, especially for sciatica, sometimes for the treatment of epilepsy, seasickness, etc., and its pharmacological effects are similar to atropine. However, its clinical application is less because of its toxicity. The clinical applications of anisodamine are treating infectious toxic shock, vascular disorders, various neuralgia, smooth muscle spasms, vertigo, fundus disorders and sudden deafness, and other diseases. It has definite curative effect and is widely used in clinical in China. Its synthetic product is called "654-2," which now still is an effective drug to treat infectious shock and other vascular diseases. While anisodine is used to treat vascular headache, retinal vasospasm, ischemic optic neuritis, cerebrovascular disease, acute paralysis, central dysfunction caused by carbon monoxide poisoning, tremor, paralysis, bronchial asthma, motion sickness, organophosphorus pesticide poisoning, and so on [4–6].

History of R&D

Scopolamine is the main active ingredient of *Scopolia japonica*, *Datura metel* L., and other Solanaceae plants. As early as 1892, the chemist E. Schmidt first isolated it from the *Scopolia japonica*, so named it as scopolamine. Due to its obvious deficiencies in solubility and usage, the researchers modified its structure as scopolamine butylbromide, which was synthesized and used scopolamine and bromo-n-butane with heating refluxing method. Preparation method of scopolamine butylbromide is simple, and it's easy to be synthesized. There are about 170 domestic enterprises producing its raw materials currently.

European Pharmacopoeia (9th ed.), United States Pharmacopeia (36), and the *Japanese Pharmacopoeia* (16th ed.) contained bromine scopolamine; British Pharmacopoeia (2015) contained the raw materials and preparations of scopolamine butylbromide, including tablets and injections. *Pharmacopoeia of the People's Republic of China* (2015) contained its raw materials, injection, and capsules.

Pharmacology

Scopolamine butylbromide, a modified scopolamine, is a peripheral anti-choline drug; it not only can smooth muscle spasm but also block the ganglion and neuromuscular junction and has the weak effect on center. This drug has stronger effects on intestinal smooth muscle spasm than atropine and anisodamine; it can selectively relieve gastrointestinal tract, biliary tract, and urinary tract smooth muscle spasm and inhibit its peristalsis. Scopolamine butylbromide has less effects on the heart, pupil, and salivary gland, so it rarely induces central nervous excitement, dilation, inhibition of saliva secretion, and other adverse reactions, which is unlike atropine.

Scopolamine butylbromide is an antagonist of the M cholinergic receptor, its pharmacological effects are similar to atropine, but it has its own characteristics that have an obvious advantage in improving blood microcirculation.

Oral absorption of scopolamine butylbromide is not easy. It produces efficacy at 2–4 min after intravenous injection, 8–10 min after subcutaneous or intramuscular injection, and 20–30 min after oral administration, which could maintain about 2–6 h [6].

Clinical Applications

Scopolamine butylbromide is used to treat various causes-induced gastrointestinal motility hyperthyroidism or cramps, biliary colic, renal colic, etc. It is characterized by having stronger antispasmodic effects, quicker onset, and less adverse reactions than atropine and anisodamine. The adverse reactions of this drug are mostly related to its anticholinergic characteristics; there could be thirst, vision disorders, drowsiness, palpitations, facial flushing, nausea, vomiting, dizziness, headache, and other reactions. Scopolamine butylbromide could reduce and lower esophageal sphincter pressure and increase muscle and gastroesophageal reflux. There are also allergic reactions. As it can increase intraocular pressure, glaucoma patients are hanged. It is widely used in clinical since its selectivity is higher than atropine, which causes less adverse reactions [7].

Discussion

As the pharmacological mechanism of scopolamine butylbromide is clear, and the selectivity is strong, it has a good application prospect. It also played a positive role in promoting the treatment theory of clinical shock through the application of this drug. We believe that, with the deeper understanding of this drug, clinical application will be given more scientific and reasonable guidance.

Fig. 1 *Scopolia japonica*

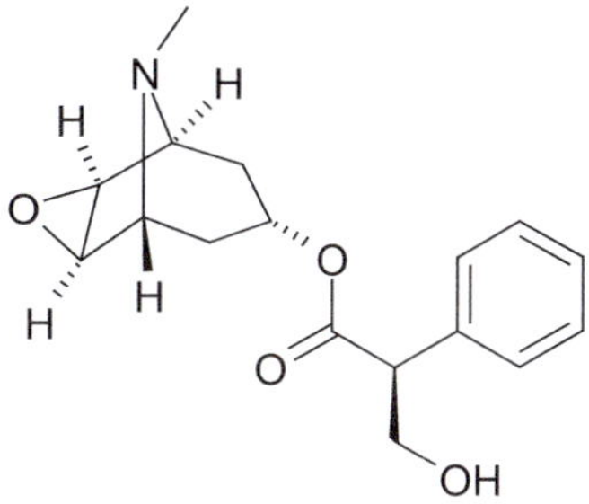

Fig. 2 The structure of scopolamine

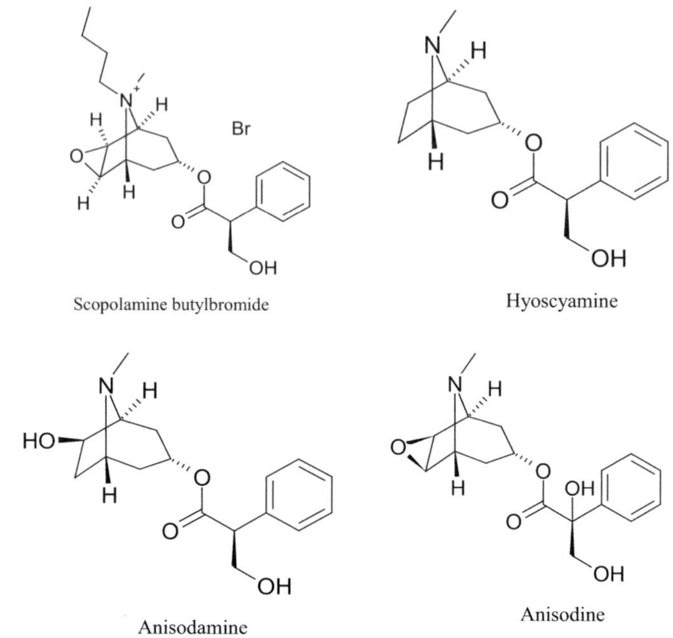

Fig. 3 The structures of the derivatives of scopolamine

References

1. Drevets WC, Zarate CA Jr, Furey ML. Antidepressant effects of the muscarinic cholinergic receptor antagonist scopolamine: a review. Biol Psychiatry. 2013;73(12):1156–63.
2. National Pharmacopoeia Commission. Chinese pharmacopoeia 2015 edition 2. Beijing: China Medical Science and Technology Press; 2015.
3. Li Y, et al. Neuropsychiatric disease rational drug use. 2nd ed. Beijing: People's Health Publishing House; 2009.
4. Institute of Pharmacology, Chinese Academy of Medical Sciences. Pharmacological effects of anisodamine. Chin J Med Sci. 1973;53(5):269–73.
5. Guo HY, Liu TE, Cao Q, et al. Anisodamine (654-2) inhibits the binding of M-, alpha1- and alpha2-receptor (brief). J Chin Acad Med Sci. 1986;8(3):170.
6. State Administration of Traditional Chinese Medicine Information Center Station. Handbook of effective ingredients of botanicals. Beijing: People's Health Publishing House; 1986.
7. Yuan H, Wang Y, Liu Y. Advances in pharmacology of anisodamine. Mod J Integr Tradit Chin West Med. 2004;13(16):2210–1.

Securinine

Dan-Shu Wang, Lian-Hua Fang, and Guan-Hua Du

Abstract Securinine is derived from the leaf of securinega suffruticosa Rehd., in which securinine is the most important alkaloid. Securinine mainly has nerve pharmacological activity; it also has antitumor, anti-inflammatory, and antivirus effects, improves bone marrow hematopoietic function, and reduces blood pressure. It has been widely used in clinical practice, mainly for the treatment of polio sequelae and facial nerve palsy. However, there are some adverse reactions in the process of clinical treatment, such as liver damage, elevated serum transaminase, or mild muscle tremor and other nerve stimulation phenomena. Therefore, it is necessary to explore alternatives that have minor side effects and to develop new drugs by applying advanced technology to regulating the clinical application of securinine.

Keywords Securinine · Alkaloid · Nerve · Clinical application

Alias: Foliaceous bead alkali
Origin: Securinega suffruticosa Rehd. (Fig. 1)
Chemical name (Fig. 2)

(6S,11aR,11bS)-9,10,11,11a-Tetrahydro-8H-6,11b-methanofuro[2,3-c]-pyrido [1,2-a]azepin-2(6H)-one

Molecular formula, $C_{13}H_{15}NO_2$; **MW**, 217.26; **CAS**, 610-40-2

D.-S. Wang · L.-H. Fang (✉) · G.-H. Du (✉)
Beijing Key Laboratory of Drug Targets Identification and Drug Screening, Centre for Pharmaceutical Screening, Institute of Materia Medica, Chinese Academy of Medical Sciences and Peking Union Medical College, Beijing, China
e-mail: fanglh@imm.ac.cn; dugh@imm.ac.cn

G.-H. Du, *Natural Small Molecule Drugs from Plants*,
https://doi.org/10.1007/978-981-10-8022-7_54

Properties

Appearance: yellow crystalline powder, slightly bitter. *Solubility*: soluble in anhydrous ethanol or chloroform, insoluble in water. *Melting point*: 139–140 °C *Specific optical rotation*: (D) −1042°(C = 1, EtOH).

Dosage Forms

Nitrate injection

Indications

Poliomyelitis sequela and facial paralysis, neurasthenia, hypotension, autonomic dysfunction, and others

Literature

Securinine is derived from the leaf of Securinega suffruticosa Rehd., which was first separated by V.I. Murav'eva in 1956 [1]. Securinega suffruticosa Rehd., a kind of subshrub plant, is widely distributed in temperate and subtropical regions [2]. It has strong adaptability and can be planted in most parts of China. Moreover, it is rich in natural resources. Securinega suffruticosa Rehd. is a folk medicine whose root is the main component, with the effects of activating circulation, spleen, and stomach to treat rheumatism waist pain, limb numbness, infant malnutrition, and other diseases. The roots, stems, leaves, flowers, and skin of Securinega suffruticosa Rehd. contain alkaloids, in which securinine is the main ingredient.

History of R&D

Securinine was first isolated by a Soviet scholar from the Wusuli region, but its chemical structure was separated and finally determined by Chinese scholars from local sources [3]. The main structural feature is a tetracyclic compound containing an indolizine, pyrrole elidine, or quinolizine ring and an α, β-unsaturated pentahydrin ring whose basic structural skeleton type is as shown in what follows (Fig. 3) [4].

Owing to the complexity of the chemical reaction of securinine, reports on its chemical reactivity and structural modification have been relatively scarce since securinine was reported on in 1956. However, the research on chemical total synthesis and biological activities of securinine has made a certain amount of progress.

Securinine has a rigid molecular structure containing four rings and three chiral centers, which makes synthesis difficult. From 1974 to 1978, three research groups from Japan, the USA, and Canada studied the biosynthesis of securinine by feeding animals with isotope-labeled S. suffruticosa [5]. Five more researchers, Horii, Liang, Honda, Liras, and Albes reported on a method of total synthesis. In other routes, securinine is synthesized into a racemate, except in Horii and Albes's report. In addition, Horii first achieved the total chemical synthesis of securinine in 1967, they used 1,2-cyclohexanedione as the B ring precursor of securinine to condense with its A ring precursor which used 2-lithium-pyridine, then constructed the mother nucleus of securinine [4].

Norsecurinines, another major class of structures, have unstable chemical properties and is easy to polymerize under alkaline conditions, making it more difficult to synthesize. At present, four groups are researching its total synthesis, two cases of which are racemates and the other two are optical pure norsecurinines [6]. The study of total chemical synthesis guides the direction for structural modification and activity optimization of securinine. In recent years, some scholars have studied the chemical reactions related to securinine, and the latest research shows that securinine is involved in 1,3-dipolar cycloaddition reaction, [2 + 2] photochemical cycloaddition reaction, and Baylis-Hillman reaction to obtain the corresponding rigid dimer. Part of the amino acid ethyl ester and amide derivatives can occur Michael addition reaction with securinine so as to get the corresponding amino acid conjugate. The light dimerization selectivity of securinine has no relation to certain external factors such as solvent polarity, reaction time, and substrate concentration but is related to the complexity of its structure, the reversibility of the reaction, and the stability of the dimerization products [7].

For its pharmacological activity, early studies suggest that securinine has a significant excitatory effect on the spinal cord, similar to strychnine, a typical spinal cord excitatory, soon after the nitrate securinine is used for the clinical treatment of poliomyelitis convalescence, paralysis, facial nerve palsy, and neurasthenia, but its mechanism on exciting the central nervous system was found by Beutler by 1985. Beutler found in ligand binding experiments that the effect of securinine and its analogs on the GABA receptor is selective. Electrophysiological experiments on neuronal cells also showed that securinine can adversely affect the inhibition of GABA. Among the derivatives, securinine can interact with the GABA receptor, whereas allosecurinine and virosecurinine have no pharmacological activities. Later, scholars at home and abroad have carried out further research on clinical and pharmacological aspects and found that right-handed securinine can significantly enhance the activity of choline acetylcholine acetylchotransferase and reduce the awareness caused by β-amyloid obstacles. Securinine mainly has nerve pharmacological activity. Recently some scholars have found that it also has antitumor, anti-inflammatory, and antivirus effects, improves bone marrow hematopoietic function, and reduces blood pressure.

Pharmacology

The pharmacological effect of securinine is mainly manifested as a central nervous system excitatory effect. As a GABA receptor inhibitor, it has an excitatory effect similar to that of strychnine on the spinal cord. A low dose of securinine can improve the excitability of brain reflection, while a high dose of securinine will cause febrile seizures, and at the same time securinine can strengthen the conditioned reflex of the cerebral cortex, shorten the latency period, and promote learning and memory capability [8], and so it is expected to be a promising drug for the treatment of Alzheimer's disease.

Securinine can improve the hematopoietic environment of patients with aplastic anemia, promote cell proliferation, have a synergistic anti-tumor effect in combination with cyclophosphamide (CTX), and can antagonize bone marrow suppression caused by CTX. It has an inhibitory effect on cell proliferation of human leukocyte cell K562 and four other kinds of tumor cells. The mechanism may be correlated to the induction of apoptosis, the promotion of external calcium influx, the increase of intracellular calcium concentration, and the down-regulation of Bcl-2 gene expression [8]. In addition, Weenen et al. found that securinine isolated from the root bark of Margaritariadiscoidea (Baill.) Webster has significant antimalarial activity, which may be because the α, β-unsaturated carbonyl group is susceptible to Michael addition reaction with parasite's nucleic acid composition [9]. Some scholars have also found that securinine has antitype I herpes simplex virus, antibacterial, and anti-inflammatory activity, which can completely inhibit the spore germination of *Alternaria alternata*, Curvularia leaf spot, spore bacteria, and curvular fungi [10]. The latest reports suggest that the differentiation of gliomas is closely related to the mutant IDH1 of isocitrate dehydrogenase. According to the key role of IDH1 and securinine on the differentiation of neuronal cells, it is presumed that securinine is likely to be a candidate drug for suppressing glioma-targeting mutant IDH1 [7].

Clinical Application

In recent years, securinine has been widely used in clinical practice, mainly for the treatment of polio sequelae and facial nerve palsy. It also has a certain effect on neurasthenia, hypotension and dizziness, tinnitus, and deafness caused by autonomic dysfunction. The nitrate and hydrochloride of left-handed securinine are mainly used clinically. In addition, securinine eye drops are found to have a remarkable effect on the treatment of herpes simplex keratitis through initial clinical observation. However, there are some adverse reactions in the process of clinical treatment. Larger doses over a long period of time may produce liver damage, elevated serum transaminase or mild muscle tremor, hand and foot numbness, and other nerve stimulation phenomena, but most of these conditions will return to normal upon drug withdrawal, so one should pay attention to usage and dosage in clinical use.

Discussion

At present, research on the bioactivity of securinine, especially the mechanism by which it excites the central nervous system, has made significant progress. More and more new pharmacological effects have been discovered gradually. The clinical application of securinine are not limited to neurological diseases; it also has the potential to treat high blood pressure, Alzheimer's disease, and tumors. However, research on its chemical reactions and structural modification is still scarce, and more extensive research needs to be conducted.

Fig. 1 Securinega suffruticosa Rehd.

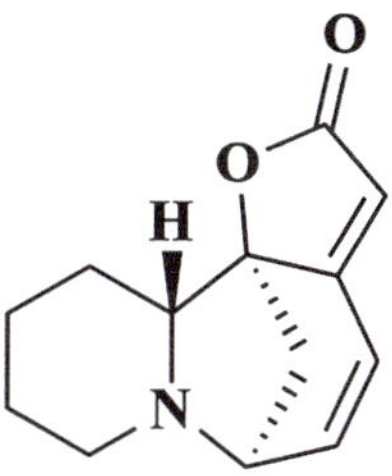

Fig. 2 The chemical structure of securinine

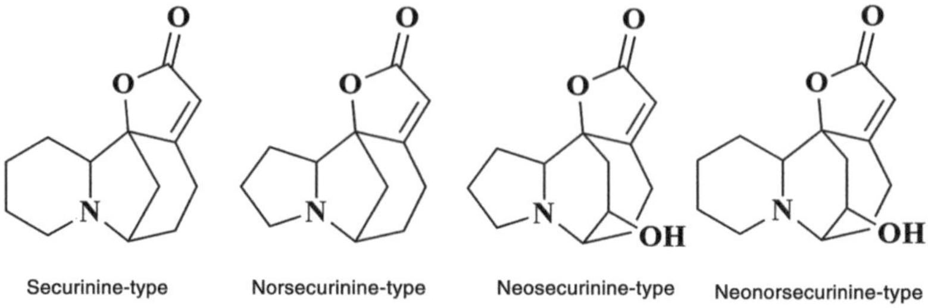

Fig. 3 Basic structural skeleton type of securinine

References

1. Murav'eva VI, Ban'kovskii AI. Chemical studies on alkaloids from Securinegasemishrub:Securinegasuffruticosa (Pall.)Rehd.Dokl.Akad. NaukSSSR. 1956;110:998–1000.
2. Yuan W, Lu Z, Liu Y, et al. Three new podocarpane-type diterpenoids from callus of Securinegasuffruticosa. Chem Pharm Bull (Tokyo). 2005;53(12):1610–2.
3. Liu Y, Yue ZH, Zhang N, et al. Development of research on Securinine. Chin Pharm Aff. 2009;23(8):817–8.
4. Yuan W. Biosynthesis and biotransformation of Securinega alkaloids. Beijing: Chinese academy of medical sciences, Beijing union medical college; 2005.
5. Sanakawa U, Ebizuka Y, Yamasaki K. Biosynthesis of securinine, the main alkaloid of Securinegasuffruticosa. Phytochemistry. 1977;16:561–3.
6. Han G, LaPorte MG, Folmer JJ, et al. Total syntheses of the Securinega alkaloids(+)-14,15-dihydronorsecurinine, (−)-norsecurinine, and phyllanthine. J Org Chem. 2000;65(20):6293–306.
7. Hu YQ. Chemical reactivity of Securinine and synthesis of dimers and amino acid conjugates of Securinine. Guangdong: Jinan University; 2015.
8. Lu XJ. Studies on chemical constituents of the leaves of Securinega suffruticosa (Pall.) Rehd. Jilin: Jilin University; 2010.
9. Weenen H, Nkunya MH, Bray DH, et al. Antimalarial compounds containing an alpha, beta-unsaturated carbonyl moiety from Tanzanian medicinal plants. Planta Med. 1990;56(4):371–3.
10. Singh AK, Pandey MB, Singh UP. Antifungal activity of an alkaloid allosecurinine against some fungi. Mycobiology. 2007;35(2):62–4.

Strychnine

Li Gao, Li-Da Du, Xue-Mei Qin, Jin-Hua Wang, and Guan-Hua Du

Abstract Strychnine, first discovered in 1818, was the first alkaloid to be identified in plants of *maqianzi* (*Strychnos nux-vomica* Linn.). Pharmacological effect of strychnine is mainly manifested in the central nervous system, cardiovascular system, digestive system and antitumor effects. Strychnine is commonly used in the treatment of flaccid paralysis, ENT diseases, diabetes, neurasthenia, acute myelitis, peripheral neuritis, and aplastic anemia in clinic. Due to the relatively high toxicity of strychnine, study of poisoning mechanism and toxicokinetics of strychnine is of great significance for the clinical safety.

Keywords Strychnine · Alkaloid · Toxicity

Origin: Nux vomica (Fig. 1)
Chemical name (Fig. 2)

(4aR,5aS,8aR,13aS,15aS,15bR)-4a,5,5a,7,8,13a,15,15a,15b,16-Decahydro-2H-4,6-methanoindolo[3,2,1-ij]oxepino[2,3,4-de]pyrrolo[2,3-h]quinoline-14-one

Molecular formula, $C_{21}H_{22}N_2O_2$; **MW**, 334.41; **CAS**, 57-24-9

L. Gao · X.-M. Qin
Modern Research Center for Traditional Chinese Medicine, Shanxi University, Taiyuan, China

L.-D. Du
School of Biomedical Science, the Chinese University of Hong Kong, Hongkong, China

J.-H. Wang · G.-H. Du (✉)
Beijing Key Laboratory of Drug Targets Identification and Drug Screening, Centre for Pharmaceutical Screening, Institute of Materia Medica, Chinese Academy of Medical Sciences and Peking Union Medical College, Beijing, China
e-mail: dugh@imm.ac.cn

G.-H. Du, *Natural Small Molecule Drugs from Plants*,
https://doi.org/10.1007/978-981-10-8022-7_55

Properties

Appearance: prismatic orthorhombic crystal with very bitter tasting (ethanol). *Solubility*: soluble in boiling water, slightly soluble in water and ethanol, and insoluble in ether. *Melting point*: 286–288 °C. *Specific optical rotation*: +52.5° to +53.0°.

Derivatives (Fig. 3)

Dosage Forms

This product is recorded in *Pharmacopoeia of the People's Republic of China* (2005) and *European Pharmacopoeia* (8.7th ed.).

The current formulations for clinical use include strychnine tablets and strychnine nitrate injections.

Indications

Strychnine is clinically used for the treatment of barbiturate poisoning, hemiplegia, amblyopia, blood diseases, and streptomycin toxicity. Strychnine is used to treat aplastic anemia.

Literature

Strychnine was the first alkaloid to be identified in plants of *maqianzi* 马钱子 (*Strychnos nux-vomica* Linn.). Traditional Chinese medicine *Strychnos* is dried mature seeds of *Strychnos nux-vomica* Linn. The *Compendium of Materia Medica* recorded *Strychnos* seems as wood turtle, the core is less than the wooden turtle and the color is white. The *Compendium of Materia Medica* recorded that *Strychnos* is very bitter tasting. The birds and dogs would die if they eat *Strychnos*. If humans eat *Strychnos* by mistake, they would be subject to limb spasm.

Strychnos was recorded as the commonly used Chinese medicine by the original work of *Traditional Chinese Medicine Daquan*. The *Compendium of Materia Medica* recorded that *Strychnos* could "treat typhoid fever, throat bitten pain, eliminate the lumps." "Traditional Chinese Medicine Will" recorded that *Strychnos* could eliminate heat, poison, and cure ulcer, evil sore.

Strychnos is grown in the tropics and distributed in India, Vietnam, Myanmar, Thailand, Sri Lanka, and other places. In ancient India, the toxicity and medicinal efficacy of *Strychnos* are also known. The seeds of *Strychnos* (rich in strychnine) are imported into Europe as a poison for killing rodents and small carnivores. The main active ingredient of *Strychnos* is alkaloids, accounting for 1.5–5%, of which the content of strychnine is the maximum, accounting for 35–50% of the total alkaloids [1]. The structure of strychnine was determined until the nineteenth century.

History of R&D

Strychnine was first discovered from a kind of bean by French chemists Joseph Bienaimé Caventou and Pierre-Joseph Pelletier in 1818. At the same time, brucine, a derivative of strychnine, was separated from the plants of *Strychnos nux-vomica* Linn. The toxicity of brucine was lower than strychnine.

Historic records indicated that preparations containing strychnine (presumably) had been used to kill dogs, cats, and birds in Europe as far back as 1640. The structure of strychnine was first defined in 1946 by Sir Robert Robinson, and this alkaloid was synthesized in a laboratory by Robert B. Woodward in 1954. This is one of the most famous syntheses in the history of organic chemistry. Both chemists won the Nobel Prize (Robinson in 1947 and Woodward in 1965) [2].

Pharmacology

The pharmacological effect of strychnine is mainly manifested in the central nervous system, cardiovascular system, digestive system and antitumor effects.

Studies have shown that the effects of strychnine on the nervous system are mainly manifested in the following areas: (1) Strychnine can improve the excitability of the spinal cord. In the recovery period of acute myelitis patients, the nerve recovery is very slow. Strychnine can stimulate the reflex function of the spinal cord, improve the function of the sensory centers of cerebral cortex, and significantly speed up the recovery of nerve function of acute myelitis patients. (2) Strychnine can excite the medulla oblongata and improve the excitabilities of vascular movement center and respiratory center. (3) Strychnine can excite the cerebral cortex. The therapeutic dose of strychnine can excite the cerebral cortex, wake up the patients in the state of inhibition, and improve the functions of hearing, vision, taste, touch, and other receptors of patients [3].

The treatment effects of strychnine on the arrhythmia. Strychnine is able to block the voltage-gated sodium channel Nav1.5, which is a major subtype of the voltage-gated sodium (Nav) channel. Nav1.5 exists in cardiomyocytes of mammalian species and plays an important role in maintaining excitability of cardiomyocytes and excitability conduction. Studies have shown that strychnine can inhibit the Nav1.5 current in a concentration-dependent manner; therefore, it has a potential advantage in the treatment of arrhythmias [3].

Strychnine can promote the digestive function through reflex. Strychnine can increase the gastric secretion through stimulation of the sensory receptors and promote digestion and appetite. In addition, 0.5 mM and 1 mM strychnine can significantly induce cell shrinkage after HepG2 exposure to strychnine for 72 h [4] and can inhibit the proliferation of HepG2 cells and promote cell apoptosis, suggesting that strychnine may possess antitumor effects [5].

Clinical Application

Strychnine is commonly used in the treatment of flaccid paralysis, ENT diseases, diabetes, neurasthenia, acute myelitis, peripheral neuritis, and aplastic anemia in clinic. However, due to the relatively high toxicity of strychnine, excessive or long-term use of strychnine may lead to poison. Studies have shown that the combination of *Strychnos* and *Glycyrrhiza* or cinnamon can reduce the toxicity of strychnine [6, 7]. Therefore, further study of poisoning mechanism and toxicokinetics of strychnine is of great significance for the clinical safety.

Discussion

Due to the great toxicity and narrow therapeutic window of strychnine, oral intake or injection of strychnine may cause toxic reactions. Clinically, strychnine has been rarely used as a central stimulant. However, efforts should be intensified in clinical setting for the treatment of barbiturate poisoning, hemiplegia, amblyopia, blood diseases, and streptomycin toxicity.

Fig. 1 Nux vomica

Fig. 2 Chemical structure of strychnine

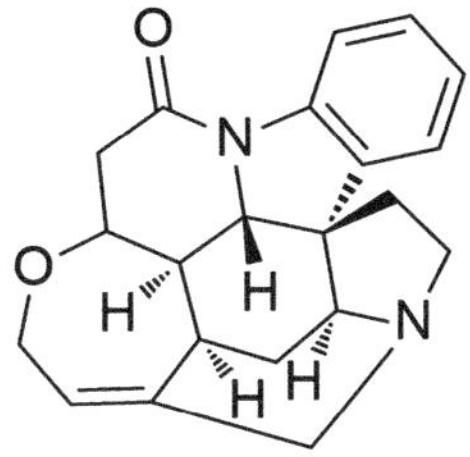

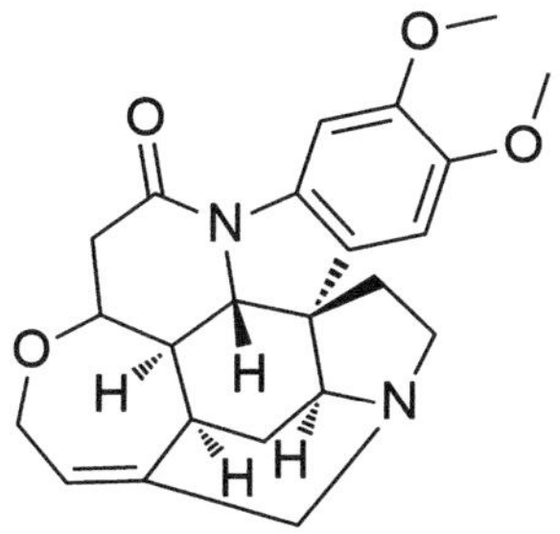

Brucine

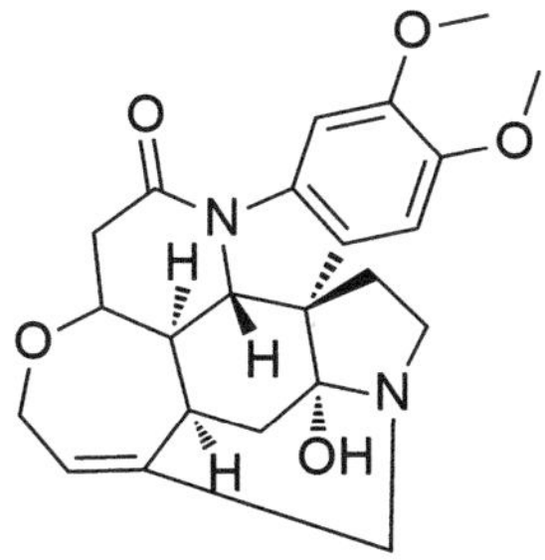

Pseudobrucine

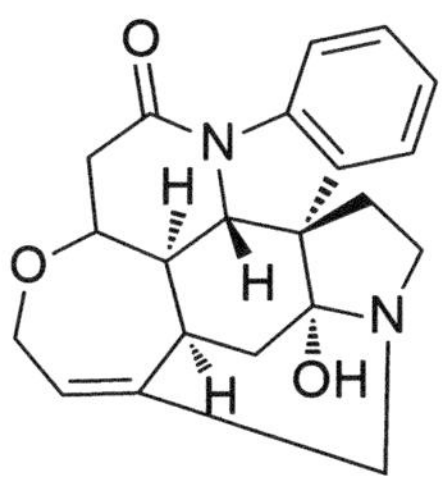

Pserdostrychnine

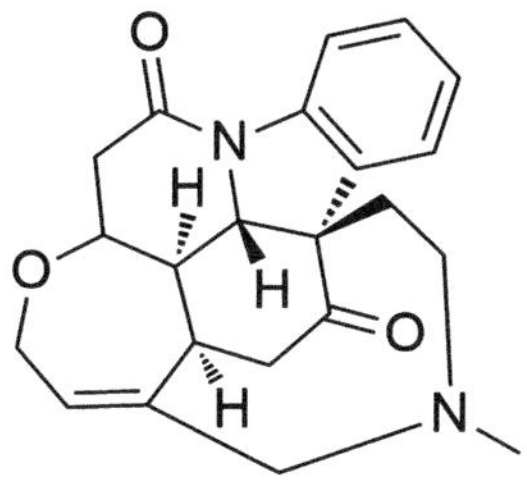

Icajine

Fig. 3 Chemical structures of derivatives of strychnine

References

1. Huang XR, Cao D. Research advancement of Semen Strychni. Shanghai J Tradit Chin Med. 2005;39(1):62–5.
2. Wikipedia. https://en.wikipedia.org/wiki/Strychnine.
3. Zhao XY. Effects of strychnine on potassium ion channels of neurons. Dalian: Dalian University of Technology; 2013.
4. Deng XK, Yin W, Li WD, et al. The anti-tumor effects of alkaloids from the seeds of Strychnos nux-vomica on HepG2 cells and its possible mechanism. J Ethnopharmacol. 2006;106(2):179–86.
5. Qu YG, Chen J, Cai BC. Advances on studies of strychnine. Chin J Exp Tradit Med Formul. 2011;17(24):247–51.

6. Yan XS, Zhu JW, Jiang B, et al. Analysis on strychnine and brucine in Strychnos Nux-vomical L. uncombined and combined with Glycyrrhiza Uralensis Fisch. Asia-Pac Tradit Med. 2009;5(10):18–9.
7. Yan XS, Zhu JW, Jiang B, et al. Analysis on strychnine and brucine in Strychnos Nux-vomical L. uncombined and combined with Cinnamomum cassia Presl. Chin J Exp Tradit Med Formul. 2010;16(6):77–8.

Tubocurarine

Shou-Bao Wang, Xiu-Ying Yang, and Guan-Hua Du

Abstract Tubocurarine is a kind of alkaloid isolated from various plant extract alkaloid arrow poisons which has a common name curare. The active dextrorotatory form was first purified by R. Boehm in 1879. It was first used in clinical practice in 1942, and it was the first typical non-depolarizing muscle relaxant. It binds the N2 cholinergic receptor on the motor nerve endplate and competitively blocks ACh-mediated depolarization, thus relaxing skeletal muscle. The drug is difficult to absorb under oral administration. For the intravenous injection, it is clinically used for anesthesia adjuvant drugs such as tracheal intubation and thoracoabdominal surgery. This drug also blocks ganglion and the release of histamine, causing a decline in heart rate and blood pressure, bronchial spasm, increased saliva secretion, etc. Artificial respiration and use of neostigmine are needed when large doses cause respiratory muscle paralysis. Due to the limitations of its plant source and more adverse reactions, this drug is now less used.

Keywords Tubocurarine · Cissampelosime methiodide · Non depolarizing muscle relaxant · N2 cholinergic receptor

Alias: Tubocurare, Curarine, Tubarine, Tubocurarine chloride
Origin: *Antiaris toxicaria* (Fig. 1)
Chemical name (Fig. 2)

7′,12′-Dihydroxy-6,6′-dimethoxy-2,2′,2′-trimethyltubocuraranium

Molecular formula, $C_{37}H_{41}N_2O_6$; **MW**, 609.73124; **CAS**, 57-94-3
Derivatives (Fig. 3): Dimethyl-L-curine dimethochloride; Cissampelosime methiodide (cissampareine)

S.-B. Wang · X.-Y. Yang · G.-H. Du (✉)
Beijing Key Laboratory of Drug Targets Identification and Drug Screening, National Centre for Pharmaceutical Screening, Institute of Materia Medica, Chinese Academy of Medical Sciences and Peking Union Medical College, Beijing, China
e-mail: dugh@imm.ac.cn

G.-H. Du, *Natural Small Molecule Drugs from Plants*,
https://doi.org/10.1007/978-981-10-8022-7_56

Properties

Appearance: white or slightly yellow crystalline powder. *Solubility*: it can be dissolved 50 mg/ml (22 °C) in water; easily soluble in methanol and ethanol; insoluble in ether, pyridine, chloroform, benzene, and acetone; and dissolved in sodium hydroxide solution. *Specific optical rotation*: +210 to +224°. *Melting point*: anhydrous 274–275 °C (decomposition).

Dosage Forms and Indications

This drug is referred in the *Pharmacopoeia of the People's Republic of China* (2015), the *British Pharmacopoeia* (2012), the *United States Pharmacopeia* (40), the *Japanese Pharmacopoeia* (15th ed.), the *European Pharmacopoeia* (7th ed.), and *The International Pharmacopoeia* (5th ed.).

The major formulation is injection. It is often injected intravenously and mainly used for abdominal surgery. It was once clinically used for the treatment of tremor paralysis, tetanus, rabies, and strychnine poisoning.

Literature

Tubocurarine is a toxic alkaloid historically known for its use as an arrow poison and obtained from the South American plant *Chondrodendron tomentosum*. It was originally thought that the dextrorotatory form was active, and its levo isomer isolated by the Chinese scientists from *Cyclea hainansis* and *Cyclea barbata Miers* (Menispermaceae) was proven to be active. Later, another derivative cissampelosime methiodide was also isolated from a medicinal herb *Yahulu* (*Cissampelos pareira*) used in Dai nationality.

History of R&D

Tubocurarine is a kind of alkaloid isolated from various plant extract alkaloid arrow poisons originating from Central and South America, with a common name curare. The active dextrorotatory form was first purified by R. Boehm in 1879. The drug was first used in clinical practice in 1942, and it was the first typical non-depolarizing muscle relaxant [1, 2].

In the 1970s, Chinese scientists isolated the levo isomer of tubocurarine from *Cyclea hainansis* and *Cyclea barbata Miers* (Menispermaceae). Its diiodomethane salt showed better relaxation on striated muscle. Another derivative dimethyl-L-curine dimethochloride further enhanced significantly the muscle relaxation efficacy [3].

Cissampelosime methiodide is another kind of muscle relaxant independently developed in China, which is isolated from the Dai medicine *Yahulu* (Menispermaceae plant *Cissampelos pareira*). It was mainly formulated into injection and exhibited significant striate muscle relaxation as its Chinese name means [4]. The discovery of Cissampelosime methiodide led into the innovative development of traditional Dai medicine which was later incorporated into the *Pharmacopoeia of the People's Republic of China* (1977).

Pharmacology

The dextroisomer of tubocurarine has pharmacological activity. It is classified into a non-depolarizing muscle relaxant and also known as competitive muscular relaxant. It binds the N2 cholinergic receptor on the motor nerve endplate and competitively blocks ACh-mediated depolarization, thus relaxing skeletal muscle.

The drug is difficult to absorb under oral administration. For the intravenous injection, onset time is 4–6 min. Upon administration of the drug, muscles used in rapid exercise such as eye muscle first relax, and then the muscles in the limbs, neck, and trunk relax too, followed by intercostal muscle relaxation and abdominal breathing. If the dose is increased, it can ultimately cause diaphragmatic paralysis until the breathing stops. The order of muscle relaxation recovery is contrary to that of muscle relaxation, i.e., the diaphragm is the fastest recovered. This drug is clinically used for anesthesia and adjuvant drugs such as tracheal intubation and thoracoabdominal surgery.

This drug also blocks ganglion and the release of histamine, causing a decline in heart rate and blood pressure, bronchial spasm, increased saliva secretion, etc. Artificial respiration and the use of neostigmine are needed when large doses cause respiratory muscle paralysis. Contraindications are myasthenia gravis, bronchial asthma, and severe shock.

Clinical Applications

The drug known for the muscle relaxants is mainly used for abdominal surgery and was once used for the treatment of tremor paralysis, tetanus, rabies, poison, and so on. For adults, the amount of one intravenous injection is 6–9 mg and can increase to 3–4.5 mg if necessary (the amount should be reduced to 1/3 in ether anesthesia). The action lasts for 20–40 min. The injection can be repeated according to the length of the operation time and muscle relaxation needs, and the dose is half of the first. For electrical shock, a dose of 0.165 mg/kg every time was administrated in 30–90 s. For diagnosis of myasthenia gravis, a dose of 0.004–0.033 mg/kg every time was used. However, attention must be paid that the drug can lead to the risk of paralysis of the respiratory muscles; emergency medicine and equipment must be prepared before. Oxygen supply, endotracheal intubation, and artificial respiration or injection of neostigmine at the same time (or phenolic ammonium chloride) can be carried out to counteract breathing stopping. It is contraindicated for the patients with myasthenia gravis. In addition, depolarizing muscle relaxants such as succinylcholine antagonizes non-depolarized muscle relaxant tubocurarine, and the clinical combination should be avoided.

Discussion

Due to the limitations of its plant source and more adverse reactions, the drug is now less used. As the substitutes of tubocurarine, other new non-depolarizing muscle relaxants such as vecuronium, ammonium, atracurium, cis-atracurium, and short-acting muscle relaxant mivacurium chloride are more preferably used currently.

Fig. 1 *Antiaris toxicaria*

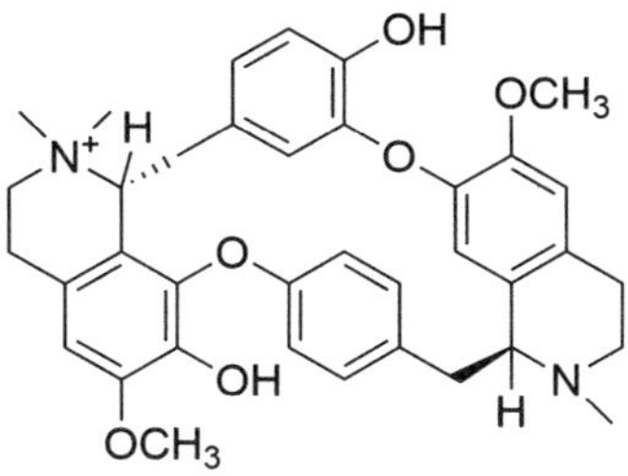

Fig. 2 Chemical structure of tubocurarine

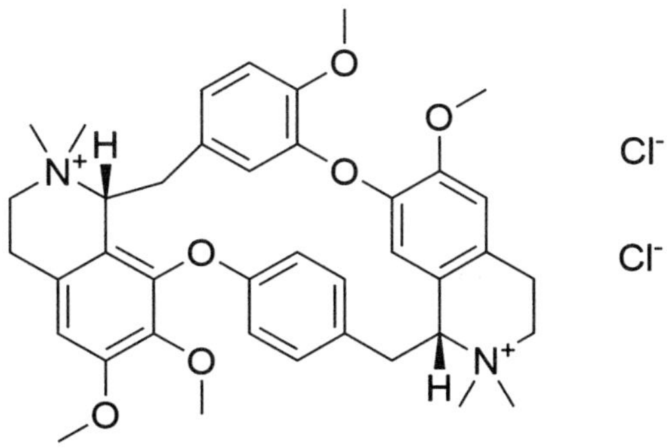

Fig. 3 Chemical structure of dimethyl-L-curine dimethochloride

References

1. Griffith HR, Johnson GE. The use of curare in general anesthesia. Anesthesiology. 1942;3(4):418–20.
2. Gray TC, Halton J. Technique for the use of d-tubocurarine chloride with balanced anaesthesia. Br Med J. 1946;2:293–5.
3. Shanghai Pharmaceutical Industry Research Institute, Haikou, Hainan Province, Guangdong Province, Haikou Hospital. Chloride left arrow poisoning trial production. China Med Ind J. 1979;5:22–41.
4. Zhiqiang M, Miao M, Suzhen Z, et al. Application of new Dai muscle relaxation clinical observation. Kunming Med Coll J. 1981;3–4:26–8.

Vanillin

Xiao-Bo Wang, Li-Da Du, Shu-Mei Wang, and Guan-Hua Du

Abstract Vanillin is extracted from gastrodin and has been used in the treatment of epilepsy. Vanillin is known as one of the first synthetic spices. It can be used as intermediate for the synthesis of a variety of drugs, such as berberine and antihypertensive drug L-methyldopa, methoxy-pyrimidine, and heart disease drug papaverine. It has an antiepileptic and anticonvulsion effect and antibacterial action ability. As an important pharmaceutical intermediate, vanillin was used in the production of drugs. The main methods used for the synthesis of vanillin are nitrosation and glyoxylic acid method. These stable and high-yield processes would promote the development direction of vanillin.

Keywords Gastrodin · Epilepsy · Synthetic

Alias: Vanillin, vanillin natural, vanillic aldehyde
Origin: Gastrodin (Fig. 1)
Chemical name (Fig. 2)

4-Hydroxy-3-methoxybenzaldehyde

Molecular formula, $C_8H_8O_3$; **MW**, 152.15; **CAS**, 121-33-5

X.-B. Wang · S.-M. Wang
Guangdong Pharmaceutical University, Guangdong, China

L.-D. Du
School of Biomedical Sciences, Chinese University of Hong Kong, Hong Kong, China

G.-H. Du (✉)
Beijing Key Laboratory of Drug Targets Identification and Drug Screening, Centre for Pharmaceutical Screening, Institute of Materia Medica, Chinese Academy of Medical Sciences and Peking Union Medical College, Beijing, China
e-mail: dugh@imm.ac.cn

G.-H. Du, *Natural Small Molecule Drugs from Plants*,
https://doi.org/10.1007/978-981-10-8022-7_57

Properties

Appearance: white or light yellow needle crystal or crystal powder, with a strong aroma. The relative density is about 1.060. *Solubility*: It is not only soluble in ethanol, chloroform, ether, carbon disulfide, glacial acetic acid, and pyridine but also in oil, propylene glycol, and hydrogen peroxide in alkaline solution. It can slowly oxidize in the air, can be unstable under illumination, and should be stored in a dark condition. *Melting point*: the melting point is 81 °C.

Derivatives

This product is included in the revised National Standard of Chemical Drugs (eighth). Vanillin tablets are used in clinical application at present.

Indications

It can be used to treat various types of epilepsy and attention deficit hyperactivity disorder and vertigo.

Literature

Vanillin is found in many plants, such as the tuber of *Rhizoma Gastrodiae* (Tian Ma), the whole herb of *Equisetum* (Mu Zei), *Ulva pertusa* (Kong Shi Chun), and sugar beets, vanilla beans, *Peru* balsam, and so on [1, 2].

History of R&D

Vanillin is known as one of the first synthetic spices. In the perfume industry, it is known as vanillic aldehyde. As early as 1858, French chemist Gby (Nicolas-Theodore Gobley) obtained pure vanillin for the first time [3] by the method of rectification. Due to less production yield of natural vanillin, it spurred the search for a chemical synthesis method of vanillin production. In 1874, German scientist

M. Haarman and co-workers deduced the chemical structure of vanillin and discovered a new way to produce vanillin with abietene as the raw material [4]. In 1965, Chinese scientists found that vanillin has antiepileptic effect and accomplished a study on the pharmacology and toxicology of vanillin from edible to officinal. They also found that vanillin has certain antibacterial activity, making it a suitable drug formulation for the treatment of skin disease. Vanillin can be used as intermediate for synthesis of a variety of drugs, such as berberine and antihypertensive drug L-methyldopa, methoxy-pyrimidine, and heart disease drug papaverine [5].

Pharmacology

Antiepileptic and anticonvulsion effect: the effect of vanillin on nervous system diseases such as depression, seizures, and epilepsy and other neurological and psychiatric diseases is mainly by promoting brain blood circulation and brain protection role. As a means of preventing and treating depression by olfactory pathway, it has been applied in clinics [6, 7].

Antibacterial action: the antibacterial mechanism of vanillin is not clear and may be related to the bacterial cell membrane integrity, enzyme activity, and activity of genetic material [8].

Clinical Application

Vanillin tablet has been used in the treatment of epilepsy and has a better therapeutic effect. Some patients have a minor dizziness response occasionally in the clinic.

Discussion

Vanillin is an important pharmaceutical intermediate; about 45% vanillin was used in the production of drugs in the world and 30% in China. The main methods used for the synthesis of vanillin are nitrosation and glyoxylic acid method. These stable and high-yield processes would promote the development direction of vanillin.

Fig. 1 Gastrodin

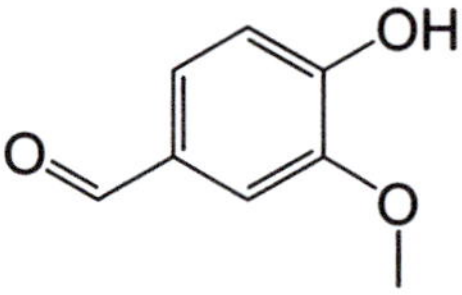

Fig. 2 Chemical structure of vanillin

References

1. Sun BG. Perfumes and essences. China Petrochemical Press; 2000: 39.
2. Handbook of active ingredients in medicinal plants. Beijing: People's Medical Publishing House; 1986.
3. Gobley N-T. Recherches sur le principe odorant de la vanille|Clc. J Pharm Chim. 1858;34:401–5.
4. Tiemann F, Haarmann W. Ueber das Coniferin und seine Umwandlung in das aromatische Princip der Vanille. Ber Dtsch Chem Ges. 2006;7(1):608–23.
5. Wangduoren. Development and application of vanillin. Beijing Daily Chem. 2000;3:13.
6. Zhang Y, Xi GM, Zhou SH. Effects of Gastrodia elata and its components on nervous system. Int J Trad Chin Med. 2006;28:268–71.
7. Yang LL, Li FA, Wei QJ. Study of clinical application of notopterygium. Qinghai Med J. 2009;7:91–3.
8. Fitzgerald DJ, Stratford M, Gasson MJ, et al. Mode of antimicrobial action of vanillin against Escherichia coli, lactobacillus plantarum and Listeria innocua. J Appl Microbiol. 2004;97(1):104–13.

Part III
Natural Small Molecule Drugs for Preventing and Treating Immune and Inflammatory Diseases

Overview

Immune and inflammatory diseases are common, serious disorders that cause people to suffer. The poor living condition can accelerate the rising incidence of immune and inflammatory diseases. It has been a long time since the search for the strategies for preventing or treating the immune and inflammatory diseases. Although the therapy for immune and inflammatory diseases made great progress by a century of efforts, they are still lacking of effective drugs.

For small molecule drugs in treating immune and inflammatory diseases, it is worth mentioning the structure modification of salicylic acid and clinical application of aspirin, which is one of the great contributions of nature products to human health. Salicylic acid's structural modification and aspirin's clinical application are important contributions to the treatment of autoimmune and inflammatory diseases with small molecule drugs. Salicylic acid is an active ingredient in willow and found in a wide variety of traditional herbs. Salicylic acid has been used to treat arthritis and other inflammatory diseases with its antipyretic, analgesic and anti-inflammatory effect. The ancient Greece and Egyptians started to use the extract of willow bark to treat pain and fever more than 2000 years ago. The traditional herbs containing salicylic acid were used to treat inflammation in almost all of the traditional medicines. Especially, they are used to clear away heat and toxic material in the traditional Chinese medicine. Salicin was successfully extracted from willow bark in 1829. Sodium salicylate was used as an analgesic in Europe in 1875. Although the active component of the extract was identified as salicylic acid, the severe gastric side effects associated with the use of sodium salicylate limited its clinical application. The severe gastric side promoted that the German medicine chemists modified the structure of salicylic acid and synthesized acetylsalicylic acid. In 1899, acetylsalicylic acid was mass-produced under the commercial name of 'Aspirin' by the German company Bayer for the treatment of fever, pain and inflammation. Aspirin was widely used in treating a variety of diseases caused by pain, such as headache, muscle pain, inflammatory pain, etc., and has been the important drug for the treatment of rheumatoid arthritis as it showed significant anti-inflammatory actions. Aspirin was extraordinarily effective for pain relief, so it was widely taken up in the world soon.

The study on mechanism of aspirin promotes the development of antipyretic-analgesic drugs. Since the middle of the twentieth century, new antipyretic-analgesic anti-inflammatory drugs with similar efficacy and fewer side effects have been developed, such as acetaminophen, indomethacin, ibuprofen, nimesulide, rofecoxib and so on. These drugs have different properties and are commonly known as non-steroidal anti-inflammatory drugs. Steroidal anti-inflammatory drugs, mainly glucocorticoids, are also natural small molecule compounds that are obtained from animal, which are not discussed here.

The treatment strategies for cardiovascular disease were also accelerated by the study on antiplatelet and antithrombotic effects of aspirin. In the 1970s, scientists found that the action mechanism of aspirin involved the reduction of prostaglandin release. It was well known that prostaglandins contributed to not only pain, inflammation and fever but also cardiovascular disease. A variety of clinical studies have proven that the low doses of aspirin can reduce the risk of recurrent myocardial ischaemia event for patients with myocardial infarction or mild ischaemia and demonstrated that serious adverse effect rarely occurs.

In addition to non-steroidal analgesic-antipyretic and anti-inflammatory drugs, there are also some natural product-based small molecule drugs, which had strong activity and certain tissue selectivity on different inflammatory diseases, such as berberine, matrine, andrographolide, potassium dehydroandrographolide succinate, sinomenine, etc. Berberine yield good therapeutic effect for enteritis in China. Potassium dehydroandrographolide succinate was proven effective and relatively safe in treatment of the upper respiratory tract infection.

At present, some drugs' mechanism has not been fully understood although they are still applied in clinic. Basic and clinical study will further understand the properties of their roles.

Till now, many immune and inflammatory diseases have to face the challenge of rare effective drugs. The rational use and systematic study of the natural small molecule drugs will not only provide information but also promote the discovery and development of novel anti-inflammation immune drugs.

Aconitine

Rong Yan, Li Li, and Guan-Hua Du

Abstract *Aconitum* plants (Wu Tou, Ranunculaceae family) have been widely used to treat various diseases, such as rheumatism, knee pain, herpes zoster, scabies, and other disorders in China for thousands of years. Aconitine was previously used as an antipyretic and analgesic. However, the clinical application of aconitine is limited by its high toxicity.

Keywords *Aconitum* · Aconitine · Antipyretic · Analgesic

Alias: Acetylbenzoylaconine, *wutoujian*
Origin: *Aconitum* (Fig. 1)
Chemical name (Fig. 2)

(1α, 3α, 6α, 14α, 16β)-8-(acetyloxy)-20-ethyl-3,13,15-trihydroxy-1,6,16-trimethoxy-4-(methoxymethyl)aconitan-14-yl benzoate

Molecular formula, $C_{34}H_{47}NO_{11}$; **MW**, 645.74; **CAS**, 302-27-2

Properties

Appearance: solid. *Solubility*: barely soluble in water and soluble in organic solvents such as chloroform or diethyl ether. Its solubility in water and ethanol are 0.3 mg/mL and 35 mg/mL, respectively. *Melting point*: 203–204 °C (397–399 °F; 476–477 K). *Optical rotation: [α]D+17.3°*.

R. Yan · L. Li · G.-H. Du (✉)
Beijing Key Laboratory of Drug Targets Identification and Drug Screening, Centre for Pharmaceutical Screening, Institute of Materia Medica, Chinese Academy of Medical Sciences and Peking Union Medical College, Beijing, China
e-mail: dugh@imm.ac.cn

G.-H. Du, *Natural Small Molecule Drugs from Plants*,
https://doi.org/10.1007/978-981-10-8022-7_58

Dosage Forms

Aconitine is contained in the fourth volume of National standards for chemical drugs. The current clinical use of dosage form is tablets at the dose of 0.3 mg (SFDA H61022876, H11021980).

Indications

Aconitine was previously used as an antipyretic and analgesic. However, the clinical application of aconitine is limited by its high toxicity. Its lethal dose 50% (LD_{50}) for mice is 1.8 mg/kg (orally) and 0.308 mg/kg (intraperitoneally) [1].

Literature

Aconitum plants (Wu Tou, Ranunculaceae family) have been widely used to treat various diseases, such as rheumatism, knee pain, herpes zoster, scabies, and other disorders in China for thousands of years. It was first described in *Shen Nong's Herbal Classic* (the earliest classical work of Chinese herbs). Aconitine is the main active component in *Aconitum* plants [2].

History of R&D

Preparations of *Aconitum* roots are employed in Chinese medicine for analgesic, antirheumatic, and neurological indications. In the Ming Dynasty, Chinese people had extracted the aconitine from the *Aconitum* plants. Aconitine is synthesized by the *Aconitum* plants via the terpenoid biosynthesis pathway (MEP chloroplast pathway).

Pharmacology

Aconitum was found to have neurotoxin. It could activate tetrodotoxin-sensitive Na^+ channels, inducing presynaptic depolarization and blocking the nerve action potential and, in turn, blocking the release of neurotransmitters and decreasing the end plate potential at the neuromuscular junction. Aconitine was also found to block norepinephrine reuptake.

In the heart, aconitine could induce ventricular tachycardia. In cultured ventricular myocytes, aconitine could induce the appearance of early after depolarization by increasing the duration of the action potential.

Research with mouse nerve-hemidiaphragm muscle preparation found that aconitine at low concentrations (<0.1 μM) could increase the electrically evoked acetylcholine release causing an induced muscle tension [3]. Action potentials were generated more often at this concentration. While at higher concentrations (0.3–3 μM), aconitine could decrease the electrically evoked acetylcholine release, resulting in a decrease in muscle tension. Under the circumstance with high concentration (0.3–3 μM) of aconitine, the sodium-ion channels were constantly activated, and transmission of action potentials was suppressed, leading to the formation of non-excitable target cells or paralysis.

Clinical Application

In clinic, it can be used for treating and alleviating the symptoms caused by arthritis, lumbocrural pain, and herpes zoster; however, it is utilized infrequently in clinical practice due to its obvious side effects.

According to a review of different reports of aconite poisoning in human beings, the following clinical features such as neurological, cardiovascular, and gastrointestinal features were observed [4]. The first symptom of aconitine poisoning appeared approximately 20 min–2 h after oral intake, and they were paresthesia, sweating, and nausea, which led to severe vomiting, colicky diarrhea, intense pain, and then paralysis of the skeletal muscles. Following the onset of life-threatening arrhythmia, including ventricular tachycardia and ventricular fibrillation, death finally occurred as a result of respiratory paralysis or cardiac arrest [5].

Discussion

Aconitine was previously used as an antipyretic and analgesic and still has some limited application in herbal medicine, although the narrow therapeutic index makes calculating its appropriate dosage difficult.

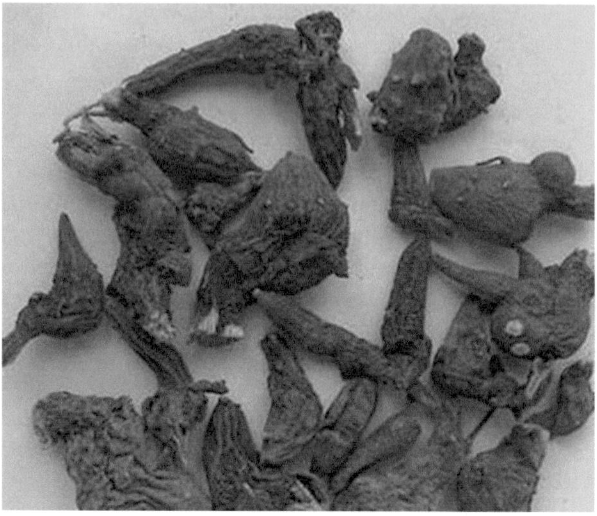

Fig. 1 *Aconitum*

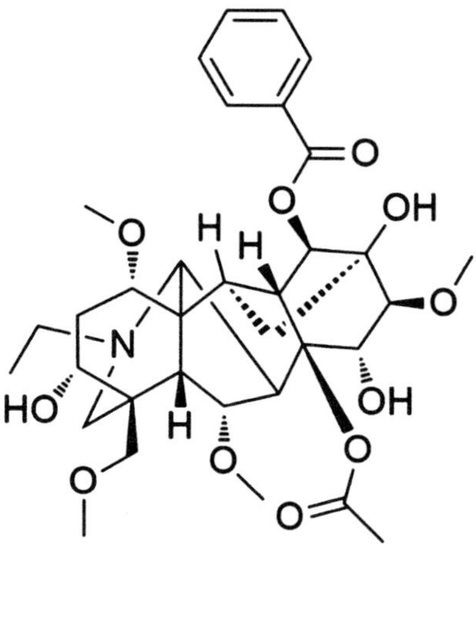

Fig. 2 Chemical structure of aconitine

References

1. Ohno Y. Experimental approach to murder by aconite poisoning from the viewpoint of medico-legal toxicology. Nihon Hoigaku Zasshi. 2006;60(2):101–9.
2. Chan TY. Aconitum alkaloid poisoning related to the culinary uses of aconite roots. Toxins (Basel). 2014;6(9):2605–11.
3. Okazaki M, Kimura I, Kimura M. Aconitine-induced increase and decrease of acetylcholine release in the mouse phrenic nerve-hemidiaphragm muscle preparation. Jpn J Pharmacol. 1994;66(4):421–6.
4. Chan TY. Aconite poisoning. Clin Toxicol (Phila). 2009;47(4):279–85.
5. Beike J, et al. Determination of aconitine in body fluids by LC-MS-MS. Int J Legal Med. 2004;118(5):289–93.

Anabasine

Ning-Bo Gong, Li-Da Du, and Yang Lu

Abstract Anatabine is the most important alkaloid found in the root, fibrous roots, or bark of *Alangium chinense* (Lour) Harms. *Alangium chinense* contains alkaloids, sugars, saponins, steroids, triterpenes, anthraquinones and their glycosides, and other ingredients. *Alangium chinense* shows great potential for treating rheumatoid arthritis. Anatasine is one of the main active ingredients in this plant, and it acts as a noncompetitive muscular relaxant. In clinics, this plant is used with other medicinal plants for treating acute soft tissue injury and degenerative arthritis, herpes zoster, rheumatic heart disease, chronic rheumatoid arthritis, pulmonary hemoptysis, breast abscess, and so on. This plant-based formulation has been widely used.

Keywords Anabasine · Alkaloid · *Alangium chinense* · Muscular relaxant

Alias: Neonicotine
Origin: *Alangium chinense* (Fig. 1)
Chemical name (Fig. 2)

2-(3-pyridyl) piperidine

Molecular formula, $C_{10}H_{14}N_2$: **MW**, 162.23; **CAS**, 13078-04-1

N.-B. Gong · Y. Lu (✉)
Beijing Key Laboratory of Polymorphic Drugs, Institute of Materia Medica, Chinese Academy of Medical Sciences and Peking Union Medical College, Beijing, China
e-mail: luy@imm.ac.cn

L.-D. Du
School of Biomedical Sciences, Chinese University of Hong Kong, Hong Kong, China

G.-H. Du, *Natural Small Molecule Drugs from Plants*,
https://doi.org/10.1007/978-981-10-8022-7_59

Properties

Appearance: Colorless to yellow liquid. *Solubility*: Soluble in water and common organic solvents. *Boiling point*: 110 °C. *Refractive index*: 1.5418. *Density*: 1.0516 g cm^{-3}.

Dosage Forms and Indications

Anabasine is available in *Pharmacopoeia of the People's Republic of China* (1977).

1. Fengshiding capsule (Pulinsong) for treating rheumatoid arthritis, rheumatoid arthritis, cervical rib neuralgia, and sciatica
2. Fengshiding tablet (Qiangzheng, Junbisha) for treating rheumatoid arthritis, rheumatoid arthritis, cervical rib neuralgia, and sciatica.
3. Jingulian capsule for treating joint swelling and pain caused by rheumatoid arthritis

Literature

Anatabine is an alkaloid found in the root, fibrous roots, or bark of *Alangium chinense* (Lour) Harms. The herb was recorded in "Jianyi Bencao," "Bencao Gangmu Shiyi," and "Zhi Wu Ming Shi Tu Kao." *Alangium chinense* is widely distributed in the east and south of China. It is also called "Bailongxu" in Shaanxi, Yunnan, and Guizhou provinces, "Baijintiao" in Guangxi province, and "Bajiaowutong," "Bajiaojinpan," or "Laolongxu" in Jiangxi and Hubei provinces. This plant is commonly used as a traditional Chinese medicine. The *Alangium* species includes more than 30 plants which are distributed in Asia, Oceania, and Africa. Nine of these plants are distributed in China. The root and the stem parts of Alangium plants are toxic.

History of R&D

Alangium chinense contains alkaloids, sugars, saponins, steroids, triterpenes, anthraquinones and their glycosides, and other ingredients. In 1974, dl-anabasine was isolated from the fibrous roots of *Alangium chinense* [1]. Four isoquinoline alkaloids including two new alkaloids (−)-10-O-dimethyl-eugenol base and 10-O-dimethyl-eugenol base were isolated. P-amyrin acetate, triacontanol, and β-sitosterol were then found in its leaves [2]. Seven glycosides were isolated from the water-soluble fraction of dried leaves of *Alangium chinense*, and eight glycosides were also isolated from the n-butanol soluble part [3]. Three new lignin compounds including 2-O-(β-apio-furanosyl)-β-glucoside, E ferulic acid ester, and Z ferulic acid ester were extracted from *Alangium chinense*, and their structures were determined. Nakamoto et al. extracted 7-O-acetylmaleic acid from *Alangium chinense* and determined the structure. There are also long-chain fatty acids and short-chain alkanes in *Alangium chinense* leaves. The volatile oil was then analyzed, and 59 components are identified by GC-MS. The total alkaloid content in the *Alangium chinense* was found to be fibrous root > fine root > coarse root > branch wood > leaf.

Pharmacology

Alangium plants have wide pharmacological effects including muscle relaxation, central inhibition, anticancer, antibacterial, and so on. The "Jisong-II" injection has presynaptic and postsynaptic effects on the neuromuscular junction, and it acts as noncompetitive muscular relaxant [4]. In 1979, the respiratory suppression experiments showed that *Alangium chinense* is a central nervous system depressant [5]. The extract of *Alangium chinense* significantly inhibits the growth of P388 lymphoblastic leukemia and Gardner lymphosarcoma in mice, but it has no effect on Gross virus-induced leukemia, Warner myelomonocytic leukemia, and B16 melanoma [6]. Another study reported that the anabasine could excite respiratory similar to nicotine and lobeline. Anabasine also induces high blood pressure and bradycardia. The components in the *Alangium chinense* extract can bind to the M, 5-HT and dopamine receptors indicating that *Alangium chinense* affects the central nervous system.

Clinical Application

The *Alangium chinense* was used for treating rheumatoid arthritis.

Discussion

In short, *Alangium chinense* shows great potential for treating rheumatoid arthritis. Anatasine in one of main active ingredient in this plant, and it acts as a noncompetitive muscular relaxant. This compound is toxic. In clinics, this plant is used with other medicinal plants for treating acute soft tissue injury and degenerative arthritis, herpes zoster, rheumatic heart disease, chronic rheumatoid arthritis, pulmonary hemoptysis, breast abscess, and so on. This plant-based formulation has been widely used.

Fig. 1 *Alangium chinense*

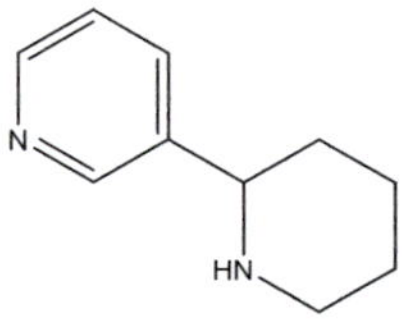

Fig. 2 Structures of anabasine

References

1. Zhejiang Medical University and Shanghai Institute of Materia Medica. Study on chemical constituents of Alangium chinense (Lour) harms. J Tradit Chin Med. 1974;410:26–8.
2. Sakurai N, Nakagawa-Goto K, Ito J, et al. Cytotoxic alangium alkaloids from Alangium longiflorum cytotoxic alangium alkaloids from Alangium longiflorum. Phytochemistry. 2006;67:894–7.
3. Itoh A, Tanahashi T, Ikejima S, et al. Five phenolic glycoside from Alangium chinense. J Nat Prod. 2000;63(1):95–8.
4. Cang Z. Study on the Chinese herbal muscle relaxant Alangium chinense (Lour) harms. China J Chin Mater Med. 1981;5:34–6.
5. Xue K. Experimental research on the effects of resibufo-genin and neostigminum on the respiratory depression caused by anabasine. Acta Pharm Sin. 1979;14(12):738–41.
6. Li H. Study on the anti-mouse lymphoma effect of Acanthopanax alanium vitiense extraction. Foreign Med Plant Drugs. 1980;1:40–1.

Andrographolide

Yu Yan, Lian-Hua Fang, and Guan-Hua Du

Abstract Andrographolide is a diterpene lactone compound extracted from *Andrographis paniculata* (Burm. F) Nees, which is commonly used as traditional Chinese medicine. Andrographolide has effects of anti-inflammatory, antibacterial, antivirus, antitumor, and immune regulation and is used in treatment of cardiovascular-cerebrovascular diseases and protection of the liver and gallbladder. Andrographolide water solubility is poor and its bioavailability is low; its pharmacological effects are extensive but weak, so the preparation requirements are strict.

Keywords Andrographolide · Diterpene lactone · Anti-inflammatory · Antibacterial

Origin: *Andrographis paniculata* (Fig. 1)
Chemical name (Fig. 2)

3-(2-(Decahydro-6-hydroxy-5-(hydroxymethyl)-5,8a-dimethyl-2-methylenenaphthyl)ethylidene)dihydro-4-hydroxyfuran-2(3H)-one

Molecular formula, $C_{20}H_{30}O_5$; **MW**, 350.44; **CAS**, 5508-58-7
Andrographolide derivatives (Fig. 3)

Y. Yan · L.-H. Fang · G.-H. Du (✉)
Beijing Key Laboratory of Drug Targets Identification and Drug Screening, Centre for Pharmaceutical Screening, Institute of Materia Medica, Chinese Academy of Medical Sciences and Peking Union Medical College, Beijing, China
e-mail: dugh@imm.ac.cn

G.-H. Du, *Natural Small Molecule Drugs from Plants*,
https://doi.org/10.1007/978-981-10-8022-7_60

Properties

Appearance: white square prism or flaky crystals (ethanol or methanol) and odorless, with a bitter taste. *Solubility*: dissolved in boiling ethanol, slightly soluble in methanol or ethanol, very slightly soluble in chloroform, and almost insoluble in water. Since andrographolide has an ester structure, it is easy to hydrolysis, open loop and isomerization in the aqueous solution to affect the drug stability. At lower temperature, the stability of andrographolide is better; it is unstable in alkaline conditions, and its instability increased with the increase in alkaline strength. The most stable pH value is 3–5. Andrographolide is more stable in chloroform.

Dosage Forms and Indications

This product is included in the *Pharmacopoeia of the People's Republic of China* (2015).

It comes in the form of andrographolide tablets, andrographolide drop-pills, and andrographolide capsules. This product is often used for detoxification, sore throat, and mouth and tongue sores.

Literature

Andrographolide is a diterpene lactone compound extracted from *Andrographis paniculata* (Burm. F) Nees. *Andrographis paniculata* is also known as lanhelian, yijianxi, zhanshecao, kucao, kudancao, etc. Since this product originates in India, it is also known as Indian grass. This product is dry ground part of Jupiteraceae plant *Andrographis paniculata* (*conus*) may originate in South Asia. It is commonly cultivated in China such as in Fujian, Guangdong, Hainan, Guangxi, and Yunnan and also in Australia and was also introduced in Jiangsu and Shaanxi.

Andrographis paniculata is commonly used in traditional Chinese medicine. Its ability of relieving "snake venom, and internal injuries cough" was recorded in "Lingnan herbal medicine records." "Quanzhou Materia Medica" described it as "heat-clearing and detoxifying, anti-inflammatory, detumescence and inhibition throat inflammation, dysentery and high fever."

The above ground part of *Andrographis paniculata* contains andrographolide, neo-andrographolide, 14-deoxygenated andrographolide, etc., and the root contains mainly flavonoids.

History of R&D

In 1911, Gorter firstly isolated a crystalline substance from *Andrographis paniculata*, identified it as a diterpene lactone compound, and named it andrographolide [1].

Andrographolide structure is complex and difficult to artificially synthesize, so it is often extracted from the plant. The extraction of andrographolide includes water extraction, alcohol extraction, and other methods, but there are issues such as time-consuming, cumbersome operation and loss of active ingredients which need to be modified. In recent years, new technology processes such as enzymatic pretreatment and microwave- and ultrasonic-assisted extraction have been used in the extraction of andrographolide, which have significantly shortened the time and improved the extraction efficiency [2, 3].

Andrographolide water solubility is poor and its bioavailability is low; its pharmacological effects are extensive but weak, so the preparation requirements are strict. Therefore, from the 1970s, drugs and organic chemical researchers have done a lot of work in the modification and transformation of andrographolide, mainly concentrated in the α, β-unsaturated lactone double bond Michael addition, redox, selective esterification of hydroxyl groups, oxidation and substitution reactions, intramolecular cyclization, replacement of lactone rings, etc. They had obtained a number of new derivatives. In the research of pharmacological effects, the main study of the andrographolide derivatives focuses on antitumor, antivirus, and other biological activities and has made some progress [4, 5].

The study of andrographis developed in not only andrographolide oral preparations but also injectables of andrographolide derivative compounds such as potassium dehydroandrographolide succinate and potassium sodium dehydroandrographolide succinate, which has now become a commonly used drug (see potassium dehydroandrographolide succinate, potassium sodium dehydroandrographolide succinate).

So far, the methods of andrographis preparations have been more than of 600 kinds [6]. The andrographolide preparations approved by the SFDA are 14 kinds of andrographolide tablets; 2 kinds of andrographolide capsules, andrographolide capsules, and sodium nitrite andrographolide; and 5 kinds of andrographolide soft capsule and one kind of andrographolide tablet [7].

Pharmacology

Modern pharmacological studies have shown that andrographolide has effects of anti-inflammatory, antibacterial, antivirus, antitumor, and immune regulation and can be used in treatment of cardiovascular-cerebrovascular diseases and protection of the liver and gallbladder. Andrographolide can treat the fever caused by pneumococcus and hemolytic *Streptococcus mutans*, mainly by inhibiting the hypothalamus PGE2 and cAMP content to exert its antipyretic, nonetheless reducing the chemotaxis (fMLP) CD^{+}_{11b} and CD^{+}_{18} may be the main mechanism of its anti-inflammatory effect. Andrographolide has antagonistic effects on Hong Kong virus (HKV), Ebola virus (EBOV), and respiratory syncytial virus (RSV) and has been shown to inhibit HIV, SARS, and viral myocarditis ST2 in vitro.

Andrographolide can also ameliorate myocardial ischemia and serve as protection from ischemia-reperfusion injury. It also has protection effect of vascular endothelial cells as well as can regulate lipid disorder, lower blood pressure, and exert effect of anti-atherosclerosis. It can also prevent angiogenesis after restenosis and improve blood rheology. Its mechanisms are mainly due to its free radical scavenging and antiplatelet aggregation properties. Andrographolide can increase the bile flow, bile salt, bile acid, and deoxycholic acid in experimental rats and guinea pigs and can reverse the reduction of bile and cholic acid and other secretions caused by paracetamol and improve liver function [8].

Clinical Application

This product is commonly used clinically for dysentery, leptospirosis, meningitis, pneumonia, upper respiratory tract infection, and enhancing adrenal cortical function. In view of the requirement for clinical viral infection emergency, it is important to introduce the hydrophilic group in different lactone structure, enhance its water solubility, and improve efficacy. There are a variety kinds of andrographolide injection, including potassium dehydroandrograpolide succinate needle, potassium sodium dehydroandrographolide succinate needle. Lianbizhi is the representative of this class of drugs in current clinical application. Oral administration of andrographolide may induce bitter vomiting. Oral administration of this product and other preparations in large doses can cause epigastric discomfort and loss of appetite. It had been reported of drug eruption, upper abdominal pain, and anaphylactic shock caused by intramuscular injection. Severe symptom response emerged such as chest tightness, shortness of breath, pallor, lip bruising, cold sweats, weak pulse, decreased blood pressure, etc. Light reactions include abdominal pain, vomiting, asthma, urticaria, pimples, dizziness, head swelling, sneezing, chest pain, etc. The reaction is real time and also 5–20 min after injection and gradually improves 5–45 min after rescue and individual recovered by 24 h. There are also reports of acute amniotic fluid blockage caused by weaning lotion exchange. In addition, taking into account the adverse effects of andrographolide on reproduction, it is recommended to be used cautiously by fertility couples and pregnant women.

Discussion

Andrographolide is widely used. It is rapidly absorbed in animals and has a long efficacy and high bioavailability with no obvious side effects. In recent years, clinical application is still focused on antibacterial and anti-inflammatory; with the development of research, andrographolide shows the prevention and treatment of cardiovascular, liver, and gallbladder diseases and enhanced immunity and antitumor and other multiple effects, making it a promising Chinese medicine, worthy of further research and development.

Fig. 1 *Andrographis paniculata*

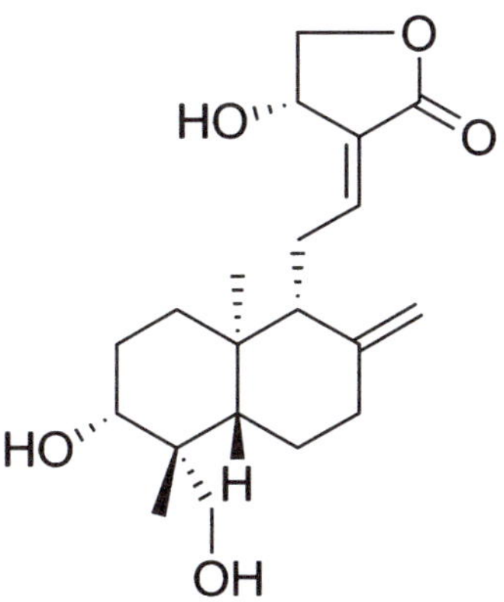

Fig. 2 The structure of andrographolide

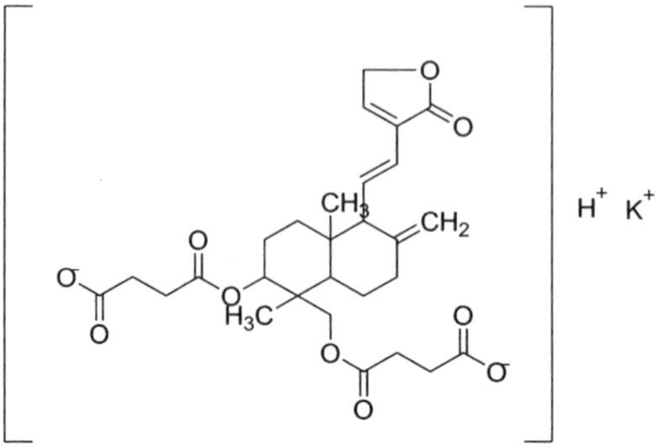

14-dehydro-11,12-didehydro-andrographolide-3,19-disuccinic acid half ester monopotassium salt (Potassium Dehydroandrograpolide Succinate)

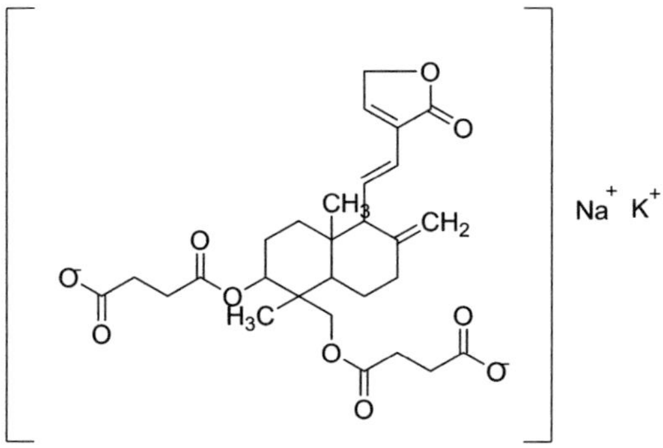

14-dehydro-11,12-didehydro-andrographolide-3,19-disuccinic acid half ester potassium sodium salt (Potassium Sodium Dehydroandrographolide Succinate)

Fig. 3 The structure of andrographolide derivatives

References

1. Amos BS, Bruce HT, Patrick JC, et al. Andrographolide: an X-ray crystallographic analysis. J Crystallogr Spectrosc Res. 1982;12(4):309–19.
2. Yu BT, Zhang ZR, Liu WS, et al. Study on stability in vitro of Andrographolide. Chin Tradit Patent Med. 2002;24(5):331–3.
3. Lin QS. Chinese herbal medicine effective component. Beijing: Chemical Industry Press; 1965. p. p624.
4. Aromdee C. Andrographolide: progression in its modifications and applications-a patent review (2012–2014). Expert Opin Ther Pat. 2014;24(10):1129–38.
5. Liu GZ, Xu HW, Sun K, et al. Development in synthesis of andrographolide derivatives. Chin J Organic Chem. 2008;28(2):201–9.
6. Wang LL, Yu J. Andrographolide and its pharmacological action and clinical research progress of preparation. China Pharm. 2003;12(10):72–3.
7. http://app1.sfda.gov.cn/datasearch/face3/base.jsp?tableId=25&tableName=TABLE25&title=国产药品&bcId=124356560303886909015737447882
8. Liu GL, Liu YX. The research progress of Andrographolide. Herald Med. 2006;25(1):48–50.

Baicalin

Li Li, Li Zhang, and Guan-Hua Du

Abstract Baicalin, a flavone glycoside, is the glucuronide of baicalein. As one of the effective components from radix scutellariae, baicalin has a variety of pharmacological activities such as antiviral, antivirus, removal of oxygen free radicals, antioxidant, antipyretic, analgesic, anti-inflammatory, antitumor, cardiovascular protecting, cerebrovascular and neuron protecting, liver protecting, and prevention or treatment of diabetes and its complications. It can be used alone or in combination with other drugs for the treatment of the disease such as acute and chronic hepatitis and persistent hepatitis. However, due to the low bioavailability of baicalin, a large number of studies have shown that baicalein is the ingredient with more prominent activities compared with baicalin. Therefore, it is a potent valuable drug and deserves further research and development.

Keywords Baicalin · Baicalein · Radix scutellariae · Hepatitis

Alias: Baicalein 7-O-glucuronide
Origin: *Scutellaria baicalensis* Georgi (Fig. 1)
Chemical name (Fig. 2)

5,6-Dihydroxy-4-oxygen-2-phenyl-4H-1-benzopyran-7-beta-D-glucopyranose acid

Molecular formula, $C_{21}H_{18}O_{11}$, **MW**, 446.36; **CAS**, 21967-41-9

L. Li · L. Zhang · G.-H. Du (✉)
Beijing Key Laboratory of Drug Targets Identification and Drug Screening, Centre for Pharmaceutical Screening, Institute of Materia Medica, Chinese Academy of Medical Sciences and Peking Union Medical College, Beijing, China
e-mail: dugh@imm.ac.cn

G.-H. Du, *Natural Small Molecule Drugs from Plants*,
https://doi.org/10.1007/978-981-10-8022-7_61

Properties

Appearance: light yellow crystalline powder at room temperature. *Solubility*: insoluble in methanol, ethanol, and acetone; slightly soluble in chloroform and nitrobenzene; almost insoluble in water; soluble in hot acetic acid. *Melting point*: 202–205 °C.

Dosage Forms

Baicalin is recorded in the tenth volume of *National Standards for Chemical Drugs*, the *British Pharmacopoeia* (2017), and the *European Pharmacopoeia* (8.7th ed.). Its dosage forms are tablets and capsules.

Indications

Baicalin is mainly used for the adjuvant therapy of acute and chronic hepatitis and persistent hepatitis.

Literature

Baicalin is one of the main active ingredients obtained from the roots of huang qin (*Scutellaria baicalensis* Georgi). The *Pharmacopoeia of the People's Republic of China* (2015) stipulates that the content of baicalin in radix scutellariae with dry goods shall not be less than 9.0%. The medicinal radix scutellariae distribution in China are now Yunnan radix scutellariae (*S. amoena* C. H. Wright), sticky hairs radix scutellariae (*S. viscidula* Bunge), Gansu radix scutellariae (*S. rehderiana* Diels), Lijiang radix scutellariae (*S. likiangensis* Diels), Sichuan radix scutellariae (*S. hypericifolia* Lev l.), radix scutellariae (*S. tenax* W. W. Smith var. patentipilosa G. Y. Wu), etc., which contain a certain amount of baicalin. Radix scutellariae medicinal has a long history in China. Listed as goods in *Shen Nong's Classic of Materia Medica*, it has been used clinically to treat diseases with symptoms such as "heat jaundice, intestinal dysentery, edema, amenorrhea, malignant sore, and scleritis" for about 2000 years. *Scutellaria* is recorded in the *Pharmacopoeia of the People's Republic of China* to have the effects of "clear heat and wet, purging fire to detoxify, stop bleeding and tocolysis." It is one of the commonly used traditional Chinese herbs, which is clinically used alone or with other Chinese medicine compatibility for the treatment of respiratory infections, acute dysentery, viral hepatitis, allergic disease and gynecological disease, and so on.

History of R&D

The chemical research of the genus *Scutellaria* began in 1889. Baicalein (scutellarein) is the first flavonoid isolated from Vietnam radix scutellariae (*S. altissima*) in 1910. In 1922, Shibata Gui Tai and his collaborators isolated and obtained baicalin, baicalein, wogonin, and benzoic acid from *Scutellaria baicalensis* [1].

Among the flavonoids in *Scutellaria baicalensis*, the content of baicalein is the highest. Its official name is baikeli. Baicalin is formed by the combination of baicalein and one molecule of glucuronic acid. Both baicalein and baicalin exist in *Scutellaria baicalensis* [2]. Studies showed that baicalein could be transferred into baicalin and other metabolites in the blood. However, baicalin by oral is hardly absorbed. Baicalein can be absorbed and also rapidly converted into baicalin [3].

At present, there are a variety of mature extraction methods for obtaining baicalin and baicalein from *Scutellaria baicalensis*. Due to the poor water solubility, oral preparation of baicalin is mostly used in clinical practice. Research showed that baicalin had a significant first pass effect, which led to its low bioavailability. Baicalin-metal complexes formed by the combination of baicalin and metal ions were found to enhance bioavailability and increase pharmacological activities [4]. In addition, the preparation of ester-type prednisone was easily hydrolyzed by esterase and could improve the lipid solubility of baicalin.

Pharmacology

Pharmacological studies showed that both baicalein and baicalin had a variety of pharmacological effects such as antibacterial, antiviral, removal of oxygen free radicals, antioxidant, antipyretic, analgesic, anti-inflammatory, antitumor, cardiovascular protecting, cerebrovascular and neuron protecting, liver protecting, and prevention or treatment of diabetes and its complications [5].

1. Antibacterial and antiviral effects: *Scutellaria baicalensis* was reported to have growth inhibitory effect on a variety of Gram-negative bacteria, Gram-positive bacteria, and spirochetes, showing a broad antibacterial spectrum. Baicalein was regarded as the main ingredient in *Scutellaria baicalensis* to play the antibacterial effect. The IC_{50} of baicalein on *Escherichia coli* was about 0.29 $mmol{\cdot}L^{-1}$, and the minimum inhibitory concentration (MICs) of *Candida albicans* was 264 $\mu g{\cdot}mL^{-1}$. The survival time of mice infected with influenza virus was significantly extended by the gastric irrigation of baicalin at a dose of 0.96–1.5 $g{\cdot}kg^{-1}$.
2. Antitumor effect: *In vitro* and *in vivo* experimental results showed that both baicalein and baicalin had obvious antitumor activity, and the mechanism involved the inhibition of tumor cell proliferation, invasion and metastasis, induction of tumor cell apoptosis, suppression of neovascularization, and enhanced tumor cell chemotherapy sensitivity. Baicalein was found to inhibit the

proliferation of rat bladder cancer MBT-2 cells and induce apoptosis, and its IC_{50} was 0.43 μmol·L^{-1}. In bladder cancer model formed by injection of MBT-2 cells into C3H/HeN nude mice, the formed tumor in mouse given baicalein 0.05–0.1 mg·day^{-1} for ten consecutive days was significantly smaller than that of the control group. A 50–200 μmol·L^{-1} baicalein could inhibit the proliferation of human prostate cancer DU2145 and PC3 cells, human umbilical vein endothelial cell proliferation, and the formation of buds and vascular structure in a dose-dependent manner. Baicalein could also inhibit human breast cancer cell MDA-MB-231 adhesion, metastasis, and invasion, in a dose-dependent manner in the range of 2–50 μmol·L^{-1}. Moreover, a dose of 100 μg·mL^{-1} could also reverse the resistance of ovarian cancer cell line A2780/ADM. This reversal may be associated with a decrease in P-gp drug efflux and increased intracellular drug concentration.

3. Antipyretic, analgesic, and anti-inflammatory effects: Both baicalein and baicalin can play antipyretic, analgesic, and anti-inflammatory effects by interfering with the arachidonic acid metabolism pathway, inhibiting the activity of nuclear factor, and suppressing the secretion and release. In rats, intraperitoneal injection of baicalin at a dose of 4.5 mg·kg^{-1} could exert antipyretic effect; the body temperature of the endotoxin-induced fever rabbits was heavily reduced after intravenous administration of baicalin at a dose of 20 mg·kg^{-1}; baicalein at a dose of 20 mg·kg^{-1} (iv) could improve the hemodynamics and heart rate and reduce the mortality and the leukocyte infiltration into the liver and lung tissues in endotoxic shock rats (LPS 10 mg·kg^{-1}, iv).
4. Scavenging oxygen free radicals and antioxidant effects: Both baicalein and baicalin were found to have strong scavenging effects on hydroxyl radicals, superoxide anion (O_2^-), alkane peroxy radicals, and DPPH free radicals. A 25–100 μmol·L^{-1} baicalein can inhibit H_2O_2-induced RAW264.7 cell apoptosis and diploid formation.
5. Liver protection: Both baicalein and baicalin were found to have protective effect on the liver injured by a variety of causes, and the mechanism was related to their function of antioxidant and inhibition of the secretion of inflammatory mediators. It was found that intraperitoneal injection of baicalin at a dose of 50–200 mg·kg^{-1} had protective effect on CCl_4-induced acute liver injury mice. Intraperitoneal injection of baicalin at a dose of 70 mg·kg^{-1} could improve CCl_4-induced chronic liver fibrosis in mice. In addition, gavage of baicalin to mice at a dose of 50–150 mg·kg^{-1} significantly reduced the immune liver injury caused by D-GalN and LPS.
6. Treatment or prevention of diabetes and its complications: Baicalein was found to have effects of protecting the islet tissue and improving kidney function and retinopathy in diabetic rats, and the mechanism might ascribe to its function on reducing the inflammatory response and inhibition of oxidative stress injury. Streptozotocin-induced diabetic rats were found to have decreased blood glucose concentration and the improvement of islet tissue damage after gavage of baicalin at a dose of 25–100 mg·kg^{-1}. After gavage of baicalin at a dose of 80–120 mg·kg^{-1} for 12 weeks, an obvious protective effect on the kidney tissue

of diabetic rats was observed. In addition, the administration of baicalein at a dose of 150 $mg \cdot kg^{-1}$ day^{-1} was found to ameliorate diabetic retinopathy.

7. Cardiovascular and cerebrovascular protective effects: *In vitro* experiment results showed that *Scutellaria baicalensis* had cardiovascular protective effects such as dilating blood vessels, counteracting high blood pressure, protecting heart and endothelial cells, and anti-atherosclerosis. Baicalein did not affect normal blood pressure in rats but could reduce blood pressure in hypertensive rats. Baicalin at concentration of 0.5 and 2 $\mu mol \cdot L^{-1}$ was found to decrease the contractile tension of isolated rat aorta induced by norepinephrine, KCl, and $CaCl_2$, shifting the reaction curve from the left to the right side and reducing the maximal effect. It could also significantly inhibit the norepinephrine-dependent contraction induced by intracellular and outside Ca^{2+}. Baicalein had protective effect on myocardial ischemia and reperfusion injury, and this effect was stronger than baicalin. Intravenous injection of baicalein at a dose of 10 $mg \cdot kg^{-1}$ could improve the contraction of myocardium in endotoxic shock rats.
8. Neuroprotective effect: Baicalein had a good neuroprotective effect. It could suppress nerve injury and apoptosis by inhibiting the inflammatory response induced by glial cell, anti-free radicals, and protection of mitochondria. Administration of baicalein at a dose of 200 $mg \cdot kg^{-1}$ could relieve 6-OHDA-induced muscle tremor injury in rats. Gavage of baicalein at a dose of 200 $mg \cdot kg^{-1}$ could improve the recovery of motor function in Parkinson model mice induced by 1-methyl-4-phenyl-1,2,3,6-tetrahydropyridine (MPTP). Intraperitoneal injection of baicalein at a dose of 30 $mg \cdot kg^{-1}$ reduced the neurological deficit score of MCAO model rats, decreasing brain water content and cerebral infarction volume, which indicated that baicalein had a protective effect on the neurological function of rats with focal cerebral ischemia.
9. Other effects: Baicalein and baicalin were also reported to have other pharmacological effects such as enhancing immune function.

Clinical Application

In clinical, baicalin is mainly used as the adjuvant therapy for acute and chronic hepatitis and persistent hepatitis. Baicalin was shown to reduce the expression of hepatitis B surface antigen, e antigen and the core antigen, and inhibited hepatitis B virus DNA replication. Reduced serum alanine aminotransferase level was observed in hepatitis patient treated with baicalin, and other liver function indexes were also found to be improved. No adverse reactions about baicalin have been reported. In addition, baicalin was reported to treat early diabetic nephropathy and alleviate the symptoms of diabetic neuropathy.

As one of the antimicrobial components of *Scutellaria baicalensis*, eye drops containing 3% baicalin are used clinically in the treatment of trachoma, with the similar curative effect as rifampicin.

Moreover, baicalein is used clinically for the treatment of enteritis and dysentery.

Discussion

Radix scutellariae as a traditional Chinese medicine has a long history of use. Baicalin, as one of the effective components from radix scutellariae, has a variety of pharmacological activities. It can be used alone or in combination with other drugs for the treatment of the disease. However, due to the low bioavailability of baicalin, its pharmacological effects are not prominent, and its clinical application is limited. A large number of studies have shown that baicalein is the ingredient with more prominent activities compared with baicalin. Therefore, it is a potent valuable drug and deserves further research and development.

Fig. 1 *Scutellaria baicalensis* Georgi

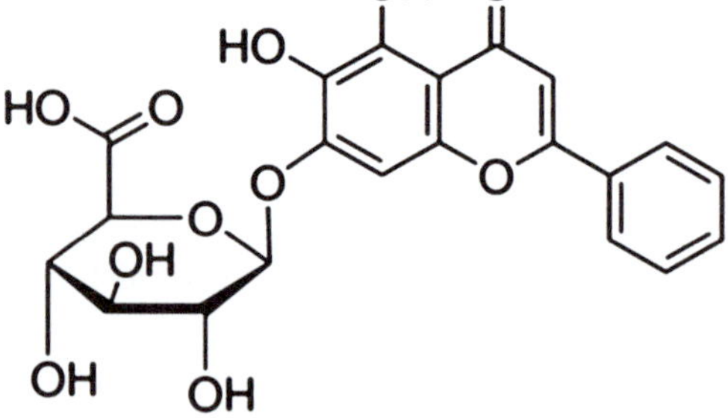

Fig. 2 Chemical structure of baicalin

References

1. Zeng GF, Zhang ZR. Studies on the yellow alkaloids in traditional Chinese medicine V, studies on chemical constituents of Scutellaria baicalensis (first report) new extraction method of Carbosil and its new methylation. Acta Pharm Sin. 1957;5(1):47–57. (In Chinese)
2. Baicalin. The free encyclopedia. In: Wikipedia, from https://en.wikipedia.org/wiki/Baicalin. Retrieved 20:01, 7 June 2016.
3. Wang WY, Dai JY, Sun SJ, et al. Advances in pharmacokinetics of baicalein. Modernization Tradit Chin Med Mater Med World Sci Technol. 2011;13(6):1018–21. (In Chinese)
4. Guo M, Wu ZL, Gao XY. Epid toxicity of baicalin-metal complexes on human hepatocellular carcinoma SMMC-7721 cells and its association with DNA interactions in hepatocellular carcinoma cells. Chin J Pharmacol Toxicol. 2014;28(4):536–49. (In Chinese)
5. Xin WY, Song JK, He GR, et al. Advances in pharmacological effects and mechanism of baicalein and baicalin. Chin J New Drugs. 2013;22(6):647–59. (In Chinese)

Berberine

Lian-Hua Fang, Jin-Hua Wang, and Guan-Hua Du

Abstract Berberine is a quaternary ammonium salt that is found in some plants such as huanglian (*Rhizoma coptidis*), huangbo (*Phellodendri Chinensis Cortex*), and sankezhen (*Berberidis Radix*). Berberine hydrochloride has extensive pharmacological effects, such as antibacterial, antiviral, anti-inflammatory, analgesic, anticancer, hyperglycemic, antilipidemic, antihypertensive, anti-arrhythmic, anti-heart failure, and so on. Therefore, berberine hydrochloride is mainly used to treat gastroenteritis, bacterial diarrhea, intestinal infection, conjunctivitis, and suppurative otitis media. In the future, it is necessary for researchers to focus on its structural optimization and mechanisms of action, and a large number of long-term clinical studies are needed to further confirm its role in the clinical treatment process and curative effect of berberine hydrochloride.

Keywords Berberine · Alkaloid · Antibacterial · Diarrhea

Alias: Berberine, berberine hydrochloride, umbellatine
Origin: *Coptis chinensis* (Fig. 1)
Chemical name (Fig. 2)

Benzo[G]-1,3-benzodioxolo[5,6-α]quinolizinium, 5,6-dihydro- 9,10-dimethoxy

Molecular formula, $C_{20}H_{18}NO_4$; **MW**, 336.36; **CAS**, 2086-83-1
Derivatives

Berberine hydrochloride, $C_{20}H_{20}NO_4Cl$; MW, 371.82; CAS, 633-65-8

L.-H. Fang · J.-H. Wang · G.-H. Du (✉)
Beijing Key Laboratory of Drug Targets Identification and Drug Screening, Centre for Pharmaceutical Screening, Institute of Materia Medica, Chinese Academy of Medical Sciences and Peking Union Medical College, Beijing, China
e-mail: dugh@imm.ac.cn

G.-H. Du, *Natural Small Molecule Drugs from Plants*,
https://doi.org/10.1007/978-981-10-8022-7_62

Properties

Appearance: Berberine is odorless and yellow crystalline powder. *Solubility*: Berberine is soluble in hot water, slightly soluble in water or ethanol and in chloroform, and insoluble in ether. *Melting point*: Melting point of berberine is about 204–206 °C. Berberine is heat labile.

Dosage Forms and Indications

This product was recorded in the *Pharmacopoeia of the People's Republic of China* (2015), the *British Pharmacopoeia* (2017), the *Japanese Pharmacopoeia* (17th ed.), the *European Pharmacopoeia* (9th ed.), and the *Korea Pharmacopoeia* (10th ed.).

Berberine hydrochloride is commonly used in clinic as tablet and capsule forms. It was mainly used to treat gastroenteritis, bacterial diarrhea, intestinal infection, conjunctivitis, and suppurative otitis media.

Literature

Coptis chinensis was widely used in China as a folk medicine by Shennong around 3000 BC. *Coptis chinensis* was firstly described in the ancient Chinese medical book *The Divine Farmer's Herb-Root Classic*. *Coptis chinensis* was used to treat intestinal bacterial infections and antipyretic analgesic for thousands of years ago.

Coptis chinensis also called zhilian, chuanlian, weilian, jizhualian, shanglian, and xuanlian in Chinese history. *Coptis chinensis* was accepted by most physicians, and the Chinese pharmacopoeia also uses *Coptis chinensis* as its official name [1]. *Coptis chinensis* mainly grows in Anhui, Hunan, Sichuan, and Yunnan and has been cultivated in Sichuan since the Ming dynasty, which has a long history of cultivation. Other species of *Coptis chinensis* from different places were used as medicine. However, commodity circulation of wild *Rhizoma coptidis* has not been formed [1]. *Coptis chinensis* is national three level protection plants at present and majorly produced in Shizhu of Chongqing, West Hubei, Shanxi, and Gansu.

Berberine is a quaternary ammonium salt from the protoberberine group of isoquinoline alkaloids. It is found in some plants such as huanglian 黄连 (*Rhizoma coptidis*), huangbo黄柏 (*Phellodendri Chinensis Cortex*), sankezhen三颗针 (*Berberidis Radix*), and so on.

The components of *Coptis chinensis* which have antibacterial and anti-inflammatory effects are original alkaloid berberine class, including berberine, coptisine, palmatine, epiberberine, columbamine, jatrorrhizine, worenine, and magnoflorine, with berberine having the highest content (5–8%).

History of R&D

Berberine is often used in the form of quaternary ammonium alkali. The solubility of berberine in water is lower, for example, berberine hydrochloride is 1:500, and berberine sulfate is 1:30. In 1926, berberine was first separated from bark of *Zanthoxylum clava*. Modern pharmacology study showed that berberine has defined structure and is a monomer of traditional Chinese medicine. Berberine can be obtained from many sources and is used in clinics, with reliable pharmacological effects and various and unique mechanisms.

Berberine sulfate injection was created by Sichuan medical pharmaceutical factory in1958, but the intramuscular dose is too low and cannot achieve intravenous bacteriostatic concentration; at the same time, intravenous injection of berberine can induce Adams-Stokes syndrome and kidney failure. The injection was eliminated because the curative effect was not accurate and was replaced by other products by the Ministry of Public Health of China on September 4, 1982 [2].

At present, berberine can be synthesized by industrial biosynthesis. A series of derivatives can also be synthesized by structure modification, and the pharmacological activities of these derivatives have been tested. A lot of information of structure-activity relationship was obtained. These results indicated that berberine derivatives had many pharmacological activities, such as anti-Alzheimer's disease, antibacterial, antitumor, and antiviral [3]. The berberine derivatives included tetrahydroberberine, dihydrogen isoquinoline berberine, four hydrogen isoquinoline berberine, and phenethylamine berberine. The derivatives of berberine have many kinds of biological activity through a variety of targets. However, none of them was approved as a drug.

Pharmacology

Berberine hydrochloride has extensive pharmacological effects, such as antibacterial, antiviral, anti-inflammatory, analgesic, anticancer, hyperglycemic, antilipidemic, antihypertension, anti-arrhythmic, anti-heart failure, and so on. Experimental study and clinical reports demonstrate that berberine has therapeutic effect on the endocrine system, circulatory system, nervous system, digestive system, and respiratory system and other diseases.

The clinical indication of berberine is intestinal bacterial infectious diarrhea, which is confirmed by years of clinical application. Berberine hydrochloride exerts effect on intestinal infection, eye conjunctivitis, and suppurative otitis media induced by *Shigella dysenteriae*, *Escherichia coli*, and *Staphylococcus aureus* and ameliorates gastritis and combined gastric and duodenal ulcers. Berberine hydrochloride also has curative effect on acute lung injury, pneumonia, and other respiratory diseases; peptic ulcer, colitis, and other digestive system diseases; pregnancy, urinary, and reproductive system infections; and other urinary tract and reproductive system diseases.

Recent research found that berberine has many pharmacological effects. However, it wasn't approved as drug. Berberine (560 mg/kg/day, 7 days) reduced body weight and caused a significant improvement in glucose tolerance without altering food intake in db/db mice. Similarly, berberine (80 mg/kg/day, for 2 weeks, i.g.) reduced body weight and plasma triglycerides and improved insulin action in high-fat-fed Wistar rats. Berberine downregulated the expression of genes involved in lipogenesis and upregulated those involved in energy expenditure in adipose tissue and muscle. Berberine (5 μg/L) treatment resulted in increased AMP-activated protein kinase (AMPK) activity in 3T3-L1 adipocytes and L6 myotubes, increased GLUT4 translocation in L6 cells in a phosphatidylinositol 3′ kinase–independent manner, and reduced lipid accumulation in 3T3-L1 adipocytes [4].

In cultured human liver cells, berberine (7.5 μg/ml) enhanced the identification of insulin receptor (InsR) messenger RNA (mRNA) and protein expression in a dose- and time-dependent manner. Berberine increased InsR expression in the L6 rat skeletal muscle cells as well. Berberine induced InsR gene expression through a protein kinase C (PKC)-dependent activation of its promoter. Inhibition of PKC abolished berberine-caused InsR promoter activation and InsR mRNA transcription. In animal models, treatment of type 2 diabetes mellitus rats with berberine (75 or 150 mg/kg/day, twice a day, 2 weeks) lowered fasting blood glucose and fasting serum insulin, increased insulin sensitivity, and elevated InsR mRNA as well as PKC activity in the liver [5].

Berberine hydrochloride has antilipidemic, antihypertensive, anti-arrhythmic, anti-heart failure, antiplatelet aggregation, and other cardiovascular system effects. In patients with hyperlipidemia taking berberine (0.5 g each time, twice a day, for 3 months), cholesterol was decreased by 29%, triglyceride was decreased by 35%, and low-density lipoprotein was decreased by 25% [6], the mechanism of which includes raising liver low-density lipoprotein receptor, reducing the synthesis of triglyceride and cholesterol and inhibiting adipocyte differentiation. The mechanism of antihypertensive is to block alpha receptors, causing vasodilatation and enhancing stimulation of acetylcholine receptors.

Berberine plays a role in blood lipid regulation. Berberine significantly upregulates the expression of low-density lipoprotein receptor (LDLR) in hepatocytes in vitro. Further studies have shown that berberine plays a role at posttranscriptional levels through the activation of cells outside the extracellular signal-regulated kinase, and the mechanism is totally different from statin. Clinical applications showed that curative effect of berberine used in the treatment of hyperlipidemia patients was good, and security of berberine in treating patients with liver dysfunction is good without the side effects of statins. The results of the study were confirmed by many European and American research institutes and hospitals, which made berberine a promising drug for reducing cholesterol.

A low (25 mg/kg per day) or a high dose of berberine (100 mg/kg per day) were administered in a 2-month-old TgCRND8 mice by oral gavage until 6 months old, and the treatment significantly ameliorated learning deficits, long-term spatial memory retention, as well as plaque load [7]. Berberine ameliorates β-amyloid pathology, hyperphosphorylation of tau, the formation of tangles, and anti-inflammatory, antioxidant, and inhibiting activities of AchE and MAO via the PI3K/AKT/GSK3 signaling pathway.

Berberine has drawn more attention for its antineoplastic effects. It can suppress the growth of a wide variety of tumor cells, such as in ovarian cancer, endometrial cancer, cervical cancer, breast cancer, lung cancer, liver cancer, colorectal cancer, renal carcinoma, bladder cancer, and prostatic cancer. The mechanisms included inducing cell apoptosis, affecting the activity of COX-2 and NF-κB, suppressing the production of PGE2 and expression of IL-8, inhibiting the activity of telomerase, downregulating bcl-2 expression, and upregulating bax expression.

Similar compounds as berberine contains jatrorrhizine, coptisine, palmatine, and other isoquinoline alkaloids, which all have quaternary ammonium groups. The main pharmacological effects include antibacterial, antiviral, antifungal, anti-inflammatory, antipyretic and analgesic, anticancer, hypoglycemic, lipid-lowering, blood pressure-lowering, anti-arrhythmic, and anti-heart failure.

Bioavailability of berberine is low, and is not easy to be absorbed after oral administration. Absorption rate of the intestinal wall is only 5%, and intestinal p-glycoprotein can increase the efflux effect of alkaloids [8]. After the injection berberine immediately distribute to various organs and tissues, but the concentration of plasma was maintained for a short time. The concentration of plasma of berberine after intramuscular injection is lower than the minimum bacteriostatic concentration. Drug distribution is wide, and the concentration in the heart, bone, lung, and liver is high. Retention time in tissues is short, most of the drugs were metabolized and cleared after 24 h, and the drug that exits the body unchanged only accounts less than 5% of dosage. In the atrioventricular model in rats, the non-conjunction type of berberine active transport to bile, metabolized by p450 enzyme system in the liver, the first stage is demethylation, and the second stage is glucosidation. In a rat model, four main metabolites are all glucuronide, such as berberrubine, thalifendine, demethyleneberberine, and jatrorrhizine.

Clinical Application

Rhizoma coptidis, as the digestive tract disease medication, has a history of more than 3000 years in China and India. Berberine, as a cathartic nonprescription drug, is mainly used in the treatment of intestinal infection clinically. Clinical research showed that berberine has hypoglycemic effect and has very good prevention and treatment for diabetic patients with complications such as hypertension, hyperlipidemia, thrombosis, and inflammation.

There are few oral side effects of berberine hydrochloride, accidentally appears nausea, vomiting, rash, and fever, which can disappear after withdrawal of drug. In patients with hemolytic anemia and lack of glucose-6-phosphate dehydrogenase, it was forbidden to be used. Berberine if used intravenously is toxic, but is only suitable for oral drug delivery [9].

Discussion

Berberine hydrochloride has extensive pharmacological activity and therapeutic effects for each system diseases in clinic. At the same time, berberine has been widely used as a medicine source which has the advantages of low cost, less adverse reaction, and the value of popularization and application, but the mechanism of berberine is still not clear. In recent years, researches were mainly focused on anti-tumor, Alzheimer's disease, diabetes, and cardiovascular complications. But at present, the study on antitumor mechanism of berberine is confined to the cellular and molecular level of modern medicine. In the future, it is necessary for research to focus on its structural optimization, mechanism of action, or targets of the research around the above three aspects, and a large number of long-term clinical studies are needed to further confirm the role of the clinical treatment process and curative effects of berberine hydrochloride.

Fig. 1 *Coptis chinensis*

Fig. 2 Chemical structure of berberine

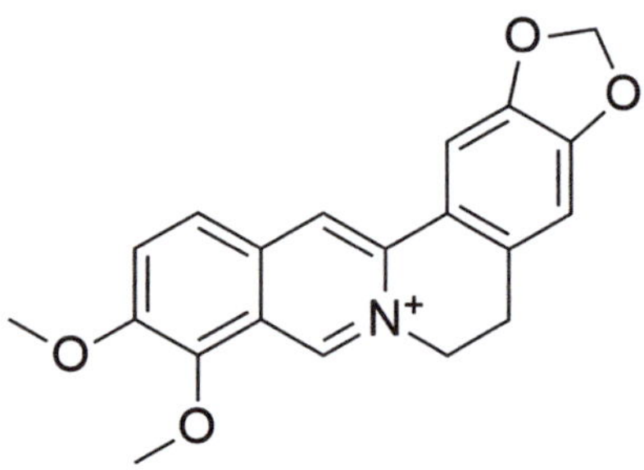

References

1. Zhao BL, Liu XY. Literatures of rhizoma coptidis. J Chin Med Mater. 2013;36(5):832–5.
2. Editorial office of Chinese journal of hospital pharmacy. Answers to questions about the drug elimination. Chin J Hosp Pharm. 1983;3(4):187–8.
3. Huang ZJ, Zeng Y, Lan P, et al. Advances in structural modifications and biological activities of berberine: an active compound in traditional Chinese medicine. Mini Rev Med Chem. 2011;11(13):1122–9.
4. Lee YS, Kim WS, Kim KH, et al. Berberine, a natural plant product, activates AMP-activated protein kinase with beneficial metabolic effects in diabetic and insulin-resistant states. Diabetes. 2006;55(8):2256–64.
5. Kong WJ, Zhang H, Song DQ, et al. Berberine reduces insulin resistance through protein kinase C-dependent up-regulation of insulin receptor expression. Metabolism. 2009;58(1):109–19.
6. Kong W, Wei J, Abidi P, et al. Berberine is a novel cholesterol-lowering drug working through a unique mechanism distinct from statins. Nat Med. 2004;10(12):1344–51.
7. Durairajan SS, Liu LF, Lu JH, et al. Berberine ameliorates β-amyloid pathology, gliosis, and cognitive impairment in an Alzheimer's disease transgenic mouse model. Neurobiol Aging. 2012;33(12):2903–19.
8. Pan GY, Wang GJ, Liu XD, et al. The involvement of p-glycoprotein in berberine absorption. Pharmacol Toxicol. 2002;91(4):193–7.
9. Qiu F, Zhu Z, Kang N, et al. Isolation and identification of urinary metabolites of berberine in rats and humans. Drug Metab Dispos. 2008;36(11):2159–65.

Bergenin

Lv-Jie Xu, Ai-Lin Liu, and Guan-Hua Du

Abstract Bergenin is a kind of lactone mainly extracted from the Chinese cabbage plant. It has antipyretic, analgesic, antitussive, expectorant, and anti-inflammatory effects. And its clinical applications are treatment of chronic bronchitis and chronic gastritis. As the chemical composition and pharmacological activity of the natural active compounds become clearer and the related products in the clinical application become ever-widening, bergenin made a very valuable contribution for human health.

Keywords Bergenin · Lactone · Chronic bronchitis · Chronic gastritis

Alias: Cuscutin
Origin: *Bergenia purpurascens* (Fig. 1)
Chemical name (Fig. 2)

(2R,3S,4S,4aR,10bS)-3,4,8,10-tetrahydroxy-2-(hydroxymethyl)-9-methoxy-3,4,4a,10b-tetrahydropyrano[3,2-c]isochromen-6(2H)-one

Molecular formula, $C_{14}H_{16}O_9$; **MW**, 328.27; **CAS**, 477-90-7

Properties

Appearance: white, loose needle-like crystal or crystalline powder with light odor and bitter taste and discolored when encountered with light or heat. *Solubility*: dissolved in methanol and slightly soluble in water or ethanol. *Melting point*: 232–240 °C. *Specific optical rotation*: −38° to −45°.

L.-J. Xu · A.-L. Liu · G.-H. Du (✉)
Beijing Key Laboratory of Drug Targets Identification and Drug Screening, Centre for Pharmaceutical Screening, Institute of Materia Medica, Chinese Academy of Medical Sciences and Peking Union Medical College, Beijing, China
e-mail: dugh@imm.ac.cn

G.-H. Du, *Natural Small Molecule Drugs from Plants*,
https://doi.org/10.1007/978-981-10-8022-7_63

Dosage Forms and Indications

Bergenin is included in the *Pharmacopoeia of the People's Republic of China* (2015).

It has been developed into solid dispersant, suppositories, capsules, clathrate, and some compound preparations for clinical use. The content of bergenin in compound bergenin tablets is 125 mg/tablets. Bergenin is mainly used for the treatment of chronic bronchitis and chronic gastritis. And it is also effective on the treatment of stomach and duodenal ulcer.

Literature

The plant resources of bergenin are rich, mainly from Rock cabbage (*Bergenia purpurascens*) and thick leafy rock cabbage (*Astilbe macroflora*) from *Saxifragaceae* and the roots, stems, and leaves of bai liang jin (*Ardisia crispa*) from *Ardisia japonica (Thunb.) Blume.* In Chinese medicine Rock cabbage (*Bergenia purpurascens*) is the common folk medicine, with nourishing, stanching bleeding, relieving a cough, and other effects. Its earliest record is contained in the *classified herbal* in Qing Dynasty [1].

There are ten kinds of Rock cabbage (*Bergenia purpurascens*) in Asia, mainly growing in East Asia, north of South Asia, and southeastern Central Asia. There are seven species (including three endemic species) in China, mainly distributed in Shaanxi (Qinling), Xinjiang, Sichuan, Yunnan, and Tibet [2].

History of R&D

In 1958, Hay and Haynes first reported the complete synthesis of bergenin, using 4-methoxy gallate and A-D-bromo-2, 3, 4, 6-tetraacetylglucose as raw materials to synthesize bergenin successfully [3].

In 1987, Chen Wendou determined the content of bergenin as 5.8% in the alcohol extract of *Astilbe chinensis (Maxim.), Franch. et Savat.(Luoxinfu)*, and Rock cabbage (*Bergenia purpurascens*) by using high-performance liquid chromatography. In 1991, Wang Junping determined the content of bergenin in the alcohol extract of *Rodgersia podophylla* as 19.17% with spectrophotometry.

By doing thin layer qualitative analysis, Liu Nunian discovered that bergenin could be found in seven kinds of plants including sour moss (*Ardisia solanacea*), Luosan tree (*Ardisia quinquegona Bl.*), Cinnabar root (*Ardisia crenata Sims*), Zijinniu (*Ardisia japonica*), Lianzuozijinniu (*Ardisia primulaefolia*), Jiujielong (*Ardisia pusilla*), and Xinyezijinniu (*Ardisia maclurei*) from *Ardisia*. In addition, Zhang Yi determined the content of bergenin in the medicinal plants from *Ardisia* with high-performance liquid chromatography, and results showed that there were four species of plants with the content of bergenin of more than 1%, having great resource utilization value. There are 11 species of plants with the content of bergenin less than 1%, such as Xiaoqiaomuzijinniu (*Ardisia arborescens*), twist fruit, fine umbrella (*Ardisia affinis Hemsl.*), and so on. They had some useful value as well [4].

In recent years, it has been found that both 8, 10-dimethylca bergenin and 11-0-gallate betaine had weak inhibitory effects on the tumor in mice, and their effects were stronger than that of bergenin. The effect of acetylated bergenin was stronger than that of bergenin on liver injury induced by chloroform in mice. The new bergenin derivatives from the *Ardisia gigantifolia stapf* had a certain degree of free radical-scavenging effect and anti-HIV viral activity [5].

Pharmacology

1. Cough-relieving effect

Bergenin had a significant relieving effect on coughs caused by electrical stimulation of the laryngeal nerve and ammonia spray. The dose of its cough-relieving effect was equivalent to 1/7–1/4 dose of codeine. No tolerance was found after a 23-day consecutive administration of bergenin due to its selective inhibition on cough center [6].

2. Expectorant effect

Dwarf tea (*Japanese Ardisia Herb*) decoction was found to have a clear expectorant effect after intragastric administration to mice at the dose of 25 g/kg, and its effect strength is equivalent to that of *Platycodon grandiflorum* at the same dose.

3. Anti-inflammatory and anti-ulcer effect

In the model of chronic bronchitis caused by sulfur dioxide fumigation in rats, 80 mg/kg/day of bergenin was given once a day for 10 days. The results showed that the tracheal goblet cell number from the rats in the treatment group decreased, suggesting that the amount of sputum reduced. Inflammatory cell infiltration was found to alleviate, and the degree of emphysema and lung collapse were also reduced. Okada found that bergenin had a therapeutic effect on experimental gastric ulcer rats, consistent with the clinical application of bergenin in the treatment of gastric ulcer, duodenal ulcer, and chronic gastritis [7].

Recent studies have shown that bergenin had certain analgesic and anti-inflammatory properties and had the potential to control inflammatory pain through inhibiting the production of IL-β and TNF-α [4]. Bergenin preparations have been widely used clinically for the treatment of chronic bronchitis, gastritis, and gastric and duodenal ulcers.

4. Antipyretic effect

The alcohol extracts of *Ardisia crispa* had a strong antipyretic effect on the fever caused by the injection of cholera typhoid, paratyphoid, TAB vaccine, and tetanus toxin in the ear of rabbit.

5. Antiviral effect

Rodgersia podophylla alcohol extract at the dose of 0.017–0.034 mg/ml cannot only inhibit the replication of DNA virus but also the replication of RNA virus. The cells with inactivated virus can continue to split and proliferate, indicating that the treatment is effective and with no obvious cell toxicity [7].

6. Liver protection

Bergenin was found to improve the liver injury induced by carbon tetrachloride in mice. It cannot only reduce the release of glutamate acetyltransferase and sorbitol dehydrogenase from mice liver but also reduce the glutathione reductase and increase the glutathione content, indicating that bergenin protected the liver through the regulation of glutathione and inhibiting the release of free radicals [8].

7. Cardiovascular system effects

Bergenin had some therapeutic effects on the arrhythmia in mice.

8. Toxicity test

Mice were intraperitoneally injected with bergenin; the minimum lethal dose is 10 g/kg; no toxic reaction was found after intragastric administration of bergenin at the dose of 12 g/kg. No effects were found in the growth and development, liver function, and ECG of young rats after the intragastric administration with bergenin at the dose of 2.5 g/kg for consecutive 60 days, and no toxic performance was found in the heart, liver, kidney, lung, spleen, stomach, intestine, brain, and other organs by pathological biopsy.

Clinical Application

Bergenin was recorded in *People's Republic of China Pharmacopoeia* as an antitussive expectorant, used for the treatment of chronic bronchitis, emphysema, bronchial asthma, and other respiratory diseases. Used alone, or in compound preparation, such as compound bergenin tablets, Qingjin syrup, Compound Hu ear tablets, Qingfei antitussive syrup, liver poison net particles, Kechuanping oral solution, and silicone lung oral solution, it is mainly used for the treatment of chronic bronchitis. Its efficacy was proven to be relatively stable and improved in the compound preparations after a wide range of clinical validation.

Discussion

As a high-yield natural active compound, the clinical application of *Bergenia* plants has a long history, and its chemical composition and related pharmacological activity are clear. A number of bergenin derivatives have been obtained through the structural modification, and their activity research is carrying out. However, as the clinical application of the related *Bergenia* plant products gradually widened, the problem of the contradiction between supply and demand is gradually highlighted and needed to be solved.

Fig. 1 *Bergenia purpurascens*

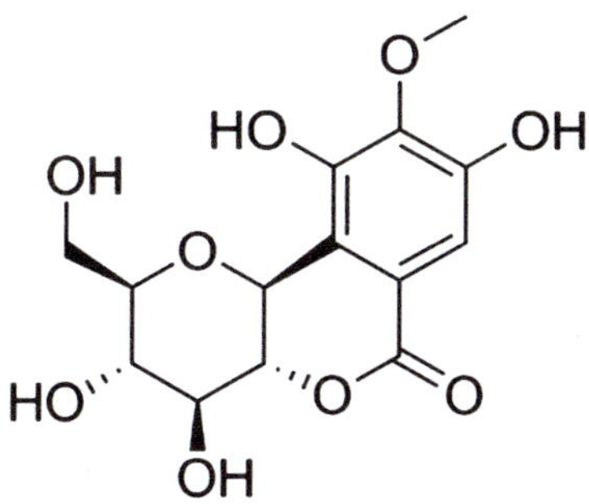

Fig. 2 Chemical structure of Bergenin

References

1. Lv XM, Wang JX. Advances in studies on Bergenia Moench. J Chin Med Mater. 2003;26(1):58–60.
2. Pan JT, Gu CZ, Douglas ES, et al. Saxifragaceae. Flora China. 2001;8:269–452.
3. Hay JE, Haynes LJ. Bergenin, a C-glycopyranosyl derivative of 4-O-methylgallic acid. J Chem Soc. 1958:2231–8.
4. Ai JL, Suo JL. Research progress in medicinal plant genus Bergenia Moench. J Baoji Univ Arts Sci (Nat Sci). 2011;31(1):46–50.
5. Lim HK, Kim HS. Protective effects of acetylbergenin against carbon tertra- chloride induced hepatotoxicity in rats. Arch Pharm Res. 2001;24(2):114–8.
6. Jiangsu New School of Medicine. Dictionary of Chinese medicine. Shanghai People's Publishing House.1999;359,2359.
7. Li BH, Wang JP. Research on Bergenin. Northwest Pharm J. 1990;5(3):45–6.
8. Wang G, Ma BJ. Overview of studies on Bergenin. J Anhui Tradit Chin Med Coll. 2002;21(6):59–62.

Brucine

Wan Li, Cui Yang, and Guan-Hua Du

Abstract Brucine is extracted from the dry and mature seeds of Loganiaceae Evergreen tree plants including *Strychnos nux-vomica* L. and *Strychnos pierriana A.W. Hill*. Brucine has many pharmacological effects, including analgesic, anti-tumor, anti-pathogen, anti-inflammatory, and antibacterial effects and regulation of the immune system. In clinical practice, brucine is used as compatibility prescription or compound medicine in treating rheumatism, rheumatoid arthritis, stroke hemiplegia, dementia, retinopathy, and orthopedic and other surgical diseases. However, the toxicity of brucine limits its wide usage in the clinic. Therefore, studies of brucine on increasing its efficiency and decreasing its toxicity will improve the safety of its clinical use.

Keywords Brucine · Analgesic effects · Rheumatism · Toxicity

Alias: Brucine, strychnine
Origin: *Strychnos nux-vomica L.* and *Strychnos pierriana A.W. Hill* (Fig. 1)
Chemical name (Fig. 2)

10,11-Dimethoxystrychnine

Molecular formula, $C_{23}H_{26}N_2O_4$; **MW**, 394.4; **CAS**, 357-57-3
Derivatives (Fig. 3)

W. Li · G.-H. Du (✉)
Beijing Key Laboratory of Drug Targets Identification and Drug Screening, Centre for Pharmaceutical Screening, Institute of Materia Medica, Chinese Academy of Medical Sciences and Peking Union Medical College, Beijing, China
e-mail: dugh@imm.ac.cn

C. Yang
Yunnan Minzu University, Yunnan, China

G.-H. Du, *Natural Small Molecule Drugs from Plants*,
https://doi.org/10.1007/978-981-10-8022-7_64

Properties

Appearance: white crystalline powder; tastes very bitter. *Solubility*: slightly soluble in water; soluble in ether, chloroform, ethanol, methanol, and other organic solvents [1]. *Melting point*: 175–178 °C. *Specific optical rotation*: −118°.

Dosage Forms and Indications

This drug is reported in the *British Pharmacopoeia* (2017) and *European Pharmacopoeia* (8.7th ed.).

Fourteen Chinese medicines containing brucine are recorded in the *Chinese Pharmacopoeia* 2010 edition, which are primarily in the form of tablets, pills, and capsules [2]. At present, the clinical usage of brucine is mainly in the form of compound drugs including Tong Bi Ling tablets and Lumbago ling capsule. The former is mainly used for the treatment of ankylosing spondylitis, and the latter is mainly used for the treatment of lumbar disc herniation, lumbar hyperplasia, sciatica, lumbar muscle strain, lumbar muscle fiber, and chronic rheumatoid arthritis.

With the development of pharmaceutical preparation technology, a variety of new formulations of strychnine are emerging, such as brucine liposomes, brucine nanoparticles, brucine total alkaloid vesicles, brucine compound spray, and brucine compound gel. However, further studies are needed to evaluate the therapeutic effects of these formulations [3].

Literature

As a Chinese medicine, brucine was firstly recorded in *Compendium of Materia Medica*. The Chinese name for brucine is "Maqian" because it looks like ancient coins with horse icon, where "Ma" refers to "horse" and "qian" refers to "coins." This traditional Chinese herb has cold, bitter, and toxic properties. It can be used to treat typhoid fever, throat pain, and lumps [4]. Brucine is mainly extracted from the traditional Chinese medicine "Maqian." In 1819, Pelletier and Caventou firstly isolated the alkaline brucine from the seeds and bark of *Strychnos nux-vomica* L. Brucine is the dry and mature seeds of Loganiaceae Evergreen tree plants including *Strychnos nux-vomica* L. and *Strychnos pierriana A.W. Hill* [5].

Pharmacology

1. Analgesic effects: Brucine is reported to exert analgesic effects through both central and peripheral pathways. Possible mechanisms of the analgesic effects include blocking the voltage-gated sodium channel, inhibiting the synthesis of PGE2, acting on the adrenergic receptor and the L-arginine-NO pathway. Brucine (0.48 mg/kg) can significantly elevate the pain threshold in mice [6].
2. Anti-tumor effects: Studies have shown that brucine inhibits the cell proliferation in breast cancer, liver cancer, leukemia, and Ehrlich ascites tumors by inducing tumor cell apoptosis, inhibiting angiogenesis, reversing tumor multidrug resistance, and regulating the expression of various cytokines. In addition, at the dose of 320 μg/ml, the in vitro hepatocarcinoma cell growth inhibition rate of brucine is nearly 100% [7].
3. Anti-pathogen effects: In vitro study shows that 0.1% brucine can completely inhibit the growth of *Haemophilus influenzae*, *Streptococcus pneumoniae*, and *alpha Streptococcus* and *Mlicrococcus catarrhalis*. At the dose of 500 μg/ml, the inhibitory rate of brucine on human immunodeficiency virus reverse transcriptase is above 30% [8].
4. Other pharmacological effects: Anti-inflammation, regulation of the immune system, regulation of the cardiovascular system, antitussive, expectorant, and antiasthenic and antibacterial effects.

Clinical Applications

The applications of brucine include reducing swelling; resolving mass; activating meridians to stop pain; attenuating rheumatism, stubborn paralysis, and numbness paralysis; alleviating bruises; etc. It has been used for nearly a thousand years and achieved prominent clinical efficacy. In the clinical practice, brucine is used as a compatibility prescription or compound medicine in treating rheumatism, rheumatoid arthritis, stroke hemiplegia, dementia, retinopathy, and orthopedic and other surgical diseases [9, 10].

Brucine is the main active ingredient in *Strychnos* and plays an important role in treating diseases. However, brucine is also a toxic ingredient, which limits the scope of its clinical application. Further studies of its pharmacological effects and toxicity will improve its clinical application.

Discussion

Brucine is the main ingredient *Strychnos* plants, which has many pharmacological effects. At present, mechanism studies of brucine are focused on the anti-inflammatory, analgesic, and anti-tumor effects, whereas the anti-pathogen and other effects are less studied. However, the toxicity of brucine limits its wide use in the clinic. Therefore, studies of brucine on increasing its efficacy and decreasing its toxicity will improve the safety of its clinical use.

Fig. 1 *Strychnos nux-vomica L.*

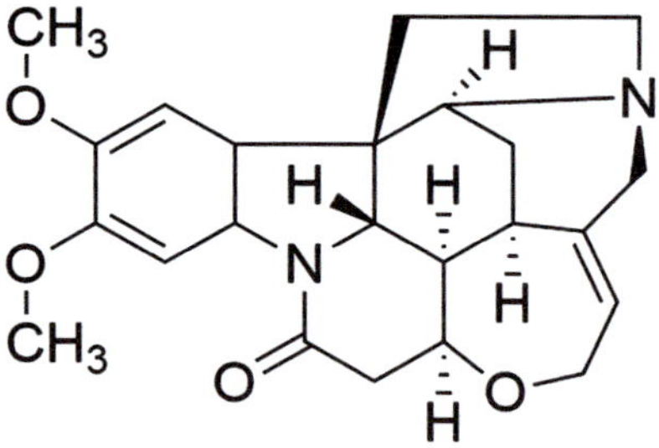

Fig. 2 Chemical structure of brucine

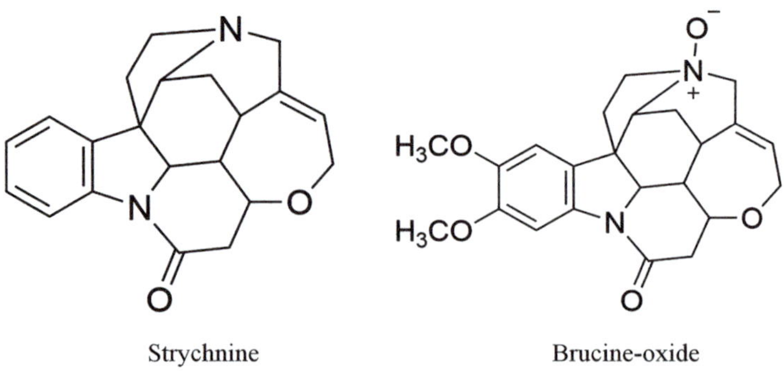

Fig. 3 Chemical structure of brucine derivatives

References

1. Xu JH, et al. Advances in studies on brucine. Chin J New Drugs. 2009;18(3):213–21.
2. An analysis of prepared TCM medicine containing Maqianzi in Chinese pharmacopoeia 2010 edition. Clin J Chin Med. 2014;6(23):9–10.
3. Zhao YL, et al. Research progress on pharmacokinetics of Strychnos nux-vomica and its preparation. Chin Tradit Herb Drug. 2015;46(11):1710–4.
4. Zhao LM, et al. Anti-tumor effect of brucine. Chin J Cancer Prev Treat. 2013;20(11):877–9.
5. Li XX, et al. Advances in studies on anti – tumor and attenuated efficacy of brucine. Guid China Med. 2013;11(20):486–8.
6. Tang DJ, et al. Comparative study on the pharmacodynamic effects on anti-inflammation and analgesia by brucine, semen strychnine and jiu-fen san. Inf Tradit Chin Med. 2014;4:1–3.
7. Qin JM, et al. The effect of brucine on hepatocellular carcinoma cell lines in vitro. Chin J Gen Surg. 2011;26(3):219–21.
8. Qin XS, et al. Study on the inhibitory effect of Chinese herbal extract on human immunodeficiency virus reverse transcriptase. Acta Acad Med Sinicae. 1990;12(6):391–5.
9. Zhu HX, et al. Discussion on the clinical application of Chinese herbal medicine brucine. Heilongjiang J Tradit Chin Med. 2008;37(3):54.
10. Lin CS, et al. Pharmacological study and clinical application of brucine. Tradit Chin Drug Res Clin Pharmacol. 2006;17(2):158–60.

Bulleyaconitine A

Wei-Qi Fu, Cui Yang, and Guan-Hua Du

Abstract Bulleyaconitine A can be extracted from *Aconitum bulleyanum* Diels and other aconitum plants. Bulleyaconitine A shows excellent anti-inflammatory, analgesic, and immunomodulatory effects, especially in the treatment of rheumatological disorders. Moreover, bulleyaconitine A has little psychological dependence and organ toxicity. Therefore, bulleyaconitine A shows considerable advantages in the treatment of rheumatic immune disease and chronic pain. Taken together, bulleyaconitine A may have good prospects for development and application after being systematically investigated.

Keywords Bulleyaconitine A · Alkaloid · Analgesic · Anti-inflammatory · Immune regulation

Alias: Crassicauline A
Origin: *Aconitum bulleyanum* Diels (Fig. 1)
Chemical name (Fig. 2)

[(1-α,6-α,14-α,16-β)-8-(acetyloxy)-20-ethyl-13-hydroxy-1,6,16-trimethoxy-4-(methoxymethyl)aconitan-14-yl](4-methoxyphenyl)-methanone

Molecular formula, $C_{35}H_{49}NO_{10}$; **MW**, 643.77; **CAS**, 107668-79-1
Derivatives (Fig. 3)

W.-Q. Fu · G.-H. Du (✉)
Beijing Key Laboratory of Drug Targets Identification and Drug Screening, Centre for Pharmaceutical Screening, Institute of Materia Medica, Chinese Academy of Medical Sciences and Peking Union Medical College, Beijing, China
e-mail: dugh@imm.ac.cn

C. Yang
Yunnan Minzu University, Yunnan, China

G.-H. Du, *Natural Small Molecule Drugs from Plants*,
https://doi.org/10.1007/978-981-10-8022-7_65

Properties

Appearance: white powder. *Solubility*: soluble in methanol, ethanol, chloroform, and ether; insoluble in water; and easily dissolved in dilute hydrochloric acid or dilute sulfuric acid [1]. *Melting point*: 160–165 °C

Dosage Forms and Indications

This product is included in the *Pharmacopoeia of the People's Republic of China* (2015).

Bulleyaconitine A preparation forms are tablets, powder injection, capsule, and injection. It is mainly used to treat rheumatic immune disease and chronic pain.

Literature

Bulleyaconitine A is an aconitine-type two terpene alkaloid, and its molecular formula is $C_{35}H_{49}NO_{10}$. As early as 1980, bulleyaconitine was separated and extracted from a traditional analgesic and anti-inflammatory herb in the west of Yunnan called *dianxi dula* (or *dianxi wutou*) (*Aconitum bulleyanum* Diels), by the Kunming Institute of Botany, Chinese Academy of Sciences [2]. At present, bulleyaconitine A can be separated from *Aconitum* plants, for example, *dianxi wutou* (*Aconitum bulleyanum* Diels), *cujing wutou* (*Aconitum crassicaule* W.T. Wang), *changhui wutou* (*Aconitum georgei* Comber), and *zhiyuan wutou* (*Aconitum transsectum* Diels).

History of R&D

In 1983, it was confirmed that bulleyaconitine A shows obvious analgesic and anti-inflammatory effects by preclinical pharmacological tests [2]. Clinical trials were carried out in 29 hospitals in 9 provinces in 1984. Bulleyaconitine A passed technical appraisals in May 1985 [2] and was first formally put into production by Yunnan Honghe Pingbian pharmaceutical company, where it obtained the trade name "Bulleyaconitine A." Bulleyaconitine A has been listed into the *Pharmacopoeia of the People's Republic of China* (2015). In recent years, bulleyaconitine A has shown a great market growth and competitive power, owing to its exact curative effect and rapid clinical promotion in chronic disease and rheumatic immune disease.

Pharmacology

1. Anti-inflammatory and analgesic effect. Bulleyaconitine A plays a significant anti-inflammatory effect by inhibiting the release of prostaglandins. The analgesic effect may be related to the rivalry of 5-HT and the inhibition of inflammatory chemokines in the brain, because inhibition of PGE_2 release may lead to disinhibition of β-endorphin [3]. Further studies showed that bulleyaconitine A achieved the analgesic effect by effectively reducing sodium ion current, the electrophysiological basis on which pain conduction is dependent [4]. Zhanguo Niu et al. found that bulleyaconitine A could be used to treat mild and moderate burn pain, with less adverse reactions and no addiction [5]. Thus, bulleyaconitine A may be suitable for long-term use as a choice of analgesic therapy for burn pain. Recently, through clinical observation on the effect of bulleyaconitine A treatment in acute gouty arthritis, some scholars discovered that bulleyaconitine A brought about good effect and could effectively improve the clinical symptoms of patients by taking 0.4 mg bulleyaconitine A orally three times in daily [6].
2. Immune regulation. Bulleyaconitine A has obvious inhibitory effects to immune cells involved in joint inflammation, which may have a close relationship to its clinical anti-inflammatory effect. Ye Lu et al. studied the influence of bulleyaconitine A on partial immune functions of BALB/c mice and found that 0.32 mg/kg bulleyaconitine A inhibits some immune functions of BALB/c mice [7]. Moreover, it was reported that bulleyaconitine A significantly inhibited the phagocytic function and the ability to secrete nitric oxide (NO) of macrophages [8].
3. Other pharmacological effects. An injection of 0.1 mg/kg bulleyaconitin A in fasting rat can inhibit the formation of corneal neovascularization induced by alkali burn [9]. Bulleyaconitine A also has local anesthesia and antipyretic effect. Besides, bulleyaconitine A can be used for the treatment of lumbar muscle strain, scapulohumeral periarthritis, and sprain of limbs.

Clinical Application

1. Rheumatic immune disease. Since listing, bulleyaconitine A has been applied in clinical treatment for rheumatoid arthritis and osteoarthritis, receiving good curative effects.
2. Chronic pain. Bulleyaconitine A does not belong to NSAID, and it exerts its effects by regulating the sodium ion channel. So there is little psychological dependence and organ toxicity, avoiding gastrointestinal/cardiovascular/renal adverse reactions and drug dependence or other potential dangers caused by NSAID and opioid analgesics.

Discussion

Bulleyaconitine A has excellent anti-inflammatory, analgesic, and immunomodulatory effects and is suitable for long-term use in the treatment of chronic pain, which shows considerable advantages in the treatment of rheumatological disorders. Therefore, bulleyaconitine A can be systematically investigated from many aspects, including dosage form, clinical observation, clinical indications, and adverse reactions.

Fig. 1 *Aconitum bulleyanum* Diels

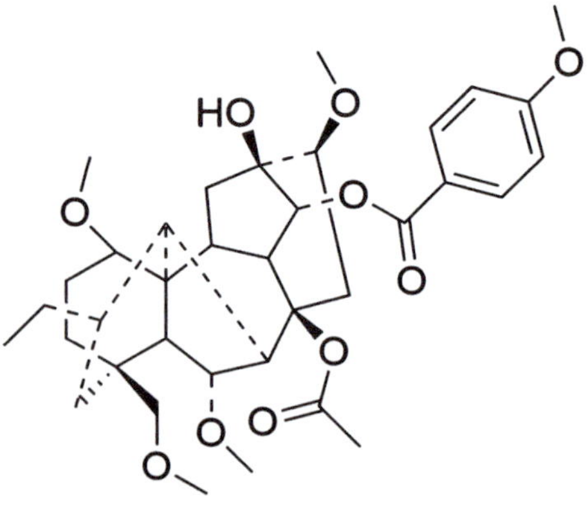

Fig. 2 Structural formula of bulleyaconitine A

Fig. 3 Derivatives of bulleyaconitine A

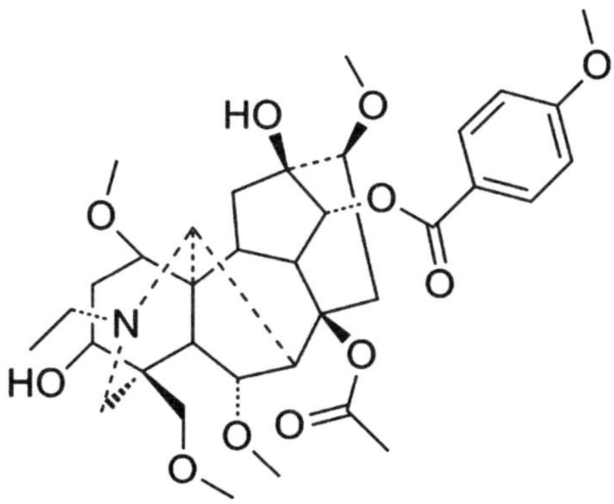

Franchetine

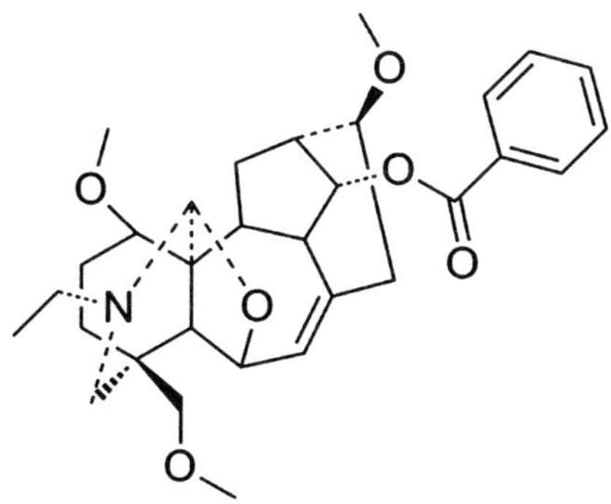

Yunaconitine

References

1. Bulleyaconitine. Pharmacopoeia of the People's Republic of China (2015). Part 2: 2015,711.
2. Tang X. Huperzine A, a novel analgesic and anti-inflammatory drug. Chin J New Drugs Clin Remedies. 1986;5(2):120–1.
3. Yang X, et al. Research progress. World Clin Drugs. 2013;34(2):80–3.
4. Gerner P, et al. Bulleyaconitine A isolated from aconitum plant displays long-acting local anesthetic properties in vitro and in vivo. Anesthesiology. 2007;107(1):82–90.
5. Niu Z, et al. Clinical observation of Bulleyaconitine A tables on analgesic effect of burn wound. Jilin Med J. 2014;35(31):7001–2.
6. Yuanxin Huang et al. Clinical curative effect observation on Bulleyaconitine A in the treatment of acute gouty arthritis. Chin J Clin Ration Drug Use. 2016;9(3A).
7. Lu Y, et al. Inhibitory effect of bulleyaconitine A on some immune function in Balb/c mice. Chin J New Drugs Clin Remedies. 2007;26(10):755–8.
8. Tsai SH, et al. Suppression of nitric oxide synthase and the down-regulation of the activation of NFκB in macrophages by resveratrol. Br J Phamacol. 1999;126(3):673–80.
9. Ye Y. Inhibitory effect of bulleyaconitine A on corneal neovascularization induced by alkali burn. J North Pharm. 2015;12(5):129.

Capsaicin

Xiu-Ying Yang and Guan-Hua Du

Abstract Capsaicin is extracted from chili pepper (*Capsicum*). It is one of the main pungent components in *Capsicum* fruits, which is one of the most popular and widely used spices in the world. Pharmacological research shows that pepper has beneficial effects on the digestive, cardiovascular, and endocrine system, as well as analgesia. Capsaicin can act on sensory nerve C primary afferent fibers, bind the end of the neuronal TRPV1 receptor. In clinical practice, capsaicin is mainly used for topical administration. In 2009, the US FDA approved the first capsaicin prescription patch with high concentration for the treatment of postherpetic neuralgia. At present, most countries have capsaicin prescription drugs on market, but clinical indications are limited to the treatment of pain. The capsaicin drug's R&D is relatively slow. If less irritated irritant capsaicin derivative can be obtained from structure modification, its development prospects would be attractive.

Keywords Capsaicin · Capsicum · Chili pepper · Analgesia

Alias: Capsaicine, Axsain, Capsidol, Capzasin, Capsin
Origin: Chili pepper (Fig. 1)
Chemical name (Fig. 2)

8-Methyl-*N*-vanillyl-6-nonenamide

Molecular formula, $C_{18}H_{27}NO_3$; **MW**, 305.42; **CAS**, 404-86-4
Derivatives (Fig. 3)

X.-Y. Yang · G.-H. Du (✉)
Beijing Key Laboratory of Drug Targets Identification and Drug Screening, Centre for Pharmaceutical Screening, Institute of Materia Medica, Chinese Academy of Medical Sciences and Peking Union Medical College, Beijing, China
e-mail: dugh@imm.ac.cn

G.-H. Du, *Natural Small Molecule Drugs from Plants*,
https://doi.org/10.1007/978-981-10-8022-7_66

Properties

Appearance: crystalline white powder, with highly volatile and pungent odor. *Solubility*: freely soluble in alcohol, ether, benzene, and chloroform; slightly soluble in carbon disulfide, petroleum, and hydrochloric acid; insoluble in water. *Melting point*: 65 °C.

Dosage Forms and Indications

This product is collected in the *British Pharmacopoeia* (2017), *United States Pharmacopeia* (40), and *European Pharmacopoeia* (9.0th ed.).

At present, the pharmaceutical dosage forms in clinical practice are external use gel and ointment. Clinical indications are mainly used to relieve pain, muscle soreness, and so on.

Literature

Capsaicin is one of the major active components of chili pepper辣椒 (*Capsicum*). Chili pepper is the fruit of annual herb of the genus *Capsicum*. Its species are native to tropical regions of central and south Americas, where they have been cultivated for thousands of years. In late fifteenth century, after Columbus discovered America, he brought the peppers to Europe and then became cultivated worldwide. Pepper was introduced in China in the Ming dynasty and then became to be the most popular and widely used spice in China.

Capsicum is not only widely cultivated as a vegetable worldwide but also used as a medicine. Capsicum was recorded in the *Traditional Chinese Medicine Classics*. For example, in *Compendium of Materia Medica Supplements*, it records that spicy eggplant hot and scattered can also dispel dampness. Modern pharmacological research shows that pepper has beneficial effects on the digestive, cardiovascular, and endocrine system. In addition, capsaicin can also be used for analgesia, reducing symptoms of cold and seasickness, alcohol addiction, and preventing cancer.

History of R&D

In 1816, Christian Friedrich Bucholz (1770–1818) first carried out the purification of capsaicin. He obtained incompletely purified capsaicin and named it "capsaicin" [1]. In 1876, John Clough Thresh (1850–1932) further purified and got pure capsaicin and named it capsaicin [1]. In 1919, Nelson identified the capsaicin structure

[2]. In 1930, E. Spath and S. F. Darling used chemical synthesis method for the first time to successfully synthesize capsaicin [3]. In 1961, the Japanese chemists S. Kosuge and Y. Inagaki isolated similar substances from capsicum and named them capsaicinoids [4, 5].

More than 14 capsaicin homologues have been found till now, the structures are similar with capsaicin, and the typical structure is H3CO(HO)-C6H3-CH2-NH-CO-R, which is only different from R group. Capsaicin is the highest content in chili pepper. Capsaicin and dihydrocapsaicin are the most potent active substances in capsicum. Capsaicin and its analogues have been synthesized in large quantities in practice and can be used not only in medicine but also in agriculture and industry.

Although there is a long history of pharmacological and chemical studies on capsaicin, until November 16, 2009, the US Food and Drug Administration (FDA) approved the first containing high concentration capsaicin prescription Qutenza (8% capsaicin) patch for the treatment of postherpetic neuralgia. At present, most countries have capsaicin prescription drugs on market, but clinical indications are limited in the treatment of pain.

Pharmacology

Hbgyes, A. (1878) first reported that capsaicin has a strong stimulating effect [6], which is the first pharmacological study on capsaicin. And then, a serial of pharmacological effects were discovered. The discovery of capsaicin receptor further explains the mechanism of capsaicin. The capsaicin receptor, also known as transient voltage receptor cation channel V1 (TRPV1), is a nonselective ligand-gated cation channel. TRPV1 is widely distributed in the body, mainly in sensory neurons, and can also be activated by heat and friction damage.

Studies on the analgesic effect of capsaicin were carried out earlier and more thoroughly. Capsaicin can act on sensory nerve C primary afferent fibers, bind the end of the neuronal TRPV1 receptor. Capsaicin (1 μM) can result in inward calcium influx [7], cell depolarization, neuronal excitation, and glutamate release. The sustained neuron excitement and then failure can result in analgesic and antipruritic effects.

The mutation of the capsaicin receptor can not only induce obesity but also may be associated with the occurrence of diabetes [8]. Capsaicin can activate and recruit brown fat to prevent obesity [9]. Brown fat can produce non-shivering heat in cold environment and participate in energy consumption. A 10–130 mg daily capsaicin can significantly increase the body's energy and fat consumption [10].

Capsaicin also has a protective effect on the cardiovascular system. Treatment of capsaicin with rats at a dose of 15 mg/kg can not only promote animal blood circulation and strengthen the cardiovascular function but also reduce the blood pressure [11], the serum cholesterol, and triglyceride levels.

The study result of capsaicin on tumor is still controversial. The epidemiology and basic research have suggested that capsaicin can not only be used as a carcinogen but also can prevent cancer. It has been shown that capsaicin can induce cancer cell apoptosis, and animal experiments have shown that prolonged use of capsaicin on the skin can induce skin cancer.

Capsaicin is able to scavenge free radicals and inhibit oxidative stress. Capsaicin can promote gastric secretion, increase appetite, relieve flatulence, improve digestive function, and also prevent gastrointestinal infection and diarrhea. Capsaicin can improve the performance of sports and anti-fatigue. Capsaicin is capable of thinning lung mucus, in favor of sputum discharging, enhancing lung tissue perfusion, and preventing and treating emphysema. Capsaicin is also beneficial on psoriasis, frostbite, cold, etc. In addition, capsaicin is also used for the paralysis of peripheral nerve function for hypertension treatment.

Capsaicin can be absorbed by the intestine and skin and is able to pass through the blood-brain barrier.

Clinical Applications

In clinical practice, capsaicin is mainly used for topical administration, such as in the treatment of osteoarthritis and rheumatoid arthritis pain, diabetic nerve pain, pain after surgery, chemotherapy- or radiotherapy-induced oral pain, psoriasis, etc.

Capsaicin irritates the mucous membrane to cause sneezing, nose bleeding, coughing, mucus secretion, tears, bronchoconstriction, breathing difficulties, and other symptoms. The main adverse effects of capsaicin preparations are contact dermatitis, skin inflammation or blisters, and in severe situation burn-like lesion.

Discussion

Chili pepper exists widely in human diet and is a main ingredient. Capsaicin was found more than a hundred years with variety of pharmacological activities. However, the capsaicin drug's R&D is slow and only a few medicines are on market. The main reason is that capsaicin has irritation reaction, which limits the drug development of capsaicin. If less irritant capsaicin derivative can be obtained from structure modification, its development prospect would be attractive.

Fig. 1 Chili pepper

Fig. 2 Chemical structure of capsaicin

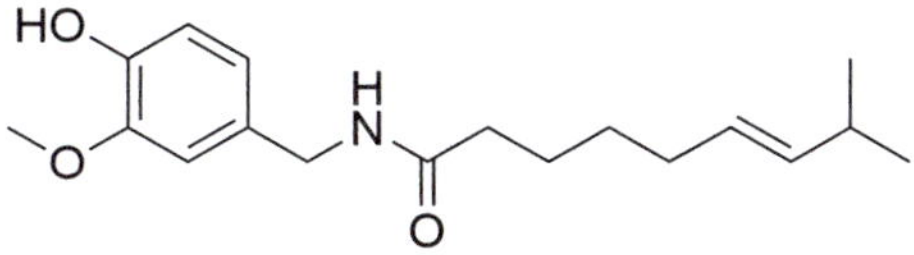

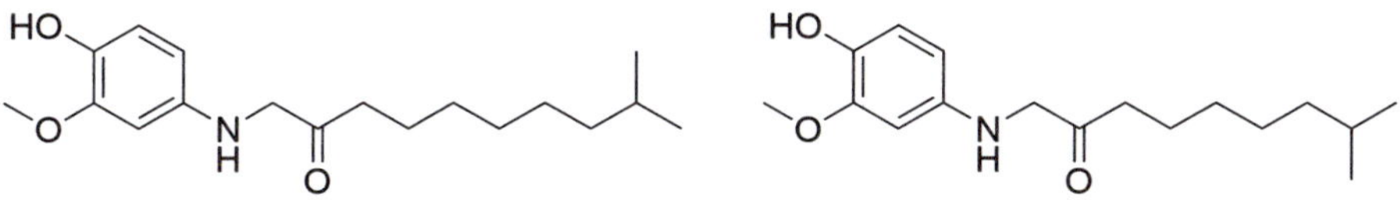

Dihydrocapsaicin

Nordihydrocapsaicin

Fig. 3 Chemical structure of capsaicinoids

References

1. Wikipedia contributors. "Capsaicin." Wikipedia, The free encyclopedia. Wikipedia, 15 Oct. 2015.
2. Nelson EK. The constitution of capsaicin, the pungent principle of capsicum. J Am Chem Soc. 1919;41(1):1115–21.
3. Späth E, Darling SF. Synthese des capsaicins. Ber Dtsch Chem Ges. 1930;63(3):737–43.
4. Kosuge S, Inagaki Y, Okumura H. Studies on the pungent principles of red pepper. Part VIII. On the chemical constitutions of the pungent principles. Nippon Nogei Kagaku Kaishi(J Agric Chem Soc). 1961;35:923–7.

5. Kosuge S, Inagaki Y. Studies on the pungent principles of red pepper. Part XI. Determination and contents of the two pungent principles. Nippon Nogei Kagaku Kaishi(J Agric Chem Soc). 1962;36:251.
6. Toh CC, Lee TS, Kiang AK. The pharmacological actions of capsaicin and analogues. Br J Pharmacol Chemother. 1955;10(2):175–82.
7. Caterina MJ, Schumacher MA, Tominaga M, Rosen TA, Levine JD, Julius D. The capsaicin receptor: a heat-activated ion channel in the pain pathway. Nature. 1997;389(6653):816–24.
8. Suri A, Szallasi A. The emerging role of TRPV1 in diabetes and obesity. Trends Pharmacol Sci. 2008;29(1):29–36.
9. Saito M, Yoneshiro T, Matsushita M. Food ingredients as anti-obesity agents. Trends Endocrinol Metab. 2015;26(11):585–7.
10. Whiting S, Derbyshire E, Tiwari BK. Capsaicinoids and capsinoids. A potential role for weight management? A systematic review of the evidence. Appetite. 2012;59(2):341–8.
11. Yang D, Luo Z, Ma S, et al. Activation of TRPV1 by dietary capsaicin improves endothelium-dependent vasorelaxation and prevents hypertension. Cell Metab. 2010;12(2):130–41.

Glycyrrhetinic Acid

Wen Zhang, Qi-Meng Zhou, and Guan-Hua Du

Abstract Glycyrrhetinic acid is extracted from liquorice, which is an important compound in liquorice. Glycyrrhetinic acid plays a role in anti-inflammatory, anti-ulcer activity, and antioxidation. Glycyrrhetinic acid also has an antiviral effect, antitumor effect, and antiarrhythmic effects. In addition, glycyrrhetinic acid functions as an anticholinesterase, anticoagulant, anti-tetanus toxin; it also improves inner ear hearing and improves absorption of insulin. However, its pseudoaldosteronism effect restricts clinical application, and it needs further study to develop new drugs.

Keywords Glycyrrhetinic acid · Liquorice · Pseudoaldosteronism

Alias: Arthrodont; Enoxolone; Biosone; Glycyrrhetin; Subglycyrrhelinic acid; α-Glycyrrhetinic acid
Origin: Liquorice (Fig. 1)
Chemical name (Fig. 2)

(2S,4aS,6aS,6bR,8aR,10S,12aS,12bR,14bR)-10-hydroxy-2,4a,6a,6b,9,9,12a-heptamethyl-13-oxo-1,2,3,4,4a,5,6,6a,6b,7,8,8a,9,10,11,12,12a,12b,13,14b-icosahydropicene-2-carboxylic acid

Molecular formula, $C_{30}H_{46}O_4$; **MW**, 470.68; **CAS**, 471-53-4

W. Zhang · Q.-M. Zhou · G.-H. Du (✉)
Beijing Key Laboratory of Drug Targets Identification and Drug Screening,
Centre for Pharmaceutical Screening, Institute of Materia Medica, Chinese Academy of Medical Sciences and Peking Union Medical College, Beijing, China
e-mail: dugh@imm.ac.cn

G.-H. Du, *Natural Small Molecule Drugs from Plants*,
https://doi.org/10.1007/978-981-10-8022-7_67

Properties

Solubility: insoluble in water; it exists in crystal with methanol and chloroform. *Melting point*: the compound melts at 292–295 °C. *Specific optical rotation*: under the condition of 20 °C, 589.3 nm, and 1 dm, polarized light rotates at 68° when it passes through the chloroform with a concentration of 64 mol/L. Both 18α-glycyrrhetic acid and 18β-glycyrrhetic acid are chiral isomers of glycyrrhetic acid.

Derivatives (Fig. 3)

In recent years, scientists have modified the structure of 18β-glycyrrhetinic acid. By adding the appropriate acyl group or glycosyl group in the position of 3-hydroxyl group, the activity of anti-virus, anti-inflammation, anti-ulcer, and anti-tumor can be enhanced. It also shows the activities of antiulcer and antitumor. In the position of C-30, if metal ions, alkoxy, amino acid, or glycoside ligand is combined, the antiulcer and anti-inflammatory property will be improved. Besides, if glycoside ligand and other groups can be added in the position of 3 and 30 at the same time, the therapeutic effect on ulcer and inflammation is totally different [5]. The antibacterial activity is enhanced when the hydroxyl group is introduced in the position of 2 of A ring, and electron-withdrawing group is equipped for antitumor. Five-membered ring takes the place of A ring, hydroxyl group oxidation, or double bonds introduced in the position of 1and 4 which leads to the decreased activity. The compound with triazoles recognizes specifically and binds with Hg^{2+}. If carbonyl group in C-11 of C ring were deoxidized, it would reduce the effect like steroids. Forming double bonds in C-9, C-11 can increase antiulcer effect [6].

Dosage Forms

Glycyrrhetic acid tablet

Indications

Treatment of Addison's disease, deoxycorticosterone

Literature

Extracting from liquorice (*Glycyrrhiza uralensis Fisch*, Gan Cao), glycyrrhetic acid also can be detected in other plants, such as *Abrus cantoniensis Hance* and *Herba Abri fruticulosi*. As one of traditional Chinese medicines, liquorice has been applied clinically for a long period. Due to its extensive usage, it plays an extremely important role in traditional Chinese medicine formula mainly as "guide" drug. As is recorded in *Shen Nong's Herbal Classic* and later in pharmaceutical monographs, liquorice is able to strengthen bones and muscles and enhance metabolism and detoxification. Also, abnormal symptoms of the body and wound can be improved. Glycyrrhetic acid, the most important and potent ingredient of liquorice, has been recorded in *Pharmacopoeia of the People's Republic of China*.

History of R&D

Glycyrrhetic acid originates from hydrolysis of glycyrrhizin, which has a therapeutic effect on disease. Dating back to the 1930s, the chemical structure of glycyrrhetic acid was demonstrated [1–3]. Subsequently, the discovery of antiulcer activity promotes following research [4]. The ramification of glycyrrhetic acid, carbenoxolone sodium, has a therapeutic effect on ulcer. In 2010, followed by the approval of raw materials, batches of tablets and capsules were approved in 2009, respectively. In foreign countries, 18β-glycyrrhetic acid was studied for anti-inflammatory effect on arthritis, rheumatoid disease, and periodontitis in BioNetWorks. The company applied for the patent of 18β-glycyrrhetic acid in 1999. Also, after joining the leading worldwide market in 2006, phase III clinical trials would be carried out in 2007. However, the progress was hindered in 2008.

To detecting more indications, its carbenoxolone sodium was studied by other three companies: after conducting phase III clinical trials in the UK, the project of RB intending to improve nonspecific inflammatory bowel disease was given up in 1992. York Pharma expected to make progress in psoriasis with gel or cream; however, the project has been in a standstill after phase II clinical trials was conducted from 2005 to 2009. Canada pharmaceutical company, Oxalys Pharmaceuticals, research it for treating Huntington's disease, and it was included in the orphan drug list by the USA in 2014. Till now, phase I clinical trials are still continuing.

Pharmacology

Thirty percent of glycyrrhetinic acid can be effectively used by the body; both 18α-glycyrrhetic acid and 18β-glycyrrhetic acid reduce by half in 2.24 h and 11.5 h separately. CYP3A promotes metabolism with hydroxyl added to 22α and 24α [7].

There are lots of pharmacological activities [8, 9]: it plays an anti-inflammatory role by inhibiting the activity of phospholipase A2 and lipoxygenase to reduce mediators of inflammation; the compound promotes antiulcer activity through the production of more PGE2 and secretion of gastric mucus; it also provokes proliferation of gastric cell to protect the mucosa from ulceration. The complex which consists of glycyrrhetinic acid and carotenoid plays antioxidation by scavenging free radical.

Glycyrrhetinic acid inhibits the replication of viral DNA to achieve an antiviral effect at the concentration of 4×10^{-5} mol/L; it also inhibits proliferation of tumor cell and promotes apoptosis and differentiation. The decreasing ability of invasion exerts an antitumor effect. Glycyrrhetinic acid is considered to have extensive antiarrhythmic effects through inhibition of L-type calcium channel. By decreasing 11-beta-HSD oxidase and reductase IC_{50} $((2.9 \pm 0.5) \times 10^{-9}$ mol/L and $(1.3 \pm 0.1) \times 10^{-8}$ mol/L)), respectively, 18β-glycyrrhetinic acid can be used as an inhibitor of 11-beta-HSD oxidase and reductase. 18α-Glycyrrhetinic acid is similar with 18β-glycyrrhetinic acid (11-beta-HSD oxidase and reductase IC_{50} $(2.9 + 0.5) \times 10^{-9}$ mol/L and $(1.3 + 0.1) \times 10^{-8}$ mol/L)). Both of them exert a mineralocorticoid effect on disease. The complex plays a role in immune regulation by inhibiting activation of lymphocyte and macrophage. In addition, glycyrrhetinic acid functions as an anticholinesterase $(1.7 \times 10^{-5}$ mol/L), anticoagulant, and anti-tetanus toxin; it also improves inner ear hearing (100 mg/kg, intramuscular injection) and improves absorption of insulin.

Clinical Application

Glycyrrhetic acid has not been applied in clinical treatment till now. Meanwhile, the ramification has come into the market for the property of antiulcer. However, with large doses and long-term usage, the drug gives rise to hypertension, sodium retention, and hypokalemia. When renin-angiotensin-aldosterone system fails to function properly, liquorice-induced pseudoaldosteronism threatens human health [10].

Discussion

In view of the above adverse reaction, especially for liquorice-induced pseudoaldosteronism, the application of glycyrrhetic acid in clinical treatment has been restricted. However, the study and application of carbenoxolone sodium indicate that glycyrrhetic acid which acts as a leading compound should be further studied.

Fig. 1 Liquorice

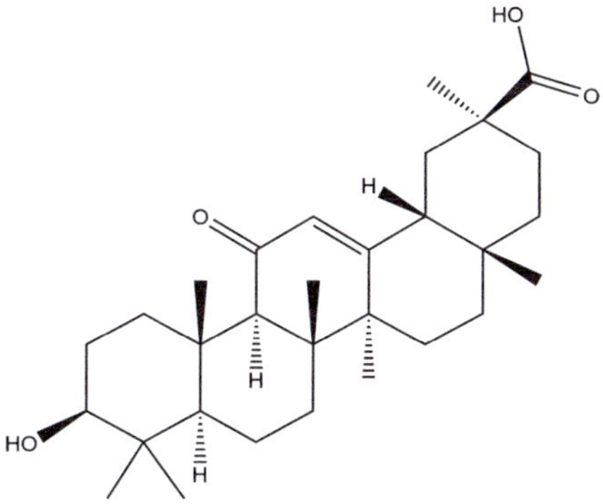

Fig. 2 Chemical structure of glycyrrhetic acid

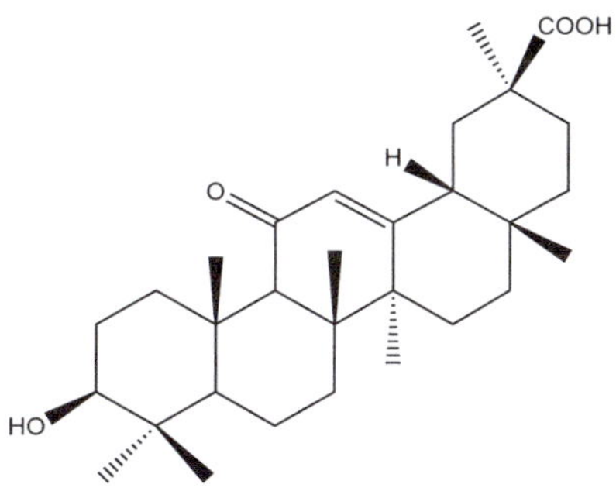

Fig. 3 Chemical structure of 18β- glycyrrhetinic acid

References

1. Ruzicka L, Leuenberger H, Schellenberg H. Polyterpene und Polyterpenoide CXVIII. Katalytische Hydrierung der α, β-ungesättigten Ketogruppe in der Glycyrrhetinsäure und dem Keto-α-amyrin. Helv Chim Acta. 1937;20(1):1271–9.
2. Doppelbindungen EV. Zur Lage der Carboxylgruppe bei der Glycyrrhetinsäure. Helv Chim Acta. 1942;25:775.
3. Ruzicka L, Jeger O, Ingold W. Neuer Beweis für die verschiedene Lage der Carboxylgruppe bei der Oleanolsäure und der Glycyrrhetinsäure. Helv Chim Acta. 1943;26:2278–82.
4. Rovers F. Effect of antiulcer of glycyrrhetic acid. Tijdschr Geneesk. 1946;90(12):135.
5. Li B, Jiang T, Wan SB. Progress in chemical modification and structure transformation of glycyrrhetic acid. Fine Chem. 2006;23:643–8.
6. Kang L, Li XQ, Wang FR. Advances in studies on structural modifications and biological activities of 18β-glycyrrhetinic acid. Chin Tradit Herb Drug. 2012;43:1430–42.
7. Gao K, Wei Y, Yang J. CYP3A1/2 and 2C9/10 in rat liver microsomes mediate 22α-and 24-hydroxylation of 18α-glycyrrhetic acid. Chin J Clin Pharmacol Ther. 2008;12:1255–60.
8. Xie SR, Zhao J, Liu R. Study and expectation of glycyrrhetic acid. J Dalian Univ. 2005;26:85–8.
9. Jin M, Wu HJ. Advancement on pharmacological action of glycyrrhetic acid. Med Recapitulate. 2009;15:1712–5.
10. Yang JN, Zhu M. Review of pharmacology of glycyrrhetinic acid and derivatives. Chin Pharmacol Bull. 1997;13:110–4.

Hemsleyadinum

Yang Lu, Li Zhang, and Guan-Hua Du

Abstract Hemsleyadinum is the mixture of cucurbitacin IIa and cucurbitacin IIb, which are extracted from the *Cucurbitaceae, Chrysanthemum*, and *Hemsleya*, having the function of clearing heat, removing toxicity, antisepsis, and anti-inflammation. Hemsleyadinum are known as the "natural antibiotics" and have the advantage of having ideal therapeutic effect under low dosages, no drug resistance, and so on. Thus, further study should be performed.

Keywords Cucurbitacin IIa · Cucurbitacin IIb · Natural antibiotics

Alias: Mixture of cucurbitacin IIa and cucurbitacin IIb
Origin: *Hemsleya amabilis* (Fig. 1)
Chemical name (Fig. 2)

Cucurbitacin IIa, 23,24-dihydroxycucurbitacin F-25-acetate; cucurbitacin IIb, 23,24-dihydroxycucurbitacin F

Molecular formula, Cucurbitacin IIa, $C_{30}H_{42}O_6$; cucurbitacin IIb, $C_{30}H_{48}O_7$
MW, Cucurbitacin IIa, 498.65; cucurbitacin IIb, 520.71
CAS, Cucurbitacin IIa, 60137-06-6; cucurbitacin IIb, 50298-90-3

Properties

Appearance: rhombohedral or flaky crystals, *Solubility*: soluble in methanol, ethanol, and acetone, slightly soluble in chloroform, insoluble in water.

Y. Lu · L. Zhang · G.-H. Du (✉)
Beijing Key Laboratory of Polymorphic Drugs, Research Center of Polymorphic Drugs, Institute of Materia Medica, Chinese Academy of Medical Sciences and Peking Union Medical College, Beijing, China
e-mail: dugh@imm.ac.cn

G.-H. Du, *Natural Small Molecule Drugs from Plants*,
https://doi.org/10.1007/978-981-10-8022-7_68

Dosage Forms and Indications

Its dosage forms and indications are available at the *Pharmacopoeia of the People's Republic of China* (1977).

The current clinical use of hemsleyadinum includes tablets and capsules, which are suitable for bacillary dysentery, enteritis, bronchitis, and acute tonsillitis.

Literature

Cucurbitacin IIa and cucurbitacin IIb are the characteristic chemical constituents of the hemsleya. Hemsleyadinum are extracted from the Cucurbitaceae, *Chrysanthemum*, and *Hemsleya*, which have the function of clearing heat and removing toxicity and antisepsis and anti-inflammation. There are 31 species of *Hemsleya* in the world, most of which are distributed in China's Yunnan, Guizhou, and Sichuan provinces, besides the 2 species distributed in India and Vietnam. The main species are *H. emiensis*, *H. pengxianensis*, *H. amabilis*, *H. dolichocarpa*, *H. graciliflora*, and *H. gigantha* [1]. Hemsleya plants are commonly used as Chinese herbal medicines.

History of R&D

Hemsleyadinum and hemsloside inhibit *Shigella dysenteriae*, *Salmonella typhimurium*, and *Salmonella choleraesuis* in vitro. Intravenous injection of cucurbitacin IIa has mild antipyretic effect caused by intraperitoneal injection of typhoid and paratyphoid bacteria but invalid to those with normal body temperature and nonbacterial fever. Xuedansu tablets have significant antitussive effect and show inhibitory effect against p-xylene-induced mouse ear swelling and carrageenan-induced rat toe swelling. Cucurbitacin IIa and cucurbitacin IIb also have anti-HIV-1 activity, which plays a vital role in HIV-1 virus infection [2]. Hemsleyadinum has significant protective effect against acute liver injury caused by CCl_4, especially at the high dosage [3]. In recent years, a large number of reports on the chemical composition of hemsleyadinum have been made by domestic and foreign scientists, and a variety of active ingredients were identified. Studies found that most of the hemsleya plants are rich in cucurbitacin IIa (up to 6.018 mg/g) [4], cucurbitacin IIb, hemsloside, and so on. *Hemsleya amabilis* tubers contain 15 active ingredients. It is reported that the content of hemsleyadinum varies from 0.38% to 1.7% [5] in different hemsleya plants. In 2008, the structure of the cucurbitacin IIa and IIb was analyzed.

Pharmacology

The main active ingredient in Cucurbitaceae plants is cucurbitacins, and cucurbitacin IIa and IIb have been used in clinics. Their indications include bacillary dysentery, enteritis, bronchitis, acute tonsillitis, and so on. Cucurbitacin IIa also has anticancer effect [6]. Cucurbitacin IIa (25–100 μM) dose-dependently inhibits proliferation of human lung adenocarcinoma A549 and mouse adenocarcinoma. It blocks the G0/G1 and S phase of A549 cells and thus promotes cell apoptosis [7]. The anticancer mechanism of cucurbitacin IIa is to prevent cancer cell spreading by destroying cytoskeleton of actin and leading to PARP-induced apoptosis by inhibiting production of downstream surviving of JAK/STAT3.

Clinical Application

At present, hemsleyadinum mainly are used as antibacterial and anti-inflammatory drugs, suitable for treatment of upper respiratory tract infection, acute tonsillitis, mumps, pharyngitis, acute and chronic bronchitis, pneumonia, acute urinary tract infection, acute and chronic uterine attachment inflammation, pelvic inflammation, acute gastritis, enteritis, bacillary dysentery, sinusitis, tympanitis, conjunctivitis, keratitis, skin and soft tissue infections, periodontitis, etc. Generally speaking, hemsleyadinum and its preparations have less side effects, such as diarrhea, abdominal pain, and other gastrointestinal reactions. But these side effects can be ignored. Cardiac patients should be used with caution. Xuedansu tablets and Xuedansu capsules are now put into commercial production. Xuedansu tablets are clinically used for the treatment of bacillary dysentery, enteritis, bronchitis, and acute tonsillitis while Xuedansu capsules for the treatment of respiratory infections, gynecological infections, and various infections. In addition to inflammation treatment, there are some preparations containing hemsleyadinum. Jinzeguanxinpian are used for treating heart disease and angina, while Miao Ethnomedicine Weichangdan is effective for treating gastric ulcer, duodenal ulcer, and gastrointestinal disorders.

Discussion

Hemsleya plants have complex and diverse chemical composition and a wide range of pharmacological activities. In addition, hemsleyadinum has a good broad-spectrum antimicrobial activity, and it has been clear that some of the oleanane-type compounds and cucurbitacin in this genus have a strong anticancer activity. In addition, it also has anti-HIV-1 activity. Despite this, their pharmacological effects, mechanisms involved, and material basis are still not clear. Through molecular docking design and drug virtual screening, lead compounds of this aspect are expected to be developed. In addition, hemsleya plants are rich in resources, and most of the hemsleya tubers are rich in hemsleyadinum. Therefore systematic research of each species will enrich the drug source, expand the range of antimicrobial active ingredients, improve its economic value, and promote the development of Chinese medicine, especially the development of traditional national medicine. Hemsleyadinum, known as the "natural antibiotics," have the advantage of having ideal therapeutic effect under low dosages, no drug resistance, and so on. Its market demand is huge. There are broad economic benefits and development prospects.

Fig. 1 *Hemsleya amabilis*

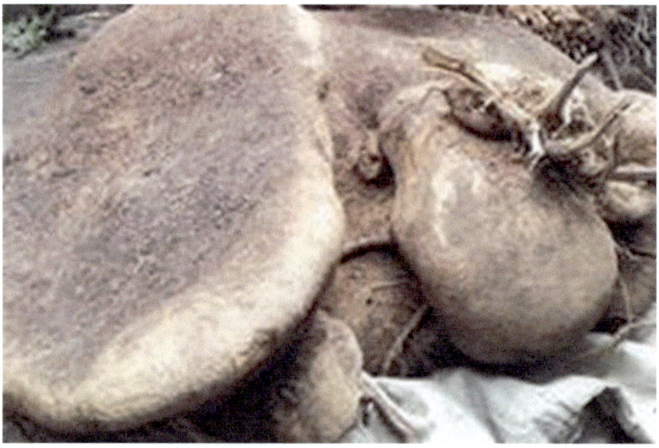

Fig. 2 Chemical structure of hemsleyadinum

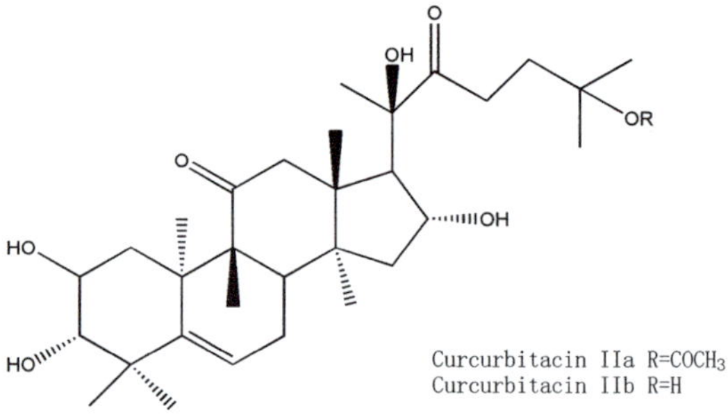

References

1. Xu XT, Deng ZP, Zhong H. Research progress of hemsleyadinum. Qilu Pharm. 2012;31:600–1, 4.
2. Tian RR, Chen JC, Zhang GH. Anti HIV-1 activity of cucurbitacin IIa and cucurbitacin IIb in vitro. Chin J Nat Med. 2008;6(3):214–8.
3. Liu AH, Shi M, Bao KF. Protective effect of hemsleyadinum to experimental liver injury. J Kunming Med Univ. 2012;9:8–10.
4. Shi YQ, Yang PQ, Tai YM. Content determination of cucurbitacin IIa of Hemsleya dolichocarpa W. J. in different harvest periods using R-HPLC. Chin J Chin Mater Med. 1996;21(5):276–7.
5. Nie RL, Chen ZL. Historical and current situation of research of hemsleya plants. Acta Bot Yunnanica. 1986;8(1):115–24.
6. Tao CY, Yi YH, Lin HW. Research of anti-cancer active ingredients from hemsleya roots. Acad J Second Mil Univ. 1999;20(5):337–9.
7. Gao S, Yu MK, Wei JH. Inhibition effect of cucurbitacin IIa to proliferation of human non-small cell lung cancer. Chin J Biol. 2012;25(1):69–71.

Houttuynin

Bi-Yu Hou, Li Zhang, and Guan-Hua Du

Abstract Houttuynin is one of the main ingredients in the volatile oil derived from *Houttuynia cordata* Thunb. Houttuynin and its derivatives have strong antibacterial effects. *Houttuynia* injection preparations are commonly used in clinical antimicrobial drugs and have exact effects on infectious diseases, respiratory diseases, and gynecological inflammation. However, there are side effects associated with the use of these injection preparations, mainly involving the cardiovascular system, gastrointestinal system, circulatory system, skin or kidney. It has been a direction of efforts to obtain the new derivatives with high efficiency and low toxicity.

Keywords Houttuynin · Injection preparations · Side effects

Alias: 3-oxo-dodecanal, 3-oxo-dodecanal
Origin: *Houttuynia Cordata* (Fig. 1)
Chemical name (Fig. 2): decanoyl acetaldehyde
Molecular formula, $C_{12}H_{22}O_2$; **MW**, 198.3; **CAS**, 56505-80-7
Houttuynin derivatives (Fig. 3)

Properties

Appearance: aromatic oil with a volatile odor, yellow oily liquid. *Solubility*: soluble in methanol, ethanol, ether, and petroleum ether; insoluble in water; easily polymerizable

---

B.-Y. Hou · L. Zhang · G.-H. Du (✉)
Beijing Key Laboratory of Drug Targets Identification and Drug Screening, Centre for Pharmaceutical Screening, Institute of Materia Medica, Chinese Academy of Medical Sciences and Peking Union Medical College, Beijing, China
e-mail: dugh@imm.ac.cn

G.-H. Du, *Natural Small Molecule Drugs from Plants*,
https://doi.org/10.1007/978-981-10-8022-7_69

Dosage Forms and Indications

Houttuynin is recorded in the *Pharmacopoeia of the People's Republic of China* (1995).

Houttuynin preparations include *Houttuynia* injection, sodium neo-houttuyfonate injection, sodium neo-houttuyfonate and sodium chloride injection, sodium houttuyfonate tablets, and tabellae natrii houttuyfonatis. These preparations are mainly used to treat chronic bronchitis, pediatric pneumonia, and other respiratory inflammatory diseases; cervicitis, annex inflammation, pelvic inflammatory disease, chronic cervicitis, and other gynecological inflammation; and also other skin diseases such as psoriasis. Houttuynin preparations also include sodium houttuyfonate suppository and compound *Houttuynia* tablets. The former is used to treat cervical erosion, and the latter is used to treat acute pharyngitis and acute tonsillitis.

Literature

Houttuynin is one of the main ingredients in the volatile oil derived from *jicai* 蕺菜 (*Houttuynia cordata* Thunb), which belongs to the family Saururaceae, and is also named as *yuxingcao* 鱼腥草 because of its fishlike smells when grated. *H. cordata* has been widely used in traditional Chinese medicines and is recorded in Grand Compendium of Materia Medica. *H. cordata* is used in the treatment of skin diseases, hemorrhoids, and certain diseases of women, relieving fever, resolving toxicity, reducing swelling, draining pus, and promoting urination. *H. cordata* is available in China and India.

History of R&D

In 1952, houttuynin was firstly found in the volatile oil extracted from *Houttuynia* by steam distillation. The chemical structure was named decanoyl acetaldehyde [1]. However, houttuynin is easy to polymerize and became invalid. In 1971, Chinese scholars synthesized sodium houttuyfonate (houttuyninum, HOU), which is an addition compound of sodium bisulfite and houttuynin. HOU is more stable and maintains the original antibacterial activity [2–4]. HOU possessed broad-spectrum antibacterial and anti-inflammatory activities, diuretic swelling, rapid antifebrile, antivirus and antitumor proliferation, increased immunity, and other pharmacological effects.

Neo-houttuyninum as the other derivative of houttuynin possesses more obvious inhibitory effect on a variety of bacteria and is used for the treatment of infectious diseases.

Since houttuynin was found 60 years ago, *Houttuynia* injection preparations have been developed, such as *Houttuynia* injection, neo-houttuyfonate sodium chloride injection, neo-houttuyfonate injection, etc. These drugs are commonly used in clinical antimicrobial drugs. However, these injection preparations were withdrawed for serious adverse reactions in 2006, and continued to be used in clinic since 2008 [5].

Pharmacology

Houttuynin is reported as a major antibacterial constituent in the volatile oil of *H. cordata*. Houttuynin exhibits significant inhibitory effects on *Staphylococcus aureus*, pneumococcus, and other bacteria as well as certain fungi and also has certain antiviral and anti-inflammatory effects. It is reported that houttuynin could reduce cardiomyocyte cross-section area and collagen deposition. Houttuynin attenuates ventricular remodeling in coronary artery ligation rats by restricting the excessive activation of rennin-angiotensin-aldosterone system (RAAS) and peroxidation [6]. Houttuynin also shows hypolipidemic and anti-atherosclerotic effects [7].

Houttuyninum has a significant inhibitory effect on *Candida albicans*, *Cryptococcus neoformans*, and *Brochothrix thermosphacta*. Houttuyninum has anti-inflammatory and analgesic effects, and the inhibition of inflammation is related to NF-κB and MCP-1 [8]. In recent years, studies have shown that intragastric administration of houttuyninum can increase apoptosis of the rheumatoid arthritis synovial cell and reduce joint index and inflammation factor levels in rheumatoid arthritis rat model [9].

Clinical Application

Houttuynin and its derivatives are used to treat milk moths, lung carbuncle, leaching disease, carbuncle swollen sore, and other infectious diseases. Currently it is used mainly for the treatment of gynecological diseases and respiratory system, especially upper respiratory tract infection. However, there are side effects associated with the use of these injection preparations, mainly involving the cardiovascular system, gastrointestinal system, blood system, and skin or kidney.

Discussion

Houttuynin and its derivatives have exact effects on infectious diseases, respiratory diseases, and gynecological inflammation. Although houttuyninum and neo-houttuyninum are more stable than houttuynin, they emerge the same allergic reactions in the clinical application. Until now, it has been a direction of efforts to obtain the new derivatives with high efficiency and low toxicity.

Fig. 1 *Houttuynia cordata*

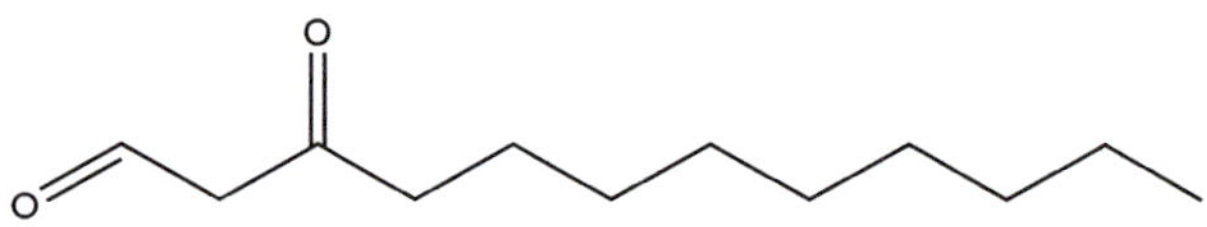

Fig. 2 The structure of houttuynin

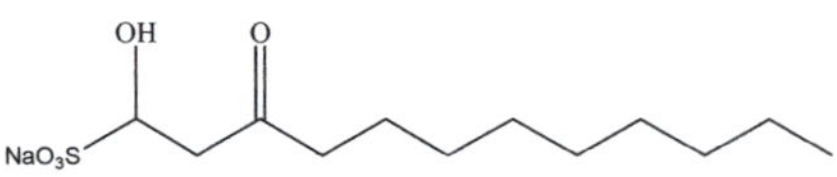

sodium houttuyfonate

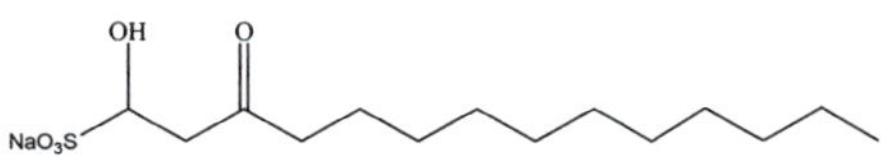

neo-houttuyninum

Fig. 3 The structures of houttuynin derivatives

References

1. Isogai H. Science papers cool of Genal. Education. 1952;2:1.
2. The Ministry of Health of People's Republic of China. The effective ingredient of synthetic Houttuynia cordata Thunb. Chin J Pharm. 1972;2:19.
3. The Ministry of Health of People's Republic of China. The clinical effect and synthetic pathway of houttuyninum. Chin J Pharm. 1972;8:5–8.
4. Qiu CQ, Xiao XX, Lei HQ, et al. Preliminary study on the synthesis and clinical efficacy of the new synthetic Houttuynia cordata. New Chin Med. 1979;10(12):601–2.
5. Yan WJ, Li LD. The general situation, existing problems and counter measures of Houttuynia injections. Chin J New Drugs. 2012;21(21):2506–10.
6. Gao Y, Gao JP, Chen CX, et al. Beneficial effects of houttuynin on ventricular remodeling induced by coronary artery ligation in rats. Eur J Pharmacol. 2014;740(5):200–8.
7. Li J, Chen CX, Gao Y, et al. Research of decanoyl acetaldehyde on anti-inflammation and anti-atherosclerosis. Chin Tradit Patent Med. 2010;32(1):26–30.
8. Pan P, Wang YJ, Han L, et al. Effects of sodium houttuyfonate on expression of NF-kappaB and MCP-1 in membranous glomerulonephritis. J Ethnopharmacol. 2010;131(1):203–9.
9. Li J, Zhao FT, Wang L, et al. Inhibition of sodium houttuyfonate on synovial proliferation in rats with rheumatoid arthritis. Cent South Pharm. 2015;13(5):502–5.

Matrine

Li Gao, Li-Da Du, Xue-Mei Qin, Jin-Hua Wang, and Guan-Hua Du

Abstract Matrine is an alkaloid extracted by organic solvents from the dry roots, plants, and fruits of *kushen* (*Sophora flavescens* Ait). Matrine has significant effects of negative frequency and positive inotropic action, achieving the role of an antiarrhythmic. Moreover, matrine has an inhibitory effect on tumor cells by inhibiting growth and proliferation and inducing apoptosis. In the clinical setting, matrine injection was used for chronic active hepatitis, and a suppository was used for trichomoniasis or Candida vaginitis and chronic cervicitis. Improvements in the modern formulation of matrine can improve its clinical application.

Keywords Matrine · Alkaloid · Antiarrhythmic · Anti-tumor

Alias: cis-Matrine, Matricaria alkali
Origin: Radix sophorae flavescentis (Fig. 1)
Chemical name (Fig. 2)

(7aS,13aR,13bR,13cS)-Dodecahydro-1H,5H,8H-dipyrido[2,1-f:3,2,1-ij][1,6]-naphthyridin-10-one

Molecular formula, $C_{15}H_{24}N_2O$; **MW**, 248.37; **CAS**, 519-02-8
Derivatives (Fig. 3)

L. Gao · X.-M. Qin
Modern Research Center for Traditional Chinese Medicine, Shanxi University, Shanxi, China

L.-D. Du
School of Biomedical Science, The Chinese University of Hong Kong, Hong Kong, China

J.-H. Wang · G.-H. Du (✉)
Beijing Key Laboratory of Drug Targets Identification and Drug Screening, Centre for Pharmaceutical Screening, Institute of Materia Medica, Chinese Academy of Medical Sciences and Peking Union Medical College, Beijing, China
e-mail: dugh@imm.ac.cn

G.-H. Du, *Natural Small Molecule Drugs from Plants*,
https://doi.org/10.1007/978-981-10-8022-7_70

Properties

Appearance: white, needle-like crystal or crystalline powder that becomes yellow and solid after long-time exposure in air. *Solubility*: soluble in water, benzene, chloroform, and methanol and slightly soluble in petroleum ether. *Melting point*: 76 °C (α-matrine), 87 °C (β-matrine), 223 °C (γ-matrine), and 84 °C (δ-matrine). *Specific optical rotation*: +31 to +36°.

Derivatives (Fig. 3)

Dosage Forms

This product is listed in the National standards for chemical drugs, Volume 1.

The current formulations for clinical use include injections, tablets, capsules, suppositories, dropping pills, pellets, and gels.

Indications

In clinical settings, matrine injection was used for chronic active hepatitis, and suppository was used for trichomoniasis or Candida vaginitis and chronic cervicitis and can also be used for senile vaginitis and pelvic inflammatory disease.

Literature

Matrine is extracted by organic solvents from the dry roots, plants, and fruits of *kushen* 苦参 (*Sophora flavescens* Ait) and also existed in *S. alopecuroides* L. and *S. subprostrata Chun et T. Chen*.

Chinese medicine Radix sophorae flavescentis have dampness eliminating insecticide and diuretic effects. *Sophora flavescens* is used for heat dysentery, hemafecia, jaundice and anuria, red leukorrhea, vulva pruritus, eczema, wet sores, skin itching, scabies, ringworm, and trichomonas vaginitis. *Sophora flavescens* is recorded in the ancient *Materia Medica*. "Southern Yunnan Materia Medica" also recorded the effects of *Saphora flavescens* [1].

Sophora flavescens is a species of *Sophora japonica* and distributed in temperate and subtropical regions. *Sophora flavescens* is widely distributed in China, mainly in the north of Hebei, the west of Henan, and southwest of Shandong, Anhui, Hubei, and Guizhou. Matrine is extracted and isolated from the Chinese medicine *Sophora flavescens* [2, 3].

History of R&D

As early as the 1930s, the Soviet Union began to study *Sophora flavescens*, and the domestic research began in 1972. *Sophora flavescens* in the chemical composition are divided into two major categories of alkaloids and flavonoids. Domestic and foreign researches are focused on the alkaloids. So far, there are 23 species of alkaloids isolated and identified in roots, stems, leaves, and flowers of *Sophora flavescens*. Most of them are the quinolizidine type, and there are few pairs of alkaloids, which include oxymatrine, matrine, sophocarpine, sophoridine, and so on.

In the matrine-type alkaloids, the maternal compound undergoes significant changes in the biological activities after oxidation. Compared with the parent, the toxicities of oxymatrine and oxysophocarpine were significantly decreased, and they exert obvious negative frequency and positive inotropic action. In addition, dehydrogenation can also reduce the toxicity of alkaloids. Sophocarpine and sophoridine are dehydrogenation and double dehydrogenation products of matrine. The toxicity of dehydrogenation is not obvious, and the toxicity of double dehydrogenation is significantly reduced. Dehydrogenation can enhance negative frequency and positive inotropic actions [4].

The main components of *Sophora flavescens* have obvious anti-cancer effects, and oxymatrine has anti-cancer and antiaging effects. Flavonoids are predominantly A-ring isoprenyl-branched. Isopentenyl flavonoids are generally thought to act as plant defensins and play an important physiological role in plant defense. These components are only distributed in plants of legumes, *Moraceae*, and *Asteraceae*. Since the 1970s, owing to the multiple pharmacological activities of isopentenyl flavonoids, the studies on these ingredients have attracted more and more attention. So far, the natural separation of these compounds has more than a thousand species. Early chemical work focused on the composition of alkaloids in *Sophora flavescens*. In recent years, more and more researches were focused on isopentenyl flavonoids, which achieved a lot of results [5].

Pharmacology

Matrine has antiarrhythmic, anti-inflammatory, anti-fibrosis, and anti-tumor effects.

Matrine has significant effects of negative frequency and positive inotropic action, which can enhance myocardial contractility, slow down the heart rate, and extend the PR and QTC interval. Matrine can promote the expression of bcl-2 gene and can improve the ratio of bcl-2/Bax to inhibit apoptosis of myocardial cell caused by myocardial ischemia reperfusion and then reduce myocardial deficiency caused by coronary atherosclerotic heart disease and myocardial infarction to achieve the role of antiarrhythmic. In addition, matrine can also play an anti-pulmonary fibrosis effect by inhibiting the proliferation of lung fibroblasts and the expression of pulmonary interstitial fibroblasts (FN). Studies have found that matrine has anti-allergy and anti-inflammatory effects [6].

Matrine has an inhibitory effect on tumor cells by inhibiting growth and proliferation and inducing apoptosis. Studies have shown that 0.8 g/L and 1.0 g/L matrine could significantly inhibit the growth, proliferation, invasion, and metastasis of human hepatocellular carcinoma SMMC-7721 cells [7]. In addition, 0.0625–0.5 mg/ml matrine significantly inhibited proliferation of HT29 cells after 48 h incubation. 1 mg/ml matrine significantly induce apoptosis of HT29 cells [8].

Oxymatrine is also an anti-cancer ingredient of *Sophora flavescens*. In recent years, oxymatrine is widely used in the clinic for treatment of leukopenia and chronic hepatitis B, hepatitis C induced by tumor radiotherapy, and chemotherapy. Oxymatrine can improve liver biochemical indicators by protecting liver cells, reducing cell necrosis, and preventing liver fibrosis. In addition, oxymatrine can treat all kinds of eczema, especially for acute and subacute eczema, and contact dermatitis [9].

Clinical Application

Matrine has multiple pharmacological effects and can treat a variety of diseases in clinical setting. At present, the researches on matrine are mainly focused on anti-liver injury and liver fibrosis and anti-tumor and anti-cardiovascular diseases. It was worthy of attention that matrine has other pharmacological activities, such as leukocyte-elevating effects and anti-virus effects, and these effects were necessary for further development and utilization. The pharmacological effects of matrine stayed in the experimental phase, and they need further development and research to promote the development of clinical practice. Improvements in the modern formulation of matrine can improve their clinical application.

Discussion

Matrine is a kind of active alkaloid extracted from traditional Chinese medicine *Sophora flavescens*, and it has a lot of pharmacological effects. *Sophora flavescens* contains a large number of alkaloids and flavonoids with broad prospects for development in the biological, pharmaceutical, healthcare, and other fields. *Sophora flavescens* is rich in wild populations and has high adaptability and artificial propagation ability, which provide a guarantee of crude drug resources for the development of medicine products of *Sophora flavescens*.

Sophora flavescens has great potential for the treatment of tumors and other diseases [10]. With the rapid development of molecular biology technology, especially the gene chip and protein chip technologies, the investigations of anti-tumor mechanisms of matrine-type alkaloids will be largely promoted. For these anti-tumor targets, novel drugs may be developed, which will bring a new approach for treatment of tumor. Therefore, further study of matrine has great significance for the development and utilization of matrine.

Fig. 1 Radix sophorae flavescentis

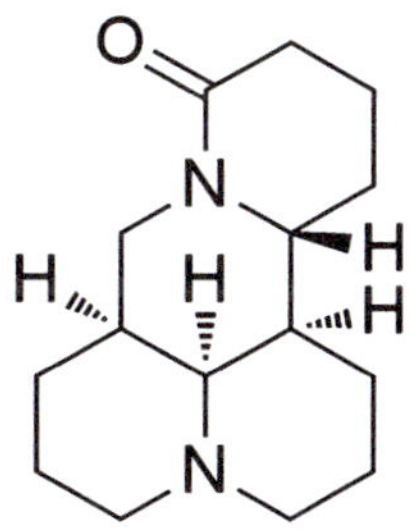

Fig. 2 Chemical structure of matrine

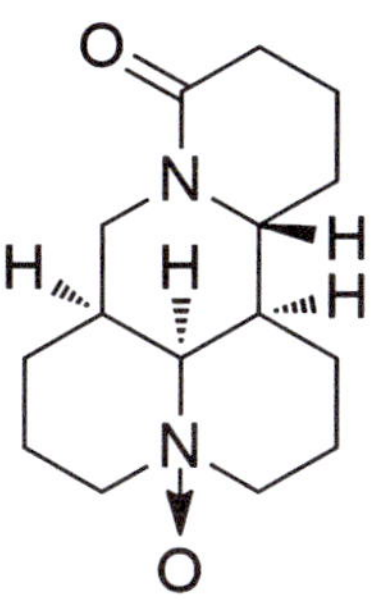

Oxymatrine

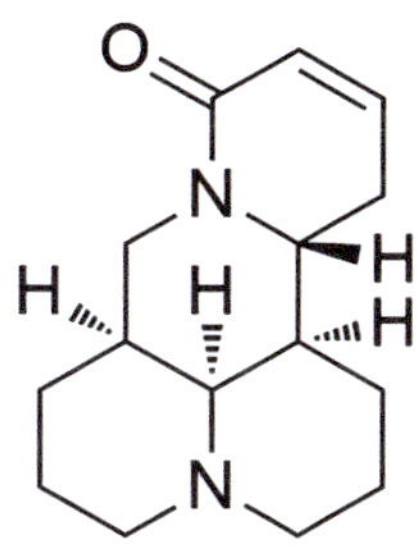

Sophocarpine

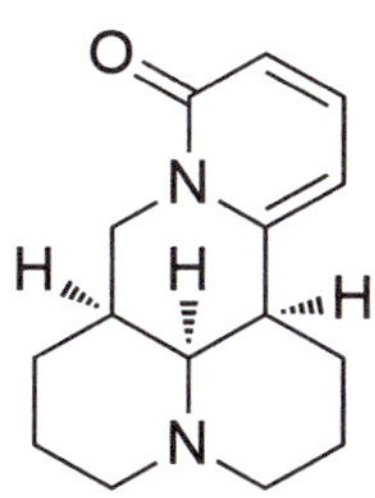

Sophoramine

Fig. 3 Chemical structures of derivatives of matrine

References

1. Qian XZ. Color illustrations of Chinese Materia Medica. Beijing: People's Medical Publishing House; 2003.
2. Editorial Committee of Chinese flora of Chinese Academy of Sciences. Flora of China. Beijing: Science Press; 2004.
3. Pharmacopoeia Commission of the Ministry of health of the People's Republic of China. Pharmacopoeia of the People's Republic of China. Beijing: Chemical Industry Press; 2015. p. 2015.
4. Zhang JT, Wang W, Duan ZH. Progress of research and application of matrine-type alkaloids. Prog Mod Biomed. 2007;7(3):451–4.
5. Zhao P, Zhang YJ, Yamamoto H, et al. Recent advance on the chemistry, bioactivity and biosynthesis of prenylated flavonoids from sophora flavecens. Nat Prod Res Dev. 2006;16(2):172–8.
6. Zhang LH, Chen BE, Pan MJ. Research progress on pharmacological action of matrine. Chin Tradit Herb Drug. 2009;40(6):1000–3.
7. Wang Y, Peng CH, Zhang GP. Study on invasion and metastasis related factors in differentiation of SMMC-7721 cells induced by matrine. J Chin Med Mater. 2003;26(8):566–9.
8. Huang J, Chen KJ, Zhang W. Effect of matrine on inhibiting proliferation and inducing apoptosis of human intestinum crassum carcinoma HT29 cells. Chin Tradit Herb Drug. 2007;38(8):1210–4.
9. Jiang HZ. Advances in pharmacological effects and preparation methods of matrine and oxymatrine. Pract Clin J Integr Tradit Chin West Med. 2007;7(1):89–90.
10. Qu ZP, Cui J, Harata-Lee Y, et al. Identification of candidate anti-cancer molecular mechanisms of compound kushen injection using functional genomics. Oncotarget. 2016;7(40):66003–19.

Neoandrographolide

Ning-Bo Gong, Li-Da Du, and Yang Lu

Abstract Neoandrographolide is a colorless column crystal that has been isolated from the stem and leaves of *Andrographis paniculata*. Neoandrographolide and andrographolide are the main active ingredients of a variety of pharmaceutical preparations containing the *Andrographis* herb. These drugs are mainly used for treating acute bacterial dysentery, acute gastroenteritis, upper respiratory tract infection, acute tonsillitis, pharyngitis, etc. Neoandrographolide can inhibit lipopolysaccharide-induced NO production in the mouse peritoneal macrophages and inhibits the release of endothelin and malondialdehyde induced by oxidized low-density lipoprotein (ox-LDL) in the porcine aortic endothelial cells. It also shows a protective effect in endotoxemia mice, and the mechanism may be through downregulation of TLR4 and NF-kB mRNA expression. Neoandrographolide may be developed as an effective antibiotic drug with less side effects in the future.

Keywords Neoandrographolide · *Andrographis paniculata* · Antibiotics

Origin: *Andrographis paniculata* (Fig. 1)
Chemical name (Fig. 2)

19-hydroxy-8(17),13-laboiadien-16,15-olide-o-β-D- glucopyranoside

Molecular formula, $C_{26}H_{40}O_8$; **MW**, 480.58; **CAS**, 27215-14-1

N.-B. Gong · Y. Lu (✉)
Beijing Key Laboratory of Polymorphic Drugs, Research Center of Polymorphic Drugs, Institute of Materia Medica, Chinese Academy of Medical Sciences and Peking Union Medical College, Beijing, China
e-mail: luy@imm.ac.cn

L.-D. Du
School of Biomedical Sciences, Chinese University of Hong Kong, Hong Kong, China

G.-H. Du, *Natural Small Molecule Drugs from Plants*,
https://doi.org/10.1007/978-981-10-8022-7_71

Properties

Appearance: Neoandrographolide is a colorless column crystal that has been isolated from the stem and leaves of *Andrographis paniculata. Melting point*: 167–168 °C. *Specific optical rotation*: −48°(pyridine); −45°(c = 1, absolute ethanol). *Solubility*: Soluble in methanol, ethanol, acetone, and pyridine and slightly soluble in chloroform and water but insoluble in ether and petroleum ether.

Dosage Forms and Indications

Neoandrographolide is available in the *Pharmacopoeia of the People's Republic of China* (1977).

Neoandrographolide and andrographolide are the main active ingredients of a variety of pharmaceutical preparations containing the *Andrographis* herb, such as andrographolide capsules, andrographolide tablets, *Andrographis* injection, potassium dehydroandrographolide succinate injection, *Andrographis* tablets, Xiaoyan Lidan tablets, and Fufang chuan xin lian tablets. These drugs are mainly used for treating acute bacterial dysentery, acute gastroenteritis, upper respiratory tract infection, acute tonsillitis, pharyngitis, etc. In addition, they are also used to treat malignant hydatidiform mole and choriocarcinoma.

Literature

Andrographic plants are derived from the aerial parts of *Andrographis paniculata* (Burm.f.) Nees and known by various vernacular names such as "chunlianqiuliu," "yijianxi," "lanhelian," "kudancao," "jinxiangcao," "jinergou," "India grass," and "bitter grass." *Andrographis paniculata* is planted in South and Southeast Asia, India, and Australia. This herb is also cultivated in Fujian, Guangdong, Hainan, Guangxi, Yunnan, Jiangsu, and Shaanxi provinces in China. *Andrographis paniculata* has been used to treat colds, sore throat, mouth sores, cough, diarrhea dysentery, and so on. *Andrographis paniculata* contains diterpene lactone compounds, and they have anti-inflammatory and antimicrobial activities. The main active ingredients include andrographolide, neoandrographolide, deoxy andrographolide, dehydrated andrographolide, and so on. *Andrographis paniculata* has been included by US Pharmacopeia.

History of R&D

The chemical constituents of *Andrographis paniculata* are mainly lactones and flavonoids, including the lactone compounds such as andrographolide and dehydrated andrographolide. In 1952, Kleipool first reported the separation of andrographolide from *Andrographis paniculata* [1]. In 1968 and 1971, W. R. Chan et al. identified the chemical structure and stereostructure of andrographolide. In the 1990s, a variety of diterpene lactones were isolated from the aerial part and the lotus leaf of *Andrographis paniculata*, such as andrographolide, neoandrographolide, dehydroandrographolide, deoxyandrographolid, deoxyandrographiside, 8-methylandrograpanin, 3-dehydrodeoxyandrographolide, andrograpanin, and 3-oxo-14-deoxy-andrographol,and also some flavonoids and sitosterol and daucosterol compounds [2, 3]. Until now, more than 40 diterpene lactone components and more than 70 flavonoids have been found from *Andrographis paniculata*. There are sterols, organic acids, diterpene alcohols, diterpene acid salts, and cycloalkene ether also found from *Andrographis paniculata* [4].

Since the 1970s, pharmacists and organic chemists have done a lot of work on the modification of andrographolide, mainly in Michael addition of α, β-unsaturated lactone double bonds and three hydroxyl groups selective esterification, oxidation and substitution reactions, redox reduction of double bonds, intramolecular cyclization, and lactone ring replacement reactions [5], and the resulting andrographolide derivatives or analogues can improve the antibacterial, anti-inflammatory, cardiovascular system, immunomodulation, and antitumor activities.

Pharmacology

Andrographis paniculata has detoxifying abilities, reduces swelling, and relieves pain. The four major diterpene lactones in *Andrographis paniculata* have antipyretic and anti-inflammatory activities against 2,4-dinitrophenol or endotoxin-induced fever and egg white-induced edema or croton oil-induced inflammatory models.

Animal experiments have shown that *Andrographis paniculata* has inhibited and delayed the body temperature elevation of persons with infections caused by pneumococcus and hemolytic streptococcus. It also can be used for treating cough and asthma and especially for treating acute bacterial dysentery, acute gastroenteritis, upper respiratory tract infection, acute tonsillitis, pharyngitis, and so on.

The active ingredients andrographolide and neoandrographolide inhibit lipopolysaccharide-induced NO production in the mouse peritoneal macrophages [6]. Andrographolide, neoandrographolide, and deoxy-didehydro andrographolide have antipyretic and anti-inflammatory activities. Among them, deoxy-didehydro andrographolide shows the strongest effect. The diterpene bicyclic and penta-lactone ring and especially the α, β-unsaturated double bond are the key for their activities.

Andrographolide inhibits the release of endothelin and malondialdehyde induced by oxidized low-density lipoprotein (ox-LDL) in the porcine aortic endothelial cells. It also inhibits the superoxide dismutase activity and reduces the production of guanosine 3′,5′-cyclic phosphate. These data show that andrographolide acts as a ROS scavenger.

Neoandrographolide shows a protective effect in endotoxemia mice, and the mechanism may be through downregulation of TLR4 and NF-kB mRNA expression [7]. Neoandrographolide at the high concentration can inhibit the macrophage outbreak caused by LPS and the proliferation of lymphocytes and can synergistically enhance the PMA stimulating effect on the respiratory outbreak. The possible mechanism is related to its ROS scavenging activity [8, 9].

Clinical Application

Neoandrographolide is one of the main active ingredients of the natural plant *Andrographis paniculata*. But there is no formulation of neoandrographolide. Andrographis injection has been clinically used, but it has several side effects.

Discussion

As an antibacterial traditional Chinese medicine, *Andrographis paniculata* attracts more and more attention by the pharmaceutical industry. One of the tasks for future development of antibiotics is to develop an effective drug with less side effects. *Andrographis paniculata* shows great potential in this area.

Fig. 1 *Andrographis paniculata* (chuan xin lian)

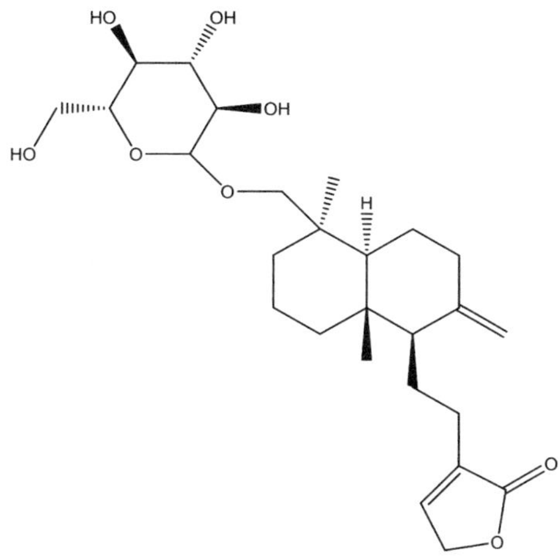

Fig. 2 The structure of neoandrographolide

References

1. Kleipool RJC. Constituents of *Andrographis paniculata* Nees. Nature. 1952;169:33–4.
2. Jiang Z. Research advances in chemical constituents of andrographis. Chin Tradit Patent Med. 2011;33(8):1382–8.
3. Jin X, Shi S, Zhang D, Zheng Z. Chemical constituents of *Andrographis paniculata*. Chin Tradit Herb Drug. 2012;4(1):47–50.
4. Chenliang Chu. Studies on chemical constituents and quality control of Andrographis Herba. Guangdong Pharmaceutical university; 2013.
5. Liu G, Xu H, Sun K, Wang J, Liu H. Development in synthesis of andrographolide derivatives. Chin J Org Chem. 2008;28(2):201–9.
6. Javzan B, Koichi H, Fumihide T, Shinji F, Ko-ichi O, Yoshiaki F. Suppression of NO production in activated macrophages in vitro and ex vivo by neoandrographolide isolated from Andrographis paniculata. Biol Pharm Bull. 2002;25(9):1169–74.
7. Fei Tan. Study about the effect and mechanism of Neoandrographolide against Endotoxemia. Southern Medical University; 2008.
8. Liu J, Wang Z. Effect of neoandrographolide on activated mouse macrophages in vitro. Chin J Nat Med. 2005;3(5):308–11.
9. Liu J, Tang Q, Wang Z. Effect of neoandrographolide on respiratory burst of macrophage RAW 264.7 and proliferation of lymphocytes in mice. Chin J New Drugs Clin Remedies. 2005;24(3):206–9.

Oleanolic Acid

Qi-Meng Zhou and Guan-Hua Du

Abstract Oleanolic acid is extracted from *Olea europaea*, which is an important triterpenoid. Oleanolic acid aids in liver protection, especially increasing the amount of metallothionein and preventing glutathione emptying, and has antimicrobial and antiviral effects. Further, oleanolic acid exerts anticancer, anti-inflammatory, anti-HIV infection, hypoglycemic, and other functions. However, the poor bioavailability limited its application range. In the history of oleanolic acid's R&D, balancing the bioavailability and the pharmacological activity in the process of drug preparation is the key to the further development of new drugs.

Keywords Oleanolic acid · Triterpenoid · Liver protection · Bioavailability

Alias: Oleanol, caryophyllia, astrantigeninc, giganteumgeninc
Origin: *Olea europaea* (Fig. 1)
Chemical name (Fig. 2)

3-Hydroxy-12-oleanen-28-oic acid

Molecular formula, $C_{30}H_{48}O_3$; **MW**, 456.70; **CAS**, 508-02-1

Properties

Appearance: *Olea europaea* exists in white needle crystal with ethanol or methanol. *Melting point*: It melts at 306–310 °C. *Solubility*: The ingredient is insoluble in water. *Specific optical rotation*: Polarized light rotates +83.3° when it passes through chloroform with concentration of 6 mol/L.

Q.-M. Zhou · G.-H. Du (✉)
Beijing Key Laboratory of Drug Targets Identification and Drug Screening, Centre for Pharmaceutical Screening, Institute of Materia Medica, Chinese Academy of Medical Sciences and Peking Union Medical College, Beijing, China
e-mail: dugh@imm.ac.cn

G.-H. Du, *Natural Small Molecule Drugs from Plants*,
https://doi.org/10.1007/978-981-10-8022-7_72

Derivatives

In recent years, there are many studies on the structure-activity relationship. Hydroxyl in the position of C-3, C-28 is the modificatory target to decrease blood glucose. The combination with glycogen phosphorylase, protein tyrosine phosphatase 1B, and α-glucosidase locus can be influenced by qualification of ring A and double bond in the position of C-12, C-13. Cyano group into the position of C-2 and modified ring A, C give rise to anti-inflammatory activity of the compound. Hydroxyl in the position of C-3, C-28 and ring A can be replaced to resist tumor [4]. People are dedicated to increase the activity of anti-HIV by modifying the hydroxyl in C-3, C-28.

Dosage Forms

Tablet and capsule

Indication

It is included in the *Pharmacopoeia of the People's Republic of China* (Part 2, volume 3), the *British Pharmacopoeia* (2017), and the *European Pharmacopoeia* (8.7th ed.).

It is mainly used for the adjuvant treatment of acute infectious icteric hepatitis and acute and chronic hepatitis. It also can be utilized in psoriasis, rheumatoid arthritis, nephritic edema, ascites, stomach with turbid, metrorrhagia, bruises, carbuncle, soreness and weakness of the waist and knees, fetal irritability and so on.

Literature

The ingredient is mainly extracted from the leaves of *muxilan* 木樨榄 (*Olea europaea*). Besides, it is found in the fruit of *ligustrum lucidum ait*, the whole grass of *swertia mileensis t.n.he et w.l.shi*, *s. mussotii franch*, the leaves and roots of *astrantia* major, the root bark and stem bark of *aralia chinensis*, the roots of *hemsleya macrosperma c.y.wu* , the roots of *hemsleya amabilis diels*, and the roots of *hemsleya chinensis cogn*. There is a long history in the herb and fruit which contain oleanolic acid. They play an important role in the treatment of various diseases. Oleanolic acid may be the most effective component in various Chinese medicines.

History of R&D

The chemical formula was first identified in 1918 [1]. In 1960, Khastgir purified the compound for the first time [2]. Corey identified the structure of oleanolic acid in 1993 [3].

Now the compound has been made into a drug and is being marketed in China. As early as 2002, the tablet form of oleanolic acid was listed. In 2010, capsule preparations came into the market, and to date, there is no other listed preparation. Sanofi treated it as antiulcer and NSAIDs candidate in abroad.

Although it has anti-inflammatory and anti-ulcer effects, Sanofi stopped the study of developing oleanolic acid into a drug in 1989.

Pharmacology

Liver protection: Pretreatment of 100 mg/kg once a day for 3 days can protect the liver and decrease the activity of enzymes [5]. It achieves by increasing the amount of metallothioneins and liver glucoside transferases to prevent glutathione emptying, to inhibit several subtypes of P450 oxidase, and to prevent combination with toxic substances.

Antimicrobial and antiviral effect: Oleanolic acid may inhibit the proliferation or production of NO induced by interferon or in the macrophages by competitively binding to the enzymes necessary for the growth and reproduction of microorganisms or virus and play an anti-inflammatory and antiviral role. EC50 of HIV-1-infected H9 cells is 3.4 mol/L.

Effects on metabolism of glucose and lipid: The treatment of 50–100 mg/kg for 4 days achieves the glucosidase activity inhibition, reduction of glucose absorption, inhibition of glycogen phosphorylase, activation of glycogen synthesis, inhibition of protein tyrosine phosphatase 1B, regulation of insulin signaling pathway, and increase of insulin sensitivity, and it may also be achieved by inhibiting the transport of glucose from the stomach to the intestine and transport activity of intestinal villi. 200 mg/kg suspension administrated by gavage for 4 weeks decreases the level of triglycerides, cholesterol, and β-lipoprotein in high-fat animal, and it contributes to increased platelet swimming rate and decreased blood viscosity.

Effects on immunologic function: 60 mg/kg intraperitoneal injection may regulate the immune system by inhibiting type I allergic reaction, promoting proliferation of lymphocyte and inhibiting enzyme C-3 in the traditional complement pathway.

Antioxidant effect: It fights against lipid peroxidation by scavenging free radicals.

Other functions: 10^{-5} mol/L oleanolic acid may inhibit the activity of DNA ligase I active site. It exerts anticancer and anti-mutation effects, and IC_{50} value is

2.2×10^5 mol/L in inhibiting ornithine decarboxylase (ODC) induced by external TPA at the transcriptional level. Oleanolic acid in polyvinylpyrrolidone suspension at the dose of 16 mg/kg by intragastric administration for 30 days also reversibly decreases quality and exercise capacity of mouse sperms [6].

It is reported that it is consistent with the single compartment model, or non-compartmental model, and may be distributed in the compound in accordance with the two compartment models [7]. Eventually, it will be eliminated by P450 oxidase in the liver.

Clinical Application

It is reported that contrast with Western medicine, the 60~90 mg/d oleanolic acid tablets supplemented with vitamin B can treat acute and chronic hepatitis, with effective rates of 94.4% and 69.81%, respectively, in clinical results [8]. It has been listed on the drug manual as adjuvant therapy for acute and chronic hepatitis.

Discussion

According to the results of current research, oleanolic acid is effective in protecting the liver. It also exerts anticancer, anti-inflammatory, anti-HIV infection, and hypoglycemic functions. Due to its poor bioavailability, balancing the bioavailability and the pharmacological activity in the process of drug preparation is the key to the further development of new drugs.

Fig. 1 *Olea europaea*

Fig. 2 Chemical structure of oleanolic acid

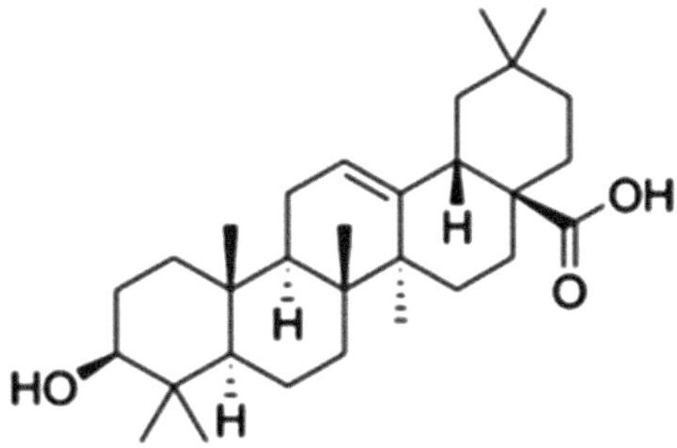

References

1. Dodge FD. The isomeric lactones, caryophyllin add urson. J Am Chem Soc. 1918;40:1917–39.
2. Khastgir H, Sengupta S, Sengupta P. Note on the constituents of the Indian medicinal plant Oldenlandia corymbosa Linn. J Am Pharm Assoc. 1960;49(8):562–3.
3. Corey E, Lee J. Enantioselective total synthesis of oleanolic acid, erythrodiol, beta.-amyrin, and other pentacyclic triterpenes from a common intermediate. J Am Chem Soc. 1993;115(19):8873–4.
4. Tang C, Chen Y, Bai S, et al. Advances in the study of structural modification and biological activities of oleanolic acid. Chin J Org Chem. 2012;33(01):46–65.
5. Tian LT, Ma L, Du NS. Survey of pharmacology of oleanolic acid. China J Chin Mater Med. 2002;27(12):884–6.
6. Mdhluli M, Horst GVD. The effect of oleanolic acid on sperm motion characteristics and fertility of male Wistar rats. Lab Anim. 2002;36(4):432–7.
7. Pei D, OuYang Z, Zhao M, et al. Pharmacokinetics studies on ursolic acid and oleanolic acid from total terpenes of Longchai recipe in rats. Tradit Chin Drug Res Clin Pharmacol. 2013;3:017.
8. Wang LX, Han GX, et al. Survey of pharmacology and chemistry of oleanolic acid. J Pharm Pract. 2001;19(2):104–7.

Paeonol

Shi-Ying Yang, Li-Da Du, and Yang Lu

Abstract Paeonol is the main active ingredient of the cortex of the *Ranunculaceae* plants including *Paeonia suffruticosa* Andr. and *Rhododendron* plant *Cynanchum paniculatum* (Bge.) Kitag. Paeonol has a wide range of pharmacological effects, such as antibacterial, anti-inflammatory, antipyretic, analgesic, and antispasmodic. Paeonol monomeric drugs are developed, and the formulations approved by SFDA include tablets, injections, and ointment. Paeonol injection was the earliest recorded in the Chinese pharmacopoeia 1977 edition, and it is the main formulation until now. Paeonol is considered to be a broad development prospects because of its significant effect and nonaddictive advantage in anti-inflammatory analgesia and other broad pharmacological activities.

Keywords Paeonol · Antibacterial · Anti-inflammatory · Analgesic

Alias: Paeonol
Origin: Moutan cortex (Fig. 1)
Chemical name (Fig. 2)

2′-Hydroxy-4′-methoxyacetopheone

Molecular formula, $C_9H_{10}O_3$; **MW**, 166.18; **CAS**, 552-41-0
Derivatives of paeonol (Fig. 3)

S.-Y. Yang · Y. Lu (✉)
Beijing Key Laboratory of Polymorphic Drugs, Research Center of Polymorphic Drugs, Institute of Materia Medica, Chinese Academy of Medical Sciences and Peking Union Medical College, Beijing, China
e-mail: luy@imm.ac.cn

L.-D. Du
School of Biomedical Sciences, Chinese University of Hong Kong, Hong Kong, China

G.-H. Du, *Natural Small Molecule Drugs from Plants*,
https://doi.org/10.1007/978-981-10-8022-7_73

Properties

Appearance: white or light yellow, shining, needlelike crystals. *Solubility*: freely soluble in ethanol and methanol; soluble in diethyl ether, acetone, benzene, chloroform, and carbon disulfide; slightly soluble in water; and soluble in hot water, practically insoluble in cold water, and volatilizing in water vapor. *Melting point*: 48–51 °C.

Dosage Forms and Indications

This drug has been recorded in *Pharmacopoeia of the People's Republic of China* (1977).

At present, the dosage forms of paeonol for clinical use include tablet, ointment, injection, and plaster. Paeonol is clinically used for the treatment of fever, headache, neuralgia, muscle pain, abdominal pain, rheumatoid arthritis and rheumatoid arthritis; its ointment can also be used for a variety of eczema, dermatitis, skin itching, mosquito bite redness, and other skin diseases, also has a certain effect for allergic rhinitis, and is used for prevention of cold.

Literature

Paeonol is the main active ingredient of the cortex of the *Ranunculaceae* plants including *Paeonia suffruticosa* Andr. (Moutan cortex, Mudanpi) and *Rhododendron* plant *Cynanchum paniculatum* (Bge.) Kitag. (Cynanchi paniculati radix et rhizome, Xuchangqing). Moutan cortex and Xuchangqing are commonly used as traditional Chinese medicine. Moutan cortex has the effects on clearing heat, cooling blood, activating blood, resolving stasis, and retreating heat; Xuchangqing also has many effects, such as dispelling wind, relieving pain and itching, and so on. Both of them have been used for more than 2000 years in China, and their functions are recorded in the ancient Chinese Materia Medica.

Moutan cortex is a traditional Chinese medicine as a medium in "Shennong's Herbal Classic of Materia Medica." The earliest record on the official use of peony is in Mawangdui unearthed cultural relics. Zhang Zhongjing in the late Han dynasty made a further discussion about the efficacy of Moutan cortex in his "Synopsis of Prescriptions of the Golden Chamber." After that, peony was recorded in several other books, such as *Mingyi Bielu*, *Wu Pu Bencao*, and *Bencaojing Jizhu* [1]. Xuchangqing was earliest recorded in "Shennong's Herbal Classic of Materia Medica" as the top grade and also in generations of Chinese Materia Medica, with as many as 50 alias, such as *Vernonia cinerea*, "Duiyue grass," "Xiaoyao bamboo," "soil *Asarum*," "hero grass," and so on [2].

Paeonol raw materials with high purity can be derived from plant extracts by chemical synthesis. The commonly used extraction methods of paeonol include boiling method, alcohol extraction, steam distillation, and CO_2 supercritical fluid extraction [3]. But due to the limitation of plant resources and the lower content of paeonol in the natural plants, there are many separation and purification difficulties existed in the extraction methods. The chemical synthesis process is relatively easy, and the final product usually can be obtained through two- to three-step reaction and separation process [4].

History of R&D

As the main active ingredient of Moutan cortex and Xuchangqing, paeonol has been used in clinical practice since ancient times. Paeonol was first isolated from Moutan cortex by Japanese scholar Changjing Changyi in 1884 [5]. With the development of modern extraction and artificial synthesis technology, paeonol monomeric drugs are developed gradually. Formulations approved by SFDA include tablets, injections, and ointment. The clinical application of its dosage form is still mainly on antibacterial, anti-inflammatory, antipyretic, and analgesic aspects.

Paeonol injection, formerly known as Xuchangqing injection, has been proven to have a significant effect on rheumatoid arthritis, visceral pain, lumbar muscle strain, and other diseases and does not show any addiction or other adverse reactions. This clinical observation was proceeded at more than 10 medical units in Shanghai and other places and lasted more than 1 year since 1971. Paeonol injection was the earliest recorded in the Chinese pharmacopoeia 1977 edition; it is the main formulation until now. But because of the poor water solubility of paeonol, it is necessary to improve its solubility to be used as injection. Ten times amount of Tween compared to paeonol was used in the injection to increase the dissolution. In the 1970s, paeonol was made into oil solution for intramuscular injection. But these are with security risks and the use of inconvenience and other shortcomings. Pharmacist obtained paeonol sodium sulfonate by sulfonated paeonol, so the problem of water solubility had been resolved, and the injection of paeonol sodium sulfonate also came into being [6]. In 2003, the quality standard of paeonol sodium sulfonate was also promoted from the local standard to the national standard.

The dosage form of paeonol for skin is mainly ointment. Paeonol ointment was first recorded in the "Anhui Province Drug Standard" 1987 edition [7]; afterwards it was published in 1995, "Drug Standard of Ministry of Public Health of the People's Republic of China" Traditional Chinese medicine prescription preparation tenth volume. It can be used for the treatment of rheumatism, stomachache, and other pain, eczema, allergic dermatitis, and so on.

With the development of preparation techniques, in recent years, some new dosage forms, such as those of beta-cyclodextrin, w/o/w-type multiple emulsion gels, phospholipid complex, liposome, and so on, are being developed to overcome the shortcomings of volatility, poor water solubility, and stability, but these are still in the research stage [8].

Pharmacology

Paeonol has a wide range of pharmacological activities, and its antibacterial, anti-inflammatory, antipyretic, analgesic, and antispasmodic pharmacological effects have been well known. With the recent further study of pharmacological effects of paeonol, the improvement of cardiovascular and cerebrovascular function, liver and kidney care, and antitumor, immune regulation, antioxidant, and other new pharmacological activities have also been reported successively.

Studies have shown that the protective effects of paeonol on cardiovascular and cerebrovascular systems are displayed in three aspects: it has antiarrhythmic activity because of the blocking effect on calcium channel current; it can affect the endothelial cell inflammatory response, thus protecting the blood vessels and improving blood circulation; and it can stabilize the cell membrane and inhibit myocardial ischemia and membrane damage.

Paeonol inhibits the proliferation of a variety of tumor cells, such as human leukemia cell line K562, human breast cancer gene T6-17, hepatocellular carcinoma cell line BEL-7404, hepatoma cell line HepA, human leukemia tumor cell line K562/ADM, cervical cancer cell line HeLa, human large intestine cancer cell line HT-29, and so on. It can also sensitize the multidrug resistance of tumor cells to the chemotherapy drugs.

Low concentration of paeonol can increase the proportion of T lymphocytes in the blood circulation and also make T lymphocytes play a stronger lymphokine separation function but also can promote the non-specific phagocytic removal of bacteria function of neutrophil. The role of systemic cellular immunity and humoral immune function has also been experimentally confirmed. In addition, paeonol has a strong effect on scavenging free radical; its antioxidant effect also enables paeonol's applications in the pharmaceutical field and food preservation industry [4, 9, 10].

Clinical Application

In clinical practice, paeonol is mainly used as a major active ingredient of multi-flavor traditional Chinese medicine. For example, paeonol is the main active ingredient and quality control index of more than ten kinds of commonly used Chinese medicine preparations, such as Feng Shi Ding tablets, Liu Wei Di Huang pills, Zheng Gu Shui water, Gui Shao Di Huang Tang pills, Mai Wei Di Huang pills, Qi Ju Di Huang tablets, Ming Mu Di Huang pills, Zhi Bai Di Huang pills, Gu Ci pills, Gu Ci Xiao Tong tablets, Fu Fang Yi Gan pills, Yang Yin Qing Fei Tang pills, Ji

Sheng Shen Qi Wan pills, and Danggui Yangxue pills. In addition, paeonol can also be administered as a monomeric drug. The listed preparations are mainly paeonol injection, paeonol tablets, and paeonol ointment, clinically for the treatment of rheumatism, stomachache, and other pains, eczema, allergic dermatitis, etc. Paeonol and its preparations have the characteristics of less side effect, no addictive, non-rebound, and high security.

Discussion

Although paeonol has the characteristics of volatility and poor water solubility, its clinical efficacy in anti-inflammatory analgesia is significant and has no addictive advantages. At the same time, its cardiovascular and cerebrovascular, immune system, and antitumor pharmacological activities have been gradually developed and reported. These are all making paeonol to have a broad development prospects. It is believed that with the improvement of the medicinal value of paeonol and the development of science and technology such as structural modification, preparation process, polymorphism, and cocrystals, new drugs of paeonol and new dosage forms will continue to flourish and further meet the clinical needs.

Fig. 1 Moutan cortex

Fig. 2 Chemical structure of paeonol

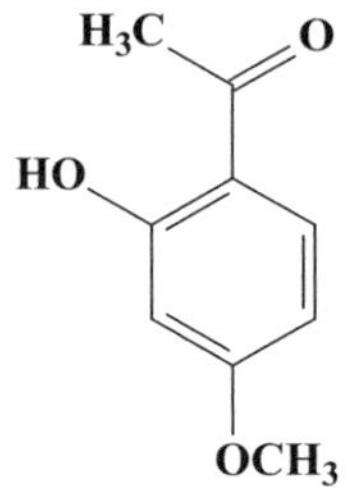

Fig. 3 Chemical structure of sodium paeonolsilate

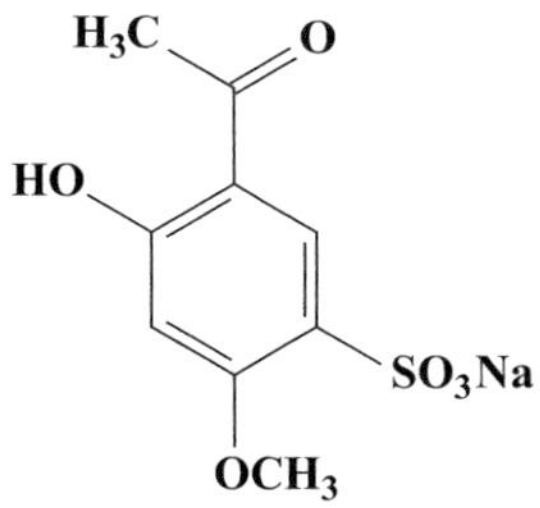

References

1. Zhou YJ, Wei ZX. A survey of Moutan cortex. China Pharmaceuticals. 2012;21(16):111–2.
2. Zhang YQ, Yan P. Textual research on Xu Changqing Materia Medica. Tradit Chin Med Res. 2005;18:54–7.
3. Fu QF, Meng FJ, Miao Q, et al. Advances in the extraction technology of paeonol from Moutan cortex. Inf Tradit Chin Med. 2010;27:108–9.
4. Hu CD, Zhang J. Advances in pharmacological activities and synthesis of paeonol. Chem Bioeng. 2009;26:16–8.
5. Xie P. A research overview of traditional Chinese medicine Moutan cortex. J Chin Med Mater Technol. 1983;4:44–6.
6. Tang HY, Yang S, Wang JB. An overview of the preparation, dosage form and clinical application of paeonol. Jiangsu J Tradit Chin Med. 2004;25:58–60.
7. Xi J, Liu SY, Liu C. Advances in research and clinical application of paeonol and its compound preparations. Lishizhen Med Mater Med Res. 2005;16:66.
8. Han Y. Transdermal delivery of paeonol using microemulsion. Shanghai: The Second Military Medical University; 2009.
9. Liu YL, Wei LC, Shen ZG, et al. Research progress on paeonol. Anhui Med Phamaceutical J. 2011;15:896–9.
10. Guo Q, Li YK, Wang ZG, et al. Advances in pharmacological studies of paeonol. Inf Tradit Chin Med. 2009;26:20–2.

Potassium Dehydroandrographolide Succinate and Potassium Sodium Dehydroandrographolide Succinate

Yu Yan, Lian-Hua Fang, and Guan-Hua Du

Abstract Potassium dehydroandrographolide succinate and potassium sodium dehydroandrographolide succinate are derivatives of andrographolide, which is the diterpene lactone compounds extracted from *Andrographis paniculata (Burm. F) Nees*. Andrographolide has effects on anti-inflammatory, antibacterial, antivirus, antitumor, immune regulation, treatment of cardiovascular-cerebrovascular diseases, and protection of liver and gallbladder but is unable to satisfy the clinical requirement due to the poor water solubility. Potassium dehydroandrographolide succinate and potassium sodium dehydroandrographolide succinate have a good effect in the wide clinical application; there is no need for skin test. However adverse reaction occurs due to its instability in the solution. Therefore, the researchers still need to study and develop preparation which is more safe, reliable, stable, and convenient to use to meet the clinical requirements.

Keywords Andrographolide · Diterpene lactone · Anti-inflammatory · Antibacterial

Y. Yan · L.-H. Fang · G.-H. Du (✉)
Beijing Key Laboratory of Drug Targets Identification and Drug Screening, Centre for Pharmaceutical Screening, Institute of Materia Medica, Chinese Academy of Medical Sciences and Peking Union Medical College, Beijing, China
e-mail: dugh@imm.ac.cn

G.-H. Du, *Natural Small Molecule Drugs from Plants*,
https://doi.org/10.1007/978-981-10-8022-7_74

Origin: *Andrographis paniculata* (Fig. 1)

Potassium Dehydroandrographolide Succinate

Chemical name (Fig. 2)

4-[[(1R,2R,4aR,5R,8aS)-2-(3-Carboxypropanoyloxy)-1,4a-dimethyl-6-methylidene-5-[(E)-2-(5-oxo-2H-furan-4-yl)ethenyl]-3,4,5,7,8,8a-hexahydro-2H-naphthalen-1-yl]methoxy]-4-oxobutanoate

Molecular formula, $C_{28}H_{35}KO_{10}{\bullet}H_2O$; **MW**, 610.68; **CAS**, 76958-99-1

Properties

Appearance: white to yellowish crystalline powder, odorless, bitter, slightly wetting. *Solubility*: slightly soluble in ethanol, insoluble in chloroform, dissolved in 1% sodium bicarbonate solution.

Dosage Forms and Indications

This product is contained in the *Pharmacopoeia of the People's Republic of China* (2010).

Injection, freeze-dried powder injection. Used for viral pneumonia and viral upper respiratory tract infection

Potassium Sodium Dehydroandrographolide Succinate (Fig. 3)

Molecular formula, $C_{28}H_{34}KNaO_{10}$; **MW**, 592.68; **CAS**, 6700-42-5

Dosage Forms and Indications

This product was included in national standards for chemical drugs.

Injection, freeze-dried powder injection. Used for viral pneumonia and viral upper respiratory tract infection

History of R&D

Andrographolide is unable to satisfy the clinical requirement due to the poor water solubility. Since the 1970s, researchers had tried to increase the water solubility through the introduction of different hydrophilic groups in the structure of lactone and had developed a variety of *Andrographis* lactone injection. Among them, potassium dehydroandrographolide succinate has the strongest anti-inflammatory and antipyretic effect. The synthetize of potassium dehydroandrographolide succinate is as follow described: the formation starts with andrographolide as an intermediate, after succinic anhydride reaction catalyzed by pyridine, dehydrated andrographolide succinate half ester was formed then reacted in dilute ethanol solution and potassium bicarbonate to form the final production [1]. Potassium dehydroandrographolide succinate is known as natural antibiotics, and its anti-inflammatory effect is the best among the 13 andrographolide injections, which has been listed as the top 10 Chinese emergency medicines by the National Chinese Medicine Administration and has been listed as the first batch of necessary drugs for the National Chinese Hospital of emergency department [2].

In 1996, Tonghua company developed potassium sodium dehydroandrographolide succinate injection drugs. It was synthetized after dehydration with andrographolide, hydroxyl esterification of potassium salt, or directly from the addition of sodium salt synthesis. It is a kind of potassium dehydroandrographolide succinate derivatives varieties [3].

Pharmacology

Potassium dehydroandrographolide succinate and potassium sodium dehydroandrographolide succinate have the same substance of active metabolites in the body (andrographolide half ester monopotassium salt). As for antipyretic, anti-inflammatory, and antiviral activity, it can promote the anterior pituitary biosynthesis and release of ACTH and alleviate inflammation through the inhibition of histamine-induced increase in capillary permeability and stimulation to pituitary-adrenal cortical function at the designated site. It can inhibit the binding of DNA with protein in the process of virus replication, as well as inhibit bacterial endotoxin-induced fever. It can inactivate adenovirus, influenza virus, respiratory virus, and other viruses and also inhibit *Staphylococcus aureus*, *Streptococcus aureus*, etc. Clinically, it is used for the treatment of upper respiratory tract infection, bronchial pneumonia, viral pneumonia, viral enteritis, and hand, foot, and mouth disease [4].

Clinical Application

Potassium dehydroandrographolide succinate is widely used in the treatment of viral pneumonia, influenza virus-induced acute upper respiratory tract infection, acute bronchitis, pediatric bronchial pneumonia, dysentery, epidemic encephalitis, and asthma attacks, especially for viral and bacterial respiratory tract infections, and has a significant effect on dysentery, especially for infant pneumonia [5].

Clinically, potassium sodium dehydroandrographolide succinate is mainly used for the treatment of upper respiratory tract infection, bronchial pneumonia, viral pneumonia, viral enteritis, and hand, foot, and mouth disease [6].

The adverse reactions of allergic reactions infusion after intravenous drip of potassium dehydroandrographolide succinate were reported most, some of them emerge of pimples, abdominal pain, vomiting, dizziness, head swelling, severe cases, however the sever of them may occur life-threatening anaphylactic shock. The reactions occurred within 20 min and usually gradually improved in 5–45 min after symptomatic treatment. There are individuals needing 24 min to start to recover. In addition, there is also chance to occur thrombocytopenia, liver damage, blood vessels to stimulate pain, difficulty in breathing, chills, fever, etc. after the use of potassium dehydroandrographolide succinate [7].

Skin allergies and diarrhea often occur in children after intravenous drip of potassium sodium dehydroandrographolide succinate, occasionally allergic shock and liver damage reports. Serious adverse reactions of potassium sodium dehydroandrographolide succinate injection are systemic damage; the main symptoms contain allergic shock, allergic reactions, chills, fever, etc., while skin and accessories damage mainly occur rash, of which 53% of patients is children under the age of 14, 38% of deaths caused by drug-induced allergic shock. According to the 23rd issue of "adverse drug reactions" published by the National Drug Adverse Drug Reaction Center, the irrational use of potassium sodium dehydroandrographolide succinate mainly includes overdose, cross-indications, and allergic constitution [8].

Discussion

Potassium dehydroandrographolide succinate and potassium sodium dehydroandrographolide succinate have a good effect in the wide clinical application, and there is no need for skin test. However, adverse reaction occurs, due to its instability in the solution. Therefore, the researchers still need to study and develop preparation which is more safe, reliable, stable, and convenient to use to meet the clinical requirements.

Fig. 1 *Andrographis paniculata*

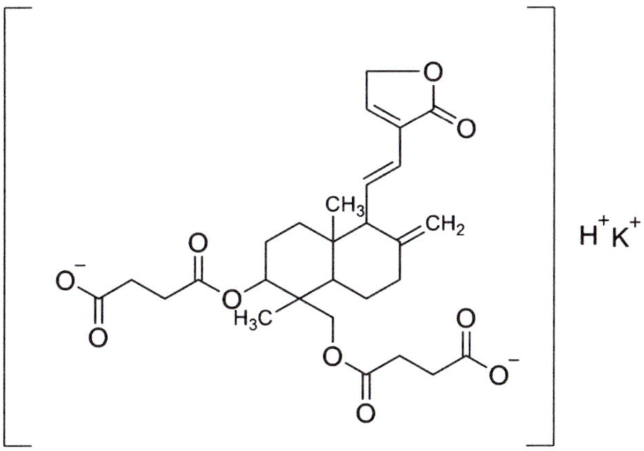

Fig. 2 The structure of potassium dehydroandrographolide succinate

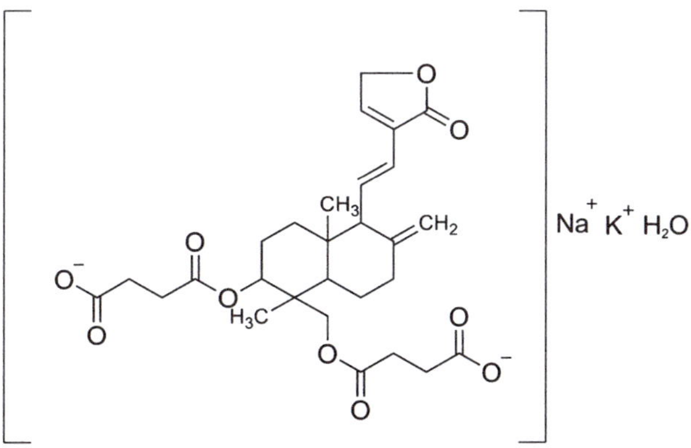

Fig. 3 The structure of potassium sodium dehydroandrographolide succinate

References

1. Andrographolide Research Team of Department of Pharmaceutical chemistry of Traditional Chinese medicine research institute of sichuan province. Research of dehydration paniculta lactone succinate half ester single potassium and its injection. Chin Herb Med Commun. 1978;8:1.
2. Xu R, Xu XL. Progress of clinical application of potassium dehydroandrographolide succinate injection. Chengdu Med J. 2005;31(4):278–80.
3. James RJ, Rui HL, Bud CT. Andrographolide derivatives to treat viral infections. USA: WO2006101537A2, 2006-09-28.
4. Zhang YT, Zhang ZJ, Sun YH. A study on the interaction of the DAS-K with bovine serum albumin by on-line ultrafiltration and chemiluminescence. Chin J Chem. 2004;24(12):1777–83.
5. Kong QF, Deng YW, Pan XF. The clinical application and adverse reactions of potassium dehydroandrographolide succinate. Chin Pharm. 2006;9(8):765–6.
6. Liang JC. The clinical application of potassium sodium dehydroandrographolide succinate [J]. Chin Med Abstr: Intern Med. 2006;27(5):435–6.
7. Yan SY, Wang LF, Lv AQ. The clinical application of potassium dehydroandrographolide succinate injection. Chin Pharm Aff. 2005;19(5):311–2.
8. Cheng F. The adverse reaction and safety analysis of potassium sodium dehydroandrographolide succinate. Chin J Clin Rational Drug Use. 2015;8(11):96–7.

Rorifonum

Xiao-Li He, Shou-Bao Wang, and Guan-Hua Du

Abstract Rorifonum is extracted from rorippa (*Rorippa Montana (Wall.) Small*) which is distributed widely in China. It is mainly used for the treatment of patients with chronic bronchitis at protraction stage. However, the pharmacological studies of rorifonum were unsystematic and the specific mechanism was unknown. At present, the clinical application of rorifonum has been reduced. As the first discovered sulfone compounds with the effects of relieving cough and eliminating phlegm, the research has promoted the study of finding innovative products from traditional medicine plant.

Keywords Rorifonum · Sulfone · Chronic bronchitis · Protraction stage

Alias: Rorifone
Origin: Rorippa (Fig. 1)
Chemical name (Fig. 2)

10-Methylsulfonyl decanenitrile

Molecular formula, $C_{11}H_{21}NO_2S$; **MW**, 231.35; **CAS**, 53078-90-3

X.-L. He
Institute of Medicinal Plant Development, Chinese Academy of Medical Sciences and Peking Union Medical College, Beijing, China

S.-B. Wang · G.-H. Du (✉)
Beijing Key Laboratory of Drug Targets Identification and Drug Screening, Centre for Pharmaceutical Screening, Institute of Materia Medica, Chinese Academy of Medical Sciences and Peking Union Medical College, Beijing, China
e-mail: dugh@imm.ac.cn

G.-H. Du, *Natural Small Molecule Drugs from Plants*,
https://doi.org/10.1007/978-981-10-8022-7_75

Properties

Appearance: white or whitish crystalline powder. *Solubility*: practically insoluble in water; slightly soluble in ether; freely soluble in ethyl acetate or chloroform; soluble in methanol and ethanol. *Melting point*: 44–47 °C.

Derivatives

Rorifamide

Dosage Forms

Rorifonum tablets are recorded in the *Pharmacopoeia of the People's Republic of China* (1977).

Indications

It is mainly used for the treatment of patients with chronic bronchitis at protraction stage.

Literature

Rorifonum is an antichronic bronchitis sulfone derived from *hancai* 蔊菜 (*Rorippa Montana (Wall.) Small*) which is distributed widely in China [1]. The plant has been used as a folk medicine to relieve cough and asthma and also as a mucolytic agent. The results of clinical trials in Shanghai showed that the water decoction of *Rorippa Montana (Wall.) Small* (hancai) had a certain effect in some kind of chronic bronchitis.

History of R&D

In the early 1970s of the last century in China, about 300 kinds of effective Chinese herbal medicines were explored for the treatment of chronic bronchitis, which was carried out as a national program [2]. Rorifonum was obtained and developed under this historical condition. Two crystals named rorifonum and rorifamide were obtained after studies on the chemical components of *Rorippa Montana (Wall.) Small* (hancai) by Shanghai Institute of Materia Medica. Using elemental analysis, spectral identification, and mass spectrometry, the chemical structure of these two compounds was finally determined [1]. Rorifonum and rorifamide are the first discovered sulfone compounds with the effects of relieving cough and eliminating phlegm.

After a series of pharmacological studies, rorifonum had been proved effective in the expectorant and antibacterial experiments. Clinical trials also revealed that rorifonum has a certain effect of relieving cough and asthma [3]. Synthetic compound of rorifonum and natural product have the same effects [4]. However, the pharmacological effect of rorifamide was not obvious. Meanwhile, in order to find a new drug with higher efficacy and lower toxicity, the researchers synthesized a series of similar compounds, some of them showing significant expectorant activity in the animal models [5].

Pharmacology

The pharmacological studies of rorifonum were unsystematic and the specific mechanism was unknown. At present, only a few pharmacological experiments were carried out including acute toxicity test, subacute toxicity test, expectorant test, antitussive experiment, and antibacterial experiment [3]. The expectorant test with phenol red in mice showed rorifonum had the effect of expectorant, while the antitussive experiment by sulfur dioxide showed there was no obvious effect.

Clinical Application

It is mainly used for the treatment of patients with chronic bronchitis at protraction stage. At this stage, the effect of a single drug is weak and should be used in combination with other drugs. At present, the clinical application of rorifonum has been reduced.

Discussion

Rorifonum and rorifonum tablets were included in the 1977 edition of *Pharmacopoeia of the People's Republic of China* but are no longer commercially available now. As the first discovered sulfone compounds with the effects of relieving cough and eliminating phlegm, the research on them has promoted the study of finding innovative products from traditional medicine plant.

Fig. 1 Rorippa

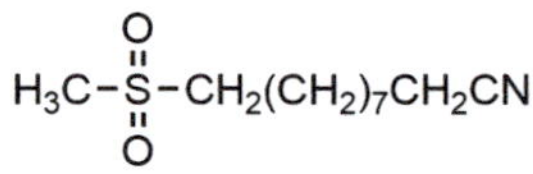

Fig. 2 The structure of rorifonum

References

1. Tang ZJ, Chen Y, Xi GL. Studies on the active principles of Hancai. Sci Chin. 1974;1:15–20.
2. The prevention and control of chronic bronchitis office. Research progress of prevention and treatment of chronic bronchitis drugs. People's Mil Surg. 1976;12:89–2.
3. Experimental study of Rorifonum and clinical observation. Med Res Commun. 1972;3:19–20.
4. Shanghai Institute of Materia Medica. New compounds for the treatment of chronic bronchitis-Rorifonum. Nati Medi J Chin. 1973;2:73.
5. Xi GL, Gu KJ, Zhang HL, et al. Syntheses of some analogs of Rorifone. Sci Chin. 1974;6:585–90.

Salicylic Acid

Yu-Cai Chen, Gui-Fen Qiang, and Guan-Hua Du

Abstract Salicylic acid is extracted from willow and Rosaceae, which was modified to aspirin to reduce the irritation of the digestive tract. Low-dose aspirin (75–300 mg/day) has antiplatelet aggregation effect, middle dose of aspirin (0.5–3 g/day) has antipyretic analgesic effects, and high doses of aspirin (more than 4 g/day) has anti-inflammatory and antirheumatic effects. However, aspirin also has adverse effects, mainly gastrointestinal symptoms such as nausea, vomiting, upper abdominal discomfort or pain. As a nonsteroidal anti-inflammatory drug (NSAID), aspirin was listed in July 1899 and has been used for more than 100 years. It is one of the three classic drugs in the history of medicine and the world's most widely used antipyretic, analgesic and anti-inflammatory drugs.

Keywords Salicylic acid · Aspirin · NSAID

Alias: 2-Hydroxybenzoic acid; O-hydroxybenzoic acid; Salonil; O-Carboxyphenol; 2-Carboxyphenol
Origin: Willow bark (Fig. 1)
Chemical name (Fig. 2 salicylic acid, Fig. 3 aspirin)

2-Hydroxybenzoic acid

Molecular formula, $C_7H_6O_3$; **MW**, 138.12; **CAS**, 69-72-7

Molecular formula, $C_9H_8O_4$; **MW**, 180.16; **CAS**, 0-78-2

Y.-C. Chen · G.-F. Qiang · G.-H. Du (✉)
Beijing Key Laboratory of Drug Targets Identification and Drug Screening, Centre for Pharmaceutical Screening, Institute of Materia Medica, Chinese Academy of Medical Sciences and Peking Union Medical College, Beijing, China
e-mail: dugh@imm.ac.cn

G.-H. Du, *Natural Small Molecule Drugs from Plants*,
https://doi.org/10.1007/978-981-10-8022-7_76

Properties

Salicylic acid. *Appearance*: white crystalline powder. *Solubility*: Absolutely soluble in ethanol, soluble in ether and chloroform, slightly soluble in water and anhydrous ether. *Stability*: Stable at room temperature, discomposes into phenol and carbon dioxide after rapidly heated. It's partially acidic.

Aspirin. *Appearance*: white crystal and decomposes at 136–140 °C. *Melting point*: 136 °C. Aspirin is the acetyl derivative of salicylic acid with weak acidity. Its acidity coefficient is 3.5 at 25 °C. *Stability*: Aspirin decomposes rapidly in ammonium acetate, alkali metal of acetate, carbonate, citrate or hydroxide solutions. There are two crystal forms of aspirin including crystal form I and II.

Dosage Forms

Salicylic acid was included in *Pharmacopoeia of the People's Republic of China* (2015), *British Pharmacopoeia* (2017), *U.S. Pharmacopeia* (40), *The Japanese Pharmacopoeia* (17th ed.), *European Pharmacopoeia* (9.0th ed.) and *The International Pharmacopoeia* (5th ed.).

Aspirin is included in *Pharmacopoeia of the People's Republic of China* (2015), *British Pharmacopoeia* (2017), *U.S. Pharmacopeia* (40), *The Japanese Pharmacopoeia* (17th ed.) and the *Indian Pharmacopoeia* (2010). The clinical application of salicylic acid is aspirin, the acetyl derivatives of salicylic acid. The dosage forms include ordinary tablet, enteric-coated tablets, enteric capsule, effervescent tablets, suppository and dispersant.

Indications

The clinical applications contain two main parts. One is for the pain relief, anti-inflammation and treatment of rheumatic disease. Another is to prevent coronary artery and cerebrovascular thrombosis and other complications of postoperative thrombosis. The different doses are used depending on the different indications.

Literature

Salicylic acid is the chemical found in willow and Rosaceae. Salicylic comes from salix, the Latin name of the willow. According to *Ebers Papyrus*, the ancient Egyptian book, willow bark was used to deal with many ailments about 1500 B.C. Around 300 B.C., it was recorded that Hippocrates, the father of the Western medicine, used willow bark to alleviate the pain during delivery [1]. In the early nineteenth century, Karl Jacob Lovig extracted spirsaure from Spiraea, an acid with fever and pain relief effect, which was proved to be salicylic acid later.

History of R&D

Willow is China's native tree species, which has more than 2000 years of cultivation history in China. There are about 520 species all over the world, and China has more than 250 species. *Spiraea* is distributed in Mongolia, Japan, North Korea, the Soviet Union Siberia and southeastern Europe. In China, they are cultivated in Liaoning, Inner Mongolia, Hebei, Shandong, Shanxi and other places. They grow on riversides, wetlands, open places and ravines.

In 1763, at Oxford University's Wadham College, British pastor Edward Stone found the active ingredient, salicylic acid, from the willow bark for the first time [1, 2]. Due to the severe side effects of salicylic acid, most patients were reluctant to take it. In 1853, the French professor of chemistry, Charles Gerhardt, tried to connect the acetyl group to the hydroxyl group of salicylic acid, thereby greatly reducing the acidity of salicylic acid. However, due to the limited technical conditions, the purity of the product was very low. At that time, no attention was paid to this problem [3]. In 1897, Felix Hoffmann, the German company Bayer's chemist, successfully synthesized aspirin after synthesis of salicin from the spinach (Latin: *Spiraea ulmaria*). It was less irritating to the digestive tract than salicylic acid [4]. Aspirin was listed in July 1899 and has been used for more than 100 years.

In 1971, John Vann, a pharmacist at the Royal College of Surgeons, found that aspirin could inhibit the synthesis of prostaglandins and cause vasodilatation, and then, they further studied the effect of prostaglandin on muscle contraction. Six years later, Vane and his student Moncada confirmed that aspirin could inhibit the production of prostaglandins and thromboxane [5, 6].

In 1982, John Vane and two Swedish scholars, Bergstrom and Samuelsson, were co-awarded with the Nobel Prize in Physiology and Medicine, due to the achievements of prostaglandins [7].

For a long time, people only know that aspirin has one kind of crystal form. Since the 1960s, scientists began to suspect that it may have another crystal form. In 2005, Bond found another new crystal form of aspirin, crystal form II, when making common crystallization of aspirin and levetiracetam in the hot acetonitrile [8, 9].

Pharmacology

Aspirin is a nonsteroidal anti-inflammatory drug (NSAID). The main pharmacological effect is to inhibit prostaglandin metabolism and thromboxane synthesis by inhibiting prostaglandin metabolism-required cyclooxygenase, via irreversible acetylation of 530 serine residues in the hydroxyl of COX-1 polypeptide chain, which results in COX-1 inactivation, blocks the conversion of arachidonic acid into thromboxane A2 pathway and then inhibits the platelet aggregation.

Prostaglandin is a hormone produced locally in the body. It can pass the pain to the brain, regulate body temperature in the hypothalamus and cause inflammation. Inhibition of prostaglandin synthesis can have antipyretic, analgesic, anti-inflammatory and antirheumatic effects. The adverse effects of aspirin are mainly gastrointestinal symptoms such as nausea, vomiting, upper abdominal discomfort or pain. It can also cause allergic reactions, cardiotoxicity, liver and kidney damage and Wright's syndrome. In addition, high doses of aspirin can cause salicylic acid reactions such as headache, dizziness, tinnitus, hearing loss and other central nervous system symptoms.

Clinical Application

The clinical application of aspirin varies with the therapeutic dose. Low-dose aspirin (75–300 mg/day) has antiplatelet aggregation effect and can be used to prevent and treat the coronary and cerebrovascular thrombosis and other postoperative thrombosis. The middle dose of aspirin (0.5–3 g/day) has antipyretic analgesic effects, so it is commonly used in the treatment of fever, headache, toothache, neuralgia, muscle pain and menstrual pain. High doses of aspirin (more than 4 g/day) have anti-inflammatory and antirheumatic effects for the treatment of acute rheumatic fever and rheumatoid arthritis. In addition, aspirin is used for the treatment of skin and mucous membrane lymphadenopathy (Kawasaki disease) in paediatric.

Discussion

So far, aspirin has been used for more than 100 years. It is one of the three classic drugs in the history of medicine and the world's most widely used antipyretic, analgesic and anti-inflammatory drugs. It has been established as a classic drug for the treatment of acute myocardial infarction, unstable angina and secondary prevention of myocardial infarction. This drug has been included in the list of essential medicines standards of the World Health Organization and is one of the most important and essential medicines. Recent studies have shown that aspirin has a wide range of effects in stroke, pregnancy complications, colon cancer, diabetes and senile dementia. Therefore, the study of aspirin will continue in future.

Fig. 1 Willow bark

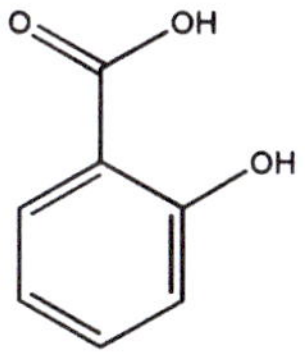

Fig. 2 Chemical structure of salicylic acid

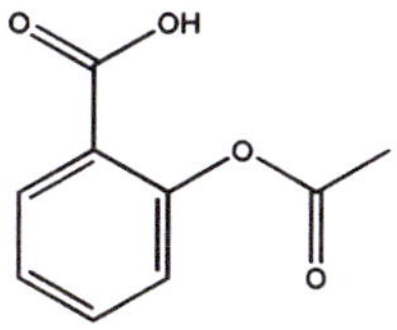

Fig. 3 Chemical structure of aspirin

References

1. Mueller RL, Scheidt S. History of drugs for thrombotic disease. Discovery, development, and directions for the future. Circulation. 1994;89(1):432–49.
2. Stone E. An account of the success of the bark of the willow in the cure of agues. Philos Trans. 1763;53:195–200.
3. Gerhardt CH. Untersuchungen über die wasserfreien organischen Säuren [Investigations into anhydrous organic acids]. Annalen der Chemie und Pharmacie. 1853;87(2):149–79.
4. Singer H. Ueber aspirin. Archiv für die gesamte. Physiologie des Menschen und der Tiere. 1901;84:527–46.
5. Vane JR. Inhibition of prostaglandin synthesis as a mechanism of action for aspirin-like drugs. Nat New Biol. 1971;231(25):232–5.
6. Vane JR, Botting RM. The mechanism of action of aspirin. Thromb Res. 2003;110(5–6):255–8.
7. The Nobel Prize in Physiology or Medicine 1982.
8. Vishweshwar P, McMahon JA, Oliveira M, et al. The predictably elusive form II of aspirin. J Am Chem Soc. 2005;127(48):16802–3.
9. Bond AD, Boese R, Desiraju GR. On the polymorphism of aspirin: crystalline aspirin as intergrowths of two "polymorphic" domains. Angew Chem Int Ed Engl. 2007;46(4):618–22.

Sinomenine

Zhao Zhang, Ling-Lei Kong, and Nai-Hong Chen

Abstract Sinomenine is derived from Caulis Sinomenii, which is mainly used for the treatment of rheumatoid arthritis and other types of rheumatism and arrhythmia in clinical practice. Sinomenine has anti-inflammatory, immunosuppressive, analgesic sedation, antiarrhythmic, detoxification, and other pharmacological effects, while the half-life of sinomenine is short. In addition, sinomenine also causes a strong histamine release, which leads to rash, gastrointestinal reactions, and other side effects, limiting its wide range of clinical applications. At present, a variety of formulations improve its clinical efficacy and reduce adverse reactions to a certain extent. It may be more effective to modify the sinomenine to increase its stability from the viewpoint of pharmacochemistry. In addition, sinomenine has a chemical structure somewhat similar to that of morphine, but it is not addictive. Research on the relationship between drug addiction and the structure and activity of sinomenine may lead to the discovery of highly effective and low-toxicity natural drugs.

Keywords Sinomenine · Rheumatism · Arrhythmia · Histamine release · Morphine

Origin: *Caulis Sinomenii* (Fig. 1)
Chemical name (Fig. 2)

7, 8-didehydro-4-hydroxyl-3, 7-dimethoxy-17-methylmorphinan-6-one

Molecular formula, $C_{19}H_{23}NO_4$; **MW**, 329.38; **CAS**, 115-53-7

Z. Zhang · N.-H. Chen (✉)
State Key Laboratory of Bioactive Substances and Functions of Natural Medicines, Institute of Materia Medica & Neuroscience Center, Chinese Academy of Medical Sciences and Peking Union Medical College, Beijing, China
e-mail: chennh@imm.ac.cn

L.-L. Kong
Beijing Key Laboratory of Drug Targets Identification and Drug Screening, Centre for Pharmaceutical Screening, Institute of Materia Medica, Chinese Academy of Medical Sciences and Peking Union Medical College, Beijing, China

G.-H. Du, *Natural Small Molecule Drugs from Plants*,
https://doi.org/10.1007/978-981-10-8022-7_77

Properties

Appearance: acicular crystals (crystallized from benzene). *Solubility*: soluble in ethanol, acetone, chloroform, and dilute alkali; slightly soluble in water, ether, and benzene. *Melting point*: 219–221 °C. *Specific optical rotation*: – 71° ($c = 2.1$, ethanol). Its hydrochloride, crystallization (water or ethanol), decomposed at 278 °C. Its hydroiodide, needle crystal (crystallized from water), decomposed at 272 °C. Its picrate, which is yellow needle crystal, decomposed at 176 °C. Sinomenine is sensitive to light and heat to decompose.

Dosage Forms

This product is contained in the eighth volume of national standards for chemical drugs. The current clinical use of dosage forms includes sinomenine hydrochloride enteric-coated tablets, sinomenine hydrochloride injection, and sinomenine sustained-release agent.

Indications

It is mainly used for the treatment of rheumatoid arthritis and other types of rheumatism and arrhythmia in clinical.

Literature

Sinomenine is derived from the roots and stems of the *Sinomenium actum* Rehd. et wils, the leaves of the *Menispermun dauricum* DC. Qingfeng rattan is distributed in the Yangtze River Basin and its southern district (its south is around the north of Guangdong), which originates in forest, forest edge, ditch edge, or shrub, climbing on trees or rocky mountain. Cut the cane of Qingfeng rattan in June to July, remove the thin branches and leaves, then moisten with water, and cut into sections and dry.

Qingfeng rattan, with lukewarm nature, tastes bitter. It can be used to relieve rheumatism, to dredge meridians, to increase the passing of water, and to relieve pain. Qingfeng rattan can also be used, for example, in the treatment of rheumatoid arthritis, joint swelling, and limb pain numbness. Qingfeng rattan was initially contained in "Figures of Herbal Medicines" of the Song Dynasty. It has been used singly or in a compound in oral or topical treatments for rheumatic diseases in traditional Chinese medicine for more than a thousand years [1, 2].

The stem and rhizome of Qingfeng rattan have many chemical constituents, including sinomenine, sinoacutine, tuduranine, acutumine, N-acutumidine, michelalbine, stepharine, disinomenine, magnoflorine, sinactine, isosinomenine, and so forth, of which sinomenine is the main active ingredient, with a content as high as 2%.

History of R&D

The extraction of sinomenine hydrochloride is commonly used in the reflux method [3] (the content of sinomenine is greater than 95%) and ultrasound method [4] (the content of sinomenine is greater than 98.73%).

The chemical structure of sinomenine is composed of four rings, A, B, C, and D (Fig. 3), similar to the structure of morphine. Ring A is a benzene ring, and ring B is a half-chair-shaped, six-member ring. The C ring is a twisted-chair-type, six-member ring that has an a, β-unsaturated ketone structure attached to the B ring. The D ring is a nitrogen-containing, six-member ring under the B ring. Its structure is shown below; the current structural modification of sinomenine is mainly focused on the A/C active group.

Based on the transformation of the A ring, it was found that the 1-substituted formyl derivative of sinomenine showed the strongest inhibitory effect on the inflammatory response of the mouse ear. The 4-substituted p-chlorobenzoyl-sinomenine has the strongest anti-inflammatory and analgesic activity. The biotransformation and chemical synthesis were also used to prepare the di-sinomenine derivatives linked by carbon and carbon, which was stronger than that of sinomenine and had a strong inhibitory effect on cell inflammatory factors. Sinomenine derivatives of the C ring with a pyrazine ring have a strong inhibitory effect on T, B lymphocyte proliferation reaction, which can be used for the preparation of immunomodulatory drugs. The transformation of C ring carbonyl yielded a series of shift alkali derivatives, with strong anti-inflammatory and analgesic effects. These attempts are important for the development of new drugs.

Pharmacology

Sinomenine has anti-inflammatory, immunosuppressive, analgesic sedation, antiarrhythmic, detoxification, and other pharmacological effects, while the half-life of sinomenine is short. In addition, sinomenine also causes a strong histamine release, which leads to rash, gastrointestinal reactions, and other side effects, limiting its wide range of clinical applications.

1. Effects on central nervous system

(a) Analgesic effect: The chemical structure of sinomenine and morphine are similar (Fig. 3). They both act on the central nervous system with a significant analgesic effect, but the mechanisms are different. It has been shown that the analgesic effect of sinomenine has nothing to do with the release of histamine.

(b) Sedative effect: Sinomenine has an inhibitory effect on the central nervous system. The sedative effect works by inhibiting the excitement of advanced neurological activity. Sinomenine can also eliminate the “anger” response of mice caused by electrical stimulation, showing a stabilizing effect. In addition, although, like morphine, it had mainly a sedative effect on the central nervous system, sinomenine also has some excitatory effects on some parts of the central nervous system, especially the spinal cord.

(c) Antitussive effect: Sinomenine has a significant antitussive effect. The antitussive potency of mice and cats is close to that of codeine, while the titer for guinea pigs is about one-fourth that of codeine. Promethazine can enhance the antitussive effect of sinomenine.

(d) Detoxification effect: Long-term subcutaneous injection of sinomenine in monkey causes no abnormal reaction after withdrawal for 72 h, indicating that sinomenine is not addictive. Sinomenine has a significant inhibitory and blocking effect on the isolated ileum of withdrawal morphine-dependent guinea pigs. Its inhibition of enkephalin receptor activity is more than 75%, suggesting that the withdrawal symptoms of morphine might be treated through the opioid receptor. Another study showed that sinomenine acting on the M cholinergic receptor might also be related to its treatment of morphine withdrawal.

(e) Other effects on the central nervous system: Sinomenine may, to a slight degree, induce vomiting. Sinomenine also has local anesthetic effects on frog nerve endings and rabbit cornea that could be applied for local infiltration anesthesia.

2. Effects on peripheral nervous system

Sinomenine can reversibly block the neuromuscular transmission, which showed a concentration-dependent inhibitory effect. Sinomenine had no significant effect on nerve stem excitability and conductivity.

3. Impact on cardiovascular system

Sinomenine has a significant antihypertensive effect. Sinomenine also has a significant antagonistic effect on ischemic arrhythmia.

4. Anti-inflammatory and antiallergic effects

The anti-inflammatory effect of sinomenine is significant. As a Cox-2 inhibitor, sinomenine can inhibit inflammation by reducing local PGE synthesis and release. In addition, sinomenine may affect the pituitary-adrenal system through the hypothalamus to have an anti-inflammatory effect. Sinomenine also has a certain inhibitory effect on acute rejection of rat kidney transplantation, and its mechanism may be due to the inhibition of IL-2 production by Th1 cells.

5. Histamine release

Sinomenine is one of the strongest histamine release agents currently known in plant components. Sinomenine granules can cause guinea pig peritoneal cells to hypertrophy by 90%.

6. Metabolism

Wistar rats were orally treated with 30 mg/kg sinomenine once; T_{max} was 1.083 ± 0.204 h, its maximum plasma concentration was 5.235 ± 0.390 ng/mL (C_{max}), and the half-life was 4.875 ± 0.635 h [5]. It mainly distributed in the large circulation and blood-flow-rich organs. According to the distribution concentration, the order was liver, heart, kidney, lung, and brain. No sinomenine was detected in testes.

Clinical Application

The treatment of rheumatism and rheumatoid arthritis is one of the most important clinical applications of sinomenine. Sinomenine is particularly suitable for the treatment of arrhythmia caused by organic heart disease. Sinomenine is used for the treatment of glomerular disease, which can reduce urinary protein and relieve hematuria symptoms, and the side effects were significantly lower than that of tripterygium glycoside tablets, which are commonly used in clinical practice. In addition, sinomenine can significantly inhibit renal interstitial fibrosis and the production of tissue growth and growth factor TGF-β1. Sinomenine can significantly delay the development of chronic renal failure and effectively treat ankylosing spondylitis. In addition, in recent years there have also been clinical reports on the use of sinomenine treatment of chronic nephritis, cancer, and drug addiction.

Sinomenine has a mild effect on liver cells, with no significant effect on other organs. Sinomenine has no mutagenic activity on somatic cells. Sinomenine has no physical or mental dependence and is not addictive. Adverse reactions caused by high doses of injections are burning pain at the injection spot, itching in head and upper limbs, edema in lips and eyelids, transient headache, and other, which can be relieved by H1 receptor blockers.

Discussion

Sinomenine has significant anti-inflammatory and antiarrhythmic effects. However, its half-life is short, and the histamine release promotion effect easily leads to rash, gastrointestinal disorder, and other adverse reactions. At present, a variety of formulations, such as sustained release, controlled release, gel, spray, and other new formulations, improve its clinical efficacy and reduce adverse reactions to a certain extent. It may be more effective to modify the sinomenine to increase its stability from the viewpoint of pharmacochemistry. In-depth studies of sinomenine may lead to the development of new Cox-2 inhibitors. In addition, sinomenine's chemical structure is similar to that of morphine, but it is no addictive. Research on the relationship between drug addiction and the structure and activity of sinomenine may lead to the discovery of new, highly effective, and low-toxicity natural drugs.

Fig. 1 *Caulis Sinomenii*

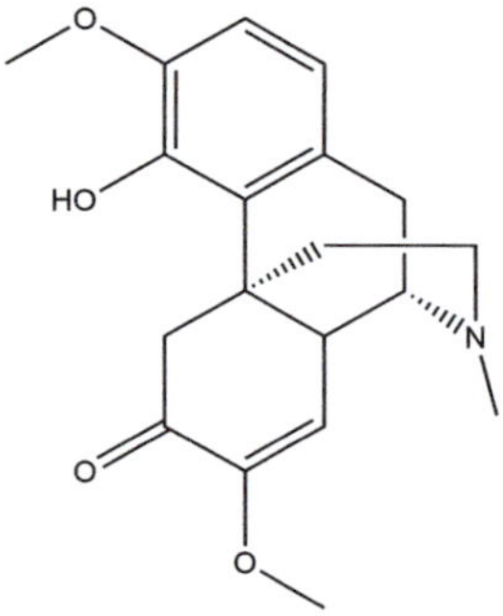

Fig. 2 The chemical structure of sinomenine

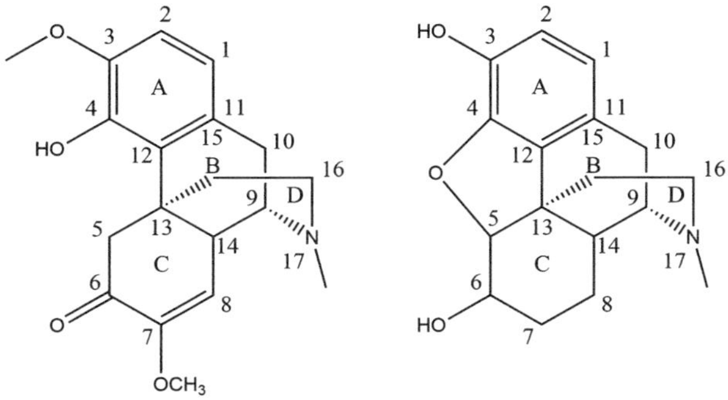

Sinomenine Morphine

Fig. 3 Chemical structures of sinomenine and morphine

References

1. Xiao PG. Chinese Academy of Medical Sciences Institute of Chinese medicine. Traditional Chinese medicine zhi. People's Health Publishing House, China; 1959.
2. Jiangsu New School of Medicine. Traditional Chinese medicine dictionary (volume 1). Shanghai: Shanghai Science and Technology Press; 1982. p. 1234.
3. Wang XL, Wen PH, Feng LM. Optimization of extraction technology of Sinomenine. Chin Trad Med. 1997;28(4):247–9.
4. Pan WJ, Li XN, Zhang ZT, et al. Extraction of sinomenine from Cynanchum paniculata by ultrasonic method. Nat Prod Res Dev. 2003;15(2):127–9.
5. Zhang MF, Zhao Y, Jiang KY, Han L, Lu XY, et al. Comparative pharmacokinetics study of sinomenine in rats after oral administration of sinomenine monomer and Sinomenium acutum extract. Molecules. 2014;19:12065–77.

Theophylline

Li-Li Gong, Li-Da Du, and Guan-Hua Du

Abstract Theophylline was firstly extracted from tea leaves and naturally found in tea and cocoa beans. Theophylline is a drug used for respiratory diseases such as chronic obstructive pulmonary disease (COPD) and asthma. It inhibits phosphodiesterase (PDE) and thereby prevents the intracellular breakdown of cAMP. Side effects are related to plasma concentrations and include nausea, vomiting, and headaches due to PDE inhibition. Higher concentrations cause cardiac arrhythmias and seizures due to adenosine A1-receptor antagonism. Theophylline has been used much less in developed countries at present. However, low dosage of theophylline shows anti-inflammatory and immunomodulatory effect, which raises the reconsideration of it for the treatment of severe asthma, smoking asthma, and COPD.

Keywords Theophylline · Asthma · Chronic obstructive pulmonary disease

Alias: Dioxane purine, 1,3-dimethylxanthine
Origin: Tea (Fig. 1)
Chemical name (Fig. 2)

1,3-Dimethyl-3,7-dihydro-1H-purine-2,6-dione

Molecular formula, $C_7H_8N_4O_2$; **MW**, 180.17; **CAS**, 58-55-9

L.-L. Gong
Beijing Chao-Yang Hospital, Capital Medical University, Beijing, China

L.-D. Du
School of Biomedical Sciences, Chinese University of Hong Kong, Hong Kong, China

G.-H. Du (✉)
Beijing Key Laboratory of Drug Targets Identification and Drug Screening, Centre for Pharmaceutical Screening, Institute of Materia Medica, Chinese Academy of Medical Sciences and Peking Union Medical College, Beijing, China
e-mail: dugh@imm.ac.cn

G.-H. Du, *Natural Small Molecule Drugs from Plants*,
https://doi.org/10.1007/978-981-10-8022-7_78

Properties

Appearance: white, crystalline powder, odorless, with a bitter taste. *Solubility*: freely soluble in solutions of alkali hydroxides and in ammonia; sparingly soluble in alcohol, in chloroform, and in ether; slightly soluble in water. Water solubility, 7.36 g/L (20 °C); *density*, 1.62 g/cm^3; *melting point*, 270–274 °C; *boiling point*, 390.1 °C (760 mmHg); *flash point*, 189.7 °C; *vapor pressure*, 2.72E-06 mmHg (25 °C).

Dosage Forms

Theophylline is included in the *Pharmacopoeia of the People's Republic of China* (2015), *British Pharmacopoeia* (2017), *United States Pharmacopeia* (40), *Japanese Pharmacopoeia* (17th ed.), and *European Pharmacopoeia* (9.0th ed.).

The dosage forms clinically available are theophylline injection, theophylline sustained-release tablets, theophylline sustained-release capsules, and theophylline controlled-release capsules.

Indications

Theophylline was used for bronchial asthma, asthmatic bronchitis, obstructive emphysema, and other symptoms of wheezing. It can also be used for asthma caused by cardiogenic pulmonary edema.

Literature

Theophylline is naturally found in tea and cocoa beans. Tea is likely originated in southwest China during the Shang dynasty as a medicinal drink. *Cha Jing* written by Lu Yu in Tang dynasty (AD 618–907) is an early professional work on tea.

Tea is mainly grown in China, India, Indonesia, Sri Lanka, Kenya, Uganda, and Zimbabwe. There are many types of teas, such as green tea, white tea, oolong tea, black tea, etc. The herb tea has diuretic, antibacterial, anti-inflammatory, and anti-diarrheal effects. In addition to as drink, tea is used in many Chinese medicines, such as Lingrui laxative tea and green tea antihypertensive tablet.

History of R&D

Theophylline was firstly extracted from tea leaves and chemically identified by the German biologist Albrecht Kossel [1]. A cup of tea contains about 1 mg/mL theophylline. In 1895, a chemical synthesis of theophylline starting with 1,3-dimethyluric acid was described by Emil Fischer and Lorenz Ach [2]. Theophylline was synthesized by Wilhelm Traube in 1900. Aminophylline, a derivative of theophylline ethylenediamine, is widely used due to its greater aqueous solubility.

Theophylline was firstly used clinically as a diuretic in 1902. Twenty years later it was firstly reported by D.I. Macht and G.C. Ting for asthma treatment in pig bronchial smooth muscle. The first successful clinical use of theophylline in bronchial asthma was reported in 1922 by S. Hirsch, who described that four patients responded well to the rectal administration of a mixture of 66.7% theophylline and 33.3% theobromine. He also tested the combination of theophylline with theobromine on bovine bronchial smooth muscle strips and noted smooth muscle relaxation. Thus he concluded that dimethylxanthines act by producing relaxation of bronchial smooth muscle. In 1937, two concurrent but independent clinical trials reported that methylxanthines were efficacious in asthma. The Food and Drug Administration approved the use of theophylline for asthma in the USA in 1940.

There are more than 300 derivatives of theophylline. The main derivatives include aminophylline, dihydroxypropyl theophylline, and oxtriphylline.

1. Aminophylline: the combined mixture of theophylline and ethylenediamine; used as a bronchodilator, diuretic agent, and smooth muscle relaxant. It is more soluble in water than theophylline.

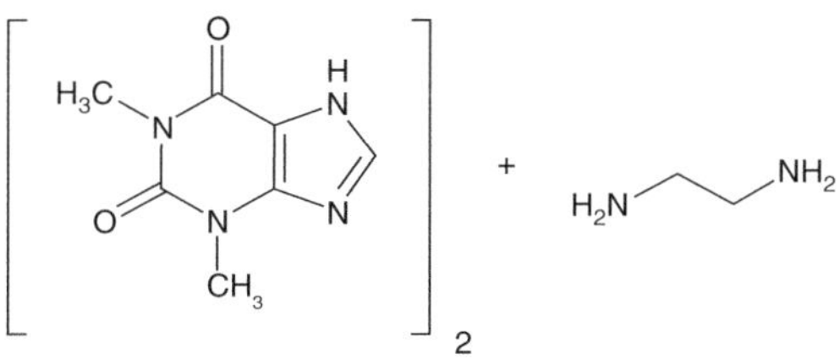

2. Doxofylline: 7-(1,3-dioxalan-2-ylmethyl) theophylline. It has antitussive and bronchodilator effects [3]. In animal and human studies, it has shown similar efficacy to theophylline but with fewer side effects [4]. Related research has showed that the effect of doxofylline on airway relaxation is 10–15 times that of aminophylline.

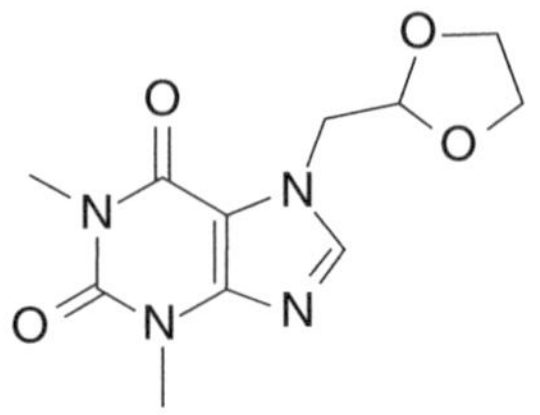

3. Diprophylline: 7-(2,3-dihydroxypropyl)-1,3-dimethyl-3,7-dihydro-1H-purine-2,6-dione. Diprophylline is the neutral preparation of theophylline. It causes less of nausea and gastric irritation.

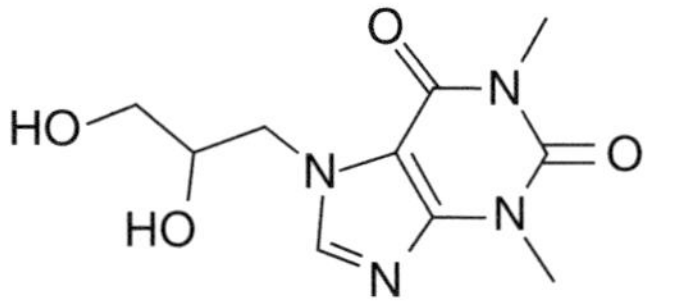

4. Oxtriphylline: choline theophyllinate; administered orally. Oxtriphylline is five times more soluble than aminophylline.

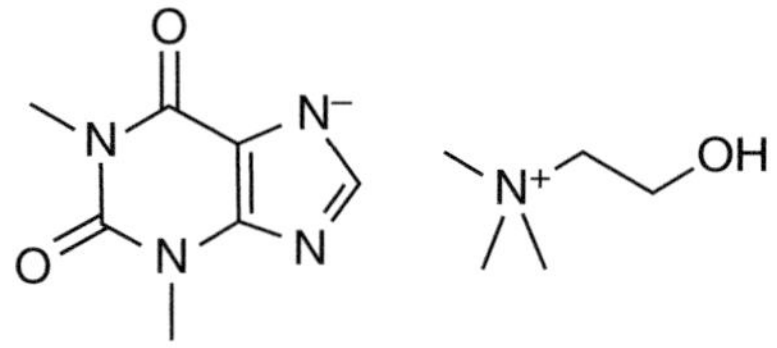

Pharmacology

Theophylline is a drug used for respiratory diseases such as chronic obstructive pulmonary disease (COPD) and asthma. It inhibits phosphodiesterase (PDE) [5] and thereby prevents the intracellular breakdown of cAMP. The increase of intracellular cAMP reduces smooth muscle tone resulting in the dilation of the airways. IL-10 release is also increased by theophylline, which may be mediated via PDE inhibition [6]. Theophylline prevents the nuclear translocation of the pro-inflammatory transcription factor nuclear factor-κB (NF-κB) [7]. Theophylline treatment results in an increase of activated circulating CD^{4+} and CD^{8+} T cells, but a decrease of these cells in the airways suggests that it may reduce the trafficking of activated T cells into the airways.

Theophylline sustained-release tablets: 0.1–0.2 g, bid, daily dose should not exceed 0.9 g. Theophylline sustained-release capsules: 0.2–0.3 g, bid. Aminophylline injection: aminophylline injection may be given by slow intravenous injection (0.125–0.25 g, bid) or intravenous infusion (0.2–0.5 g, bid) in glucose injection or sodium chloride injection.

Theophylline is rapidly and completely absorbed after oral administration with peak concentrations occurring within 1–2 h of a single dose. The bioavailability of theophylline is 96–100% in adult and 80% in newborns. Theophylline is distributed in extracellular fluid, placenta, mother's milk, and central nervous system. It is extensively (>70%) metabolized in the liver by N-demethylation to form 1,3-dimethyluric acid, 3-methylxanthine, and 1-methylxanthine.

The effect of theophylline is closely correlated with concentration of the drug in serum. The therapeutic range for theophylline is 10–20 μg/mL in adults. Toxic effects usually occur at concentrations above 20 μg/mL in adults, although mild symptoms can occur above 15 μg/mL. These effects include anorexia, nausea, vomiting, headaches, and nervousness. Severe side effects, such as increased cardiac rate, arrhythmia, cerebral seizures, and respiratory or cardiac arrest, usually occur at concentrations above 40 μg/mL but may also occur at lower concentrations.

Clinical Application

Theophylline is usually used as an add-on therapy in asthma and COPD patients. Side effects are related to plasma concentrations and include nausea, vomiting, and headaches due to PDE inhibition. Higher concentrations cause cardiac arrhythmias and seizures due to adenosine A1-receptor antagonism. Theophylline has been used much less in developed countries at present. However, low dosage of theophylline shows anti-inflammatory and immunomodulatory effect, which raises the reconsideration of it for the treatment of severe asthma, smoking asthma, and COPD [8].

Discussion

Theophylline has been used to treat airway diseases for over 70 years. It was originally used as a bronchodilator, but the required high dosages are associated with frequent side effects. With the wide application of inhaled β2-agonists, its clinical prescription declines. Compared with long-acting inhaled β2-agonists and leukotriene modifiers, slow-release theophylline preparations are much cheaper, which justifies the choice of it as an add-on therapy for asthma control. Low dosage of theophylline may find an important place in the management of patients with severe asthma and all patients with COPD.

Fig. 1 Tea

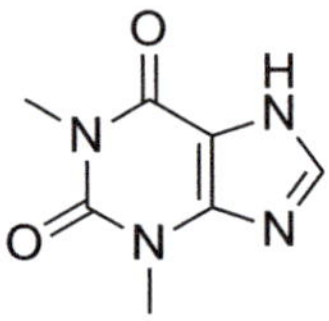

Fig. 2 Chemical structure of theophylline

References

1. Somerville LL. Theophylline revisited. Allergy Asthma Proc. 2001;22(6):347–51.
2. Schultze-Werninghaus G, Meier-Sydow J. The clinical and pharmacological history of theophylline: first report on the bronchospasmolytic action in man by s. R. Hirsch in Frankfurt (main) 1922. Clin Allergy. 1982;12(2):211–5.
3. Poggi R, Brandolese R, Bernasconi M, et al. Doxofylline and respiratory mechanics. Short-term effects in mechanically ventilated patients with airflow obstruction and respiratory failure. Chest. 1989;96(4):772–8.
4. Sankar J, Lodha R, Kabra SK. Doxofylline: the next generation methylxanthine. Indian J Pediatr. 2008;75(3):251–4.
5. Rabe KF, Magnussen H, Dent G. Theophylline and selective pde inhibitors as bronchodilators and smooth muscle relaxants. Eur Respir J. 1995;8(4):637–42.
6. Mascali JJ, Cvietusa P, Negri J, et al. Anti-inflammatory effects of theophylline: modulation of cytokine production. Ann Allergy Asthma Immunol. 1996;77(1):34–8.
7. Ichiyama T, Hasegawa S, Matsubara T, et al. Theophylline inhibits nf-kappa b activation and i kappa b alpha degradation in human pulmonary epithelial cells. Naunyn Schmiedeberg's Arch Pharmacol. 2001;364(6):558–61.
8. Pauwels RA, Joos GF. Characterization of the adenosine receptors in the airways. Arch Int Pharmacodyn Ther. 1995;329(1):151–60.

Part IV
Antitumor Drugs

Overview

In the history of human struggle against disease, malignant tumor is one of the most threatening diseases to human health. Especially with the deepening of modern medical knowledge of tumor and the lack of therapeutic treatments, tumor is not only a disease that destroys human health but also a disease that disrupts the patient's mental defense. It has become a common phenomenon for people to turn pale at the mention of cancer because of the disappointment with the effectiveness of cancer treatments.

The application of modern small molecule antitumor drugs began in the early twentieth century. War toxicants represented by alkylating agents were used in cancer treatment and achieved the effects of killing tumor cells. From then on, people began a long war to kill tumor. In the process, various antitumor drugs have been discovered. So far, antitumor drugs are still the focus of medical research, and they still target to kill tumor cells. This treatment pattern, practiced for nearly a century, has exposed problems. People began to explore new treatment modalities and drugs, such as immunotherapy drugs, whose appearance is an important progress in research and development of antitumor drugs.

In the process of developing antitumor drugs, people realized the severe adverse reactions of the war toxicant showed on patients. Then, they started to look for drugs with higher efficiency but lower adverse reactions. Natural products thus become an important resource for drug discovery. Through the active efforts of researchers, a batch of antitumor natural products have been applied in clinical trials, including small molecules found in plants and microorganisms, and the latter is commonly referred to as antineoplastic antibiotics (not discussed in this book). The application of these drugs has promoted the research and development of antitumor drugs and the treatment of malignant tumors.

After a long period of effort, a batch of antitumor drugs have been discovered from natural small molecule compounds, which also have achieved obvious therapeutic effects in clinical applications. The discovery and application of natural small molecule drugs not only provide effective therapeutic agents for clinical treatment but also promote the study of modern antitumor drugs. It is noteworthy that among the successfully developed antitumor natural small molecule drugs, most are from medicinal plants of Chinese medicine with a long history of application. It suggests that the therapeutic experience and extensive information accumulated in the long history of clinical application of traditional Chinese medicine have extremely important scientific values.

Harringtonine and homoharringtonine, clinically used in the treatment of acute nonlymphocytic leukemia and malignant lymphoma, are alkaloids extracted from *Cephalotaxaceae* plants *Cephalotaxus fortunei* and their congeneric plants. These compounds can inhibit the activity of DNA polymerase α and the synthesis of nucleoprotein, leading to dyssynthesis of DNA. They become typical cell cycle non-specific drugs for their strong killing effects on S phase cells and a certain influence on G_0 phase cells.

Vincaleukoblastinum and vincristine, extracted from the *Apocynaceae* plant *Catharanthus roseus* (*Vinca rosea*), can inhibit spindle formation and prevent mitosis. They are cell-cycle-specific drugs, acting mainly on proliferating tumor cells at M phase. Besides, they can also inhibit synthesis of RNA and lipids. Vincristine shows more curative effect than vincaleukoblastinum and has a wide spectrum of antitumor activity. Vincristine displays remarkable therapeutic effect in acute leukemia and malignant lymphoma and is also effective in other malignant tumors, such as chorionic epithelioma, breast cancer, nephroblastoma, neuroblastoma, Ewing's sarcoma, leiomyosarcoma, lung cancer, cervical carcinoma, and malignant melanoma.

Another cell-cycle-specific drug is colchicine, an alkaloid extracted from *Colchicum autumnale* and *Iphigenia indica* Kunth. Colchicine may block cell mitosis at M phase and eventually lead to cell death. The most vigorously dividing cells are the first to be affected by colchicine. This medicine may cause many adverse reactions, and it also has diverse functions in nervous system. Colchicine can be used in skin cancer, breast cancer, cervical carcinoma, leukemia, Hodgkin's disease, and many other kinds of cancers. Colchicine is also an effective drug to prevent and control acute gout attack at present.

Paclitaxel is an antineoplastic drug and has wide clinical applications. Possessing a unique action mechanism, paclitaxel is a novel antineoplastic drug that affects the function of microtubules. Paclitaxel can promote tubulin dimer to assemble into microtubules and then stabilize microtubules by inhibiting the depolymerization process, thus inhibiting normal dynamic reorganization of microtubule network and interfering with cell cycle. Paclitaxel has significant radiosensitizing effects, which may be related to the action of paclitaxel to interrupt cell activity at the stage of radiation-sensitive phases G_2 and M.

Camptothecin and hydroxycamptothecin, alkaloids extracted from *Camptotheca acuminata*, are cell-cycle-specific antineoplastic drugs. They mainly act on the S phase of cell cycle and can inhibit DNA topoisomerase I, thereby inhibiting DNA replication and transcription. They are suitable for primary liver cancer, gastric cancer, bladder cancer, rectal cancer, non-small cell lung cancer, head and neck epithelial cancer, leukemia, and other malignant tumors.

Seen from the aforementioned drugs with distinctive features, natural small molecule drugs show obvious inhibitory and killing effects on many kinds of tumors and have become important antineoplastic drugs in clinical applications. Meanwhile, the discovery and research, especially the research on action mechanism and chemical structure modifications of these drugs, deepen our understanding about tumor nosogenesis and mechanism of antitumor drug action and also promote the research and development of antitumor drugs.

In addition to the few drugs listed above, in clinical applications, there are many antitumor drugs derived from natural products. They also play unique roles, such as indirubin in the treatment of leukemia and podophyllotoxin in the inhibition of cell mitosis. Both of them have played positive roles in clinical treatments.

It is particularly noteworthy that the emergence of a number of new antitumor natural drugs will provide new insights into cancer therapy. For example, the clinical application of ginsenoside Rg3 in the treatment of malignant tumors is of great significance for the development of new antitumor drugs and the exploration of new therapies for malignant tumors. Today, malignant tumor is still a major disease which seriously threatens human health and life. It is a long-term task to study new drugs. The research and development of natural products still have a bright future.

Anethole

Bao-Xi Zhang, Li-Da Du, and Yang Lu

Abstract Anethole was obtained from dried and ripened fruits of *Illicium verum* Hook. f. or *Foeniculum vulgare* Mill. *Illicium verum* possesses a variety of pharmacological effects, such as increasing the white blood cellular activity, bacteriostatic effect, antiviral effects, and so on. With the deepening research, anethole will make a greater contribution to human health.

Keywords Anethole · Seasonings · Antiviral effects · Bacteriostatic effect

Alias: Para-methoxyphenylpropene
Origin: *Illicium verum* (Fig. 1)
Chemical name (Fig. 2)

1-Methoxy-4-(1-propenyl)benzene

Molecular formula, $C_{10}H_{12}O$; **MW**, 148.2; **CAS**, 104-46-1

Properties

Appearance: this compound shows a colorless or light-yellow liquid appearance. *Melting point*: 20–21 °C. *Solubility*: dissolve in chloroform and ether in unlimited amount; soluble in benzene, ethyl acetate, acetone, carbon disulfide, petroleum ether, and alcohol; insoluble in water. It has a sweet smell.

B.-X. Zhang · Y. Lu (✉)
Beijing Key Laboratory of Polymorphic Drugs, Research Center of Polymorphic Drugs, Institute of Materia Medica, Chinese Academy of Medical Sciences and Peking Union Medical College, Beijing, China
e-mail: luy@imm.ac.cn

L.-D. Du
School of Biomedical Sciences, Chinese University of Hong Kong, Hong Kong, China

G.-H. Du, *Natural Small Molecule Drugs from Plants*,
https://doi.org/10.1007/978-981-10-8022-7_79

Dosage Forms and Indications

It is included in the *British Pharmacopoeia* (2017), *United States Pharmacopeia* (40), and *European Pharmacopoeia* (8.7th ed.). The available formulation is capsule.

It is used for treating leukocytopenia due to tumor chemotherapy, radiotherapy, and other causes.

Literature

Anethole was obtained from dried and ripened fruits of *Illicium verum* Hook. f. or *Foeniculum vulgare* Mill. *Illicium verum* also known as star anise is mainly distributed in Guangxi, Guangdong, and Yunnan provinces in China. It can be used to treat cold hernia, abdominal pain, lumbago, etc. [1]. Its dry and mature fruit contains 5–8% of aromatic oil and 22% of fat oil, as well as protein, resin, etc. This plant has been widely used in food and spice industry and traditional Chinese medicine [2].

Fennel is an annual or perennial herb with a strong smell and is originally grown in the Mediterranean region and also in most provinces of China [3].

History of R&D

In China, anise has been used as a traditional food spice and seasoning. In the anise plant, the main biological active ingredient is volatile oils such as trans-anethole. Till the end of last century, several methods to obtain the pure anethole have been developed, including:

1. Cooling, crystallizing, and recrystallizing after the distillation of the anise oil.
2. Heating p-methoxyphenyl crotonic acid at 220–240 °C.
3. Heating and dehydrating the derived product of anisaldehyde and C_2H_5MgX.
4. Heating anisaldehyde together with propionic anhydride and sodium propionate.
5. Adding concentrated hydrochloric acid and phosphoric acid to the mixture of anisole and propionaldehyde at 0 °C and then heating the product with pyridine to remove hydrogen chloride.
6. Prepare Grignard reagent using parabromoanisole, then react with allyl bromide to produce p-methoxyphenylpropylene, then heat with potassium hydroxide, and finally anethole was obtained after isomerization.
7. Using crystalline ferric chloride to catalyze the reaction of p-propenyl phenol and methanol [4].

Anethole is easily oxidized when exposed in the air, especially in the presence of heat, light, or catalyst [5]. Therefore, in recent years, a series of studies have been carried out on the synthesis of its derivatives in order to obtain more active substances.

Pharmacology

Anethole is the main ingredient in star anise oil and possesses a variety of pharmacological effects.

1. Increasing the white blood cellular activity. Some agents such as Shengbaining and Shengxuening, whose main active ingredients were extracted from the star anise, can promote mature white blood cells in the bone marrow to spread into the surrounding blood. Due to the body's own feedback, mature and release speed of bone marrow cells were accelerated. It can also keep bone marrow cells' activity, enhancing the white blood cells (especially granulocytes).
2. Bacteriostatic effect. Star anise oil shows antibacterial effects in a variety of strains [6] including *Staphylococcus aureus*, *Escherichia coli*, *Bacillus subtilis*, *Aspergillus niger*, *Aspergillus flavus*, *Penicillium citrinum*, yeast, *Shigella*, diphtheria bacillus, and *Salmonella typhi*. The results laid a theoretical foundation for the development and utilization of novel plant-derived antifungal propenylbenzene derivatives.
3. Antiviral effects [7]. Star anise oil can act on different acyclovir-susceptible and acyclovir-resistant herpes simplex virus type 1 (HSV-1) strains.
4. Other effects. Anethole can improve the activity of anticholinesterase [8]. It shows significant inhibitory effect against acetylcholinesterase and butyrylcholinesterase, and the IC_{50} value was 39.89 ± 0.32 μg/mL and 75.35 ± 1.47 μg/mL, while the value for star anise oil was 36.00 ± 0.44 μg/mL and 70.65 ± 0.96 μg/mL, respectively. It has also been reported that anethole has an antioxidant effect.

Through structural transformation, a variety of compounds can be prepared from anethole. Anisaldehyde, produced by the oxidation reaction, was widely used in the preparation of flavors for its lasting aroma. On the other hand, it can also be used as the intermediate during the preparation of drugs such as amoxicillin.

Clinical Application

As a drug, anethole is mainly used for leukopenia caused by tumor chemotherapy.

Discussion

With the deepening research, especially applications in the field of pharmaceuticals, high-end spices, and daily necessities, the demand for high-purity anethole is huge, and the market prospects are very broad. Anethole will make a greater contribution to human health.

Fig. 1 *Illicium verum*

Fig. 2 Molecular structures of anethole

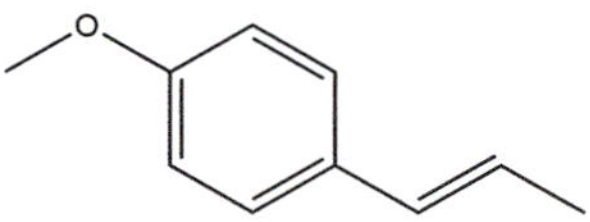

References

1. Chinese Pharmacopoeia Commission. Pharmacopoeia of the People of Republic of China, vol. I. Beijing: China Medical Science and Technology Press; 2015.
2. Lian JH, Sun GS. Research progress of star anise. Technol Dev Chem Ind. 2010;39(3):31–3.
3. Editorial Committee of flora of China. Flora of China, vol. 55. Beijing: Science Press; 1985. p. 213–4.
4. Dong Y, Zhu YJ. Synthesis of anethole with activated carbon supported ferric chloride as catalyst. Guangdong Chem Ind. 1997;5:27–8.
5. Zanardi J, Leriverend C, Aubert D, et al. A catalytic cycle for the asymmetric synthesis of epoxides using sulfur ylides. J Org Chem. 2001;66:5620–3.
6. Fu MD, Li CH. Extraction progress of star anise oil and its antibacterial and antioxidative effects. China Med Her. 2011;8(34):29.
7. Koch C, Reichling J, Kehm R, et al. Efficacy of anise oil, dwarf-pine oil and chamomile oil against thymidine-kinase-positive and thymidine-kinase-negative herpesviruses. J Pharm Pharmacol. 2008;60(11):1545.
8. Bhadra S, Mukherjee PK, Kumar NS, et al. Anticholinesterase activity of standardized extract of Illicium verum Hook. f. fruits. Fitoterapia. 2011;82(3):342.

Berbamine

Xiu-Yun Song, Ling-Lei Kong, and Nai-Hong Chen

Abstract Berbamine is an alkaloid, separated from Berberidaceae, Menispermaceae or Ranunculaceae plants. Nowadays, it mainly comes from chemical synthesis. The current clinical formulation is berbamine hydrochloride tablets. It has wide pharmacological effects, low adverse reactions, and low long-term toxicity. It is used in the prevention and treatment of patients with cancer, which is due to radiotherapy- and chemotherapy-induced neutropenia or due to benzene poisoning, radioactive substances, and drug-induced neutropenia. However, the other effects, mechanisms and clinical applications of barbamin and its derivatives, especially the using in tumor and diabetic antaract, need further research.

Keywords *Berberis* · Alkaloid · Neutropenia

Origin: *Berberis* poiretii Schneid (Fig. 1)
Chemical name (Fig. 2)

6, 6′, 7-trimethoxy-2, 2′-dimethylberbaman-12-ol

Molecular formula, $C_{37}H_{40}N_2O_6$; **MW**, 608.71; **CAS**, 478-61-5

X.-Y. Song · N.-H. Chen (✉)
Institute of Materia Medica, Chinese Academy of Medical Sciences and Peking Union Medical College, Beijing, China
e-mail: chennh@imm.ac.cn

L.-L. Kong
Beijing Key Laboratory of Drug Targets Identification and Drug Screening, Centre for Pharmaceutical Screening, Institute of Materia Medica, Chinese Academy of Medical Sciences and Peking Union Medical College, Beijing, China

G.-H. Du, *Natural Small Molecule Drugs from Plants*,
https://doi.org/10.1007/978-981-10-8022-7_80

Properties

Appearance: odorless and bitter white crystalline, powder, belongs to bisbenzylisoquinoline alkaloid. *Melting point*: 197–210 °C in petroleum ether. *Solubility*: easily soluble in dilute hydrochloric acid, sulfuric acid, ethanol, chloroform, and acetone; soluble in ether; slightly soluble in boiling water and petroleum ether; but insoluble in cold water, ammonia, sodium carbonate, and calcium hydroxide solution.

Dosage Forms and Indications

Berbamine is contained in the 13th volume of the book, *National Standards for Chemical Drugs*.

The current clinical dosage is berbamine hydrochloride tablets. It is used in the prevention and treatment of patients with cancer, which is due to radiotherapy- and chemotherapy-induced neutropenia or due to benzene poisoning, radioactive substances, and drug-induced neutropenia.

Literature

Berbamine is an alkaloid, separated from Berberidaceae, Menispermaceae, or Ranunculaceae plants. Most of Berberidaceae plants have medicinal value, including their root, root bark, and stem. The plants of *Berberis* L. and *Mahonia* Nutt. contain many alkaloids, the most important of which are berberine and a small amount of palmatine, jatrorrhizine, berbamine, oxyacanthine, and so on. These compounds have many biological activities and have been used to replace goldthread and golden cypress. Also they are well succedaneums of berberine in Chinese and Western medicine. Traditional Chinese medicine Oregon grape root has different therapeutic effects for dysentery, enteritis, children pneumonia, and chronic bronchitis. The major treatment diseases of Chinese mahonia are bacillary dysentery, acute gastritis, infectious hepatitis, pneumonia, tuberculosis, bronchitis, sore throat, and so on.

History of R&D

Berbamine is an alkaloid mainly from *Berberis julianae*, *Berberis wilsonae*, *Berberis poiretii*, and huanglumu, which belong to *Berberidaceae berberis* and *Mahonia*. Nowadays, it mainly comes from chemical synthesis. Berbamine hydrochloride is contained in the 13th volume of the book, *Hua xue yao ping di biao sheng guo biao*. As early as 1940, the United States scholars had obtained berbamine from *Mahonia* [1]. Chinese scholars began the research of berbamine extraction and pharmacological activities since the 1980s in the last century [2].

Although the berbamine study in our country later than the west, the traditional Chinese medicine *Berberi's* medicinal records can be traced back to the Tang Dynasty. The book *Tang Ben Cao* recorded that Oregon grape root could be a therapy for aphtha, kill insects, and remove hot gas in the body. In the Ming Dynasty the book *Ben Cao Gang Mu* recorded that Oregon grape root could treat metrorrhagia. Now, berbamine is widely used for leukopenia.

Early research mainly focused on the affects and clinical effects on the immune function. It is found that berbamine can reduce the antigen-presenting cell stimulation effect, resulting in immune regulation effect [3]. Further study found that berbamine and its derivatives are calmodulin antagonists, which have the effects of antitumor, antioxidation, and so on. In recent years, many derivatives of berbamine have been synthesized. Researchers showed that both the ether, esterification of berbamine, and the addition of benzene or hydroxyl to berbamine can enhance the cytotoxicity of berbamine. The results in vitro confirmed that their effect was many times to berbamine [4].

Pharmacology

Berbamine can promote hematopoietic function, stimulate myeloid proliferation.

Also berbamine can enhance the content of hematopoietic stem cell and granulocyte-colony stimulating factor (G-CSF), then increase the proliferation and differentiation of bone marrow hematopoietic stem cells and granule progenitor cell, which increase the number of peripheral blood leukocytes. Preclinical and clinical studies showed that berbamine was good for the treatment of leukopenia induced by chemotherapy. Berbamine can inhibit the proliferation of many kinds of tumor cells, such as the human high metastatic breast cancer cells MDA-MB-231, non-small cell lung cancer cells A549, liver cancer cells SMMC772l and HepG2, multiple myeloma cells KM3 and RPMl8226, chronic myeloid leukemia (CML) cells KU812, and imatinib-resistant K562 cells [5]. The mechanisms of berbamine may be through blocking the cell cycle, inducing cell apoptosis, inhibiting tumor cell invasion and metastasis, reversing MDR, affecting tumor signal transduction, and so on.

In addition, berbamine also has the pharmacological effects in antihypertension, anti-arrhythmia, anti-myocardial ischemia and anti-silicosis.

Clinical Application

Berbamine is used for the prevention and treatment of cancer patients with neutropenia due to radiotherapy, chemotherapy [6, 7], benzene poisoning, radioactive substances, and drug.

Discussion

Berbamine is a bisbenzylisoquinoline alkaloid separated from the plants of *Berberis* and *Mahonia*. It has wide pharmacological effects, low adverse reactions, and low long-term toxicity. Clinical studies show that berbamine can stimulate the formation of hematopoietic cells, improve the patient's response to imatinib, and promote neutrophil counts returned to normal for patients of chronic myeloid leukemia with imatinib-induced neutropenia [8]. At present berbamine is mainly using as leukogenic drug in clinic. However, the other effects of berbamine and its derivatives, especially as a new calmodulin antagonist used in antitumor therapy, the prevention and therapy of diabetic cataract, and the mechanisms and clinical application need further research.

Fig. 1 *Berberis* poiretii Schneid

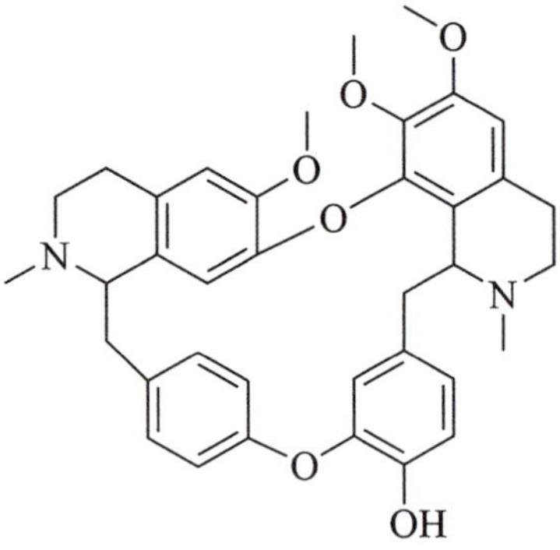

Fig. 2 The chemical structure of berbamine

References

1. Greathouse G, Rigler N. Isolation of the alkaloids, berberine and berbamine, from Mahonia Swaseyi. Plant Phydiol. 1940;15(3):563–4.
2. Liu G, Chen B, Song W, Xiao P. The comprehensive utilization of Berberis poiretii schneid – the medicinal value of berberine and berberine's content in 22 kinds of Berberis. Chin Bull Bot. 1978;20(3):255–7.
3. Ren Y, Lu L, Guo TB, Qiu J, Yang Y, Liu A, Zhang JZ. Novel immunomodulatory properties of berbamine through selective down-regulation of STAT4 and action of IFN-gamma in experimental autoimmune encephalomyelitis. J Immunol. 2008;181:1491–8.
4. Nam S, Xie J, Perkins A, Ma Y, Yang F, Wu J, Wang Y, Xu RZ, Huang W, Horne DA, Jove R. Novel synthetic derivatives of the natural product berbamine inhibit Jak2/Stat3 signaling and induce apoptosis of human melanoma cells. Mol Oncol. 2012;6(5):484–93.
5. Wang J, Feng Y, Hu Y, Zhang G, Hu R. Research progress on antitumor effect and mechanism of berbamine and its derivatives. Tianjin Med J. 2012;40(12):1273–5.
6. Lin C, Huang X. The clinical observation of Shengbai amine (berbamine) on the treatment of leukopenia induced by chemotherapy. Tradit Chin Med Patent Prescription. 1996;07:29.
7. Kapoor S. Emerging role of berbamine as an anti-cancer. Agent in systemic malignancies besides chronic myeloid leukemia. J Zhejiang Univ Sci B. 2012;13(9):761–2.
8. Zhao Y, Tan Y, Wu G, Liu L, Wang Y, Luo Y, Shi J, Huang H. Berbamine overcomes imatinib-induced neutropenia and permits cytogenetic responses in Chinese patients with chronic-phase chronic myeloicl leukemia. Int J Hematol. 2011;94(2):156–62.

Camptothecin

De Kang, Ai-Lin Liu, Jin-Hua Wang, and Guan-Hua Du

Abstract As an alkaloid derived from *Camptotheca acuminata*, camptothecin exerted its anticancer activity by specifically targeted topoisomerase I. The intolerable toxicity and poor solubility and stability in vivo of camptothecin limited its clinical application, but the sequential approval of camptothecin derivatives has provided patients with a new selection for the treatment of cancer. Camptothecin is often used in combination with other drugs in clinical application. As one of the most important anticancer drugs derived from plants, camptothecin played an important role in the treatment of cancer with its unique anticancer mechanism and therapeutic effect.

Keywords Camptothecin · Alkaloid · Topotecan · Cancer

Origin: *Camptotheca acuminata* (Fig. 1)
Chemical name (Fig. 2)

4-Ethyl-4-hydroxy-1H-pyrano-[3′,4′:6,7]indolizino[1,2-b]quinoline-3,14(4H,12H)-dione

Molecular formula, $C_{20}H_{16}N_2O_4$; **MW**, 348.35; **CAS**, 7689-03-4

D. Kang · A.-L. Liu · J.-H. Wang · G.-H. Du (✉)
Beijing Key Laboratory of Drug Targets Identification and Drug Screening, Centre for Pharmaceutical Screening, Institute of Materia Medica, Chinese Academy of Medical Sciences and Peking Union Medical College, Beijing, China
e-mail: dugh@imm.ac.cn

G.-H. Du, *Natural Small Molecule Drugs from Plants*,
https://doi.org/10.1007/978-981-10-8022-7_81

Properties

Appearance: pale yellow needlelike crystal. *Solubility*: slightly soluble in ethanol and chloroform; poorly soluble in water; camptothecin fails to generate stable salt with acid, whereas it can produce sodium salt which is soluble in water by reacting with heated sodium hydroxide solution. *Melting point*: 264–267 °C.

Camptothecin derivatives (Fig. 3)

Dosage Forms

Camptothecin is recorded in the first edition of the national standards of chemical drugs. At present, several derivatives of camptothecin are used in clinical setting for treatment of malignant tumors, with the main dosage forms being hydroxylcamptothecine injection, irinotecan hydrochloride injection, and topotecan hydrochloride injection.

Indications

It is mainly used in digestive tract tumors and has a good effect on gastric cancer, rectal cancer, and colon cancer. Besides, it can improve the surgical resection of advanced gastric cancer and also has some therapeutic effect on bladder cancer and lung adenocarcinoma. Moreover, camptothecin can be used for treatment of psoriasis, warts, acute and chronic leukemia, and hepatosplenomegaly caused by schistosomiasis.

Literature

Camptothecin is an alkaloid derived from Xi Shu (*Camptotheca acuminata*), which belongs to Nyssaceae. The traditional Chinese medicine *Camptotheca acuminata* (Xi Shu) has been collected in the Compilation of Chinese Herbal Medicine, *Chinese Materia Medica*, and Great Dictionary of Chinese Medicine. *Camptotheca acuminata* (Xi Shu) is widely distributed in the basin of Yangtze river and the southwestern provinces. The main medicinal parts of *Camptotheca acuminata* (Xi Shu) are root bark and fruit, which get rid of heat and toxic materials and eliminate the disease.

History of R&D

In 1966, Wall M E et al. [1] from the United States isolated an alkaloid from *Camptotheca acuminata* and defined its chemical structure. The in vitro anticancer tests revealed the anticancer activity of the tryptophan-terpene alkaloid, which is known as camptothecin and received widely concern. At the beginning of the 1970s, the experimental treatment of human gastric cancer was carried out, and the symptoms of some patients were relieved. However, due to the toxicity and the intolerable side effects such as nausea and vomiting, as well as the reduction anticancer activity of its water-soluble sodium salt, the study of camptothecin has entered a low ebb stage.

The chemical composition of *Camptotheca acuminata* is very complex, but the main active ingredients are alkaloids. So far, various alkaloids like camptothecin, 10-hydroxycamptothecin, methoxy-camptothecin, and venoterpine have been isolated from *Camptotheca acuminata*. In addition, the camptothecin and methoxy-camptothecin have also been found in some other plants, such as Apocynaceae *Heynea Ervatamia*, Icacinaceae smelly *Nothapodytes obtusifolia*, and Rubiaceae *Ophiorrhiza japonica*.

The natural resources of *Camptotheca acuminata* are very limited, and the content of camptothecin is very low, which limited the extraction and application of camptothecin. In 1975, Corey et al. first opened the door for the chiral synthesis of camptothecin, but the reaction step was long and the yield rate was very low [2]. It was not until 1997 that Ciufolini et al. developed a new method for the synthesis of camptothecin by five steps, with a total yield rate up to 51%. The great breakthrough in the chemical synthesis of camptothecin has made its extensive application become a reality [3].

A study in the 1970s found that the synthesis of DNA and RNA in mammalian cells can be inhibited by camptothecin, whereas the removal of the alkaloid could restore the DNA and RNA synthesis function. It is speculated that camptothecin may exert a direct effect on the S phase of cell replication.

In 1985, Hsiang et al. found that CPT can directly inhibit topoisomerase I (topo I), which is an enzyme involved in DNA replication and transcription. Topoisomerase I was closely related to cell division, and blocking its enzyme activity can inhibit growth of cancer cell [4]. The anticancer mechanism of camptothecin by targeting topoisomerase I has created a new breakthrough point and led to a new upsurge in the study of camptothecin.

Hydroxycamptothecin, as a camptothecin derivative with a hydroxyl group on the tenth carbon atom, is widely used for the treatment of various cancers. In 1969, researchers from Shanghai Institute of Materia Medica found that hydroxycamptothecin possessed potent anticancer activity and low toxicity. And this finding promoted the production and clinical application of hydroxycamptothecin, but its usage was interrupted for technology and quality [5]. In the 1980s, hydroxycamptothecin was reproduced for clinical application with an improvement in producing technology, and hydroxycamptothecin got its approval number in 1986 for clinical usage in China. In the 1990s, the US Food and Drug Administration approved the clinical application of topotecan and irinotecan, which played a significant role in the prevention and treatment of cancers [6].

Pharmacology

The pharmacology of camptothecin was mainly manifested as antitumor activity. Camptothecin specifically targeted topoisomerase I and exerted anticancer activity by inhibiting the synthesis of DNA. Camptothecin mainly influenced the S phase of cell cycle and was considered as a specific inhibitor agent of cell cycle. The results of animal experiments showed that camptothecin had some inhibitory effects on leukemia, Yoshida sarcoma, and Ehrlich ascites carcinoma.

Previous clinical trials showed that camptothecin and its analogs have therapeutic effects on bladder cancer, brain cancer, breast cancer, cervical cancer, colon cancer, neural stromal tumor, lymphoreticulosis, lung cancer, leukemia, lymphoma, melanoma, ovarian cancer, pancreatic cancer, pediatric cancer, prostate cancer, and liver cancer [7].

Injection of camptothecin (2.5 mg/ml, 5–10 mg/day) with a treatment course of 140 mg achieved effective rate of 44.8% and 38.3% for gastric cancer and colon cancer, respectively. Hydroxycamptothecin can be used for the prevention and treatment of gastric, liver, head, and neck cancer and leukemia, and the effective rate is 44% [6]. In addition, the dimethyl sulfoxide solution of camptothecin was also successfully used for treatment of psoriasis.

Clinical Application

Because of the toxicity and side effects of camptothecin, the currently used agents in clinical applications are camptothecin derivatives like topotecan, irinotecan, and hydroxycamptothecin. Topotecan, a water-soluble camptothecin derivative developed by SmithKline Beecham, was approved by FDA in 1996 for the treatment of ovarian cancer. As another water-soluble camptothecin derivative approved by FDA in 1996, irinotecan was mainly used in the treatment of advanced colorectal cancer. In addition, it was also shown to have obvious inhibitory effect on small cell lung cancer and leukemia [8]. Hydroxycamptothecin possesses a broad-spectrum antitumor activity and was clinically used for intravesical therapy of bladder cancer. In addition, it has remarkable curative effect on colon cancer, breast cancer, gastric cancer, and leukemia.

Discussion

With an increasing morbidity and mortality, cancer becomes one of the most serious diseases threatening human health. Camptothecin has been shown to exert anticancer effect by targeting topoisomerase I, thus providing a new thinking for cancer therapy. The intolerable toxicity and poor solubility and stability in vivo of camptothecin limited its clinical application, but the sequential approval of camptothecin derivatives has provided patients with a new selection for the treatment of cancer. There is no obvious cross drug resistance between camptothecin and other most commonly used anticancer drugs. Camptothecin is often used in combination with other drugs in clinical application. As one of the most important anticancer drugs derived from plants, camptothecin played an important role in the treatment of cancer with its unique anticancer mechanism and therapeutic effect.

Fig. 1 *Camptotheca acuminata*

Fig. 2 Chemical structure of camptothecin

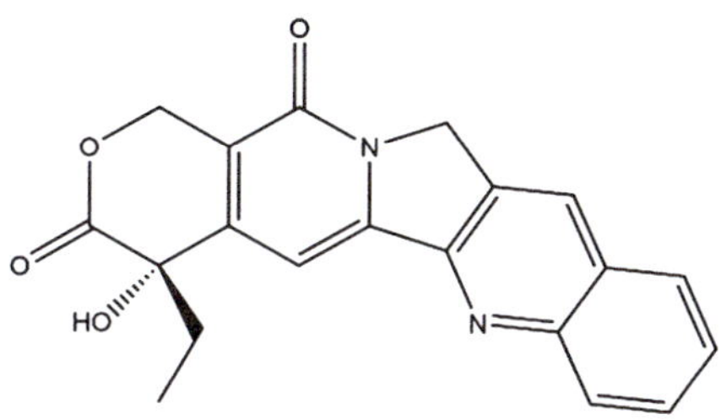

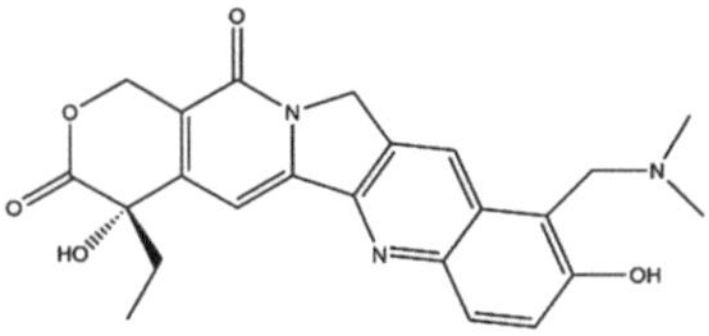

Topotecan

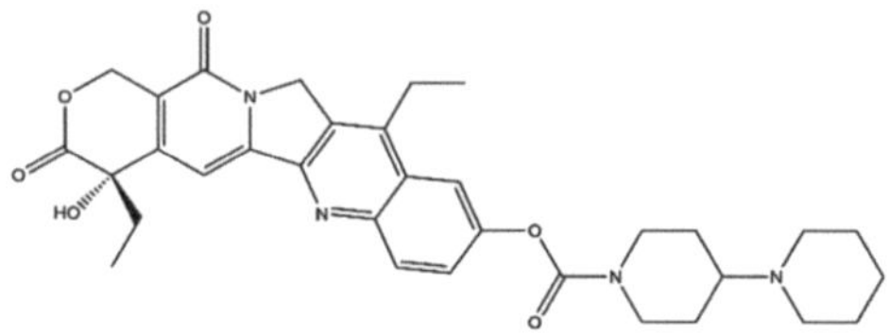

Irinotecan

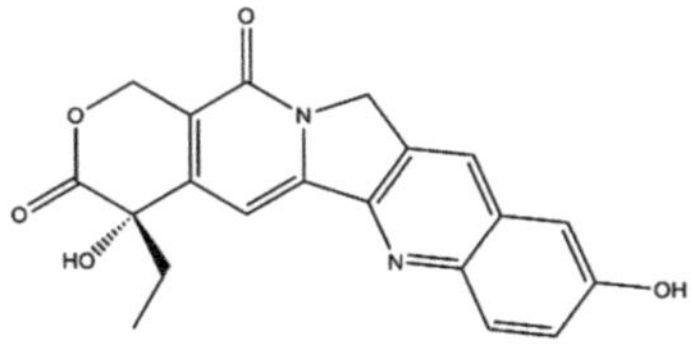

Hydroxycamptothecin

Fig. 3 Chemical structure of camptothecin derivatives

References

1. Wall ME, Wani MC, Cook CE, Antitumor agentI. The isolation and structure of camptothecin, a novel alkaloidal leukemia and tumor inhibitor from camptotheca acuminate. Am Chem Soc. 1966;88(16):3888–90.
2. Corey EJ, Crouse DN, Anderson JE. Total synthesis of natural 20(S)-camptothecin. J Org Chem. 1975;40(20):2140–1.
3. Ciufolini MA. Practical total synthesis of (+)-camptothecin: the full story. Tetrahedron. 1997;53(32):11049.
4. Hsiang YH, Hertzberg R, Hecht S. Camptothecin induces protein-linked DNA breaks via man malian DNA topoisomerase I. Biol Chem. 1985;260(27):14873–8.
5. Chinese academy of sciences, Shanghai Institute of Materia Medica. Study on antitumor effect of 10-hydroxycamptothecin. Chin Med J. 1978;6(3):42–8.
6. Peng G. Research progress of camptothecin as anticancer drugs. Acta Acad Med CPAF. 2001;10(3):255–8.
7. Jiancan F. Development and utilization of Camptotheca acuminata and Camptothecin. For Sci. 2000;36(5):100–8.
8. Xiaoyang F. Retrospection and rethinking on forty years development of camptothecin drugs. Med Philos. 2005;26(1):30–1.

Cepharanthine

Shi-Feng Chu, Ling-Lei Kong, and Nai-Hong Chen

Abstract Cepharanthine is diclofenac quinoline alkaloid isolated from rhizome of *Stephania japonica*. It could stimulate the reticuloendothelial system, then activate hematopoietic tissue, and promote myeloproliferation. The major clinical application of cepharanthine is to improve their immunity impaired by chemotherapy or radiotherapy in tumor patients. In addition, cepharanthine can increase the sensitivity of antitumor drugs. Now, it is commonly used as an adjuvant drug for anticancer treatment. In addition, it is also used as anti-pneumoconiosis drugs in China.

Keywords Cepharanthine · Myeloproliferation · Tumor · Pneumoconiosis

Alias: *Stephanotis*, metaphanine
Origin: *Stephania japonica* (Fig. 1)
Chemical name (Fig. 2)

6′,12′- dimethoxy- 2,2′- dimethyl-6,7- (methylenebis (oxy)oxyacanthan)

Molecular formula, $C_{37}H_{38}N_5O_6$; **MW**, 606.71; **CAS**, 481-49-2
Derivatives (Fig. 3)

S.-F. Chu · N.-H. Chen (✉)
State Key Laboratory of Bioactive Substances and Functions of Natural Medicines, Institute of Materia Medica & Neuroscience Center, Chinese Academy of Medical Sciences and Peking Union Medical College, Beijing, China
e-mail: chennh@imm.ac.cn

L.-L. Kong
Beijing Key Laboratory of Drug Targets Identification and Drug Screening, Centre for Pharmaceutical Screening, Institute of Materia Medica, Chinese Academy of Medical Sciences and Peking Union Medical College, Beijing, China

G.-H. Du, *Natural Small Molecule Drugs from Plants*,
https://doi.org/10.1007/978-981-10-8022-7_82

Properties

Appearance: light yellow or yellow powder. *Solubility*: soluble in acidic aqueous solution and ether, acetone, and other organic solvents; insoluble in petroleum ether. *Melting point*: 148–150 °C. *Specific optical rotation*: +277°.

Dosage Forms and Indications

Cepharanthine is recorded in the pediatric volumes for oral carcinoma of the national standard of proprietary Chinese medicine. Now, it is used in tablet in clinic with 20 mg for the treatment of agranulocytosis and leukopenia induced by chemotherapy, radiotherapy, and so on. In addition, it can also be used for prevention and treatment of silicosis and pneumoconiosis in coal workers.

Literature

Cepharanthine is diclofenac quinoline alkaloid isolated from rhizome of *Stephania japonica*, [1] which was first recorded in "Bencao shiyi" and functions as clearing heat, promoting diuresis and detumescence in traditional Chinese medicine. In this chapter, *Stephania japonica* was used to treat tuberculosis and other consumptive diseases which provided a clue for its development on lung disease [2].

History of R&D

Promoting leukocytosis of cepharanthine in tuberculosis patients was first recorded in the proceeding Institute of Chemotherapy by the Japanese scholar Yamaguchi in 1946. In 1948, Chinese scholar Yuhuang Zhao also received this compound and published it in D.M. Med. In 1955, the Japanese scholar reported that daily administration of cepharanthine (3–5 mg) relieved leukopenia in a tumor patient with radiotherapy; White blood cells showed a faster recovery and persistent efficacy in 5 mg group than that in 3 mg group. Furthermore, this effect was also found in X-ray staff.

It is thought that cepharanthine may stimulate the reticuloendothelial system, then activate hematopoietic tissue, and promote myeloproliferation. Therefore, the number of white blood cells in the peripheral blood increased significantly. Due to its good safety, cepharanthine is currently widely used in cancer patients with granulocytopenia or leukopenia embolism after radiotherapy and chemotherapy and so on.

In addition, Chinese scientists extracted five kinds of dibenzylisoquinoline alkaloids from Chinese herbal medicine in 1977, which could significantly improve the silicosis symptoms in rat [3]. In 1993, it was issued as anti-pneumoconiosis drug by CFDA and is used for delaying the progress of pneumoconiosis in clinic.

Pharmacology

The pharmacological effects of cepharanthine are very broad, mainly for the promotion of leukocytosis, antitumor, anti-inflammation, improvement of the body immunity, and other aspects of efficacy. Improvement of leukocytosis is the most widely studied among them, and the mechanism may be to stimulate the reticular endothelium system, activate hematopoietic tissue, and promote bone marrow hyperplasia [4].

The effective dose of cepharanthine is 20 mg three times a day by oral administration. It was also confirmed that cepharanthine has a certain antitumor effect when raising white blood cells. Oral administration of 20 mg/kg can significantly inhibit the deterioration and metastasis of Lewis lung cancer. In addition, the growth of Ehrlich tumor is also inhibited significantly by cepharanthine [3].

Cepharanthine is also a good antitumor drug sensitizer. It can significantly improve their efficacy and reduce their side effects when combined with other antitumor drugs, as manifested by increasing the concentration of FT-207 metabolite (5-Fu) in tumor tissue under cepharanthine and FT-207 combination therapy, which was significantly higher than that in blood [5], suggesting that cepharanthine has the potential to improve the efficacy of antitumor drugs.

Clinical Application

The major clinical application of cepharanthine is to improve their immunity impaired by chemotherapy or radiotherapy in tumor patients. In addition, cepharanthine can increase the sensitivity to antitumor drugs, and itself also has a certain antitumor effect. Now, it is commonly used as adjuvant drug for anticancer treatment.

Discussion

Cephalosporins, extracted from natural product, significantly increase the efficacy of white blood cells in patients with tuberculosis. The following study found that cepharanthine can activate the reticuloendothelial system and hematopoietic tissue, promote bone marrow hyperplasia, increase the white blood cells, and maintain the number of white blood cells in the normal range, which is to benefit for the improvement of the immunity of cancer patients and reduce the risk of infection.

Clinical reports of cepharanthine can be used for prevention and treatment in silicosis and coal workers' pneumoconiosis. Cepharanthine not only can significantly improve the respiratory symptoms but also improve the immunity of patients with silicosis and reduce the incidence of pulmonary infection. However, due to certain side effects, such as gastrointestinal symptoms, rash, pigmentation, and poor compliance during long-term administration, more and more adverse reactions were reported [6], Therefore, organ-specific mode of administration route for improving cepharanthine efficacy get more and more attention.

Fig. 1 *Stephania japonica*

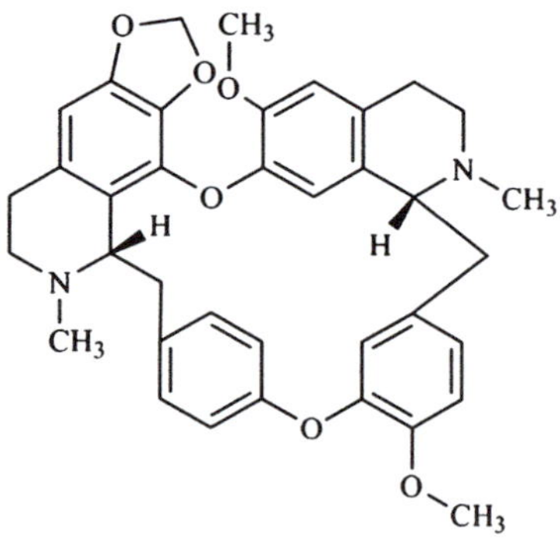

Fig. 2 Chemical structure of cepharanthine

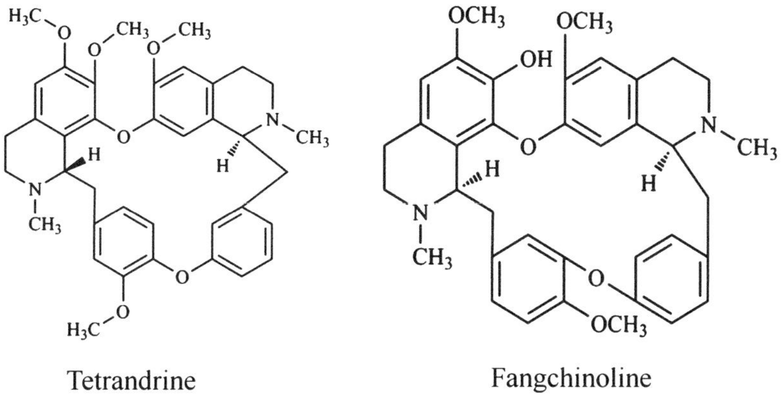

Fig. 3 The chemical structure of cepharanthine derivatives

References

1. Hasegawa S, Takahashi K. The effect of cepharanthine on pertussis. Jpn J Exp Med. 1949;20(2):229–34.
2. Huang, et al. The role of cepharanthine in the prevention and treatment of leukopenia. Foreign Med: Pharm. 1974;04:203–4.
3. Effects of five kinds of dibenzylisoquinoline alkaloids on experimental silicosis in rats. Health Res. 1977;02:92–8.
4. Cui, et al. An overview of pharmacological effects of caffeine. Chin Herb Med. 1995;09:502.
5. Ono M. Effect of cepharanthine on antitumor activity of 1-(2-tetrahydrofuryl)- 5-fluorouracil (FT-207)--5-fluorouracil delivery into tumor tissue. Nihon Gan Chiryo Gakkai Shi. 1989;24(7):1379–92.
6. Wu. Treatment of silicosis and coal workers' pneumoconiosis by thalidomide. Ind Hyg Occup Dis. 1997;23(5):297–8.

Colchicine

Wen Zhang, Qi-Meng Zhou, and Guan-Hua Du

Abstract Colchicine is the most important alkaloid extracted from colchicum. Colchicine promotes the binding of tubulin to microtubules and inhibition of tubulin polymerization. Therefore, it also influences metabolism of protein, RNA polymerase, synthesis of membrane lipid, and transportation of amino acid through cell membrane. As for colchicine, both anti-inflammatory and anti-gout effects are achieved by inhibiting the motion of neutrophil. In the long history of colchicine R&D, it was used early, but it got the FDA approval late. And its adverse effect limits the range of its clinical application.

Keywords Colchicine · Alkaloid · Tubulin · Orphan drug

Alias: Acetamide, colchineos, colchisol, colcin, colsaloid, condylon
Origin: *Colchicum* (Fig. 1)
Chemical name (Fig. 2)

N-[(7S)-1,2,3,10-Tetramethoxy-9-oxo-5,6,7,9-tetrahydrobenzo[a]heptalen-7-yl] acetamide

Molecular formula, $C_{22}H_{25}NO_6$; **MW**, 399.44; **CAS**, 364-86-8

W. Zhang · Q.-M. Zhou · G.-H. Du (✉)
Beijing Key Laboratory of Drug Targets Identification and Drug Screening, Centre for Pharmaceutical Screening, Institute of Materia Medica, Chinese Academy of Medical Sciences and Peking Union Medical College, Beijing, China
e-mail: dugh@imm.ac.cn

G.-H. Du, *Natural Small Molecule Drugs from Plants*,
https://doi.org/10.1007/978-981-10-8022-7_83

Properties

Appearance: colchicine exists in white or light-yellow crystal powder with no smell, and it is seldom prone to absorb moisture. *Melting point*: it becomes dark when it is exposed to light, and it melts at 87–89 °C. *Solubility*: this product is soluble in chloroform or ethanol and it dissolves in water. However, the semihydrate crystal can form in certain concentrations. The product is hardly soluble in ether. *Specific optical rotation*: −121° (0.9 g/100 mL, chloroform, 589.3 nm, 17 °C).

Derivatives (Fig. 3)

Colchicine has been used in clinical practice for many years. However, it was not approved by FDA until in 2006 due to lack of demonstrated randomized controlled trials. In the same year, FDA began to experiment by itself, and the tablet produced by URL Pharma which was acquired by Takeda Pharmaceutical was approved by FDA in 2009. In September 2010, the FDA banned selling the unapproved oral colchicine with single ingredient.

The structural modification of colchicine focused on the transformation of ring: both A and C rings have high affinity with tubulin, and B ring is essential for its conjunction, reaction rate and activation. B ring also regulates the pharmacokinetic properties.

Dosage Forms

Colchicine is recorded in the *Pharmacopoeia of the People's Republic of China* (2015), *British Pharmacopoeia* (2017), *United States Pharmacopeia* (40), *European Pharmacopoeia* (9.0th ed.), *The Japanese Pharmacopoeia* (17th ed.), and *The International Pharmacopoeia* (5th ed.). The dosage forms in current clinical use are tablet and capsule.

Indications

Gout, antitumor, familial Mediterranean fever, and Behcet's disease

Literature

The ingredient is extracted from lily bulbs of *Colchicum autumnale* L. (Qiu Shui Xian). It is also found in *lily tulip* bulbs of *Tulipa edulis Yili*, *Compositae Saussurea laniceps*, *Saussurea*, *Saussurea medusa* were curved flowers, *three-finger Saussurea involucrata*, the whole grass of lotus leaf oak with roots (*lotus Herba saussureae*).

History of R&D

Meadow saffron (*Colchicum*) is recorded to treat rheumatic swelling on ancient Egyptian medical papyrus in 1500 B.C. [1]. According to *De Materia Medica* written by Pedanius Dioscorides in the first century, extract of Meadow saffron is used in treating gout. *London Pharmacopoeia* in 1618 recorded that colchicine is also applied to treat gout [2].

In 1820, the ingredient was first isolated by the French chemist P.S. Pelletier and J.B. Caventou [3]. In 1833, it was purified and named by Geiger [4]. Michael Dewar guessed that there are two seven-membered rings in colchicine in 1945 [5]. Murray Vernon King et al. determined the structure of colchicine by X-ray diffraction in 1952 [6]. In 1959, Albert Eschenmoser integrated the product successfully [7].

Colchicine tablet and raw material are approved mostly in domestic in 2010. The tablet produced by Taiwan manufacturers is approved for being listed in mainland of China in 2012. The raw material made by Indian obtained the approval in 2013. There are three kinds of colchicine approved by FDA: with the combination of probenecid, it is prior to be approved. The others are tablet (2009) and capsule (2014).

Pharmacology

Colchicine promotes the binding of tubulin to microtubules and inhibition of tubulin polymerization [8]. It plays an important role in antitumor therapy with blocking mitosis. It is reported that apoptosis of various tumor cells can be induced by interfering metabolism of protein, inhibiting RNA polymerase, influencing synthesis of membrane lipid, and transport of amino acid through cell membrane. In addition, both anti-inflammatory and anti-gout effects are achieved by inhibiting the motion and cytoskeleton deformation of neutrophil, which is achieved by blocking tubulin polymerization. Since the product has no influence on formation and metabolism of uric acid, it is supposed to control the outbreak of inflammation.

Forty-five percent of colchicine can be utilized effectively. There is 34–44% binding with protein. Part of colchicine is eliminated by CYP3A4, and it remains 50% in 26.6–31.2 h. Sixty-five percent of colchicine is excreted in feces.

Clinical Application

Colchicine is used in treating acute gout patients with NSAID intolerance in clinical application. It also has antitumor activities in breast cancer or liver cancer. It is reported that breast cancer in phase II is treated with CMU therapy which combines intravenous administration to 3 mg colchicine twice a week with methotrexate and uramustine.

There are some other applications in colchicine. Familial Mediterranean fever can only be relieved by colchicine. It also plays a role in Behcet's disease for long-term anti-inflammatory. The dose ranges from 0.015 to 0.03 mg/kg. It will be better with twice a day and long-term oral administration. 0.6 mg three times a day is utilized for women with constipation and irritable bowel syndrome. Severe or persistent oral ulcer and pericarditis can be improved with colchicine [9]. It also has a certain effect on atrial fibrillation and recurrence triggered by resection of the heart tissue with 0.5 mg twice a day [10].

Discussion

It has certain and positive influence on some diseases. However, more adverse reactions can be triggered, such as gastrointestinal discomfort and decreased neutrophils caused by inhibiting mitosis. High dosage of colchicine can lead bone marrow toxicity, anemia, and hair loss. On the other hand, inhibition of mitosis contributes to peripheral neuropathy. There is no effective antidote to colchicine poisoning. Although colchicine's side effects and toxicity limits its scope of application, there is no substitute for the orphan drug status in treatment for acute gout.

Fig. 1 *Colchicum*

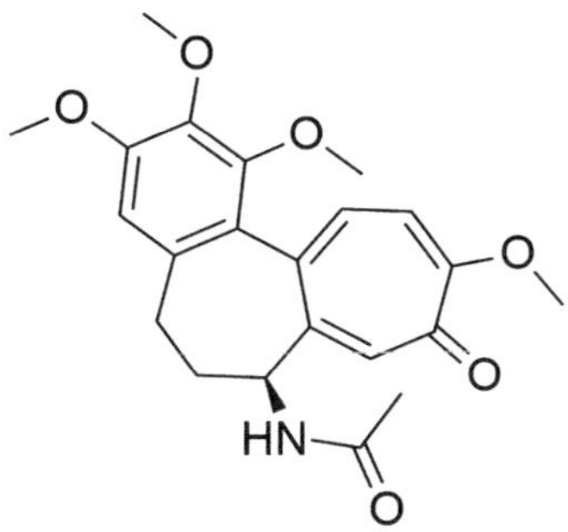

Fig. 2 Chemical structure of colchicine

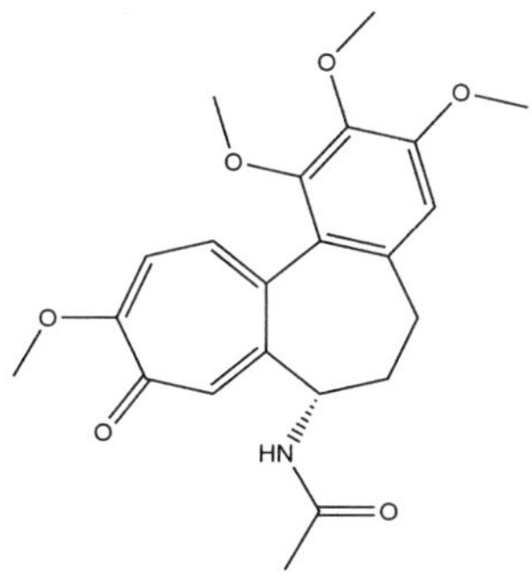

Fig. 3 Chemical structure of colchicine

References

1. Graham W, Roberts JB. Intravenous colchicine in the management of gouty arthritis. Ann Rheum Dis. 1953;12(1):16–9.
2. Hartung EF. History of the use of colchicum and related medicaments in gout: with suggestions for further research. Ann Rheum Dis. 1954;13(3):190–200.
3. Pelletier PJ, Caventou JB. Examen chimique de plusieurs végétaux de la famille des Colchicées, et du principe actif qu'ils renferment [Cévadille (veratrum sabadilla); hellébore blanc (*veratrum album*); colchique commum (*colchicum autumnale*)]. 1820.
4. Geiger PL. Ueber einige neue giftige organische Alkalien. Ann Pharm. 1833;7(3):269–80.
5. Dewar M. Structure of colchicine. Nature. 1945;155(3927):141–2.
6. King MV, Vries JD, Pepinsky R. An x-ray diffraction determination of the chemical structure of colchicine. Acta Crystallogr. 1952;5(4):437–40.
7. Schreiber J, Leimgruber W, Pesaro M. Synthese des colchicins. Angew Chem. 1959;71(20):637–40.
8. Zhang Z, Feng Y, Zhao DM. Progress in the study of tubulin inhibitors binding to the colchicine-site. Chin J Med Chem. 2014;24(3):241–9.
9. Alabed S, Cabello JB, Irving GJ. Colchicine for pericarditis. Cochrane Libr. 2013;7:1–10.
10. Deftereos S, Giannopoulos G, Efremidis M. Colchicine for prevention of atrial fibrillation recurrence after pulmonary vein isolation: mid-term efficacy and effect on quality of life. Heart Rhythm. 2014;11(4):620–8.

Ginsenoside Rg3

Zhen-Zhen Wang, Ling-Lei Kong, and Nai-Hong Chen

Abstract Ginsenoside Rg3 is a dammarane tetracyclic triterpenoid saponin found in natural ginseng, which belongs to the protopanaxadiol. It is also the natural ingredient with prominent activity extracted from red ginseng (processed ginseng). Ginsenoside Rg3 is mainly used in the clinical treatment or adjuvant therapy of cancer. In addition, it improves and prevents the common diseases in the elderly such as cardiovascular disease, coronary heart disease, limb weakness, difficulty walking, and loss of memory. It also has a certain biological activity in antiviral, radiosensitization, and treatments of chronic obstructive pulmonary disease. Ginsenoside Rg3 has a wide safety range, good oral absorption, and the characteristics of multi-target, multi-link, and multi-effect on the antitumor effect. Ginsenoside Rg3 is the active ingredient of Shenyi (参一) capsule which is a kind of traditional Chinese anticancer medicine developed in China. Further study on the antitumor mechanism of ginsenoside Rg3 will be helpful to the clinical application of this drug. Extraction of more compounds from the ginseng and red ginseng will be benefit for the findings of new important drugs to contribute to human health.

Keywords Ginsenoside · Rg3 · Antitumor · Shenyi capsule

Alias: Ginsenoside RG3,20(S); ginsenoside-RG3,20(R); ginsenoside-RG3
Origin: Ginseng (Fig. 1)

Z.-Z. Wang · N.-H. Chen (✉)
State Key Laboratory of Bioactive Substances and Functions of Natural Medicines, Institute of Materia Medica & Neuroscience Center, Chinese Academy of Medical Sciences and Peking Union Medical College, Beijing, China
e-mail: chennh@imm.ac.cn

L.-L. Kong
Beijing Key Laboratory of Drug Targets Identification and Drug Screening, Centre for Pharmaceutical Screening, Institute of Materia Medica, Chinese Academy of Medical Sciences and Peking Union Medical College, Beijing, China

G.-H. Du, *Natural Small Molecule Drugs from Plants*,
https://doi.org/10.1007/978-981-10-8022-7_84

Chemical name (Fig. 2)

(3β,12β)-12,20-Dihydroxydammar-24-en-3-yl-2-O-β-D-glucopyranosyl-β-D-glucopyranoside

Molecular formula, $C_{42}H_{72}O_{13}$; **MW**, 785.025; **CAS**, 38243-03-7 for [20(R)-ginsenoside Rg3], 14197-60-5 for [20(S)-ginsenoside Rg3]

Ginsenoside Rg3 is a chiral compound, which has two common configurations: 20(R)-ginsenoside Rg3 and 20(S)-ginsenoside Rg3. 20(R)-ginsenoside Rg3 is the dominant configuration.

Properties

Appearance: white powder. *Density*: 1.40. *Melting point*: 315–318 °C. *Solubility*: 20(R)-ginsenoside Rg3 is soluble in methanol and ethanol; low water solubility; insoluble in ether and chloroform. *Storage temperature*: 2–8 °C. The component classification belongs to three dammarane tetracyclic triterpenoid saponins.

Dosage Forms and Indications

This product is contained in "New drug standards" the 87th volume.

Ginsenoside Rg3 capsule, ginsenoside Rg3 emulsion, and ginsenoside Rg3 microemulsion are mainly used in the clinical treatment or adjuvant therapy of lung cancer, breast cancer, liver cancer, gastric cancer, colon cancer, etc. In addition, they improve and prevent the common diseases in the elderly such as cardiovascular disease, coronary heart disease, limb weakness, difficulty walking, and loss of memory. Ginsenoside Rg3 also has a certain biological activity in antiviral, radiosensitization, and treatment of chronic obstructive pulmonary disease [1–3].

Literature

Ginsenoside Rg3 is mainly derived from the dried roots of Araliaceae *Panax ginseng* C.A. Mey. and *Panax quinquefolius* L. Ginseng has a long history of medicinal uses, well known as "king of herbs." Ginseng is listed as the top grade in "Shen Nong's herbal class," with "the main security set up the five internal organs, spirit, soul, stop fright, in addition to evil, eyesight, fun puzzle" account. The traditional Chinese medicine regards ginseng as the top grade of reinforcing Qi, with Dabu strength, spleen and lungs, Sheng Jin, Anshen Yizhi, etc.

History of R&D

Ginseng is currently one of the most studied traditional medicines. Chemical researches certificate that ginseng contains a variety of ginseng saponin, ginseng polysaccharide, protein, polypeptide, amino acids, and other chemical constituents. Ginseng saponins are the main active components, accounting for about 4%. Ginsenoside is a kind of sterol compound, which has a similar basic structure. Ginsenoside contains the steroid nucleus of 17 carbon atoms arranged in 4 rings. At present, more than 40 kinds of ginsenosides have been isolated from ginseng. They are divided into four types according to their chemical structure, the protopanaxadiol, the protopanaxatriol, the ocotillol, and the oleanolic acid.

Ginsenoside Rg3 is a dammarane tetracyclic triterpenoid saponin found in natural ginseng, which belongs to the protopanaxadiol [2]. It is also the natural ingredient with prominent activity extracted from red ginseng (processed ginseng). Modern medicine has proved that Rg3 has certain biological activity. Since 1983, Rg3 has been widely studied by many scholars in the world [4]. The results showed that the content of ginsenoside Rg3 in the fresh and dried ginseng roots was less than 1 ppm while in red ginseng was 3/100,000. After several years of research, the method of extracting 20(R)-ginsenoside Rg3 from red ginseng has been successfully improved, and the industrialized production has been realized. A multicenter clinical study by eight domestic hospitals indicated that 20(R)-ginsenoside Rg3 has therapeutic effects on tumor. It has been issued the new drug certificate by the Chinese Food and Drug Administration, as a new anticancer drug approved.

Pharmacology

It is reported that the antitumor effects of ginsenoside Rg3 and Rh2 were the most significant in more than 40 kinds of ginsenosides obtained. Moreover, they enhanced the immune function of tumor patients after chemotherapy, including humoral immunity, cell immunity, and non-specific immunity [5]. This effect is consistent with the vitality nourishing and body strengthening of ginseng in the theory of traditional Chinese medicine. There is more research on the effects of 20(R)-ginsenoside Rg3 in the two isomers of ginsenoside Rg3, which has certain inhibitory effects on tumor cells of lung cancer, gastric cancer, colon cancer, breast cancer, liver cancer, etc. There is less research on the effects of 20(S)-ginsenoside Rg3. Other studies have shown that ginsenoside Rg3 also has the effects of anti-fatigue [6], antiviral [7], and treatment of chronic obstructive pulmonary disease [7].

Human tolerance test shows that ginsenoside Rg3 has a wide safety range. The recommended dose of the first phase of clinical trial (0.8 mg/kg/day) has no side effects. Ginsenoside Rg3 has good oral absorption. The drug absorption peak in the blood is detected after administration for 15–30 min, and the blood concentration reaches the peak after administration for 1.0–1.5 h. The half-life is 4.84 ± 1.41 h.

The maximum blood concentration increases with the dose within the reagent range (0.8–3.2 mg/kd/day), suggesting the pharmacokinetics of ginsenoside Rg3 the first-order kinetics process.

Clinical Application

The basic research for many years and a large number of clinical data confirm that ginsenoside Rg3 effectively inhibits the growth of lung cancer, liver cancer, gastric cancer, colon cancer, and breast cancer, significantly improves the clinical symptoms of patients, improves the quality of life, and prevents and treats cancer. Ginsenoside Rg3 effectively improves the ability of anti-fatigue, antiaging, and resistance to disease. It also improves cardiovascular function, is resistant to platelet aggregation, protects nerve cells of brain, and improves immunity.

Ginsenoside Rg3 is the active ingredient of Shenyi (参一) capsule which is a kind of traditional Chinese anticancer medicine developed in China. It prevents the metastasis and recurrence of cancer and has been an adjuvant treatment in multiple tumors [2]. Shenyi capsule improves the curative effects of primary lung cancer and liver cancer, relieves the symptoms of Qi deficiency, as well as improves immunity, when combined with chemotherapy. Further study on the antitumor mechanism of ginsenoside Rg3 will be helpful to the clinical application of this drug.

Discussion

Ginseng has a long history of consumption and medicinal use in China. Ginsenosides can be developed into natural, inexpensive, and low-toxic drugs. They are also approved by the national authorities for medicinal and edible materials. Ginsenoside Rg3 has the characteristics of multi-target, multi-link, and multi-effect on the anti-tumor effect. It will obtain treatment effect clinically. Extraction of more compounds from the ginseng and red ginseng will be benefit for the findings of new important drugs to contribute to human health.

Fig. 1 Ginseng

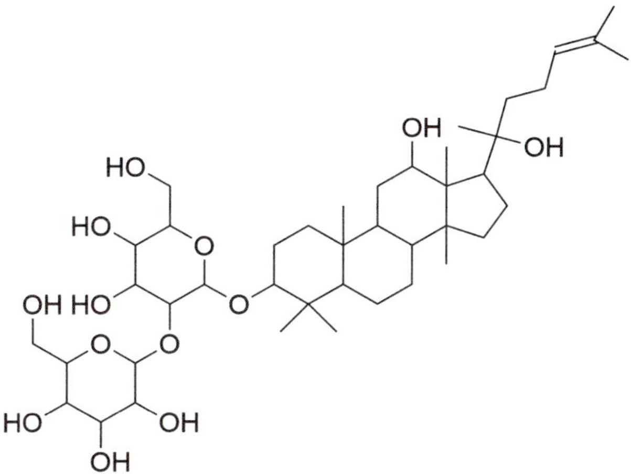

Fig. 2 The chemical constitution of ginsenoside Rg3

References

1. Mei SL, Yuan HY, Chang YP, et al. The anti-HSV-1 activity and immunoregulatory effects of ginsenoside Rg3 in vitro. J Jilin Univ (Med Sci). 2006;6:1019–22.
2. Duan LR, Xie YH, Wang SW, et al. Advances in pharmacokinetics and antitumor effects of ginsenoside Rg3. Adv Mod Biomed. 2010;4:770–3.
3. Li TZ, Liu CT. Effect of Shenyi capsule on serum immunoglobulin and cytokines in patients with chronic obstructive pulmonary disease. Chin Clin Rehabil. 2005;3:160–1.
4. Kitapwa L, Fukuda Y, Taniyama T, et al. Chemical studies on crude drugs I. Yakugaku Zasshi. 1983;6(103):612–22.
5. Zhang ZM, Jiang B, Zheng XX, et al. The immunopotentiating effects of ginsenoside Rg3 on peripheral blood lymphocytes in patients with tumor radiotherapy in vitro. Chin J Pharm Sci. 2004;4:25–8.
6. Cheng H, Song XB, Zhang LJ. Research progress of ginsenoside Rg3 and Rh2. Drug Eval Study. 2010;4:307–11.
7. Zhang NS, Zhang XH, Li WF. Research progress of ginsenoside Rg3. Med Guid. 2006;7:687–9.

Gossypol

Shi-Feng Chu, Ling-Lei Kong, and Nai-Hong Chen

Abstract Gossypol is extracted from the roots, stems, leaves, and seeds of the plant cotton. Amounting evidences showed that gossypol impaired the sperm quality significantly, and it was the world's first male contraceptive. However, due to the presence of hypokalemia and irreversible antifertility risk of gossypol, the clinical indications of male contraceptives are not approved by the Chinese Food and Drug Administration. At present, gossypol and its derivatives are mainly used for female uterine fibroids, functional uterine bleeding, and endometriosis treatment. In addition, the treatment of gossypol and its derivatives in the tumor is gradually confirmed clinically.

Keywords Gossypol · Contraceptive · Tumor

Origin: Cotton (Fig. 1)
Chemical name (Fig. 2)

2,2′-bis-(Formyl-1,6,7-trihydroxy-5-isopropyl-3-methylnaphthalene)

Molecular formula, $C_{30}H_{30}O_8$; **MW**, 518.5622; **CAS**, 303-45-7

S.-F. Chu · N.-H. Chen (✉)
State Key Laboratory of Bioactive Substances and Functions of Natural Medicines, Institute of Materia Medica & Neuroscience Center, Chinese Academy of Medical Sciences and Peking Union Medical College, Beijing, China
e-mail: chennh@imm.ac.cn

L.-L. Kong
Beijing Key Laboratory of Drug Targets Identification and Drug Screening, Centre for Pharmaceutical Screening, Institute of Materia Medica, Chinese Academy of Medical Sciences and Peking Union Medical College, Beijing, China

G.-H. Du, *Natural Small Molecule Drugs from Plants*,
https://doi.org/10.1007/978-981-10-8022-7_85

Properties

Appearance: plate or needle-like crystals with yellow color, odorless, and tasteless. *Solubility*: do not dissolve in water, slightly soluble in ethanol, and soluble in chloroform, ether, acetone, ethyl acetate, dichloroethane, carbon tetrachloride, and pyridine. Difficult to dissolve in cyclohexane, benzene, and petroleum ether. *Melting point*: 184–214 °C.

Gossypol has a chiral structure with two optical isomers in the left and right. The formic acid gossypol or gossypol acetate was commonly used.

Dosage Forms and Indications

This product as gossypol acetate is recorded in the tenth volume of national standards for chemical drugs.

Solid formulations include compound acetate gossypol tablets and gossypol potassium chloride vitamin B capsule. Gossypol was used as male contraceptive drug in clinic previously but terminated due to the induced severer hypokalemia. Now, compound acetate gossypol tablets consisted of acetate gossypol and potassium chloride, vitamins are used for the treatment of uterine functional bleeding, uterine fibroids, and menorrhagia and endometriosis.

Literature

Gossypol is mainly present in the roots, stems, leaves, and seeds of the plant cotton, and the highest content is found in cotton seeds. Its chemical structure was first determined by Adams in 1938 and was used to be studied in antitumor research.

History of R&D

The close relationship between male infertility and the consumption of cottonseed products was first found in cotton-producing areas by Shandong scientists in 1971. Further study found that gossypol impaired the sperm quality significantly in rat, rabbit, and so on. Mechanism study showed that gossypol could interact with sperm and then trigger a cascade reaction for different stages of spermatogenesis. Acetate gossypol showed similar biological activities with gossypol. The clinical trial for gossypol tablet for the male contraceptive treatment was initiated in the 1973, which was produced in Shanghai, and the raw material was provided from Shanghai, Shandong, and Zhejiang province. In 1974, a series of derivatives were developed by Shandong scientists; the activity of gossypol formate is similar with that of gossypol and gossypol acetate, while its side effects are much less than gossypol acetate, and there is no hypokalemia. To 1978, the national total volunteers reached 8806 for the male contraceptive treatment induced by gossypol, and the hypokalemia occupied 0.99% of these volunteers (occurred in Jiangsu, Shanghai, and Sichuan province) [1]. Gossypol produced by Shandong showed no hypokalemia within 2000 cases. However, irreversible azoospermia was found in 15% of these volunteers, suggesting that gossypol may induce irreversible male infertility. Due to these side effects, the clinical trial was terminated in 1980 in the national summary conference [2]. During 1981–1985, the side effect, antifertility reversibility, and safety evaluation induced by gossypol were further investigated by the national 65-35-2-5 scientific group, which was carried out by prospective clinical study in 390 volunteers, and the results showed that there is no carcinogenic, teratogenic, and mutagenic effects in gossypol [3].

Pharmacology

There are significant differences in pharmacological effects of gossypol between male and female. In males, gossypol could significantly reduce sperm quality, as manifested by the reduced sperm density or activity. Mechanism study showed that gossypol could impair spermatogenic epithelium directly, and sperm cells and spermatocytes are the most sensitive. In female, gossypol could inhibit ovarian and endometrial, muscular steroid hormone receptor, hence thinning the endometrium and muscle layer and reducing menstrual flow [4]; therefore, gossypol acetate and potassium chloride, vitamins, and other combinations were prepared for the compound acetate gossypol tablets, which were used for the treatment of uterine functional bleeding, uterine fibroids, menorrhagia, and uterus endometriosis. The recommended daily oral dose is 20 mg. Gossypol dextrorate has no effect, with levorotate as the active ingredient; therefore, the role of L-gossypol is two times that of gossypol [5].

A good antitumor effect of gossypol and its derivatives was found in recent studies. Gossypol showed significant inhibition on cancer cell proliferation in vitro studies, including tumor cells derived from lymphatic and granulocytes, adrenal glands, breast, and cervical, rectal, and central nervous system. Studies have suggested that gossypol inhibits the catalytic activity of topoisomerase II and intervenes in the stability of topoisomerase – DNA complexes. Gossypol blocks the cell in S stage by reducing the activity of DNA polymerase alpha and beta and inhibiting DNA synthesis. Gossypol is a specific DNA synthesis inhibitor that inhibits cell division. In addition, gossypol can also exhibit antitumor effect by acting as a signal pathway regulator and inhibiting the energy metabolism of tumor cells [6].

Clinical Application

Due to the presence of hypokalemia and irreversible antifertility risk of gossypol, the clinical indications of male contraceptives are not approved by the Chinese Food and Drug Administration. At present, gossypol and its derivatives are mainly used for female uterine fibroids, functional uterine bleeding, and endometriosis treatment. In addition, the treatment of gossypol and its derivatives in the tumor is gradually confirmed clinically, such as gossypol combined with docetaxel or cisplatin for the treatment of malignant non-small cell lung cancer, and is currently in phase III clinical trial [7]. Gossypol combined with rituximab can achieve better therapeutic efficacy against granulocytic leukemia and is currently in phase II clinical trials [8].

Discussion

The invention of female oral contraceptives is known as one of the 100 major inventions that affect thousands of human history. The first oral contraceptive was approved by the US FDA on May 9, 1960, but there were many side effects such as early pregnancy response, menstrual disorders, weight gain, pigmentation, and so on. In 1970, the father of contraceptive medicine Carl Djerassi proposed male oral contraceptive strategy; China researchers found that gossypol has a good male contraceptive effect, which is the world's first male contraceptive. Although there are unexplained hypokalemia and irreversible sperm injury stopped the application, the research and development of gossypol greatly promoted the development of the fields the natural medicine research, reproductive medicine and molecular biology, it has made tremendous contributions to promoting the modernization levels of science and technology in China. At present, gossypol is used for the treatment of uterine fibroids and endometriosis treatment in the form of compound gossypol tablets and achieved good therapeutic effect in China.

Fig. 1 Cotton

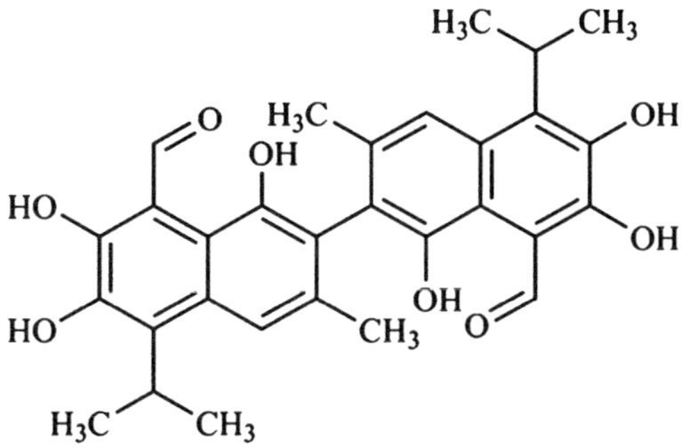

Fig. 2 Chemical structure of gossypol

References

1. Wuhan Medical College, et al. Investigation of infertility caused by the consumption of crude sesame oil. Natl Gossypol Res Exch. 1972.
2. Wang Y, Luo Y, Tang X. Study on the anti-fertility effect of cottonseed meal and gossypol. Acta Pharm Sin. 1979;14(11):662–9.
3. Wang Q, Guan M, Lei H. (−) and (+) gossypol on male fertility in male rats. Acta Pharm Sin. 1984;19(12):932–4.
4. Li J, Li Z. Study on the reproductive behavior of gossypol in women. Adv Mod Obstet Gynecol. 1994;3(4):367–71.
5. Yuan J. Review, thinking and prospect of gossypol contraceptive research. Chin J Tradit Med Sci Technol. 1996;3(9):5–7.
6. Dai Y, Yang S, Tang H. Study on the anti-tumor effect of L-gossypol. Chin J Pharmaceut Sci. 2009;18(19):1–4.
7. Gossypol combined with docetaxel and cisplatin scheme in advanced non small-cell lung cancers with APE1 high expression. *ClinicalTrials.gov Web Site* 2013, November 11.
8. Castro JE, Loria OJ, Aguillon RA, et al. A phase II, open label study of AT-101 in combination with rituximab in patients with relapsed or refractory chronic lymphocytic leukemia: evaluation of two dose regimens. Blood. 2007;110(11):Abst 3119.

Homoharringtonine (HHT)

Zhe Wang, Li Li, and Guan-Hua Du

Abstract In China, Cephalotaxaceae plants have a long history of drug application according to Shennong's Classic of Materia Medica. These plants are distributed in the Qinling Mountains, the Dabie Mountains, and the south of the Yangtze River. Homoharringtonine was mainly extracted from the dry leaves of *Cephalotaxus fortunei*. The harringtonine compounds have the anticancer activities, in which harringtonine and homoharringtonine have been used as anticancer drugs worldwide. Therefore, its main dosage form is homoharringtonine injection, which is mainly used for the treatment of chronic myelocytic leukemia and acute myeloid leukemia.

Keywords Homoharringtonine · Alkaloid · Chronic myelocytic leukemia · Acute myeloid leukemia · Antitumor

Alias: Omacetaxine mepesuccinate
Origin: *Cephalotaxus fortunei* Hook. f. (Fig. 1)
Chemical name (Fig. 2)

1-((1S,3aR,14bS)-2-Methoxy-1,5,6,8,9,14b-hexahydro-4H-cyclopenta(a)(1,3) dioxolo(4,5-h)pyrrolo(2,1-b)(3)benzazepin-1-yl) 4-methyl (2R)-2-hydroxy-2-(4-hydroxy-4-methylpentyl)butanedioate

Molecular formula, $C_{29}H_{39}NO_9$; **MW**, 545.6; **CAS**, 26833-87-4
Derivatives (Fig. 3)

Molecular formula, $C_{28}H_{37}NO_9$; **MW**, 531.6; **CAS**, 26833-85-2

Z. Wang · L. Li · G.-H. Du (✉)
Beijing Key Laboratory of Drug Targets Identification and Drug Screening, Centre for Pharmaceutical Screening, Institute of Materia Medica, Chinese Academy of Medical Sciences and Peking Union Medical College, Beijing, China
e-mail: dugh@imm.ac.cn

G.-H. Du, *Natural Small Molecule Drugs from Plants*,
https://doi.org/10.1007/978-981-10-8022-7_86

Properties

Appearance: an almost white or pale yellow crystalline powder or an amorphous friable solid. It has the hygroscopic nature. It darkens on exposure to light. *Solubility*: easily soluble in chloroform, ethanol and methanol, and slightly soluble in ether and water. *Melting point*: 143–147 °C.

Dosage Forms and Indications

This product is recorded in the *Pharmacopoeia of the People's Republic of China* (2015).

Its main dosage form is homoharringtonine injection, which is mainly used for the treatment of chronic myelocytic leukemia and acute myeloid leukemia.

Literature

Homoharringtonine was mainly extracted from the dry leaves of *Cephalotaxus fortunei*. There are nine species of plants in Cephalotaxaceae including *Cephalotaxus oliveri* Mast., *C. sinensis* (Rehd. et Wils.) L., *C. fortunei* var. alpina Li, *C. lanceolata* K. M. Feng, *C. hainanensis* Li, *C. sinensis* var. latifolia Cheng et L. K. Fu, *C. fortunei* Hook.f, *C. harringtonia* (Forbes) Koch, and *C. wilsoniana* Hayata. *Cephalotaxus* is widely distributed in eastern India, Thailand, China, the Korean Peninsula, and Japan. In China, these plants are distributed in the Qinling Mountains, the Dabie Mountains, and the south of the Yangtze River [1].

Cephalotaxaceae plants have a long history of drug application. It was recorded in *Shennong's Classic of Materia Medica* that, "Bi Zi (Cephalotaxaceae fruit) is sweet and warm, and it can be used to get rid of pathogenic factor, parasites, snake bite, antitoxic and infectious disease." In the *Compendium of Materia Medica • Fei Shi* (Bi Zi is Fei Shi), it also recorded that "Taken often, it has the function of curing five kinds of hemorrhoids, getting rid of parasites and treating infectious disease." The seeds of *Cephalotaxus* have the function of insect repellent and digestive care, and they are generally used in the treatment of ascariasis, hookworm disease, retention of food, etc. in China. The branches and leaves of *Cephalotaxus* have anticancer effect, and they are used for the treatment of malignant tumors. The main chemical components in *Cephalotaxus* are alkaloids, flavonoids, and sterols. Harringtonine is the main type of alkaloids. The harringtonine compounds have the anticancer activities, in which harringtonine and homoharringtonine have been used as anticancer drugs worldwide. It was found that 1 g *Cephalotaxus* compounds could be extracted from 100–150 kg *Cephalotaxus* dry leaves. However, *Cephalotaxus* is a national protected plant, and it is an endangered species. At present, methods of chemical synthesis, cell (tissue) culture, and plant endophytic fungi are all important ways to obtain homoharringtonine.

History of R&D

In China, especially in Fujian, etc., doctors have treated tumor with *Cephalotaxus harringtonia* long ago. In the early 1970s, American scientists Powell et al. isolated and identified the alkaloids of the *Cephalotaxus* plant and studied its antitumor activity. One type of the alkaloids (Cephalotaxine ester, including harringtonine, homoharringtonine, isoharringtonine, deoxyharringtonine, and pseudodeoxyharringtonine) was found to have the effect of inhibiting the proliferation of mouse leukemia cells [2, 3]. Chinese scientists also isolated a large number of harringtonine and homoharringtonine and used them to treat leukemia. Then, the United States and other developed countries' scientists carried out the Phase I and II clinical research and toxicology research of homoharringtonine.

Pharmacology

Mechanism

1. Inhibit the expression of protein genes: The antitumor mechanism of *Cephalotaxus* alkaloids is mainly related to its inhibitory effect on the initial stage of protein synthesis, the extension of peptide chain, and the protein gene expression.
2. Induce apoptosis: Homoharringtonine was found to activate HL-60 cell apoptosis through caspase-3 mediated Bcl-2-Bax, -MAPK pathway [4].
3. Induce cell differentiation: Homoharringtonine could induce HL-60 cell differentiation by downregulating the expression of CD44 gene, thereby increasing p27 and p21 expression and inhibition of cyclin E activity.

Structure-Activity Relationship (SAR) The methyl acetate group of 2′ position in the C-3 acyl side chain is an active essential group. R group plays a role in regulating the polarity of the molecule, and the size of the R group can affect the molecular activity. The olefinic carbon instead of the 2′ position chiral carbon is still active [5].

In Vivo Process Homoharringtonine mainly metabolizes in the liver and converts to a low activity of 4′-demethylation homoharringtonine by oxidation in the liver microsomes. It is absorbed slowly by intramuscular or oral administration, and it can be administered by intravenous or subcutaneous injection. Homoharringtonine is biphasic in vivo by intravenous injection. At a dose of 3–4 mg/m^2 by intravenous injection for 6 h, the α half-life is about 0.5 h, and the β half-life is about 9.3 h [6]. Six hours or longer time of sustained intravenous injection of administration led to a better efficacy than that of short-term rapid intravenous treatment, and the toxic side effects are relatively smaller too. When omacetaxine was administrated at a dose of 1.25 mg/m^2 subcutaneous, it was found to reach the mean peak blood plasma concentration within 1 h, and it was distributed widely, as evidenced by an apparent volume of distribution as 126.8 L/m^2. Plasma concentration versus time data demonstrated biexponential decay; urinary excretion of unchanged omacetaxine accounted for <15 percentage of the dose; mean steady-state terminal half-life was 7 h [7].

Clinical Application

1. Chronic Myelocytic Leukemia

Bai Houqiao et al. [8] treated 23 (17 men, 6 women) chronic myeloid leukemia patients. After hydroxyurea induction therapy, the patients were intravenously or intramuscularly injected with homoharringtonine, which resulted in a complete hematologic remission of 26.1%. Wu Chunhua [9] used homoharringtonine combined with cytarabine to treat 23 cases of chronic myeloid leukemia patients without effects under the treatment of hydroxyurea and interferon, and the efficiency was found to be up to 93.5%.

2. Acute and Chronic Myeloid Leukemia

Yang Yanli et al. [10] used homoharringtonine and cytarabine combined with granulocyte colony-stimulating factor called HAG regimen to treat 28 cases of acute myeloid leukemia, and the result was a complete remission of 78%. This program was proved to have obvious curing effects on acute myeloid leukemia, and its effects for the treatment of the elderly, secondary, or drug-resistant leukemia patients were affirmed too. In order to verify the efficacy of low-dose homoharringtonine in the treatment of chronic myeloid leukemia with refractory or recurrent chronic myeloid leukemia, Gu et al. [11] observed the efficacy in 67 patients with daily use of homoharringtonine cytidine; the efficiency was 89.5%.

Adverse Effect

The most common adverse effects of homoharringtonine in clinical studies are thrombocytopenia, anemia, neutropenia and neutropenia, diarrhea, nausea, general weakness, fatigue, injection site response, lymphopenia, etc.

Discussion

Since the 1960s, there have been many reports on the pharmacological mechanism of homoharringtonine. It has been found that the treatment effect was significantly improved by in combination of homoharringtonine and other drugs. However, the mechanism was largely unknown and need further exploration. In addition, reports of other cephalotaxine compounds such as isoharringtonine and deoxyharringtonine are relatively less, and the biological activity of these harringtonines cannot be fully developed and widely used. Homoharringtonine as an anticancer drug has remarkable effects on acute and chronic leukemia, and the combination of homoharringtonine and other drugs for the treatment of leukemia has a good application prospect.

Due to the serious shortage of plant resources of *Cephalotaxus*, and the resource plant as a national endangered plant, the important route of plant extraction is severely limited. Therefore, it is particularly important to obtain the cephalomannine compound by other effective routes. The plant endophytic fungi have a short culture cycle; its fermentation conditions are simple, easy to control, and large-scale. It will be a new route to obtain homoharringtonine.

Fig. 1 *Cephalotaxus fortunei* Hook. f

Fig. 2 Chemical structure of homoharringtonine

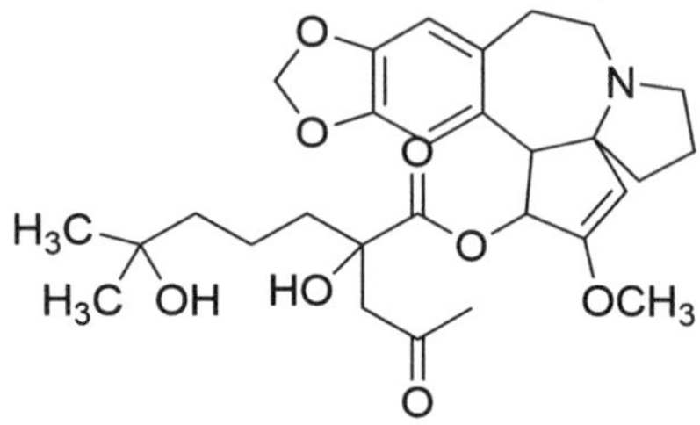

Fig. 3 Chemical structure of homoharringtonine derivatives

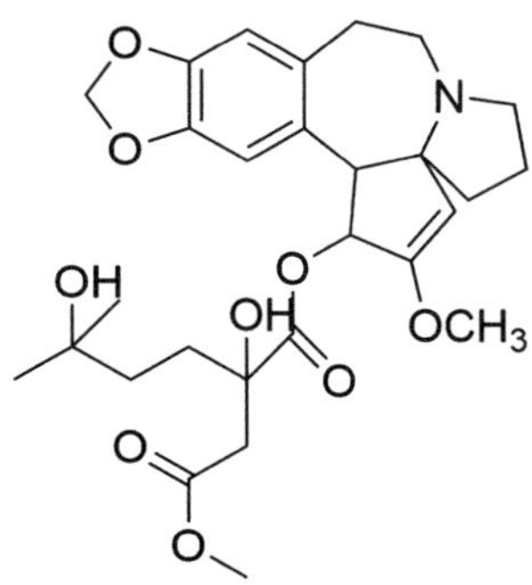

Harringtonine

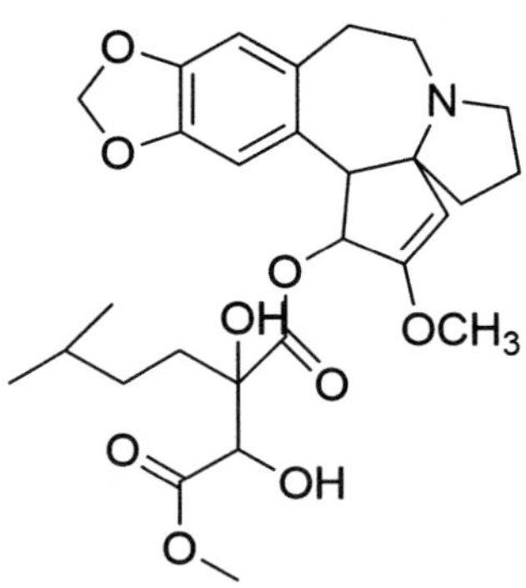

Isoharringtonine

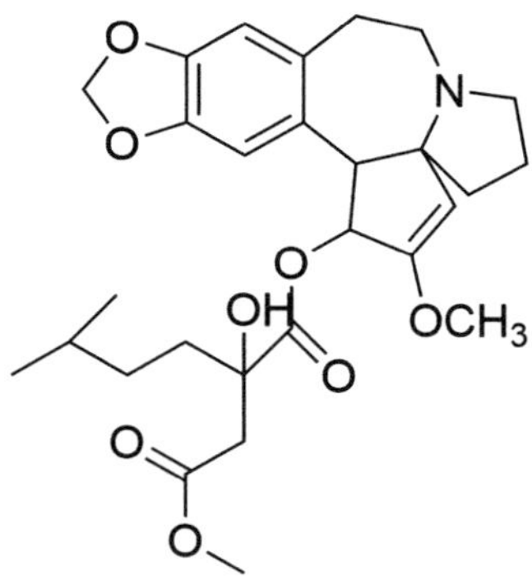

Deoxyharringtonine

References

1. Zhang YY. Research progress on sources, pharmacological activities, and clinical applications of cephalotaxines. Drugs Clin. 2011;26(5):370–4.
2. Lu DY. Biological activities and clinical utilizations of harringtonine and homoharringtonine. Nat Prod Res Dev. 2000;12(5):70–3.
3. Powell RG, et al. Antitumor alkaloids from *Cephalotaxus harringtonia* structure and activity. J Pharm Sci. 1972;61(8):1227–30.
4. Chen CY. The mechanism of apoptosis induced by homoharringtonine in HL -60 cells. Chin J Pathophysiol. 2004;20(7):1183–6.
5. Wu CW. Advances in study on Cephalotaxus fortunei and the alkaloids in it. Foreign Med Sci: Pharm. 1993;20(6):321–5.
6. Savaraj N. Clinical pharmacology of homoharringtonine. Cancer Treat Rep. 1986;70(12):1403–7.
7. Nemunaitis J. Pharmacokinetic study of omacetaxine mepesuccinate administered subcutaneously to patients with advanced solid and hematologic tumors. Cancer Chemother Pharmacol. 2013;71(1):35–41.
8. Bai HQ. The clinical analysis of 23 cases of chronic myelogenous leukemia treated with homoharringtonine. China Prac Med. 2009;4(24):151–2.
9. Wu CH. Clinical observation of low dose HA for the treatment of chronic myeloid leukemia. Pract Clin Med. 2010;11(7):30–1.
10. Yang YL. Low dose of homoharringtonine and cytarabine combined with G-CSF for treatment of acute myelocytic leukemia: report of 28 cases. J Beng bu Med Col. 2009;34(4):307–8.
11. Gu LF. Low dose of homoharringtonine and cytarabine combined with granulocyte colony-stimulating factor priming on the outcome of relapsed or refractory acute myeloid leukemia. J Cancer Res Clin Oncol. 2010;137(6):997–1003.

Indirubin

Xiu-Yun Song, Ling-Lei Kong, and Nai-Hong Chen

Abstract Indirubin comes from *Indigofera tinctoria* L., *Baphicacanthus cusia Brem.*, *Isatis tinctoria* L., and *Couroupita guianensis Aubl.* The indirubin has been artificially synthesized. Indirubin has the effects of anti-inflammatory, antibacterial, detoxification, enhancing immune function, and has anticancer. Indirubin is used for the therapy of chronic myelocytic leukemia in clinical. It has the advantages of reliable clinical curative effect, small side effects, and no significant inhibitory effect on the bone marrow. Indirubin is an effective ingredient of medicine. Therefore, indirubin and its derivatives will become a new direction of research in the field of anticancer drugs.

Keywords Indirubin · Chronic myeloid leukemia · Cancer

Origin: *Indigofera tinctoria* L. (Fig. 1)
Chemical name (Fig. 2)

2-(2-Oxo-1H-indol-3-ylidene)-1H-indol-3-one

Molecular formula, $C_{16}H_{10}N_2O_2$; **MW**, 262.27; **CAS**, 479-41-4

Properties

Appearance: dark red acicular crystal with sublimate, odorless, and tasteless. *Solubility*: slightly soluble in dimethyl sulfoxide (DMSO) or tetrahydrofuran, very slightly soluble in chloroform or acetone, and insoluble in ethanol, ethyl ether, or

X.-Y. Song · N.-H. Chen (✉)
Institute of Materia Medica, Chinese Academy of Medical Sciences and Peking Union Medical College, Beijing, China
e-mail: chennh@imm.ac.cn

L.-L. Kong
Beijing Key Laboratory of Drug Targets Identification and Drug Screening, Centre for Pharmaceutical Screening, Institute of Materia Medica, Chinese Academy of Medical Sciences and Peking Union Medical College, Beijing, China

G.-H. Du, *Natural Small Molecule Drugs from Plants*,
https://doi.org/10.1007/978-981-10-8022-7_87

water. *Melting point*: 348–353 °C. The stability of indirubin is poor that it needs to be stored away from light and thermal environment [1].

Dosage Forms and Indications

Indirubin is contained in the fourth volume of the book, "National standards for chemical drugs."

Indirubin tablets are used for the therapy of chronic myelocytic leukemia in clinical.

Literature

Indirubin comes from the leaves of *Indigofera tinctoria* L. (a leguminous plant), the rhizome of *Baphicacanthus cusia Brem.* (an Acanthaceae), the root of *Isatis tinctoria* L. (a Cruciferae), and the fruit, flower, and stem bark of *Couroupita guianensis Aubl.* (a Lecythidaceae plant). The indirubin has been artificially synthesized [2]. The inhibition effect of W-256 sarcoma and clinical effect of synthetic indirubin are similar as natural extracts.

History of R&D

The report in 1951 showed that indirubin can inhibit eosinophils in the blood of guinea pigs. But this report did not attract attention. Indirubin is the effective component of the traditional Chinese medicine Angelica aloe pill. Since 1966, the scientists in the Institute of Hematonosis, Chinese Academy of Medical Sciences Research, started the research of therapying chronic myelocytic leukemia with the TCM rules using Angelica aloe pills, which had certain effect. Angelica aloe pill consisted of 11 herbs, such as *Angelica*, *Aloe*, *Rhizoma Coptidis*, and natural indigo. It was identified that natural indigo was the effective component of Angelica aloe pill. However, the active ingredient of natural indigo was indirubin.

Indirubin is a novel antileukemia drug, which was found by the medical scientists in China in the middle of the 1970s [3]. It has the effects of antibacterial, anti-inflammatory, antitumor, and enhancing immune functions. The compound has been applied to the clinical treatment of chronic myeloid leukemia. Indirubin has the characteristics of reliable clinical curative effect, small side effects, and no obvious inhibitory effect on the bone marrow.

Pharmacology

The effect of indirubin has anti-inflammatory, antibacterial, detoxification, enhancing immune function anticancer [4]. Indirubin has moderate inhibitory effect for animal-transplanted tumor. 200 mg/kg indirubin subcutaneous or intraperitoneal injection, once daily for 6 consecutive days, had the inhibition effect to rat W256 tumor cancer sarcoma; the inhibition rates were 47–52% and 50–58%, respectively. The chemotherapy index of intraperitoneal injection was 2.23. However, the inhibition rate of 500 mg/kg indirubin by gavage was only 23–33%. Indirubin by perfusion can prolong the survival time of W526 rat. The inhibition rate of Lewis mice's lung cancer was 43% for 500 mg/kg indirubin orally, once a day for 9–10 days. Indirubin has some inhibition effects on mice breast cancer, but no obvious effect for L1212, P388, and L1210 of lymphocytic leukemia in mice [5].

The effect of indirubin on chronic myelogenous leukemia is very obvious [6, 7], which is similar to first clinical choice of Maryland. Indirubin has the advantages of fast curative effect, no obvious inhibition effect of the bone, small bone marrow toxicity, and low side effects. Indirubin may have the effect of improving adrenocorticotropic hormone. In pathological conditions, such as inflammatory diarrhea, protein metabolism disorder, kidney disease, leukemia, and other tumors, urinary excretion of indirubin increased. Indirubin can inhibit synthesis of DNA and destroy the leukemia cells. It was found by electron microscopy that juvenile cells reduced, even disappear completely, with indirubin administration. Indirubin can significantly reduce the size of spleen, increase the concentration of hemoglobin to normal level, and reduce the swelling liver. In addition, indirubin can also enhance the phagocytic ability of animal mononuclear macrophages, which play a role in body immune reaction. So the anticancer effect of this product may be related to improving the body's immunity.

Clinical Application

Indirubin is mainly used for chronic myeloid leukemia; its total effective rate was 87.3%. The effect of indirubin to decrease white blood cells is similar to Maryland. The effect of indirubin to reduce liver is better than that of Maryland. But the remission role of the blood and bone marrow is worse than Maryland, and no cross resistance with Maryland. Indirubin could be used for abnormal bone marrow hyperplasia and eosinophilia.

Discussion

At present, in the global scope, the malignant tumor has become a serious disease endangering human life and health. Anticancer drug with reliable clinical effect and low toxic side effects has gradually become the mainstream of research. Indirubin is an effective ingredient of medicine with novel structure. It has the advantages of reliable clinical curative effect, small side effects, and no significant inhibitory effect on the bone marrow, which has inhibitory effect on various human tumor cells. Therefore, indirubin and its derivatives will become a new direction of research in the field of anticancer drugs.

Fig. 1 *Indigofera tinctoria* L

Fig. 2 The chemical structure of indirubin

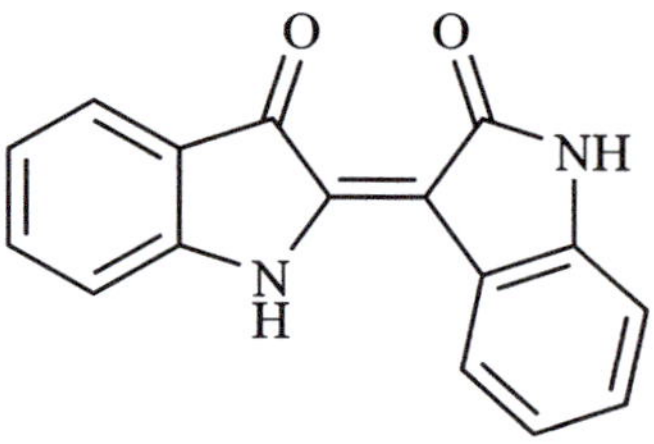

References

1. American Chemical Society. Chemical abstracts of the United States. Chemical abstracts of the United States Service, 1998.
2. Tu X. Study on the synthesis of natural products of indirubin and three phenol derivatives. Master's thesis of Northwestern University. 2010.
3. Chen H, Sun J. Research progress of Folium Isatidis. China Pharm. 2004;13(8):79–80.
4. Wu J, Ge Z, Gao Y, Zhang L. Study on the inhibitory effect of indirubin on tumor cells and the related mechanisms. Tianjin Tradit Chin Med. 2008;1:55–8.
5. Hoessel R, Leclerc S, Endicott JA, et al. Indirubin, the active constituent of a Chinese antileukaemia medicine, inhibits cyclin-dependent kinases. Nat Cell Biol. 1999;1(1):60–7.
6. Xiao Z, Hao Y, Liu B, Qian L. Indirubin and meisoindigo in the treatment of chronic myelogenous leukemia in China. Leuk Lymphoma. 2002;43(9):1763–8.
7. Nam S, Scuto A, Yang F, et al. Indirubin derivatives induce apoptosis of chronic myelogenous leukemia cells involving inhibition of Stat5 signaling. Mol Oncol. 2012;6(3):276–83.

Monocrotaline

Li Zhang, Li-Da Du, and Yang Lu

Abstract Morphine is an important alkaloid extracted from *Crotalaria sessiliflora* L. Monocrotaline showed anticancer activity in vivo, especially for treating squamous cell carcinoma, cervical cancer, and leukemia. However, since many pyrrolizidine alkaloid compounds have hepatotoxicity, further development of these compounds is restricted. Monocrotaline's toxicity is low, but its metabolite in the liver has high liver toxicity. Monocrotaline can also reduce blood pressure and inhibit cardiac contractility and heart rate and also shows mild inhibition of respiratory rate and depth.

Keywords Monocrotaline · Alkaloid · Anticancer · Hepatotoxicity

Origin: Monocrotaline (Fig. 1)
Chemical name (Fig. 2)

(3R,4R,5R,8a^{1}R,13aR)-4,5-Dihydroxy-3,4,5-trimethyl-4,5,8,8a^1,10,12,13,13a-octahydro-2H-[1,6]dioxacycloundecino[2,3,4-gh]pyrrolizine-2,6(3H)-dione

Molecular formula, $C_{16}H_{23}NO_6$; **MW**, 325.3; **CAS**, 315-22-0

L. Zhang · Y. Lu (✉)
Beijing Key Laboratory of Polymorphic Drugs, Research Center of Polymorphic Drugs, Institute of Materia Medica, Chinese Academy of Medical Sciences and Peking Union Medical College, Beijing, China
e-mail: luy@imm.ac.cn

L.-D. Du
School of Biomedical Sciences, Chinese University of Hong Kong, Hong Kong, China

G.-H. Du, *Natural Small Molecule Drugs from Plants*,
https://doi.org/10.1007/978-981-10-8022-7_88

Properties

Appearance: white prism crystal. *Solubility*: soluble in methanol, ethanol, and chloroform; slightly soluble in benzene, water, ether, and acetone; insoluble in petroleum ether. *Melting point*: 197–198 °C. *Specific optical rotation*: −54.7 ° (in chloroform).

Indications

This product is available in the *Pharmacopoeia of the People's Republic of China* (1977).

Clinical available formulations include gel and liposome transdermal preparations. It is effective in treating skin cancer, basal cell carcinoma, acute leukemia, and cervical and penile cancer.

Literature

Monocrotaline is present in the *Crotalaria sessiliflora* L. Application of *Crotalaria sessiliflora* L. was recorded in "Sheng-cao-yao-xing-bei-yao" by He Kejian in Qing dynasty, and then Wu Qijun recorded it in the "Zhi-wu-ming-shi-tu-kao." In China, it is commonly used for treating dysentery, cough, rheumatism, snake bites, infantile malnutrition, and cancers. The seeds of *Crotalaria sessiliflora* L. contain up to 0.4% of monocrotaline; thus, this plant is considered as an important resource to obtain monocrotaline [1].

Crotalaria Linn. is a Leguminosae plant. Its main active ingredient is monocrotaline [2]. The whole plant and seeds can be used as medicine. There are about 350 species of *Crotalaria* Linn. in the world, distributed in tropical and subtropical regions. There are about 34 species distributed in the south and southwest of China.

History of R&D

In 1935, W. M. Neil and L. L. Russoff isolated monocrotaline from *Crotalaria sessiliflora* L. Monocrotaline showed anticancer activity in vivo, especially for treating squamous cell carcinoma, cervical cancer, and leukemia. However, since many pyrrolizidine alkaloid compounds have hepatotoxicity, further development of these compounds is restricted.

China began the research of monocrotaline, as early as 1943. *Crotalaria sessiliflora* L. was also recorded in the 1977 edition of the *Pharmacopoeia of the People's*

Republic of China. But monocrotaline was reported for its liver toxicity; teratogenic, allergic reactions; and other side effects, limiting its further clinical applications. In 1992, the Ministry of Health stopped the clinical application of monocrotaline injection. Further pharmacology study is essential for monocrotaline [3].

Pharmacology

Monocrotaline is a pyrrolizidine alkaloid which has anticancer and anti-choline effects. In vitro study suggested that monocrotaline has significant cytotoxicity in human BEL-7402, KB cancer cells through inducing DNA alkylation [4]. Monocrotaline's toxicity is low, but its metabolites in the liver have high liver toxicity. The metabolites are highly electrophilic and can bind to enzymes, proteins, DNA, and RNA, thus causing several side effects. Monocrotaline can reduce blood pressure and inhibit cardiac contractility and heart rate and also shows mild inhibition of respiratory rate and depth.

Clinical Application

Monocrotaline injection shows well therapeutic effect for treating squamous cell carcinoma and basal cell carcinoma. In folk, the fresh pulp or dry powder of the herb has been also used for treating squamous cell carcinoma and basal cell carcinoma [5].

Fig. 1 *Crotalaria sessiliflora* L

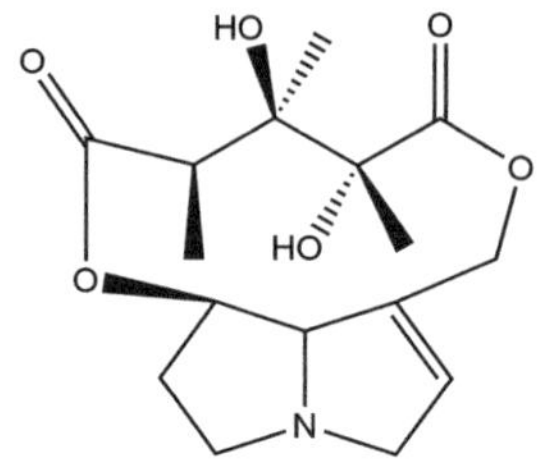

Fig. 2 Chemical structure of monocrotaline

References

1. Fan QY. Agricultural Geely in the extraction of active ingredients analysis. Guangdong College of Pharmacy Master's Degree thesis, pp 1–3.
2. Li LZH, Zhu HY, Shi JSH, et al. Chemical composition and pharmacological activities of *Crotalaria* Linn plants. Nat Prod Res Dev. 2007;19:724–30.
3. Jiang LQ, Cheng M, Gao QFTJ. Analysis of monocrotaline in seed of *Crotalaria assamica* Benth. Chin Pharm. 2010;13(9):1743–7.
4. Liu Q, Chen LL, Shuangcheng M, et al. Monocrotaline in five kinds of liver microsomes in vitro metabolism research. Chin Pat Med. 2014;36(8):1764–8.
5. Zhao QL, CHJ B, GGG Y. Prospects for development and utilization of crotalaria. Trop Agric Sci. 2008;28(4):71–4.

Paclitaxel

Lin Wang and Guan-Hua Du

Abstract Paclitaxel belongs to diterpenoid compound and the main natural source of it is yew bark. Docetaxel is the main derivative of paclitaxel and artificially synthesized on the base of the structure of paclitaxel. Paclitaxel and docetaxel showed potent suppression to most solid tumors and gained the highest sales among all the antitumor drugs in the world due to its specific mechanism. The main target of paclitaxel is microtubule in cells. Compared with paclitaxel, docetaxel shows stronger affinity to tubulin and longer retention time in cells. Paclitaxel is mainly metabolized through the liver and eliminated from the body by feces. Paclitaxel showed no obvious toxic effects on main organs, and its clinical side effects include allergic reactions, nervous system damage, hematopoietic suppression, catarrh, and cardiac toxicity.

Keywords Paclitaxel · Docetaxel · Tubulin · Cancer

Alias: Yew alcohol, violet, and purple alkanes, road tiger, tiger, Anzatax
Origin: Yew (Fig. 1)
Chemical name (Fig. 2)

(2alpha,5beta,7beta,10beta,13alpha)-4,10-bis(acetyloxy)-13-{[(2R,3S)-3-(benzoylamino)-2-hydroxy-3-phenylpropanoyl]oxy}-1,7-dihydroxy-9-oxo-5,20-epoxytax-11-en-2-yl benzoate

Molecular formula, $C_{47}H_{51}NO_{14}$; **MW**, 853.92; **CAS**, 33069-62-4

L. Wang · G.-H. Du (✉)
Beijing Key Laboratory of Drug Targets Identification and Drug Screening, Centre for Pharmaceutical Screening, Institute of Materia Medica, Chinese Academy of Medical Sciences and Peking Union Medical College, Beijing, China
e-mail: dugh@imm.ac.cn

G.-H. Du, *Natural Small Molecule Drugs from Plants*,
https://doi.org/10.1007/978-981-10-8022-7_89

Properties

Appearance: Odorless and tasteless white or kind of white crystal powder. *Solubility*: Poorly soluble in water but slightly soluble in ether. Soluble in methanol, acetonitrile, chloroform, acetone, and other organic solvents. *Melting point*: 213–216 °C. *Specific optical rotation*: −49° ($C = 1$, MeOH); Curl: [alpha] 20° to $D = 49.0$–55.0° (10 mg/mL of methanol solution) in anhydrous dry goods without solvents.

Derivatives: docetaxel (Fig. 3)

Chemical Name

5 beta, alpha 20 – epoxy – 1, 2, 4, 7 beta, 10 beta, alpha – 6 hydroxy taxane – 11-13 ene – 9 – ketone – 4, 10 – two acetate – 2 – benzoate – 13 – [(2 R, 3 S) ‘- N – benzoyl – 3 – phenyl serine ester

Molecular formula, $C_{43}H_{53}NO_{14}$; **MW**, 853.92; **CAS**, 114977-28-5

Dosage Forms

This product is included in the Pharmacopoeia of the People's Republic of China (2015), the British Pharmacopoeia (2017), the United States Pharmacopeia (40), and the European Pharmacopoeia (9.0th ed.). The current dosage form used in clinic is injection.

Indications

It's suitable for the treatment of metastatic ovarian cancer and breast cancer. It's also used in the treatment of small cell and non-small cell lung cancer, cervical cancer, malignant melanoma, head and neck cancer, and leukemia, tolerance to chemotherapy, etc.

Literature

Paclitaxel belongs to diterpenoid compound and the main natural source of it is yew bark. Yew [*Taxus chinensis* (Pilg.) Rehd] is also called the yew or red building, which is mainly scattered distributed in the northern hemisphere. However, there is no massive natural yew forest formed throughout the world. *Taxus media*, a natural hybrid, was discovered by American scholar in 1918. The female parent of *Taxus media* is northeast yew, and the male parent of it is the European *Taxus chinensis*. Taxus cultivar is divided into four groups. Among the first group, Coleana, Stovekenii, and Hicksii are preferred varieties for the extraction of taxane compounds. Especially, Hicksii has been approved by FDA as one of the varieties of *Taxus chinensis* for extracting paclitaxel [1]. From the study, baccatin III, a paclitaxel precursor compound isolated from *Taxus chinensis*, was found to have the same mother nuclear structure with paclitaxel, and the content of it is high in *Taxus chinensis* needles, although the biological activity of baccatin III is lower than paclitaxel. Moreover, paclitaxel can be obtained by four steps of chemical reaction with baccatin III, and the production rate is as high as 80%. This exciting finding made great progress for solving the new sources of paclitaxel and made it possible to produce paclitaxel in mass [2].

History of R&D

The toxic ingredients in branches and leaves of *Taxus chinensis* were separated in 1856 and named "taxine," which was identified as a kind of white alkaloid's component. In 1962, Dr. Barclay, who worked for the US Department of Agriculture, found the Pacific yew in the forest park located in Washington, D.C. Then he sent the plant samples to a research institute in Wisconsin, USA, for the activity screening and extraction of effective components. The samples from the yew tree bark extracts were found cytotoxic on KB cells in 1964. The antitumor activity of the extract was confirmed by repeated experiments later. Dr. Wall and his partners in Triangle Research Institute in North Carolina, USA, obtained a white crystalline monomer named paclitaxel from the Pacific yew tree bark in June 1967 [3].

In 1979, Horwitz, a medicine molecular pharmacologist in the Albert Einstein College, illustrated the special antitumor mechanism of paclitaxel. Paclitaxel is the first found drug which can interact with tubulin polymer [4]. In 1983, NCI applied for clinical trials for paclitaxel to the Food and Drug Administration (FDA) of the United States. In 1984, paclitaxel was approved as a new drug for phase I clinical trials in ovarian cancer. In 1992, the Bristol-Myers Squibb submitted new drug applications to FDA for the development of paclitaxel injection. In December, FDA approved paclitaxel to be listed for the treatment of advanced ovarian cancer in stage II. Currently, among all the antitumor drugs, the sale of paclitaxel becomes the first in the world as a well-recognized anticancer drug with potent broad-spectrum activity. In October of 1995, China became the second country with formal production of paclitaxel and its injection in the world. The achievement was gained under the unremitting efforts of researchers in the Institute of Materia Medica, Chinese Academy of Medical Sciences.

Pharmacology

Paclitaxel is mainly used for the treatment of ovarian cancer and breast cancer. The mechanism of it includes:

1. The effects on cell microtubules/tubulin: Inhibition of microtubule depolymerization results in abnormal micro tube bundle arrangement and makes the spindle lose normal function and then induces cell death.
2. In the absence of bird triphosphate (GTP) and microtubule associated protein (MAP), it induces cells to form microtubule lack of function.
3. It significantly sensitized cancer cells to radiotherapy through blocking the cell cycle in the stage of G2 and M [5].

Paclitaxel is mainly metabolized through the liver and enters into the intestine with bile and then eliminated from the body by the feces (90%). The renal clearance of paclitaxel only accounts for 1–8% of total removal of paclitaxel. Docetaxel (Taxotere), a similar compound of paclitaxel, is also widely used in clinical application. Docetaxel is artificially synthesized on the base of the structure of paclitaxel and exhibits the similar activity and mechanism with paclitaxel. It also showed the activity of anti-angiogenesis and inducing apoptosis. Compared with paclitaxel, docetaxel shows stronger affinity to tubulin and longer retention time in cells. Therefore, docetaxel is suitable for the treatment of local advanced non-small cell lung cancer or metastatic breast cancer. Moreover, the cytotoxicity of docetaxel is 1.3- to 12-folds of paclitaxel.

Pharmacokinetic study showed the intracellular drug concentration, and retention time of docetaxel in body was threefold of paclitaxel. In addition, drug-resistant cell lines to paclitaxel are still sensitive to docetaxel. Docetaxel is mainly metabolized into four main metabolites by the cytochrome P450 in the liver. The main metabolites are discharged from feces by the biliary system, and only 5–7% of the metabolites are ruled out from urine [6].

Clinical Application

The first indication of docetaxel for clinical application is the second-line drug treatment for the advanced ovarian cancer. The effect of paclitaxel in combination with standard chemotherapy drug therapy for early breast cancer is significant. Especially, the combination of paclitaxel and trastuzumab has been the current standard first-line therapeutic schedule for the treatment of metastatic breast cancer with overexpression of Her-2. In addition, paclitaxel is used in the second-line drug treatment for metastatic breast cancer [7].

Docetaxel showed good clinical effects for breast cancer, prostate cancer, lung cancer, gastric cancer, ovarian cancer, cervical cancer, and head and neck cancer. Kinds of multidrug resistance cancer cell lines with overexpression of P-glycoprotein and other multidrug resistance-associated proteins are found sensitive to docetaxel. Docetaxel is the first and only chemotherapy drug which was approved to exhibit clear survival benefits to hormone-independent prostate cancer patients. In 2004, FDA approved the intravenous administration of docetaxel in combination with prednisone for the treatment of advanced stage of metastatic prostate cancer. In addition, paclitaxel is the main drug in the clinic to treat advanced non-small cell lung cancer and showed great potency for the treatment of drug-resistant cancer cells [8].

Discussion

After doxorubicin and cisplatin, paclitaxel is the best anticancer drug discovered by human being with minimal drug adverse reactions. Paclitaxel and its analogue docetaxel showed potent suppression to most solid tumors through novel mechanisms. Meantime, although the clinical side effects of paclitaxel include allergic reactions, nervous system damage, hematopoietic suppression, catarrh, cardiac toxicity, etc., it showed no obvious toxic effects on the liver, kidney, lung, and other organs. In company with the constant progress and gradually penetrating in the research of related field, the clinical application of paclitaxel scope continued expanding. Paclitaxel gained the highest sales among all the antitumor drugs in the world. Around the application of paclitaxel, clinical trials have been carried out for the study of the anticancer activity on other common cancers and a variety of combination therapy. Moreover, there is a great development in the research on the methods of obtaining paclitaxel by biosynthesis, fungal fermentation, and plant tissue culture. Therefore, it is hopeful to solve the problem of inadequate materials for paclitaxel in a short term.

Fig. 1 Yew

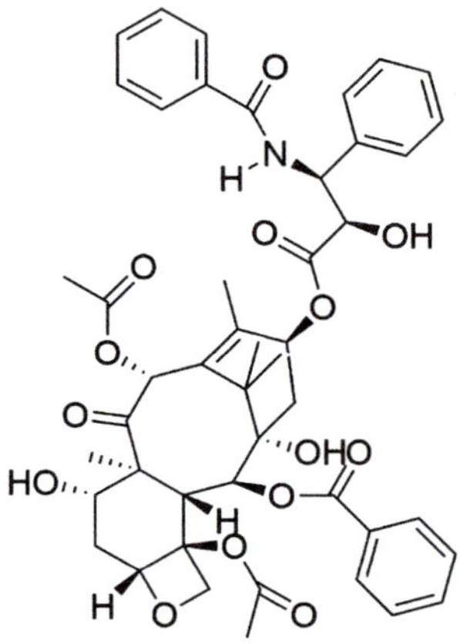

Fig. 2 Structure of paclitaxel

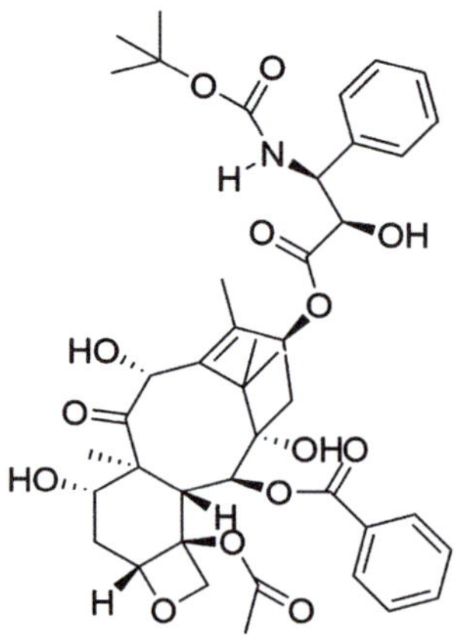

Fig. 3 Structure of docetaxel

References

1. Peng S, Zhao T-z, Wu Y. Natural ecological factors on monte and reviewing the effect of paclitaxel yield review. For Resour Manag. 2009;3:88–91.
2. Shi Q-w, Lilly, Huo, Hong C, et al. Paclitaxel anti-cancer drugs research and development process of thinking and analysis. J Med Philos. 2010;31(6):6–8.
3. Shi Q-W. Natural medicine chemistry myth: paclitaxel. Chin Herb Med, Lancet. 2011;10:1878–84.
4. Schiff PB, Fan J, Horwitz SB. Promotion of microtubule assembly in vitro by paclitaxel. Nature. 1979;277:665–7.
5. Mei XG. Dao-wei Hu: the research progress of paclitaxel pharmaceutical development. J Mod Med Clin. 1996;6:247–51.
6. Wang P-W, Dan C-J. The molecular mechanism of paclitaxel action research progress. Chem Life. 1999;6:254–6.
7. Barbuti AM, Chen ZS. Paclitaxel through the which of anticancer therapy: exploring its role in chemoresistance and radiation therapy. Working (Basel). 2015;7(4):2360–71.
8. Ma P-j. Clinical application research progress of paclitaxel. J Shanghai Med Intell Res. 2004;2:34–42.

Podophyllotoxin

Xiao-Cong Pang, Li Zhang, and Guan-Hua Du

Abstract Podophyllotoxin is isolated from *Podophyllum*. Podophyllotoxin has significant antitumor and antiviral activities. However, *Podophyllum* showed several toxicity and side effects. Podophyllotoxin derivatives, etoposide, Etopophos, amino sugar etoposide, and teniposide, have been developed as anticancer drugs. But these podophyllotoxin derivatives are not free of toxicity. Despite toxicity and side effects associated with the clinical use of podophyllotoxin and its derivatives, podophyllotoxin is still a good candidate for the design and synthesis of analogs with more potent activity and less side effects.

Keywords Podophyllotoxin · Anticancer drug · Side effects

Alias: Podophyllum toxin
Origin: *Podophyllum* (Fig. 1)
Chemical name (Fig. 2)

(5R,5aS,8aR,9R)-9-Hydroxy-5-(3,4,5-trimethoxyphenyl)-5,8,8a,9-tetrahydrofuro[3′,4′:6,7]naphtho[2,3-d][1,3]dioxol-6(5aH)-one

Molecular formula, $C_{22}H_{22}O_8$; **MW**, 414.41; **CAS**, 518-28-5
Podophyllotoxin derivatives (Fig. 3)

X.-C. Pang · L. Zhang · G.-H. Du (✉)
Beijing Key Laboratory of Drug Targets Identification and Drug Screening, Centre for Pharmaceutical Screening, Institute of Materia Medica, Chinese Academy of Medical Sciences and Peking Union Medical College, Beijing, China
e-mail: dugh@imm.ac.cn

G.-H. Du, *Natural Small Molecule Drugs from Plants*,
https://doi.org/10.1007/978-981-10-8022-7_90

Properties

Appearance: white needle crystal powder. *Solubility*: freely soluble in chloroform, acetone, ethyl acetate, and benzene; soluble in ethanol and ethyl ether; and insoluble in water. *Melting point*: after drying the melting point is 183–184 °C. *Specific optical rotation*: 132.7 °C (chloroform).

Dosage Forms and Indications

Podophyllotoxin preparations include solution, ointment, tincture, coating agent, gelling agent, etc. Podophyllotoxin has been used as antiviral agent in the treatment of condyloma acuminatum and other warts as well as anticancer agent.

Literature

Podophyllotoxin (2,3-butyl-4-aromatic naphthene) is isolated from *guijiu* 鬼臼属 (*Podophyllum*) [1]. There are two species as the main source of podophyllotoxin, *Podophyllum hexandrum* Royle and *Podophyllum peltatum* [2].

Although podophyllotoxin has significant antitumor and antiviral activities, it showed several toxicity and side effects. Podophyllotoxin derivatives, etoposide (VP-16-213), Etopophos, amino sugar etoposide (NK6l1), and teniposide (VM26), have been developed as anticancer drugs. They are used to treat small cell lung cancer, testicular cancer, acute leukemia, malignant lymphoma, etc. But podophyllotoxin derivatives are not free of toxicity. Besides the narrowing of the anticancer spectrum and low water solubility, these drugs could induce severe myelosuppression, gastrointestinal side effects, etc.

Although the synthetic and biosynthetic pathways of podophyllotoxin have been elucidated, it is still the most effective, economic, and fast way to extract podophyllotoxin from the plant.

History of R&D

Podophyllotoxin was first found in the *Podophyllum peltatum* L. The first time to isolate podophyllotoxin from podophyllin was in 1880. In 1942, it was found that venereal warts could be effectively treated by application of podophyllin.

Subsequently, podophyllotoxin was reported to inhibit the growth of the tumor through the inhibition of the microtubule formation. The chemical structure of podophyllotoxin was elucidated in 1951.

In the 1960s, two main podophyllotoxin derivatives were synthesized, etoposide and teniposide (VM-26) [3]. In 1983, etoposide was approved by FDA. Etoposide and teniposide are used in frontline cancer therapy against various cancer types, such as small cell lung cancer, testicular cancer, etc. In 1996, etoposide phosphate analog (Etopophos) was launched in America. Etopophos is the prodrug of etoposide and can be rapidly absorbed and completely converted to the parent compound in vivo. In 1990, WHO recommended 0.5% podophyllotoxin as the first-line drug for the treatment of condyloma acuminatum. Podophyllotoxin creams and gels are nowadays widely used in clinical practice.

Pharmacology

Antineoplastic and antiviral activities are the most pronounced pharmacological effects of podophyllotoxin [4]. Podophyllotoxin shows a significant inhibitory effect on the division and proliferation of epithelial cells infected by human papillomavirus (HPV), disrupts the cell cytoskeleton, and induces the necrosis and shedding of warts. It was shown that the antitumor effect of podophyllotoxin is associated with the inhibition of microtubule assembly and the induction of apoptosis. However, the antitumor effect of podophyllotoxin analogs, such as etoposide, teniposide, and Etopophos, is related to disparate mechanisms including the inhibition of DNA topoisomerase II activity and the formation of stable nucleic acid-drug-enzyme complex, which induce DNA double-strand or single-strand break and eventually lead to cell death [5–7]. It was also found that podophyllotoxin derivatives have immunosuppressive and anti-inflammatory effects.

Clinical Application

Podophyllotoxin is a useful agent for the treatment of condyloma acuminatum [8]. Podophyllotoxin and its derivatives are also widely used in the treatment of cancer, such as lymphomas and lung carcinoma. Because of the several toxicity of podophyllotoxin, for example, the irritation of skin and mucous membranes, combination therapies are used to treat condyloma acuminatum or cancer.

Discussion

Podophyllotoxin and its derivative preparations are used as antiviral and anticancer drugs in clinical practice. Despite toxicity and side effects associated with the clinical use of these preparations, podophyllotoxin continues to be a good candidate for the design and synthesis of analogs with more potent activity and less side effects.

Fig. 1 *Podophyllum*

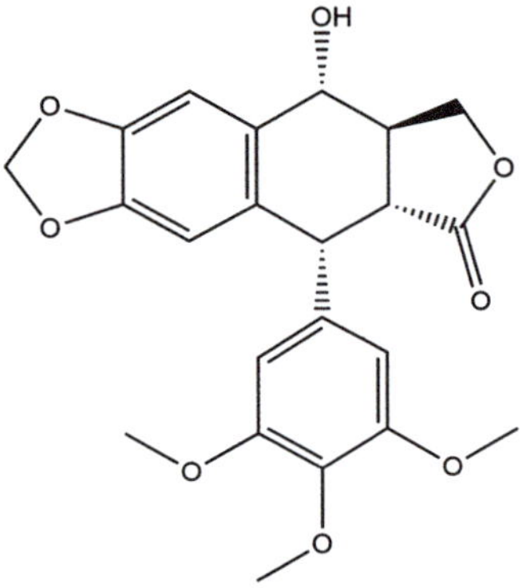

Fig. 2 The structure of podophyllotoxin

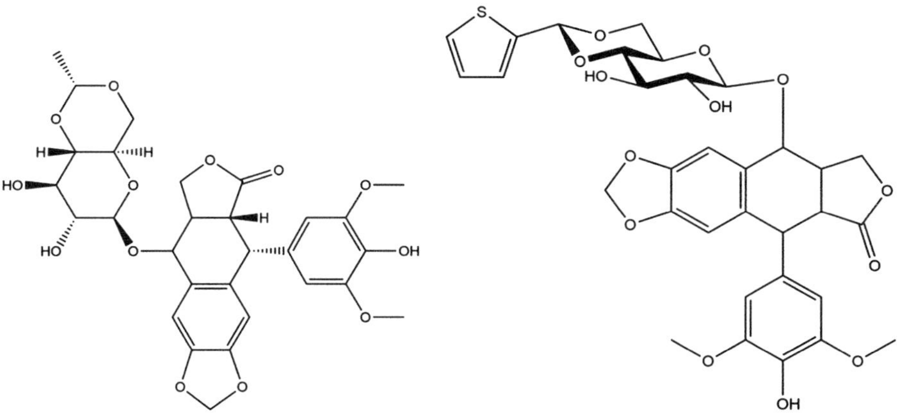

etoposide teniposide

Fig. 3 The structures of podophyllotoxin derivatives

References

1. Yang XZ, Shao H, et al. Research status of podophyllotoxin resources. Chin Herb Med. 2001;32(11):1042–4.
2. Newman DJ, Cragg GM, Snader KM. Natural products as sources of new drugs over the period 1981–2002. J Nat Prod. 2003;66:1022–37.
3. Stähelin HF, von Wartburg A. The chemical and biological route from podophyllotoxin glucoside to etoposide: ninth Cain memorial award lecture. Cancer Res. 1991;51(1):5–15.
4. Guerram M, Jiang ZZ, Zhang LY. Podophyllotoxin, a medicinal agent of plant origin: past, present and future. Chin J Nat Med. 2012;10(3):161–9.
5. Gordaliza M, Miguel del Corral JM, Angeles Castro M, et al. Cytotoxic cyclolignans related to podophyllotoxin. Farmaco. 2001;56(4):297–304.
6. Shi P. Study on the antitumor activity and mechanism of podophyllotoxin in vitro. Dissertation of East China University of Science and Technology. 2012; p 37–9.
7. Imbert TF. Discovery of podophyllotoxins. Biochimie. 1998;80(3):207–22.
8. Xiao Y, Li XY, Zeng K. Progress in the treatment of condyloma acuminatum with different forms of podophyllotoxin preparation. Chin J Dermatol Biotechnol. 2013;27(5):524–5. 534

Vinblastine and Vincristine

You-Wen Zhang, Xiang-Ying Kong, Jin-Hua Wang, and Guan-Hua Du

Abstract Vinblastine and vincristine are all extracted from *Catharanthus roseus*, which is a common Apocynaceae ornamental plant. Vinblastine and vincristine drugs have a strong inhibitory effect on monozygotic leukemia, breast cancer, liver cancer, ovarian cancer, head and neck cancer, testicular cancer, solid sarcoma, and malignant melanin in a variety of spontaneous or transplanted lymphocytic leukemia. Although the structure of vinblastine and vincristine is very similar, there are some differences in pharmacological effects, and there is no cross-resistance. The mechanism of vinblastine and vincristine is similar as the cell cycle-specific antineoplastic agents. Clinical applications of vinblastine and vincristine are very broad, especially in China. Vinblastine sulfate injection as the main application form is used intravenously to treat Hodgkin lymphoma, lymphocytic leukemia, testis tumor, and choriocarcinoma as first-line therapy in clinical practice according to ABVD method. However, antitumor effects still face challenges, and have a long way to go. The process of research and development of vinblastine drugs will provide us more meaningful revelations.

Keywords Vinblastine and vincristine · Alkaloid · Antitumor

Origin: *Catharanthus roseus* (Fig. 1)

Y.-W. Zhang
Beijing Key Laboratory of Anti-infective Agents, Institute of Medicinal Biotechnology, Chinese Academy of Medical Sciences and Peking Union Medical College, Beijing, China

X.-Y. Kong
Institute of Chinese Materia Medica, China Academy of Chinese Medical Sciences, Beijing, China

J.-H. Wang · G.-H. Du (✉)
Beijing Key Laboratory of Drug Targets Identification and Drug Screening, Centre for Pharmaceutical Screening, Institute of Materia Medica, Chinese Academy of Medical Sciences and Peking Union Medical College, Beijing, China
e-mail: dugh@imm.ac.cn

G.-H. Du, *Natural Small Molecule Drugs from Plants*,
https://doi.org/10.1007/978-981-10-8022-7_91

Vinblastine

Alias: Vinblastine
Chemical name (Fig. 2)

Dimethyl(2β,3β,4β,5α,12β,19α)-15-[(5S,9S)-5-ethyl-5-hydroxy-9-(methoxy-carbonyl)-1,4,5,6,7,8,9,10-octahydro-2H-3,7-methanoazacycloundecino[5,4-b] indol-9-yl]-3-hydroxy-16-methoxy-1-methyl-6,7-didehydroaspidospermidine-3,4 dicarboxylate

Molecular formula, $C_{46}H_{58}N_4O_9$; **MW**, 810.97; **CAS**, 865-21-4

Properties

Appearance: needlelike crystals were produced when recrystallized by methanol. *Solubility*: soluble in chloroform, acetone, and ethanol. *Melting point*: 211–216 °C. Sulfate melting point of vinblastine is 284–285 °C; hydrochloric acid melting point of vinblastine is 244–246 °C (decomposes). *Specific optical rotation*: +42°.

Dosage Forms and Indications

It was recorded in the *Pharmacopoeia of the People's Republic of China* (2015), the *British Pharmacopoeia* (2017), the *United States Pharmacopeia* (40), the *Japanese Pharmacopoeia* (17th ed.), the *European Pharmacopoeia* (9th ed.), and *The International Pharmacopoeia* (5th ed.).

Vinblastine sulfate injection is used as the first-line therapy to treat Hodgkin lymphoma, lymphocytic leukemia, testis tumor, and choriocarcinoma intravenously in clinical practice generally according to ABVD methods (Adriamycin, bleomycin, vinblastine, and dacarbazine).

Vincristine

Chemical name (Fig. 3)

(3aR,3a1R,4R,5S,5aR,10bR)-Methyl4-acetoxy-3a-ethyl-9-((5S,7S,9S)5-ethyl-5-hydroxy-9-(methoxycarbonyl)-2,4,5,6,7,8,9,10-octahydro-1H-3,7-methano[1] azacycloundecino[5,4-b]indol-9-yl)-6-formyl-5-hydroxy-8-methoxy-3a,3a1,4,5,5a,6,11,12-octahydro-1H-indolizino[8,1-cd]carbazole-5-carboxylate

Molecular formula, $C_{46}H_{56}N_4O_{10}$; **MW**, 824.96; **CAS**, 57-22-7

Properties

Appearance: needlelike crystals were produced when recrystallized by methanol. The other properties of vincristine are similar to vinblastine.

Dosage Forms and Indications

It was recorded in the *Pharmacopoeia of the People's Republic of China* (2015), the *British Pharmacopoeia* (2017), the *United States Pharmacopeia* (40), the *Japanese Pharmacopoeia* (17th ed.), the *European Pharmacopoeia* (9th ed.), and *The International Pharmacopoeia* (5th ed.).

Vincristine sulfate injection is used to treat leukemia, lymphadenoma, non-small cell lung cancer, nephroblastoma, and neuroblastoma in clinical practice generally, but the effect of vinblastine is better than vincristine in Hodgkin lymphoma treatment.

Literature

Vinblastine and vincristine are all extracted from *Catharanthus roseus*, which is a common Apocynaceae ornamental plant. Aliases are Jin Zhan Cao and Ying Tou Hong. They are originated from Mediterranean coast and Central and South America. It was introduced to China recently and widely planted in Guangdong, Guangxi. So there are few medicine records of the whole plant in China history, but it has different applications in many parts of the world. It has been recorded that the juice of vinca leaves can be used to control scurvy and hemorrhagic fever, to relieve toothache by mouthwash, and to heal wound by local application. In India and the Philippines, the herbal tea based on vinca can be used to control blood sugar, and this plant is used for the treatment of diabetes by the name of "Vin-q-lin" in the UK [1].

History of R&D

Vinblastine and vincristine were extracted from vinca, and researchers found that they are the main ingredient of vinca which have antitumor activity. This is an accidental harvest by hard work and inspiration during scientific research process.

It was firstly reported as an important discovery by *Annals of the New York Academy of Sciences*, and one report was published by Robert Noble and Charles Thomas Beer [2]. Researchers validated the antidiabetic effects on rabbits of vinca extract from West Indies. However, the following experimental results showed negative effect by researchers of Eli Lilly and Company. They found an interesting result that some components of vinca extract can cause peripheral blood leukopenia and myelosuppression. This result provided researchers an assumption whether vinca can be used to treat leukemia.

The exciting idea inspired the researchers strongly. They demonstrated that vinblastine can cause severe leukopenia. At same time researchers from Eli Lilly demonstrated that the specific components of vinca can significantly prolong the life of the rat model of lymphocytic leukemia transplantation. Since then, the two teams worked together and identified the specific components vinblastine, vincristine, and vinorelbine (vinblastine). Meanwhile, they defined the antileukemia effect of vinblastine, vincristine, and vinorelbine (vinblastine).

In 1963, vincristine (trade name, Oncovin) developed by Eli Lilly was approved by the US FDA, it provided a good choice for tumor chemotherapy program and gradually became the essential tumor chemotherapy drugs. Since then, some excellent scientists discovered a class of drugs that affect the world from the disappointing research results.

At the early stages of the study, vinblastine and vincristine are directly extracted from the vinca, but the cost is high, and the extraction efficiency is low [1]. Especially the specific chiral monomer compounds with biological activity are difficult to obtain. In the following years, the researchers used genetic engineering, biosynthesis, chemical synthesis, and other technical methods to explore and optimize the production process of vinblastine drugs. In particular, the rise of asymmetric synthesis provides a more economical choice for the synthesis of vinblastine compounds and increases productivity to more than 22% [3]. At the same time, the pharmaceutical dosage forms were developed from the beginning as simple sulfate injection to liposomal preparations and nanoparticle preparations, which improve the efficacy, reduce adverse reactions, and play better antitumor effect.

Pharmacology

Vinblastine drugs have a strong inhibitory effect on monozygotic leukemia, breast cancer, liver cancer, ovarian cancer, head and neck cancer, testicular cancer, solid sarcoma, and malignant melanin in a variety of spontaneous or transplanted lymphocytic leukemias. Although the structure of vinblastine and vincristine is very similar, there are some differences in pharmacological effects, and there is no cross-resistance [4].

The mechanism of vinblastine and vincristine is similar as cell cycle-specific antineoplastic agents. They all inhibit tubulin polymerization, interfere spindle microtubule formation, block the cell cycle in M phase, and then inhibit the proliferation of cancer cells [4]. X-ray diffraction results showed that Vinblastine targets at the hydrophobic structure on tubulin, Vinblastine inserted into the trough as the wedge, blocking the polymerization of tubulin, making a spirochete structure of tubulin itself form, and loss of biological role [5]. However, this antitumor effect can affect other rapid proliferation cells, and then small intestinal epithelial cells and bone marrow cells will be inhibited, which is one of the reasons of gastrointestinal side effects and bone marrow transplantation side effects. The effect of vincristine is better than vinblastine, and the dosage is lower, the probability and intensity of side effects are lower. Recently, it has been found that vinblastine and vincristine can also suppress growth of tumor cell by inhibiting the synthesis of RNA and protein.

Clinical Application

Although vinblastine and vincristine have been found for more than 50 years, but their clinical applications are very broad, especially in China [6]. Vinblastine sulfate injection as main application form was used to treat Hodgkin lymphoma, lymphocytic leukemia, testis tumor, and choriocarcinoma intravenously as first-line therapy in clinical practice generally according to ABVD method (Adriamycin, bleomycin, vinblastine, and dacarbazine). Vincristine can be used to treat non-Hodgkin lymphoma combine with other chemotherapy drugs, but it has been gradually replaced by ABVD method. Vincristine has a significant effect on acute lymphoblastic leukemia (ALL), especially children with acute lymphoblastic leukemia. In addition, it can also be used for the treatment of non-small cell lung cancer, nephroblastoma, neuroblastoma, breast cancer, chronic lymphocytic leukemia, and multiple myeloma.

The toxicities of vinblastine drugs includes neurotoxicity, especially for motor neurons. It always causes side effects such as slowness of reflex, numbness of the extremities, disappearance of tendons, and peripheral neuritis. Neurotoxicity is commonly produced in patients over 40 years, and the tolerant of children is better than in adults. In addition, side effects also include gastrointestinal reactions and bone marrow suppression.

Discussion

It has been 60 years since the discovery of vinblastine and vincristine; however these classic chemotherapy drugs are still used as first-line drugs for the treatment of a variety of cancer, especially Hodgkin lymphoma and lymphoblastic leukemia. We ascribe these achievements to the rigor, keenness, and efficiency of scientists. The deep understanding of the mechanisms and targets not only deepens the understanding of clinical characteristics but also provides an example and reference for the discovery and development of new anticancer drugs. In recent years, chemotherapy of cancer has changed from cytotoxic drugs to targeted drugs, antibody drugs, and immunotherapeutic drugs, which are progresses in anticancer drug research. However, the treatment of cancer still has challenges, and antitumor drugs also have a long way to go; importantly, the process of research and development of vinblastine drugs will provide us more meaningful revelations.

Fig. 1 *Catharanthus roseus*

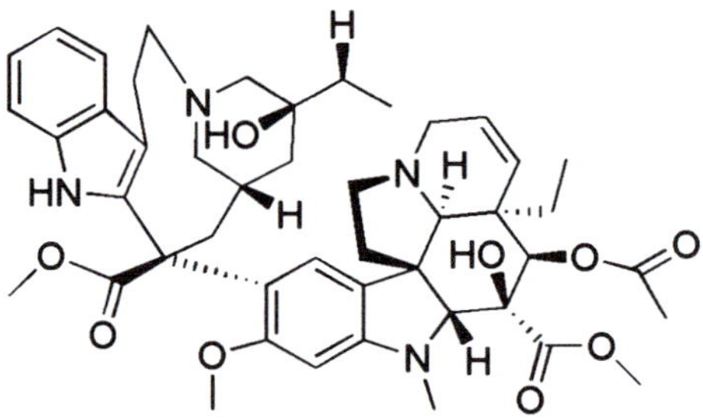

Fig. 2 Chemical structure of vinblastine

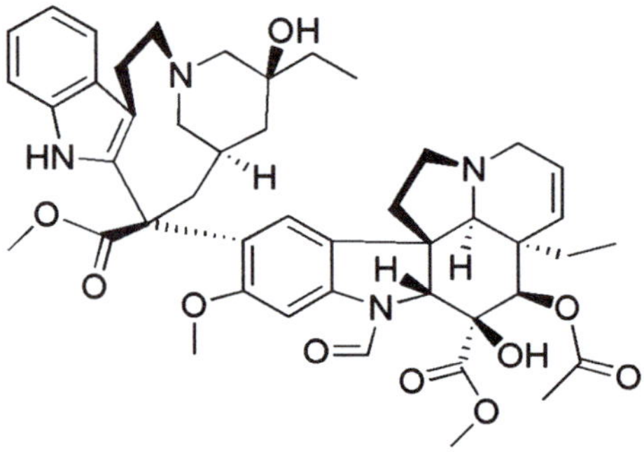

Fig. 3 Chemical structure of vincristine

References

1. Johnson IS, Armstrong JG, Gorman M, Burnett JP Jr. The vinca alkaloids: a new class of oncolytic agents. Cancer Res. 1963;23:1390–427.
2. Noble RL, Beer CT, Cutts JH. Role of chance observations in chemotherapy: Vinca rosea. Ann N Y Acad Sci. 1958;76(3):882–94.
3. Kuehne ME, Qin Y, Huot AE, Bane SL. The syntheses of 16 a′-homo-leurosidine and 16 a′-homo-vinblastine. Generation of atropisomers. J Organomet Chem. 2001;66(16):5317–28.
4. Jordan MA, Wilson L. Microtubules as a target for anticancer drugs. Nat Rev Cancer. 2004;4(4):253–65.
5. Gigant B, Wang C, Ravelli RB, et al. Structural basis for the regulation of tubulin by vinblastine. Nature. 2005;435(7041):519–22.
6. Xi LP, Song XB, Zhang LJ. Progress in study on basic and clinical application in treating liver cancer. Drug Eval Res. 2011;34(1):59–62.

Part V
Natural Small Molecule Drugs for the Treatment of Parasitic and Bacterial Infections

Overview

Parasitic infectious diseases are a class of diseases closely related to human living environment and living conditions, especially in economically underdeveloped countries and regions, where parasitic infectious diseases are usually common diseases. There are many types of parasitic infectious diseases with serious harm to human health, and some of them are even life-threatening. Some common ones have been prevalent in China, such as malaria from malaria parasite infection, lymphatic filariasis impairing the lymphatic system and the intestinal roundworm and tapeworm infections. These diseases have also been widespread in other areas of the world, and the treatment of these diseases should be widely given adequate attention.

Natural medicines play important roles in the treatment of parasitic infections, especially for the treatment of malaria parasites. One of the earlier antimalarial drugs was the well-known quinine. And in the 1980s, artemisinin was discovered, which could effectively control the spread of malaria and save millions of lives. It was found based on the traditional Chinese medicine. And in view of the contribution of artemisinin for human health, its main researcher, Chinese scientist Yoyo Tu, won the Nobel Prize in Medicine or Physiology in 2015.

In the treatment history of roundworm or tapeworm infections, people found an effective drug santonin from the natural product with a positive therapeutic effect for roundworm infection. And arecoline has a significant effect on tapeworm infection, which used to be the main drug for the treatment of this disease. In the late twentieth century, the newly developed chemical drugs with important efficacy and fewer adverse reactions gradually replaced the natural antiparasitic drugs, but these natural antiparasitic drugs have made great contributions in the history of disease treatment and own important scientific values and references to new drug discovery.

Natural compounds from plants, especially the traditional Chinese medicine, have made great contributions in the treatment of parasitic diseases. At present, parasitic infectious diseases are still threatening human health. And new drugs are especially needed for the multidrug-resistant *Plasmodium falciparum* malaria treatment. Natural products will still be an important resource for the new drugs discovery.

In addition to the treatment of parasitic infectious diseases, natural products also play important roles in the treatment of pathogenic infections. In the applications of traditional Chinese medicine, some detoxification drugs have shown good antibacterial effects. In the early and middle of the twentieth century, the researchers made a large number of studies on the antibacterial effects of natural products. Many natural products have been found to possess the function to inhibit bacterial growth or kill pathogens. Some natural products have been clinically proved to be effective in the treatment of bacterial infectious diseases, such as berberine.

However, the development of antibacterial agents from plants attracted less attention in the late twentieth century due to the discovery of antibiotics. In the early 1940s, the successful application of penicillin in the treatment of infectious diseases has led research direction to the discovery of new antibiotics. Medical scientists in the world have carried out a large number of studies and found many effective ones, such as tetracycline, chloramphenicol, and erythromycin, which have made great contributions in the treatment of bacterial diseases. In particular, with the development of pharmaceutical chemistry, a number of gratifying results and achievements have been obtained. Many powerful new antibiotics with broad spectrums were developed and used in clinical. For the good therapeutic effects of antibiotics, the discovery of antibacterial agents from natural plants was studied less at present.

Nevertheless, with the extensive use of antibiotics, the problem of drug-resistant bacteria has attracted the attention of the medical scientists. People began to reexamine the application experience, lessons, and achievements of the antibiotics. New ideas have emerged for the treatment of bacterial infectious diseases, especially for drug-resistant diseases. Based on the accumulated experience of the traditional medicine studies, it was believed that plant-derived natural products will play important roles in the treatment of these diseases, especially the antibiotic-resistant ones. The studies using natural products for the treatment can not only find new natural drugs but also can promote the theoretical research on the treatment of these diseases. The development and application of plant-derived antibiotics should be taken into consideration by medical scientists.

In addition, it is necessary to mention the role of natural small molecule compounds in the treatment of viral infectious diseases. Like the earlier antibacterial studies, the antiviral studies of small molecule compounds from plants have received extensive attention at the beginning. But due to the clinical application of the vaccine and the specificity of viral infection, the discovery of antiviral drug from natural products was hindered in the modern drug development and evaluation process. However, the practice of traditional drugs in the treatment of viral infectious diseases has shown that natural medicines can play important roles in the disease treatment. In recent years, the clinical application of traditional Chinese medicine extracts in the treatment of viral infectious diseases has achieved remarkable results.

Although there is no evidence that these drugs can directly inhibit the virus replication, the treatments using these drugs are worthy of in-depth studies.

At present, the natural small molecules in this chapter are not widely used clinically due to the lack of full therapeutic effects and mechanism studies. These drugs might have many mild medicinal effects, such as allicin with many effects but no specific clinical application feature. With the deepening understanding of these diseases and the developing progresses on thc treatment strategy, natural small molecule drug research and development will get new achievements, which will play important roles in the clinical treatment of these diseases.

Agrimophol

Gui-Min Jin, Li-Da Du, and Yang Lu

Abstract Agrimophol is the active ingredient of agrimony, the dry part above the ground of *Agrimonia pilosa* Ledeb, which has excellent anthelmintic effect, and also can be used for treating trichomonas enteritis, vaginitis, and other diseases.

Keywords Agrimophol · Anthelmintic effect · Anticancer effect

Origin: *Herba Agrimoniae* (Fig. 1)

Alias: Agrimopholum
Chemical name (Fig. 2)

6-(3-butyryl-2,6-dihydroxy-4-methoxy-5-methylbenzyl)-3,5-dihydroxy-4,6-dimethyl-2-(2-methylbutanoyl) cyclohexa-2,4-dien one

Molecular formula, $C_{26}H_{34}O_8$; **MW**, 474.54; **CAS**, 65792-05-4

Properties

Appearance: light yellow-green oblique prism. *Solubility*: soluble in chloroform and benzene and slightly soluble in methanol, ethanol, and petroleum ether but almost insoluble in water. *Melting point*: 138.5–139.5 °C.

G.-M. Jin · Y. Lu (✉)
Beijing Key Laboratory of Polymorphic Drugs, Research Center of Polymorphic Drugs, Institute of Materia Medica, Chinese Academy of Medical Sciences and Peking Union Medical College, Beijing, China
e-mail: luy@imm.ac.cn

L.-D. Du
School of Biomedical Sciences, Chinese University of Hong Kong, Hong Kong, China

G.-H. Du, *Natural Small Molecule Drugs from Plants*,
https://doi.org/10.1007/978-981-10-8022-7_92

Dosage Forms and Indications

Available in the *Pharmacopoeia of the People's Republic of China* (1977). Capsules, suitable for the treatment of tapeworm and trichomonas infection.

Literature

This substance is the active ingredient of agrimony, the dry part above the ground of *Agrimonia pilosa* Ledeb, which is distributed in most areas of China. Its same generic plants include *Agrimonia pilosa Ldb.* var. nepalensis (D.Don) Nakai, *Agrimonia nipponica Koidz.* var. occidentalis Skalicky, *Agrimonia coreana* Nakai, and *Agrimonia eupatoria* Linn. subsp. asiatica (Juzep.) Skalicky. According to ancient Traditional Chinese Medicine literature, it is astringent and bitter in property; distributes to the heart and liver, which have the function of astringency and hemostasis; prevents malaria; cures diarrhea; detoxifies and tonifies deficiency; and can be used to treat hemoptysis, hematemesis, metrorrhagia, malaria, bloody flux, and so on [1]. In recent years, the role of agrimony in the treatment of cancer diseases and other aspects of attention has gradually attracted much attention.

Agrimony also contains volatile oil, trace elements, agrimarin, agrimus lactone, tannin, sterols, organic acids, phenolic compounds, flavonoids and glycosides, and other compounds. Among them, the phenolic compounds mainly contain agrimony A, agrimony B, agrimony C, agrimony D, agrimony E, agrimony, pseudoephedrine, and so on.

History of R&D

In folk medicine, the whole agrimony was used to stop bleeding and inflammation sterilization; its rhizome was used to treat intestinal bloody flux, the radical bud to expelling tenia. The clinical application and pharmacological studies have shown that agrimony root has a broad spectrum of anthelmintic effect with high activity and low toxicity.

In the early 1970s, the Chinese scientists used lime milk extraction method or petroleum ether as a solvent to separate the extract from the grass buds and further extracted the main ingredient, agrimophol [2]. Animal experiments and clinical applications show that the agrimophol has a broad spectrum of anthelmintic effect with high efficacy, as well as the function of killing schistosomiasis.

The chemical structure of agrimophol was determined, and the synthesis was realized [3]. It is synthesized from methyl phloroglucinol.

Recent studies have shown that agrimophol has a significant anticancer effect [4].

Pharmacology

Agrimophol has excellent anthelmintic effect and also can be used for treating trichomonas enteritis, vaginitis, and other diseases. Pharmacological and clinical trials have shown that the effect of disinfecting cestode is stronger than that of niclosamide and bithionol, and the side effects are fewer.

Agrimophol can be rapidly oxidized after irradiation and release of ROS such as H_2O_2, O_2 and OH^-. OH^- which directly destroy DNA and membrane structure of tumor cells, leading to irreversible damage to the cells [5]. Compared with the monotherapy with radiotherapy, combined treatment with agrimophol further induces 25.1% more cell death of human lung adenocarcinoma cells [6]. It is reported that the mitochondria are the most prominent in the ultrastructural damage of HeLa cells after the action of agrimophol. The main manifestations are mitochondrial swelling, rupture, and even vacuolization. So the mitochondria within the tumor cells are likely to act as a specific target organelle for the action of agrimophol [7]. It was also observed that agrimophol could significantly inhibit deoxythymidine (3H-TdR) from permeating into the tumor cells after the tumor cells were irradiated. After the HeLa cells were exposed to agrimophol, the deoxyribonucleoside content of the cells was reduced by 50%, which indicated that the inhibition of 3H-TdR infiltration into tumor cells mainly affected the synthesis of deoxynucleotides in tumor cells. The synthesis of deoxynucleotides is related to nucleotide reductase. Agrimophol can inhibit the activity of nucleotide reductase in tumor cells and inhibit the synthesis of deoxynucleotides in tumor cells, thereby inhibiting the synthesis of DNA in tumor cells.

Clinical Application

Agrimophol is clinically used to expel tenia and roundworms and kill schistosomiasis. Fewer adverse effects were observed; occasional nausea, vomiting, dizziness, cold sweat, or a diarrhea after a half month even can lead to collapse reaction. Although studies have shown that agrimophol has a more pronounced antitumor activity, there is no relevant antineoplastic drug available.

Discussion

Agrimophol has been used as an effective anthelmintic drug. Further studies are essential for defining its anticancer effect.

Fig. 1 *Herba Agrimoniae*

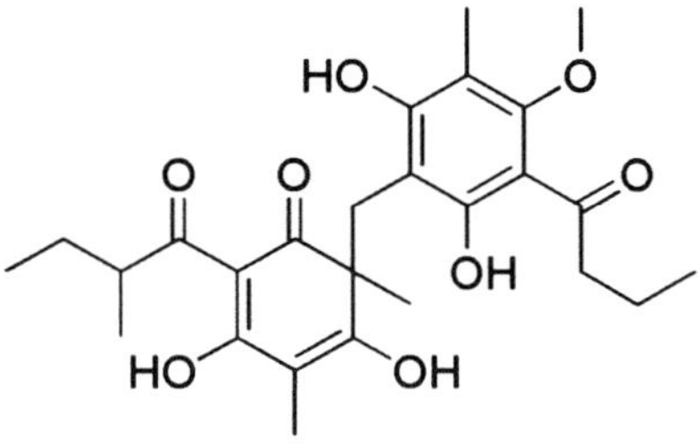

Fig. 2 The structure of agrimophol

References

1. Chinese Pharmacopoeia Commission. Pharmacopoeia of the People of Republic of China, vol. I. Beijing: China Medical Science and Technology Press; 2015.
2. Shenyang Pharmaceutical College, Department of Chinese Materia Medica; Liaoning Institute of Materia Medica. Study of the active ingredient of expelling tenia in agrimophol. J Shenyang Pharm Univ. 1973;1:1–6.
3. Shenyang Pharmaceutical College, Chemical synthetic laboratory, Liaoning Institute of Materia Medica, Hunan institute of pharmaceutical industries. The total synthesis of agrimophol. Acta Chim Sin. 1976;34(4):313–20.
4. Shraibom N. Herzelia. Molecular combinations for cancer or other disease treatment: US, 008734859B1. 2014.
5. Xue WJ, Li DH. Antitumor mechanism of phenolic compounds. Acta Pharm Sin. 1985;20(6):477–80.
6. Han JQ, Wang XW, Wang Y, et al. A clinical prospective study of radiosensitization on lung adenocarcinoma by agrimophol. Chin J Radiol Med Prot. 2003;23(5):355–7.
7. Pan YS, Yuan RX, Huang KJ, et al. Experimental study of antineoplastic of agrimophol. Chin Pharm Bull. 1983;18(7):434.

Allicin

Wei Zhou, Ai-Lin Liu, and Guan-Hua Du

Abstract Garlic is an edible and medicinal plant with a long history of application in China. Allicin is one of the main active substances of garlic, which has simple structure and is easy to be synthesized. It has a broad spectrum of pharmacological activities, such as antibacterial, antiviral, anticancer, decreasing the blood pressure, and inhibition of platelet aggregation. However, due to its chemical instability, carrying out research on allicin at molecular level has difficulties. Therefore, the full understanding of the biological activities of garlic still depends on the breakthroughs in the separation, extraction, purification, analysis, and preservation of allicin.

Keywords Allicin · Ester · Antibacterial · Antiviral

Alias: S-oxodiallyl sulfide, allicin
Origin: Garlic (Fig. 1)
Chemical name (Fig. 2)

2-propene-1-sulfinothioic acid, s-2-propenyl ester

Molecular formula, $C_6H_{10}S_2O$; **MW**, 162.273; **CAS**, 539-86-6

Properties

Appearance: Light yellow powder or light yellow oily liquid, with strong irritating odor. Unstable and light, heat, and organic solvents will easily degrade it into a variety of sulfur-containing organic compounds. *Solubility*: Its solubility is about 2.5% in the 10 °C water. Soluble in ethanol, ether, benzene, and other organic

W. Zhou · A.-L. Liu (✉) · G.-H. Du
Beijing Key Laboratory of Drug Targets Identification and Drug Screening, Centre for Pharmaceutical Screening, Institute of Materia Medica, Chinese Academy of Medical Sciences and Peking Union Medical College, Beijing, China
e-mail: liuailin@imm.ac.cn

G.-H. Du, *Natural Small Molecule Drugs from Plants*,
https://doi.org/10.1007/978-981-10-8022-7_93

solvents. *Boiling point*: 80–85 °C (0.2 kPa). Relative density of d^{20} = 1.112; refractive index n_D^{20} = 1.561.

Dosage Forms and Indications

This product conforms to the national standards for chemical drugs.

The current clinical used forms are allicin injection, garlic instant kelp capsule, etc. It is used clinically mainly for anti-pathogenic microbial infection, anti-tumor, and anti-dilation of blood vessels. It could also be used for the treatment of periodontitis and ulcerative colitis.

Literature

Garlic, the perennial underground bulbs of Liliaceae Allium garlic (*Allium sativum* L.), is an edible and medicinal plant. It was recorded in pottery of ancient Egyptian and Sumerian (Sumerian) (2600–2001 BC). It was also recorded in ancient traditional Chinese medicine books. *Compendium of Materia Medica* described that garlic benefits the body and removes cold draught out of the body. Besides, ancient Chinese also believed that garlic could exorcise evil spirits.

History of R&D

As the main active substance of garlic, garlic was separated by Cavallito and Bailey in 1944 from crushed garlic. Cavallito firstly clarified the physical properties and chemical structure of the scent of garlic [1, 2].

Fresh garlic does not contain free allicin but only its precursor material alliin. When the garlic undergoes physical mechanical crushing, garlic alliinase is activated and catalyzes the breakdown of alliin into allicin [3].

The reported preparation methods of allicin contain extraction, biosynthesis, and chemical synthesis. Chemical synthesis starts from raw material of diallyldisulfides, m-chloroperoxy benzoic acid to get crude allicin. According to the United States patent report, biosynthesis preparation of allicin comes from natural sources or synthetic, after it is made into a certain concentration of aqueous solution, it is converted into allicin by reaction with alliinase. Extraction of allicin with low boiling nonpolar solvent can acquire pure allicin, but it must be stored at −70 °C to prevent from decomposing.

Pharmacology

Allicin has a broad spectrum of pharmacological activities, such as antibacterial, antiviral, anticancer, decreasing the blood pressure, and inhibition of platelet aggregation.

1. Antibacterial and Antiviral

Allicin is known as a natural broad-spectrum antimicrobial drug. It can inhibit a variety of bacteria and is able to pass through the cell membrane of the pathogen and enter the cytoplasm. The bio-oxidation in bacteria is blocked due to lacking cysteine, which destroys normal metabolism and inhibits bacterial reproduction. In addition, allicin can also inhibit a variety of viruses, such as influenza virus B, vaccinia virus, and angular cheilitis bubble virus. The latest study reported that allicin could also inhibit the growth and reproduction of babesi in vitro [4], providing a potential way for the treatment of Babesia disease.

2. The Effect on Heart and Cerebral Vessels

Allicin exerts a protective effect on the heart and cerebral vessels through reducing plasma total cholesterol, lowering blood pressure, inhibition of platelet activity, and reducing hematocrit and blood viscosity. Allicin increases the activity of inducible nitric oxide synthetase (iNOS) and boosts the NO level, which results in the vasodilation. In vitro study also found that allicin's vasodilation is related to NO, and allicin can improve the level of iNOS and NO in platelet and placental villi as well as choriocarcinoma [5]. In addition, allicin administration effectively reverses hyperlipidemia and atherosclerosis, inhibits lipid peroxidation, and reduces the level of glutathione decomposition and superoxide dismutase and peroxidase activity in high cholesterol-fed rats [6].

3. Anticancer

Allyl sulfide, the active ingredient of allicin, shows anticancer effect. Epidemiological investigation and experimental studies have shown that allicin has a significant inhibitory effect on gastric, colon, liver, and lung cancers. Experimental results show that [7] allicin improves the cellular immune function of cancer patients. Alliin is the precursor of allicin, which also has significant antitumor activity. In the 1980s, it was found that the growth of S180 tumor in mice was significantly inhibited by alliin injection or garlic thionine. Meanwhile, alliin could selectively inhibit the reduction synthesis of reduced glutathione (GSH)-dependent prostaglandin E2 in gland cells.

Clinical Application

Allicin shows inhibitory effects on infections caused by bacteria, fungi, and virus. Clinical studies reported that allicin has cured 23 *Candida albicans*-infected patients [8]. Combining allicin with surgery had cured ten patients with maxillary sinus aspergillosis [9]. Good results for the treatment of chronic gastritis, peptic ulcer, chronic colitis, and fatty liver were achieved after allicin administration. Allicin is also capable of eliminating oxygen free radical ion. In addition, allicin can reduce the nitrite- and nitrate-reducing bacteria, and taking garlic could benefit chronic symptoms in stomach, such as stomach discomfort, fullness pain, acid reflux, heating, burning and loss of appetite, and so on.

Discussion

Garlic has a long history of application, known as the natural "medicinal plant gold." Allicin is one of main active ingredients of garlic, which has simple structure and is easy to be synthesized. Its multiple biological activities and little adverse effect make it a hot spot for research. However, due to its instability, allicin used in animal test is prepared from alliin and alliinase in advance. Foreign researchers had made intensive studies on the two key factors in the formation of allicin, as well as the detection methods and activities of allicin. Nonetheless, carrying out research on allicin at molecular level has difficulties due to its chemical instability. Therefore, the full understanding of the biological activities of garlic still depends on the breakthroughs in the separation, extraction, purification, analysis, and preservation of allicin.

Fig. 1 Garlic

Fig. 2 The structure of allicin

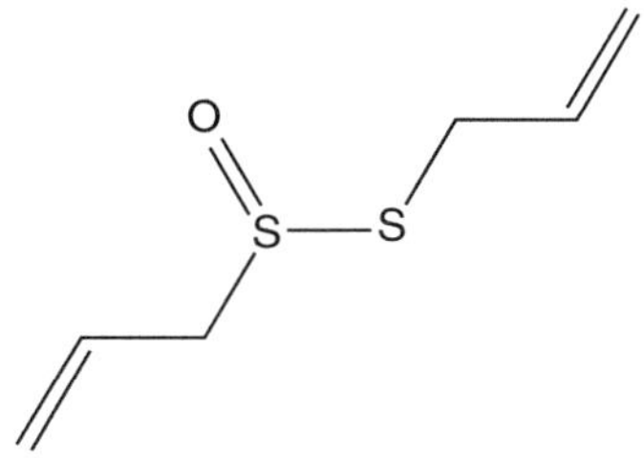

References

1. Cavallito CJ, Bailey JH. Allicin, the antibacterial principle of Allium sativum. I: isolation, physical properties and anti- bacterial action. J Am Chem Soc. 1944;66:1950–1.
2. Cavallito CJ, et al. Allicin, the antibacterial principle of Allium sativum. II: determination of the chemical structure. J Am Chem Soc. 1944;66:1952–4.
3. Feng SD, Mi XJ, Ma Y. The economic value and deodorization mechanism of garlic. North Hortic. 1998;3:37–8.
4. Salama AA, Aboulaila M, Terkawi MA, et al. Inhibitory effect of allicin on the growth of Babesia and Theileria equi parasites. Parasitol Res. 2014;213(1):273–85.
5. Nie XM, Zhou YJ, Xie Y, et al. Effect of diallyl trisulfide coated stents on iNOS protein expression and NO level in canine models of coronary restenosis. J Fourth Mil Med Univ. 2006;2(11):9752–71.
6. El-Sheakh AR, Ghoneim HA, Suddek GM, et al. Attenuation of oxidative stress, inflammation, and endothelial dysfunction in hypercholesterolemic rabbits by allicin. Can J Physiol Pharmacol. 2015;94(14):1.
7. Zhang ZM, Gao HQ, Wei Y. Effects of allicin on cellular immune function in patients with cancer. J Shan Dong Univ (Health Sciences). 2003;41(2):148–50.
8. Duan GL, Chen TD, Shi TK, et al. Treatment of allicin intravenous infusion on urinary tract infection by candida albicans. Chin J Urol. 1996;17(1):64.
9. Wang XL. Treatment of allicin on maxillary sinus aspergillosis (10 cases of clinical trail). Chin J Otorhinolaryngol Integr Med. 1998;6(2):91–2.

Arecoline

Yan Zhao, Li-Da Du, and Guan-Hua Du

Abstract Arecoline is extracted from *Areca catechu* L., and it is the most important alkaloid in it. Arecoline has the effect of antiparasite and is promoting the motility of the gastrointestinal smooth muscle. It activates the M and N cholinergic receptors, excites the nervous system, promotes the body excitability, and improves the ability of learning and memory. The application of arecoline is rare at present mainly due to its adverse effects. However, as an important traditional Chinese medicine, areca still has important value in clinical practice, especially in compatibility of herbs.

Keywords Arecoline · Alkaloid · Antiparasite · Cholinergic drugs

Origin: Areca nut (Fig. 1)
Chemical name (Fig. 2)

N-methyl-1,2,5,6-tetrahydronicotinate

Molecular formula, $C_8H_{13}NO_2$; **MW**, 55.19; **CAS**, 63-75-2

Y. Zhao
Pharmacy Department, Qingdao Municipal Hospital, Qingdao, China

L.-D. Du
School of Biomedical Sciences, Chinese University of Hong Kong, Hong Kong, China

G.-H. Du (✉)
Beijing Key Laboratory of Drug Targets Identification and Drug Screening, Centre for Pharmaceutical Screening, Institute of Materia Medica, Chinese Academy of Medical Sciences and Peking Union Medical College, Beijing, China
e-mail: dugh@imm.ac.cn

G.-H. Du, *Natural Small Molecule Drugs from Plants*,
https://doi.org/10.1007/978-981-10-8022-7_94

Properties

Appearance: oily liquid. *Solubility*: mixed with water, ethanol, or ether in any ratio, soluble in chloroform. *Density*: 1.059 g/cm^3. *Boiling point (760 mmHg)*: 209 °C. *Flash point*: 81.1 °C. *Vapor pressure (25 °C)*: 0.208 mmHg. Arecoline can be synthesized to salts with organic acids. Arecoline hydrobromide, arecoline acetarsol, and arecoline p-antimony carboxybenzoic acid (Fig. 3) are commonly used.

Derivatives (Fig. 3)

Dosage Form and Indications

This product has no source standard.

Tablets: anthelmintics, less used now. Eye drops: cholinergic drugs for glaucoma treatment.

Literature

Areca catechu L. (Bin Lang), a palm plant native to Malaysia, is mainly distributed in tropical regions of Asia and the Americas. It has been introduced to China and cultivated in China for more than 2000 years, mainly in Hainan and Taiwan province. It is also cultivated in Guangdong, Guangxi, Fujian, and Yunnan.

Arecoline is extracted from dry mature seeds of areca. The mature seeds of areca are 3–5 cm in diameter, with fibrous peel and a seed, namely, areca nut. The endosperm of areca nut is hard, with grayish-brown spots. Areca nut is harvested from August to November every year before the fruit is fully mature. The seeds are peeled, boiled, and cut into thin slices. The dried slices are brown or black, and it is an important medicine in traditional Chinese medicine. Areca nut is the raw materials of catechu. Areca nut was used as anthelmintics with arecoline as the main alkaloid in it.

Areca has a long medicinal history. It is recorded in Ben Cao Gang Mu (*Compendium of Materia Medica*) that the peels of areca *Piper betle L.* (Fu Liu Teng) and Concha Arcae (Wa Leng Zi) ash can be used together to dispel malaria. In Ben Cao Xin Bian (New compilation of Materia Medica) written by Chen Shiduo in Qing dynasty, it records areca is pungent and bitter; Qi is warm and descending. It is the Yang in the Yin. It is nontoxic. Its main entry channels are the spleen, stomach, large intestine, and lung. Its effects are invigorating spleen and stomach, expelling phlegm, relieving heart pain, killing worms, and dispelling malaria.

History of R&D

Arecoline, an alkaloid extracted from *Areca catechu* L., is first reported by Jahns in 1888. Mujumdar found that there were at least six kinds of alkaloids in areca, including arecoline, arecaidine, guavacoline, and guavacine. The content of arecoline in the fresh fruit of areca was 0.3–0.63% as determined with HPLC [1].

It has been found first that arecoline has the effect of antiparasite [2] and is promoting the motility of the gastrointestinal smooth muscle. It activates the M and N cholinergic receptors, excites the nervous system, promotes the body excitability, and improves the ability of learning and memory [3].

Pharmacology

Paralysis was the main mechanism of the antiparasite effect of arecoline.

The effect of arecoline on cholinergic receptors is similar to that of pilocarpine. It can agonize M-cholinergic receptors (M1, M2, M3, M4) [4–6] and increase the secretion of glands, especially the salivary. It can also agonize N-cholinergic receptors and activate the muscles of skeletal, ganglia, carotid body, etc. The central nervous system is also influenced by the cholinergic effect of arecoline. Intravenous injection of small dosage of arecoline can cause wake-up reaction of the cortex in cats, which is inhibited or blocked by atropine. Arecoline can cause salivation, vomiting, diuretic, lethargy, and convulsion when overdosed.

In addition, arecoline can enhance intestinal peristalsis, contract bronchus, lower heart rate, dilate blood vessels, and decrease blood pressure. While in rabbit, it induces coronary artery contraction.

Eye drops can reduce the pupil.

Pharmacokinetic parameters for arecoline after oral administration of arecoline (3 mg/kg): T_{max}, 120.07 min; C_{max}, 60.61 ng/mL; $t_{1/2}$, 69.32 min; AUC_{0-t}, 15116.86 min/ng/mL; $AUC_{0-\infty}$, 15771.37 min/ng/mL; plasma clearance, 0.19 L/min/kg [7].

Clinical Application

One percent arecoline solution can be used to reduce intraocular pressure in glaucoma, but it shows obvious irritation to the cornea and the effect lasts short time.

There were researchers who explore the possibility of developing it as an antidementia drug due to its excitement effect on the nervous system [8]. Arecoline once was used as antiparasite drug, but it has been replaced by other drugs now.

Discussion

Although there are many pharmacological effects of arecoline, its application is rare at present mainly due to its adverse effects. However, as an important traditional Chinese medicine, areca still has important value in clinical practice, especially in compatibility of herbs. In the south of China and Southeast Asia, people have the habit of chewing areca to refresh the spirit. However, it has the potential to increase the incidence of oral cancer.

Fig. 1 Areca nut

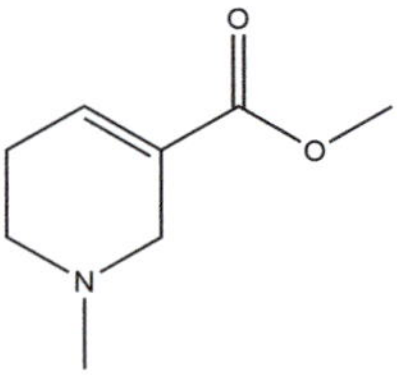

Fig. 2 Chemical structure of arecoline

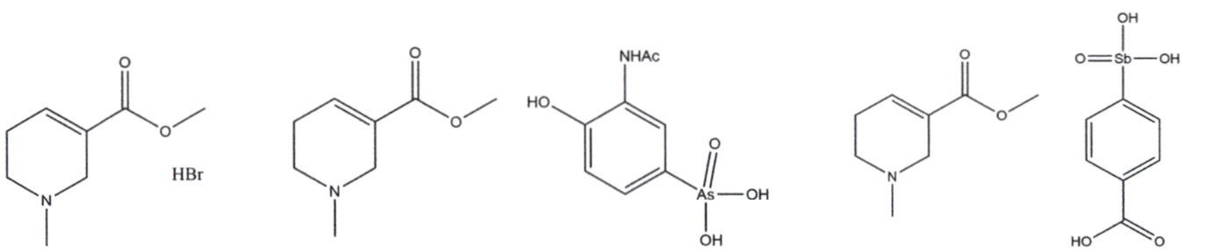

Arecoline hydrobromide   Arecoline acetylarsine   Arecoline p-antimony carboxybenzoic acid

Fig. 3 Chemical structures of arecoline derivatives

References

1. Peng W, Liu YJ, Wu N, et al. Areca catechu l. (arecaceae): a review of its traditional uses, botany, phytochemistry, pharmacology and toxicology. J Ethnopharmacol. 2015;164:340–56.
2. Zheng ZH. Modern research and application of traditional Chinese medicine. Beijing: Academic; 1999.
3. Wang G, Hu B. Research progress of arecoline. Int J Pathol Clin Med. 2010;30(2):171–5.
4. Ghelardini C, Galeotti N, Lelli C, et al. M1 receptor activation is a requirement for arecoline analgesia. Farmaco. 2001;56(5–7):383–5.
5. Yang YR, Chang KC, Chen CL, et al. Arecoline excites rat locus coeruleus neurons by activating the m2-muscarinic receptor. Chin J Physiol. 2000;43(1):23–8.
6. Xie DP, Chen LB, Liu CY, et al. Arecoline excites the colonic smooth muscle motility via m3 receptor in rabbits. Chin J Physiol. 2004;47(2):89–94.
7. Li B, Zhou XZ, Li JY, Yang YJ, Niu JR, Wei XJ, Liu XW, Li JS, Zhang JY. Determination and pharmacokinetic studies of arecoline in dog plasma by liquid chromatography-tandem mass spectrometry. J Chromatogr B Analyt Technol Biomed Life Sci. 2014;969:12–8.
8. Saikia JR, Schneeweiss FH, Sharan RN. Arecoline-induced changes of poly-adp-ribosylation of cellular proteins and its influence on chromatin organization. Cancer Lett. 1999;139(1):59–65.

Artemisinin

Ling-Lei Kong and Guan-Hua Du

Abstract Artemisinin is an antimalarial lactone derived from *qinghao*. Clinically, artemisinin is mainly used to treat malaria symptoms, malignant cerebral malaria, uncomplicated malaria, and severe malaria. Combined with different antimalarial can delay and prevent resistance of malaria parasites. Artemisinin combination therapies (ACT) have been recommended by the WHO as the first-line therapy for *P. falciparum* malaria worldwide. The discovery of artemisinin has completely changed the treatment of malaria. With the widespread use of artemisinin and its derivatives, the resistance of *Plasmodium* to these drugs has emerged, which pose a new challenge to the treatment of malaria. Therefore, a new generation of artemisinin drugs should have a broader application prospects.

Keywords Artemisinin · Malaria · Artemisinin combination therapies (ACT)

Alias: Qinghaosu
Origin: *Artemisia annua* (Fig. 1)
Chemical name (Fig. 2)

(3R,5aS,6R,8aS,9R,12S,12aR)-3,6,9-trimethyloctahydro-3,12-epoxypyrano[4,3-j]-1,2-benzodioxepin-10(3H)-one

Molecular formula, $C_{15}H_{22}O_5$; **MW**, 282.34; **CAS**, 63968-64-9
Artemisinin derivatives (Fig. 3)

L.-L. Kong · G.-H. Du (✉)
Beijing Key Laboratory of Drug Targets Identification and Drug Screening, Centre for Pharmaceutical Screening, Institute of Materia Medica, Chinese Academy of Medical Sciences and Peking Union Medical College, Beijing, China
e-mail: dugh@imm.ac.cn

G.-H. Du, *Natural Small Molecule Drugs from Plants*,
https://doi.org/10.1007/978-981-10-8022-7_95

Properties

Appearance: colorless needles or white crystalline powder. *Solubility*: practically insoluble in water, very soluble in dichloromethane, freely soluble in acetone and ethyl acetate, and soluble in glacial acetic acid, methanol, and ethanol. *Melting point*: 150–153 °C. *Specific optical rotation*: +75 to +78°.

Dosage Forms

Artemisinin is recorded in the *Pharmacopoeia of the People's Republic of China* (2015), India Pharmacopoeia (2010), and the International Pharmacopoeia (5th ed.). The current clinical use of dosage forms are tablets, suppository, and suspension injection.

Indications

Clinically, artemisinin is mainly used to treat malaria symptoms, malignant cerebral malaria, uncomplicated malaria, and severe malaria. Combined with different antimalarial can delay and prevent resistance of malaria parasites. In additional, artemisinin can also be used for systemic lupus erythematosus or discoid lupus erythematosus. Currently, artemisinin derivatives and their compound preparations are widely used in clinic.

Literature

Artemisinin is an antimalarial lactone derived from *qinghao* 青蒿 (*Artemisia annua* or sweet wormwood). *Artemisia annua* is a common herb found in many parts of the world, and the medicinal value was confirmed by Chinese herbalists for more than 2000 years. The earliest record was dated back to 200 BC, in the *Fifty-two Prescriptions* unearthed from the Mawangdui. Its antimalarial application was first described in *Zhou hou Bei ji Fang*, edited by Ge Hong. In 1596, Li Shizhen recommended a tea made from *qinghao* specifically to treat malaria symptoms in his *Compendium of Materia Medica*.

At present, the only source of artemisinin is from the wild and artificial cultivation *Artemisia annua*. In addition, only a few plants were reported to have artemisinin, such as *Artemisia apiacea*, *Artemisia Lamcea*, and *Artemisia subdigitata* Mattf. [1]. Due to limited sources and high costs of natural plants, biosynthesis, chemical synthesis, and synthesis in engineered organism have been carried out.

History of R&D

In the 1960s, the incidence of malaria increased again, and the effect of traditional drug therapy was reduced; new treatments were urgently needed. In order to support Vietnam's fight against the United States, China launched the "523" project to develop new antimalarial drugs on May 23, 1967 [2].

Scientists started to look for recipes that had been used to treat fever. They searched more than 2000 recipes of Chinese traditional herbs, compiling 640 recipes that might have some activity of antimalarial. They tested more than 200 recipes with Chinese traditional herbs and 380 extracts from the herbs in a rodent malaria model. Among the promising results, extracts from *Artemisia annua* L. (qinghao), a type of wormwood native to Asia, were shown to inhibit parasite growth by 68%. However, the activity was not stable, varying from 12% to 40% inhibition in subsequent repeats. The variation in antimalarial activity could be due to many factors, including geographic origins of the plant, seasonal variation, different parts of the plant (leaves or stem), and the methods used in extraction. A low-temperature extraction process could be used to isolate an effective antimalarial substance from the plant. The neutral extract, termed extract number 191, was tested in the mouse malaria, *Plasmodium berghei*, and achieved 100% inhibition. This finding represented the breakthrough in the discovery of artemisinin [3].

Subsequent clinical trials have shown that the *Artemisia annua* extract can get rid of malaria parasites from patients' bodies faster than chloroquine. The exciting results led to a large-scale, countrywide effort to extract large quantities of the pure ingredient to determine its structure and to chemically synthesize artemisinin. The structure and stereo absolute configuration of artemisinin was confirmed by X-single crystal diffraction method. Artemisinin is a sesquiterpene lactone with an unusual peroxide bridge, which is considered to be related to its mechanism of action [4, 5]. The discovery of artemisinin dramatically changes the landscape to combat malaria and leads to a paradigm shift in antimalarial drug development.

However, the discovery of artemisinin is the first stage; the development of artemisinin derivatives and their compound preparations is another important stage. Based on artemisinin, scientists obtained artemisinin ether derivatives by semisynthetic method. After screening of antimalarial activity, artemether was found. To further improve the solubility of artemisinin derivatives, artesunate was also found. The discovery of artesunate makes artemisinin and its derivatives much easier to promote, and more convenient dosage forms to treat malaria enriched the clinic application of artemisinin and its derivatives [6, 7].

In addition to artemisinin, several drugs used in combination with artemisinin, including lumefantrine, piperaquine, and pyronaridine, were found. Their success reflects the unique spirit of collaboration from many scientists and institutions involved in the search for antimalarial drugs. Therapies that combine artemisinin or its derivatives with some other antimalarial drug are the preferred treatment for malaria and are both effective and well tolerated in patients. Treatments containing an artemisinin derivative (artemisinin combination therapies, ACTs) are now standard treatment worldwide for *P. falciparum* malaria.

Pharmacology

The mechanism of artemisinins is not known, but the most widely accepted theory is that they are first activated through cleavage after reacting with haem and iron(II) oxide, which results in the generation of free radicals that in turn damage susceptible proteins, resulting in the death of the parasite [8, 9].

Artemisinin and its derivatives also show a good antitumor effect [10], which is mainly via (1) apoptosis, ferroptosis, or necrosis; (2) anti-angiogenesis; (3) oxidative stress; (4) tumor suppressor genes; and (5) protein targeting. In addition, artemisinin can exhibit antiarrhythmic, anti-fibrotic, and immunomodulating effects.

Clinical Application

Artemisinin and Its Derivatives [11]

Artemisinin and its derivatives are widely used to treat malaria symptoms, malignant cerebral malaria, uncomplicated malaria, and severe malaria in clinic. Especially, artemisinin suppositories are more suitable for use in primary health care for the treatment of severe malaria due to ease of use, preservation, and in situ production.

For the treatment of anti-chloroquine falciparum, based on the treatment experience in the development of antimalarial drug program, China's malaria expert advisory committee proposed that artemisinin derivatives should be used in a 5-day or 7-day course in the treatment of falciparum malaria and recommended artesunate or artemether 100 mg × 5 days or 7 days, the first dose doubled, with a total amount of 600–800 mg, and dihydroartemisinin 80 mg × 7 days, the first dose doubled, with a total amount of 640 mg.

In the treatment of severe malaria, compared with quinine, artemether or artesunate alone can remove the malaria quickly in the body and does not cause hypoglycemia.

Artemisinin Combination Therapies [12]

Although artemisinin works quickly within hours, artemisinin needs another drug in combination to obtain a cure for its short half-life. Patients recovered so quickly after taking artemisinin that they would not continue treatment after they felt better and thus were ultimately not cured. Such incomplete treatment may promote resistance of drug.

The WHO has recommended artemisinin combination therapies (ACT) to be the first-line therapy for *P. falciparum* malaria worldwide, such as artemisinin combined with mefloquine, lumefantrine, amodiaquine, piperaquine, and pyronaridine, which has a long half-time. Combinations are more effective because the artemisinin component kills most parasites at the start of the treatment, while the more slowly eliminated partner drug clears the remaining parasites.

In foreign countries, artemisinin component and mefloquine in combination was recommended. Artemether, artesunate, or dihydroartemisinin with mefloquine in the treatment of anti-chloroquine malaria can significantly reduce the reignition rate. In China, artemisinin component and lumefantrine in combination was recommended. Artesunate and lumefantrine have synergistic effect. The effect is better than any other one in reducing the dosage and course of treatment. Combination treatment can inhibit the formation and spore proliferation of the parasite, which can prevent the spread of chloroquine-resistant *Plasmodium falciparum* and reduce the spread of malaria. Artemether-lumefantrine (artemether 20 mg, lumefantrine 120 mg) in the treatment of anti-chloroquine falciparum malaria can be instantly cured, and the relapse rate is only 1.8% in 28 days.

Adverse Effects [13]

Artemisinins are generally well tolerated at the doses of treating malaria. The most frequent side effects were headache, dizziness, nausea, abdominal pain, dyspepsia, vomiting, and skin rash. A rare but serious adverse effect is allergic reaction.

Side effects are rare in patients treated with artemisinin derivatives. In a prospective study of over 3500 patients in Thailand, there was no evidence for serious side effects. Artemisinin derivatives are also safe for pregnant women.

For the artemisinin combination therapies, it will delay resistance of drug because a parasite will need time to acquire resistance to two drugs as opposed to just one. ACTs are considered the most effective treatments for *P. falciparum* malaria which have resistance to other drugs. However, the data from clinical trials showed that parasites have acquired resistance to artemisinin and some ACTs appear to lose effectiveness.

Discussion

The impact of artemisinin on public health is immediate and tremendous. Millions of lives have been saved by using artemisinin. There are few scientific discoveries that have the same scale and instant impacts on public health, human productivity, and scientific research as artemisinin has! Artemisinin represents one of the significant contributions of China to global health. The discovery of artemisinin is a gift to the mankind in its fight against malaria. The 2015 Nobel Prize in Physiology or Medicine was awarded to Professor YouYou Tu for her key contributions to the discovery of artemisinin.

Artemisinin and its derivatives have been evaluated by the World Health Organization as the only truly effective drug for the treatment of falciparum malaria and are still effective against chloroquine-resistant *Plasmodium falciparum.* The discovery of artemisinin has completely changed the treatment of malaria. Artemisinin, which is used in combination therapy, can reduce total malaria mortality by 20% and in children by 30%. With the widespread use of artemisinin and its derivatives, the resistance of *Plasmodium* to these drugs has emerged, which pose a new challenge to the treatment of malaria. Therefore, a new generation of artemisinin drugs should have a broader application prospects. Additionally, the discovery of artemisinin has led to paradigm shifts in antimalarial research and therapy.

Fig. 1 *Artemisia annua*

Fig. 2 The structure of artemisinin

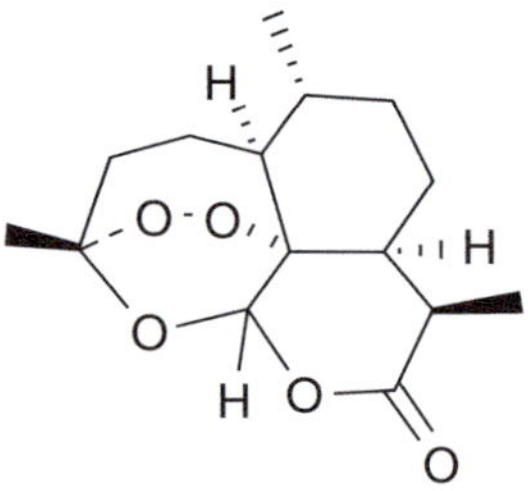

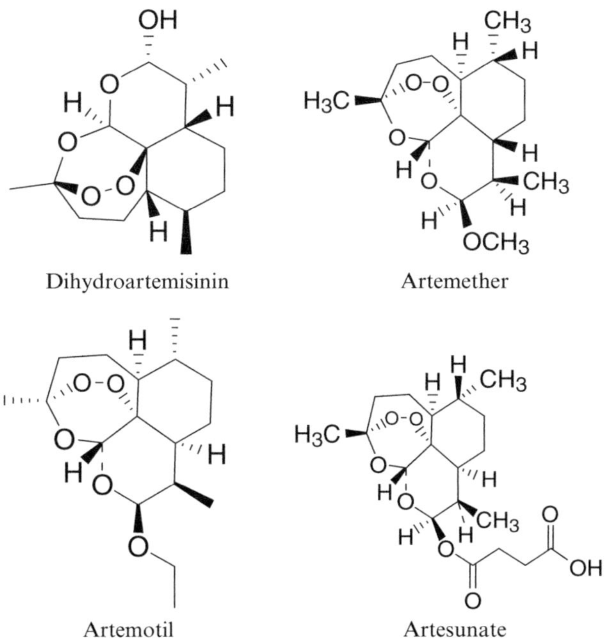

Fig. 3 The structure of artemisinin derivatives

References

1. Wang M-y. History of artemisinins development. Chin J Nat. 2012;34:44–7.
2. Miller LH, Su X. Artemisinin: discovery from the Chinese herbal garden. Cell. 2011;146(6):855–8.
3. Tu Y. The discovery of artemisinin (qinghaosu) and gifts from Chinese medicine. Nat Med. 2011;17(10):1217–20.
4. Collaboration Research Group for Qinghaosu. A new sesquiterpene lactone—qinghaosu. KexueTongbao. 1977;22:142.
5. Collaboration Research Group for Qinghaosu. Crystal structure and absolute configuration of artemisinin. Sci China. 1979;11:78–92.
6. Tu YY. Qinghao and qinghaosu derivative drugs. Beijing: Chemical Industry Press; 2009.
7. Li Y, Wu YL. How Chinese scientists discovered qinghaosu (artemisinin) and developed its derivatives? What are the future perspectives? Med Trop (Mars). 1998;58(3 Suppl):9–12.
8. Collaboration research group for Qinghaosu. Studies on new anti-malarial drug qinghaosu. YaoxueTongbao. 1979;2:49–53.
9. Pharmacology Research Laboratory of the Institute of Chinese Materia Medica, Academy of Traditional Chinese Medicine. Pharmacologic study of Qinghao. XINYIYAOXUEZAZHI. 1979;1:25–35.

10. Bhaw-Luximon A, Jhurry D. Artemisinin and its derivatives in cancer therapy: status of progress, mechanism of action, and future perspectives. Cancer Chemother Pharmacol. 2017;79(3):451–66.
11. Malaria Expert Advisory Committee of Ministry. Use of antimalarial principle and regimens in China. Chin J Parasit Dis Control. 2002;15:129–30.
12. Cui L, Su XZ. Discovery, mechanisms of action and combination therapy of artemisinin. Expert Rev Anti Infect Ther. 2009;7(8):999–1013.
13. Meshnick SR. Artemisinin: mechanisms of action, resistance and toxicity. Int J Parasitol. 2002;32(13):1655–60.

Lycorine

Xiao-Yu Bai, Li-Da Du, and Guan-Hua Du

Abstract Lycorine is extracted from lycoris radiate, which is famous because galanthamine has been approved by FDA as an anti-Alzheimer disease drug. Lycoris radiate is the only natural source for galanthamine with extremely low content (<0.02%). The main active ingredients extracted from the lycoris radiate are about 40 alkaloids with various contents and lycorines are the major effective medicinal ingredients. Lycorine and its derivatives have various pharmacological effects, including anti-inflammatory, antiviral, anti-malaria, acetylcholine esterase inhibition, cardiovascular protection, and antitumor. The in-depth studies on such compounds may fully take advantage of the natural resources in China to develop an innovative drug which is effective in the treatment of serious diseases including tumor, senile dementia, and virus infection. In conclusion, the research and development of an innovative drug relating to lycorine and its derivatives deserve much attention from both scientists and researchers.

Keywords Lycorine · Lycoris radiate · Anti-inflammatory · Cardiovascular protection · Antitumor

Alias: narcissine, galanthidine
Origin: *Lycoris radiata* (Fig. 1)

X.-Y. Bai
National Key Laboratory of Mental Health, Institute of Psychology, Chinese Academy of Sciences, Beijing, China

L.-D. Du
School of Biomedical Sciences, Chinese University of Hong Kong, Hong Kong, China

G.-H. Du (✉)
Beijing Key Laboratory of Drug Targets Identification and Drug Screening, Centre for Pharmaceutical Screening, Institute of Materia Medica, Chinese Academy of Medical Sciences and Peking Union Medical College, Beijing, China
e-mail: dugh@imm.ac.cn

G.-H. Du, *Natural Small Molecule Drugs from Plants*,
https://doi.org/10.1007/978-981-10-8022-7_96

Chemical name (Fig. 2)

1,2,4,5,12b,12c-Hexahydro-7*H*-[1,3]dioxolo[4,5-j]pyrrolo[3,2,1-*de*] phenanthridine-1,2-diol

Molecular formula, $C_{16}H_{17}NO_4$; **MW**, 287.31; **CAS**, 476-28-8
Lycorine Derivatives (Fig. 3)

Properties

Appearance: colorless prismatic crystal. *Solubility*: insoluble in water; sparingly soluble in ethyl alcohol and diethyl ether. *Melting point*: 275–280 °C (decomposition). *Specific optical rotation*: right-handed optical rotation with a specific optical rotation of −129° (98% ethyl alcohol).

Dosage Forms and Indications

Injection: 25 mg/ml, for resistance to amebic protozoa and treatment of intra-intestinal/extra-intestinal amebiasis. Subcutaneous injection: 25–50 mg/injection and 50 mg/day.

Literature

Lycoris radiate, a traditional Chinese medicine (TCM), is the bulb of the amaryllidaceous *Lycoris radiata* herb. It has been applied for clinical purposes for centuries. It is firstly recorded in *Tujing Bencao* and mainly used for the pyogenic infections. According to *A Supplement to Compendium of Materia Medica*, lycoris radiate may be used for treating acute throat trouble, phlegm node, *baihuodan*, and pulmonary abscess [1].

There are about 20 *Lycoris* species in the world, which are widely distributed in China and Japan. Lycoris radiate is an amazing horticultural plant with a graceful shape and a bright color. In TCM, it is acrid in taste and neutral in nature, with functions of detoxication, easy expectoration, and diuresis and emesis promotion. According to the modern medicine, lycoris radiate is considered to be in favor of the central nervous system and cardiovascular system [2].

The main active ingredients extracted from the lycoris herbs are about 40 alkaloids with various contents. Pharmacological tests indicate that galanthamine, lycorine, lycoramine, lycorenine, and crinine are the major effective medicinal ingredients. Lycorine may be used to treat amebic dysentery and against cancer. Moreover, galanthamine, dihydrogalanthamine, and lycoramine may be used to treat infantile paralysis and restore nerve functions and against traumatic paraplegia, etc. Lycoris radiate is famous because galanthamine has been approved by FDA as an anti-Alzheimer disease drug. Lycoris radiate is the only natural source for galanthamine with extremely low content (<0.02%) [2].

History of R&D

In 1895, Morishima successfully extracted lycorine from the bulb of Lycoris radiata. However, its structure was unidentified until in 1935. In 1959, its stereochemical structure was dissected by monocrystal [3].

The solvent extraction method, chromatographic separation, and resin absorption are commonly used for lycorine extraction. However, lycorine obtained from these techniques is not pure enough and often mixed with other alkaloids. The great differences in the efficacies of different alkaloids prevent such blending from being directly used. Furthermore, the separation and purification process so required has an adverse effect on and limits the development and utilization of the medicinal value of lycorine [4].

Pharmacology

Great progress has been made in exploration of the pharmacological activities and mechanisms of lycorine and its derivatives in recent years [5–7].

1. Effect on Central Nervous System

   Lycorine can accelerate the mice's conditioned reflex of motor and defense nature. A mouse intraperitoneally injected with lycorine at 2 mg/kg and a rabbit intramuscularly injected with lycorine at 12 or 20 mg/kg are exposed to a good sedative effect. For a mouse and a rat injected with lycorine at 12 mg/kg and 15 mg/kg, respectively, the sleep time of hexobarbital sodium, pentobarbital sodium, and miltown is extended. According to the hot plate test, the analgesic actions of morphine and rhizoma corydalis are better in a mouse intraperitoneally injected with lycorine at 12 mg/kg.
2. Effect on Cardiovascular System

   Intravenous injection of lycorine results in a slight antihypertensive effect in anesthetized dog, cat, and rabbit, while no inhibitory effect is observed in isolated heart of toad.
3. Anti-inflammatory Effect

   The lycorine may stimulate the pituitary gland-adrenocortical function, which is possibly relevant to its anti-inflammatory action. Intravenous or subcutaneous injection of lycorine at 3 mg/kg significantly inhibits formaldehyde-induced (rabbit) and albumen-induced (rat) foot swelling. Its inhibitory effect on albumen-induced foot swelling in rat was abolished when adrenal gland was removed.

4. Effect on Smooth Muscle

   Lycorine can excite the isolated uteri of guinea pig and rabbit, which is free from the counteraction by diphenhydramine. The isolated uterus of rat is excited by a small dose of lycorin but inhibited by a large dose. This effect is relevant to the inhibition by lycorine on cholinesterase.

5. Emetic Effect

   Lycorine has a good emetic action. The incubation period for emesis is similar to that of ipecine and longer than that of apomorphine, with low toxicity reported. Therefore, it is sometimes used as the emetic for food poisoning.

6. Antitumor Effect

   According to the in vivo experiment, lycorine can inhibit the anaerobic glycolysis of the ascites tumor cells of mice but has no effect on their respiration and aerobic glycolysis. The in vitro test indicates that, however, lycorine can lead to a significant inhibition on the aerobic glycolysis of tumor cells but little effect on their respiration and anaerobic glycolysis. Lycorine is capable of inhibiting adenosine triphosphatase, which is possibly relevant to its cytotoxicity.

7. Antiparasitic and anti-malaria effect

   Dihydrolycorine is better than ipecine in terms of the counteraction against amebic dysentery, with lower toxicity. Thus, it has the potentiality to be a better drug against amebic dysentery. Not only that, dihydrolycorine can be used against paragonimiasis as well.

8. Other Effects

   Upon the subcutaneous injection of a small amount of lycorine, the blood glucose is observed to be reduced slightly in the target rabbit or rat, with the epinephrine-induced hyperglycemia in the rat relieved. A large amount, however, can only result in a significant rise of blood. Lycorine, similar to SKF-525A, can inhibit drug metabolism to a weak extent. The amebic protozoa can be killed by lycorine. A rat intraperitoneally injected with lycorine at 6 mg/kg has more uric acid excreted.

Clinical Application

Dihydrolycorine, generated through the hydrogenation of lycorine, has been used clinically due to its better resistance against amebic dysentery and lower toxicity. The amine salt made of lycorine has an antitumor effect in animals.

Lycorine exposure may cause skin irritation (red and swollen) and itching. Nosebleed may be induced in case of inhalation. In case of overdose, it may cause salivation, emesis, diarrhea, bradycardia, cold hands/feet, or even death due to respiratory center paralysis. The major studies of clinical application are focused on (1) antitumor effect, (2) effect on the central nervous system, (3) effect on the cardiovascular system, (4) anti-inflammatory effect, (5) effect on smooth muscle, and (6) emetic effect.

Discussion

Lycorine and its derivatives have various pharmacological effects, including anti-inflammatory, antiviral, anti-malaria, acetylcholine esterase inhibition, cardiovascular protection, and antitumor. The in-depth studies on such compounds may fully take advantage of the natural resources in China and make a huge economic and social benefit by right of the possibility to develop an innovative drug which is effective in the treatment of serious diseases including tumor, senile dementia, and virus infection. In conclusion, the research and development of an innovative drug relating to lycorine and its derivatives deserve much attention from both scientists and researchers.

Fig. 1 *Lycoris radiata*

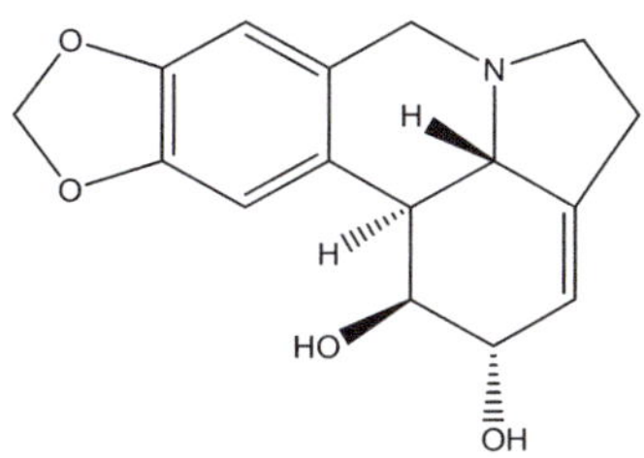

Fig. 2 Chemical structure of lycorine

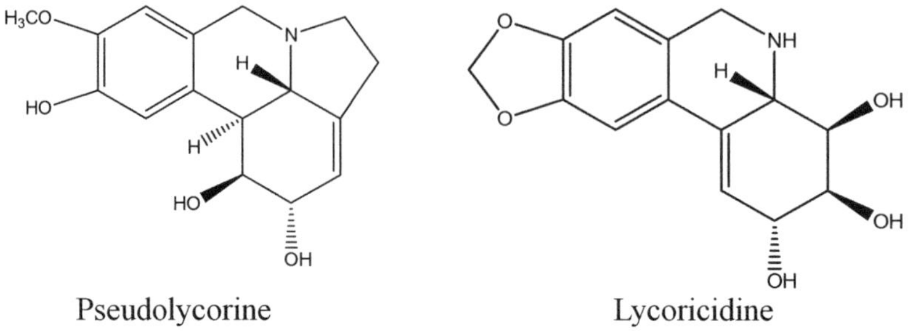

Fig. 3 Chemical structures of lycorine derivatives

References

1. Kunming Q, Lixiao, Zhao X, et al. Overview of the study on the pharmacological effects of lycorine and its derivatives. J Beijing Union Univ (JCR Sci Ed). 2009;23(1):6–10.
2. Baoxiang H, Shugen F, Zhu P, et al. Study on the lycorine of lycoris radiate. South China For Sci (Jiangxi Linye Keji). 2012;1:22–4.
3. Shanhai H, Zhengxiong C, Jingfen L. Study on the amaryllidaceae alkaloids II: the new separation method of pseudolycorine and the search for plants with high lycorine content. Acta Pharm Sin. 1962;9(12):719–24.
4. Yizheng Chen, Jianxi Li, Yonghua Luo et al. The method and process to extract the alkaloids with the *lycoris radiata* hydrobromic acid galantamine acid water. CN:1569856A, 2005-01-26.
5. Zongbin Ji. Pharmacological action and application of available composition of traditional Chinese medicine. Beijing: People's Medical Publishing House; 2011.
6. Vrijsenz R, Berghep DAV, Vlietinckg AJ, et al. Lycorine: a eukaryotic termination inhibitor? J Biol Chem. 1986;261(2):505–7.
7. Cedrón JC, Gutiérrez D, Flores N, et al. Synthesis and antiplasmodial activity of lycorine derivatives. Bioorg Med Chem. 2010;18:4694–701.

Naringin

Yue-Hua Wang, Wei-Han Li, and Guan-Hua Du

Abstract Naringin, also called naringoside, hesperidin, and isohesperidin, is a kind of flavanone compound extracted from immature or nearly mature outer layer of *C. paradisi* Macfad which belongs to *Citrus grandis* (L.) Osbeck. Naringin has many biological activities and pharmacological effects including antioxidant, anti-aging, anticancer, antimicrobial, hypoglycemic, lowering blood pressure, and prevention of atherosclerosis, and it has been used in fields of medicine and cosmetic. In addition, naringin can be used as natural colorants and flavor modifiers in food industry.

Keywords Naringin · Naringoside · Analgesic · Anti-inflammation

Origin: Grapefruit (Fig. 1)
Chemical name (Fig. 2)

(7-[[2-O-(6-Deoxy-α-L-mannopyranosyl)-β-D-glucopyranosyl]]oxy]-2,3-dihydro-5-hydroxy-2-(4-hydroxyphenyl)-4H-1-benzopyran-4-one)

Molecular formula, $C_{27}H_{32}O_{14}$; **MW**, 580.53; **CAS**, 10236-47-2

Properties

Appearance: white to light yellow crystalline powder. *Solubility*: soluble in methanol, ethanol, acetone, acetic acid, dilute alkali solution, and hot water and insoluble in nonpolar solvent like petroleum benzin, ether, benzene, and chloroform. *Melting*

Y.-H. Wang · W.-H. Li · G.-H. Du (✉)
Beijing Key Laboratory of Drug Targets Identification and Drug Screening, Centre for Pharmaceutical Screening, Institute of Materia Medica, Chinese Academy of Medical Sciences and Peking Union Medical College, Beijing, China
e-mail: dugh@imm.ac.cn

G.-H. Du, *Natural Small Molecule Drugs from Plants*,
https://doi.org/10.1007/978-981-10-8022-7_97

point: 83 °C containing 6-8 crystalline water; containing 171 °C contain 2 crystalline water. *Specific optical rotation*: 19D-82° (in ethanol solution).

Dosage Forms

This product is registered both in British Pharmacopoeia (2017) and European Pharmacopoeia (9.0th ed.).

Its main dosage forms are suppositories, lotions, injections, tablets, capsules, etc. It is used for the treatment of bacterial infections, anxiety, and cancer.

Literature

Naringin is also called naringoside, hesperidin, and isohesperidin. It is a pale yellow flavanone compound extracted from immature or nearly mature outer layer of *C. paradisi* Macfad which belongs to *Citrus grandis* (L.) Osbeck. It tastes bitter and is naturally existed in the skin and flesh of *Rutaceae* like grapefruit, mandarin orange, and orange. It is also one of the active ingredients in many traditional Chinese medicines like *Rhizoma Drynariae*, *Immature Bitter Orange*, *Fructus Aurantii*, and *Exocarpium Citri Grandis*. The contents of naringin in different plants vary greatly with the category and origin, and the content of naringin is high in immature fruits [1]. In terms of traditional Chinese medicine, the flavor of grapefruit is sweet and sour, and the nature is cold; the flavors of its peel are sweet, pungent, and bitter, and the nature of its peel is warm. Both the pulp and the peel of grapefruit have the biological function including reducing phlegm, helping digestion, relieving abdominal distention, and fast diaphragm. And they are mainly used for the treatment of cough with asthma, sense of suppression in the chest, coldness and pain in abdomen, dyspeptic retention, and hernia. Naringin was recorded in a *Chinese Materia Medica* book *Rihuazibencao* that grapefruit can be used to treat pregnant woman with symptoms of eating less and tastelessness. It could remove extra gas from stomach, help digestion, and detoxicate and is effective to remove the peculiar smell after drunk. It was also recorded in the *Compendium of Materia Medica* that the peel of grapefruit can help digestion, fast diaphragm, eliminate anger, and reduce phlegm. In *Sichuan Traditional Chinese Medicine Book*, the peel of grapefruit was recorded to be used to relieve alcoholism, treat renal edema, help digestion, reduce phlegm, and treat hernia.

History of R&D

Naringin is mainly existed in the peel of grapefruit, lime, and their varieties; it has multiple biological activities and is widely applied in the fields of medicine, food, and cosmetics. In the 1930s, Harvey and Rygg obtained naringin through the method of isolation and extraction. They also established a colorimetric method for the determination of naringin, which laid the foundation for the following researches [2]. Booth and other researchers conducted the systematic researches on the naringin metabolites [3]. In the 1960s, Hagen and other researchers established a fluorescence chromatography method for the determination naringin [4]. In addition, the biological activity of naringin was evaluated. It was observed to improve ascites, experimental pulmonary edema, peritonitis, and oxygenation [5–7]. At present, the extraction methods of naringin are hot water extraction, alkali extraction, and acid precipitation and organic solvent extraction. A series of pharmacological activity studies have been conducted and demonstrated its various biological activities [8].

Pharmacology

Analgesic and Anti-inflammatory Effects

The writhing times of mice were found to be significantly reduced after intragastric administration of naringin at the dose of 60, 120, and 180 mg/kg. The swelling degree of ear was inhibited after intragastric administration of naringin at the dose of 120, 180, and 240 mg/kg. The peritoneal exudate was significantly reduced after intragastric administration of naringin at the dose of 120, 180, and 240 mg/kg. The inflammatory swelling caused by rat paw injection of protein was inhibited after intragastric administration of naringin at the dose of 60, 120, and 180 mg/kg/day for consecutive 3 days, and the anti-inflammatory mechanism may be related to the inhibitory effects on the process of synthesizing or releasing of inflammatory mediators PGE_2 [9].

Anti-apoptosis, Anti-radiation, and Cancer Prevention

Naringin was found to inhibit the liver damage caused by TNF releasing induced by lipopolysaccharide, which may be used to reduce the damage to liver and the incidence of liver cancer [10]. The anticancer effect of naringin was a comprehensive result caused by many physiological activities, including the effect of antioxidant and anti-free radicals, inhibiting the proliferation of cancer cells, inducing the apoptosis of tumor cells, and inhibiting the expression of cancer gene, among which inducing tumor cell apoptosis was an important way of naringin to prevent cancer.

Protective Effects on Early Diabetes and Complications

Naringin was reported to promote the decomposition of sugar in the liver and decrease glycogen concentration by inhibiting the apoptosis of pancreatic B cells caused by oxidation and regulating the expressing and activity of fatty acid, cholesterol, and glucose metabolism enzyme [11].

Naringin is the important component of traditional Chinese medicine drynaria rhizome. Experiment results have proved that naringin can inhibit many kinds of inflammation effects including osteoarthritis. Besides, naringin was found to have the effects of promoting osteoblast proliferation and differentiation as well as inhibiting the activity of osteoclast.

Other Effects

Antioxidant, anticholesterol, and neuroprotection [12]

Clinical Application

It was used for the treatment of bacterial infection, calm, and cancer prevention.

Discussion

Recent research results have showed that naringin had the pharmacological action like antioxidant, antiaging, anticancer, antimicrobial, hypoglycemic, lowering blood pressure, and prevention of atherosclerosis and it has been used in fields of medicine and cosmetic. In addition, naringin can be used as natural colorants and flavor modifiers in food industry.

Fig. 1 Grapefruit

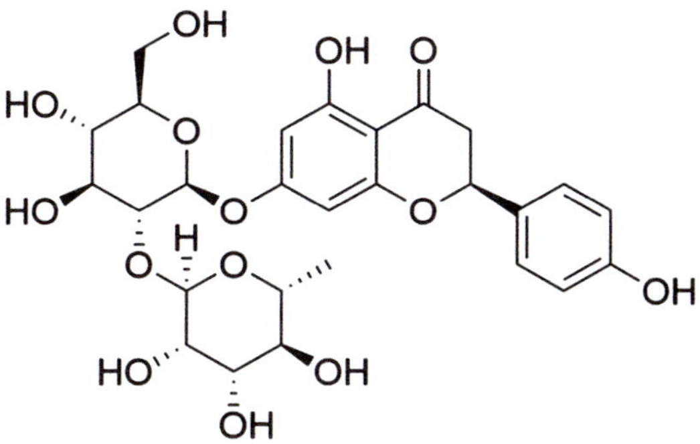

Fig. 2 The structure of naringin

References

1. Li JD, Huang QR. Research progress on extraction, isolation and pharmacological effects of naringin. J N Pharm. 2014;11(7):67–9.
2. Harvey EM, Rygg GL. Colorimetric determination of naringin. Plant Physiol. 1936;11(2):463–5.
3. Booth AN, Jones FT, Deeds F. Metabolic and glucosuria studies on naringin and phloridzin. J Biol Chem 1958; 233(2): 280–2.
4. Hagen RE, Dunlap WJ, Mizelle JW, et al. A chromatographic-fluorometric method for determination of naringin, naringenin rutinoside, and related flavanone blycosides in grapefruit juice and juice sacs. Anal Biochem. 1965;12(3):472–82.
5. Kalman SM, Clewe ER. Respiration of ascites tumor cells: effects of quercitrin and naringin. Stanford Med Bull. 1953;11(4):216–7.
6. Lambev I, Krushkov I, Zheliazkov D, et al. Antiexudative effect of naringin in experimental pulmonary edema and peritonitis. Eksp Med Morfol. 1980;19(4):207–12.
7. Lambev I, Belcheva A, Zhelyazkov D. Flavonoids with antioxidant action (naringin and rutin) and the release of mastocytic and nonmastocytic histamine. Acta Physiol Pharmacol Bulg. 1980;6(2):70–5.
8. Jia DY, Yao K, Tan M, et al. Ethanol extract processing of naringin from pummelo pericarp. Chin Tradit Herb Drug. 2002;33(9):801–2.

9. Xie RF, Wen SE, Li Y, et al. Study on the antiinflammation and analgesia of naringin. J Human Normal Univ (Med Sci). 2011;8(4):5–12.
10. Kawaguchi K, Kikuchi S, Hasegawa H, et al. Suppression of lipopolysaccharide- induced tumor necrosis factor-release and liver injury in mice by naringin. Eur J Pharmacol. 1999;368(2–3):245–50.
11. Wu K, Huang RN, You Q, et al. Protective effect of naringin on myocardium of diabetic rats. Int J Cardiovasc Dis. 2012;39(5):302–5.
12. Gopinath K, Sudhandiran G. Naringin modulates oxidative stress and inflammation in 3 nitropropionic acid induced neurodegeneration through the activation of nuclear factor erythroid 2 related factor 2 signalling pathway. Neuroscience. 2012;227:134–43.

Nevadensin

Zhao Zhang, Ling-Lei Kong, and Nai-Hong Chen

Abstract Stone *Chlorophytum* is the folk antituberculous herb. Nevadensin is extracted from *Lysionotus pauciflorus* Maxim, which has been used for the adjuvant therapy of cervical lymph node tuberculosis, tuberculosis, bone tuberculosis, and bronchitis treatment. Adverse reactions are mainly nausea, vomiting, diarrhea and abdominal discomfort, etc. With the improvement of the extraction technology of nevadensin, its biological activity is improved continuously, which will surely be expanding its application areas.

Keywords Nevadensin · Tuberculosis · Bronchitis

Alias: Lysionotin
Origin: *Lysionotus pauciflorus* Maxim (Fig. 1)
Chemical name (Fig. 2)

5 7-Dihydroxy-3 4′ 8-trimethoxyflavone

Molecular formula, $C_{18}H_{16}O_7$; **MW**, 344.32; **CAS**, 152743-19-6

Z. Zhang · N.-H. Chen (✉)
State Key Laboratory of Bioactive Substances and Functions of Natural Medicines, Institute of Materia Medica & Neuroscience Center, Chinese Academy of Medical Sciences and Peking Union Medical College, Beijing, China
e-mail: chennh@imm.ac.cn

L.-L. Kong
Beijing Key Laboratory of Drug Targets Identification and Drug Screening, Centre for Pharmaceutical Screening, Institute of Materia Medica, Chinese Academy of Medical Sciences and Peking Union Medical College, Beijing, China

G.-H. Du, *Natural Small Molecule Drugs from Plants*,
https://doi.org/10.1007/978-981-10-8022-7_98

Properties

Appearance: yellow needle crystal or light yellow powder, odorless, and tasteless. *Solubility*: slightly soluble in chloroform and in methanol, ethanol, or ethyl acetate, very slightly soluble in ether, insoluble in water, and soluble in 5% sodium carbonate solution, potassium hydroxide, or sodium hydroxide solution. *Melting point*: 197–199 °C.

Dosage Forms

Nevadensin is contained in the *Pharmacopoeia of the People's Republic of China* (1997).

Dosage: ShidiaolanPian (Xiaoheling): 0.3 g per tablet (equivalent to the primary drug 4 g), four tablets each time, oral thrice daily.

Nevadensin tablet: 50 mg nevadensin each tablet, one to two tablet(s) each time, oral thrice daily.

Indications

It can make expectoration easy, relieve a cough, soften hard lumps, and dispel nodes. It is used clinically for bronchitis, lymph node tuberculosis, chill cold, etc.

Literature

Chinese stone *Chlorophytum* is the whole herb of *Lysionotus pauciflorus* Maxim, which belongs to Gesneriaceae. Stone *Chlorophytum* is a small evergreen shrub, which is born in wet places of rock slope and clings to the stone or a tree. Cutting stone *Chlorophytum* in summer and autumn season when leafiness, removing impurities, and drying. Stone *Chlorophytum* is more in Jiangsu, Zhejiang, Anhui, Jiangxi, Hunan, Hubei, Shaanxi, Sichuan, Yunnan, Guizhou, and Guangxi.

Stone *Chlorophytum* has a long medical history in Yunnan, Guizhou, and south of Qinling region. The function was described in various previous dynasty medicine classics. Stone *Chlorophytum* tastes bitter, is flat, and goes into the lung, liver, and kidney meridians. It has been recorded in herbal formula that stone *Chlorophytum* could eliminate sputum and poison, promote digestion, nourish blood, and cure rheumatism, emphysema, dizziness, and various kinds of deficiency. In the classification of herbs, it was recorded that stone *Chlorophytum* could cure hematemesis, blood vomiting, pain of waist and knee, and contusions. It has been described in common folk medicinal herbs that stone *Chlorophytum* could clear the lung to relieve cough, cool blood, and hemostasis [1, 2].

Other plants containing this element in China include *Iva nevadensis* M.E. *Jone*, *I. acerosa* (Nutt.) Jackson, *Helianthus pumilus* L., *Gardenia gummifera* Linn.f., *Ocimum canum* Sims, etc.

History of R&D

Stone *Chlorophytum* is the folk antituberculous herb, the fat-soluble component separated from which is lysionotin. China's scientists for the first time found that lysionotin is the effective component of the antituberculosis in stone *Chlorophytum.* Xu Yin et al., who worked in Shanghai Institute of Materia Medica, Chinese Academy of Sciences, found that lysionotin is 5,7-dihydroxy-4′,6′,8-trimethoxy flavonoids by structure identification, which is the known compound nevadensin [3–5].

Farkas et al. used a 14-step synthesis to demonstrate the structure of nevadensin with complex reaction steps. The synthesis route was simpler, which was designed by Tang Cheng-fang in Shanghai Institute of Materia Medica, Chinese Academy of Sciences, and the total yield was ten times higher [6].

Nevadensin is a water-insoluble compound. Although its sodium salt is water-soluble, clinical use of injection caused pain to the patients. In order to overcome the above shortcomings, methylamine glucoside salt of nevadensin was synthesized for intramuscular injection. The effect is significant without causing pain. In order to explore its structure-activity relationship, seven chemical analogues were designed and synthesized with good activity in antibacterial test of *Mycobacterium tuberculosis*, which include 6,8-dichloro-4′-methoxyflavone, 6,7-dimethyl-4′-methoxyflavone, 6,7-dichloro-4′-methoxyflavone, dimethoxy-3′, 4′-dimethoxyflavone, 6,7-dichloro-3′, 4′-dimethoxyflavone, 6-chloro-4′-methoxyflavone, and 4-dichloro-6-(4-methoxybenzoyloxy) acetophenone (Fig. 3) [6].

Pharmacology

1. Anti-*Mycobacterium tuberculosis*, Anti-inflammatory Effect

Nevadensin 200 μg/ml in vitro has obvious anti-*Mycobacterium tuberculosis* effect; in vivo test also has a protective effect. The treatment for lymph node is significant [5]. Nevadensin has significant inhibitory effect on the experimental arthritis caused by serotonin, formaldehyde, and kaolin. It also has an inhibitory effect on the cotton ball granuloma. Furthermore, the anti-inflammatory effect does not depend on the presence of adrenal cortex [7].

2. Antihypertensive Effect

The main role of nevadensin in the central nervous system causes blood pressure reduction, which also has an impact on the peripheral nervous system. Lateral ventricle injection of nevadensin could reduce arterial blood pressure, and the effect of lowering systolic blood pressure was stronger than that of reducing diastolic blood pressure. The mechanism might be that the action of nevadensin on α-adrenergic receptor is stronger than that of the central β-adrenergic receptor.

3. Scavenging Free Radicals

Phenolic hydroxyl group of nevadensin is the main active group that scavenges free radicals (OH).

4. Metabolism

SD rat single oral nevadensin 50 mg/kg. $T_{max} = 6.5 \pm 1.6$ h. The maximal plasma concentration $C_{max} = 58.4 \pm 10.3$ ng/mL; the half-life of the SD rats was 4.8 ± 1.3 h. Nevadensin could not be detected 36 h later [8].

Clinical Application

Nevadensin has been used for the adjuvant therapy of cervical lymph node tuberculosis, tuberculosis, bone tuberculosis, and bronchitis treatment. Adverse reactions were mainly nausea, vomiting, diarrhea and abdominal discomfort, etc.

Discussion

Nevadensin is safe and effective for drug-resistant tuberculosis and refractory extrapulmonary tuberculosis (EPTB). The medical value of nevadensin is high in the traditional Chinese medicine compound. With the improvement of the extraction technology of nevadensin, its biological activity is improved continuously, which will surely be expanding its application areas.

Fig. 1 *Lysionotus pauciflorus* Maxim

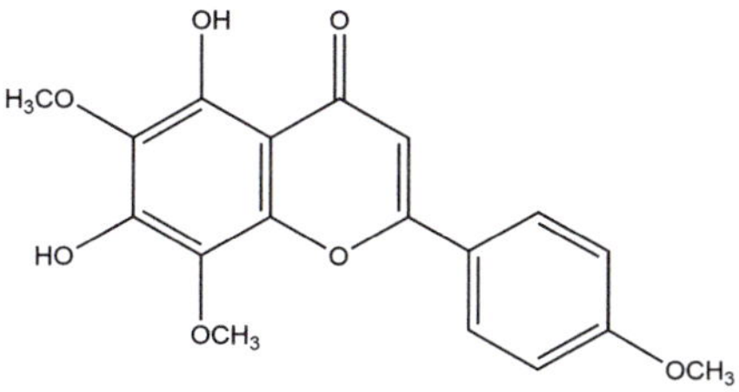

Fig. 2 The chemical structure of nevadensin

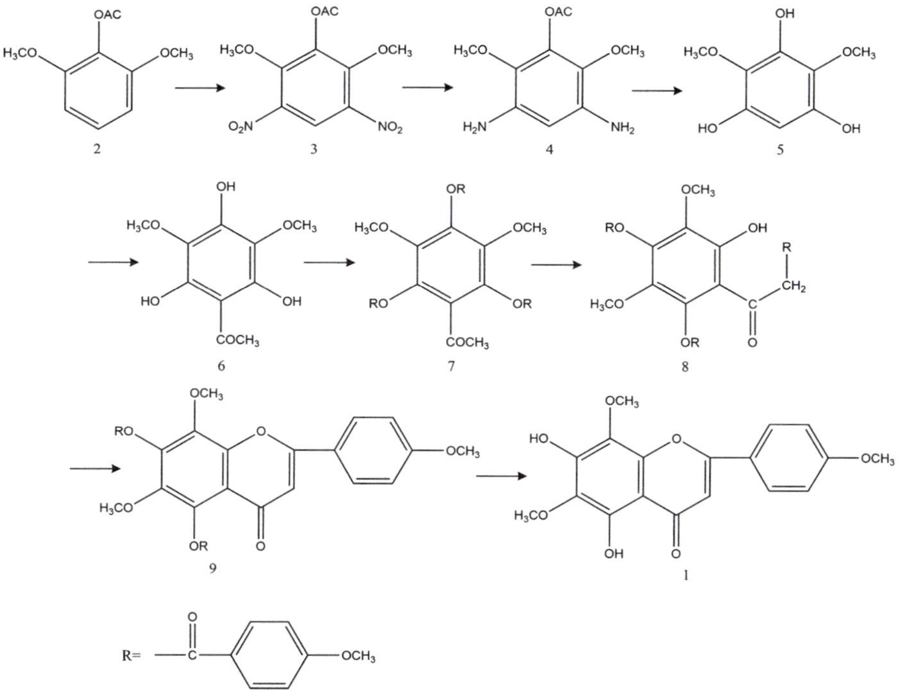

Fig. 3 The synthetic route of nevadensin [6]

References

1. China Pharmacopoeia Committee, Pharmacopoeia of the People's Republic of China. The first division of 1997 edition. Beijing: People's Medical Publishing House Press; 1977. p. 141.
2. State Administration of Traditional Chinese Medicine, Editorial Board of the Chinese materia medica. The Chinese materia medica: the 20th volumes. Shanghai: Shanghai Science and Technology Press; 1999. p. 498–500.
3. Farkas L, Nogradi M, Sudarsanam V, et al. Constituents of Iva species V. Isolation, structure, and synthesis of nevadensin, a new flavone from Iva nevadensis M. E. Jones and Iva acerosa (Nutt.) Jackson. J Org Chem. 1966;31:3228–32.
4. Mabry J. The systematic identification of flavonoids, vol. 107. New York: Springer; 1970.
5. Xu Y, Hu ZB, Feng SC, et al. The research of antituberculous active ingredients of Lysionotius pauciflorus Maxim I: the separation and identification of Nevadensin. Acta Pharm Sin. 1979;14(7):447–8.
6. Tang CF, Xi GL, Gu KJ, et al. Synthesis of tuberculostatic nevadensin and its analogs K. Chin Pharm J. 1981;16(3):55–6.
7. He XZ, Luo GY, Wang ZN. The anti-inflammatory effect of Nevadensin. Tradit Chin Med J. 1985;10(11):516–8.
8. Liu H, Yan C, Li C, et al. LC-MS/MS determination of nevadensin in rat plasma and its application in pharmacokinetic studies. J Pharm Biomed Anal. 2013;74:56–61.

Palmatine

Zi-Ru Yu and Guan-Hua Du

Abstract Palmatine is mainly extracted from the dried cane, and it is one of the main medicinal ingredients of Huangteng. Palmatine is a detoxification agent with a variety of pharmacological effects. It has broad-spectrum antimicrobial antiviral effect, insect resistance and can enhance the ability of leukocyte phagocytosis and also protect cardiovascular system. In clinical it is used mainly for gynecological inflammation, bacillary dysentery, enteritis, urinary tract infection, surgical infection, conjunctivitis, respiratory tract infections, and so on. However, the mechanism of its action and the principle of adverse reactions are still unknown. It is believed that palmatine will have a broad application prospects in clinical treatment with the continuous deepening of pharmacological research.

Keywords Palmatine · Huangteng · Gynecological inflammation

Alias: BRN 1555498; berbericinine; O,O-dimethyldemethylene berberine; palmatine; UNII-G50C034217; berbinium
Origin: *Fibraurea recisa* (Fig. 1)
Chemical name (Fig. 2)

5,6-Dihydro-2,3,9,10-teyramethoxy-dibenzo[a,g]quinolizinium

Molecular formula, $C_{21}H_{22}NO_4$; **MW**, 352.4; **CAS**, 3486-67-7

Z.-R. Yu · G.-H. Du (✉)
Beijing Key Laboratory of Drug Targets Identification and Drug Screening, Centre for Pharmaceutical Screening, Institute of Materia Medica, Chinese Academy of Medical Sciences and Peking Union Medical College, Beijing, China
e-mail: dugh@imm.ac.cn

G.-H. Du, *Natural Small Molecule Drugs from Plants*,
https://doi.org/10.1007/978-981-10-8022-7_99

Properties

Appearance: yellow needle crystal, odorless, taste very bitter. *Solubility*: freely soluble in hot water, sparingly soluble in water, slightly soluble in ethanol and chloroform, and almost insoluble in ether. *Melting point*: 205 °C. *Relative density*: 1.2 g/cm^3.

Dosage Forms

Ordinary tablets, sustained-release tablets, dispersible tablets, hard capsules, soft capsules, suppositories, injection needle, powder, and other series of preparations are mainly used in clinical treatments.

Indications

Palmatine is included in the *Pharmacopoeia of the People's Republic of China (2015)*. As an antibacterial drug, palmatine is used to treat upper respiratory tract infections, tonsillitis, enteritis, dysentery, urinary tract infections, surgical and gynecological bacterial infections such as inflammation, and so on. Locally, it can treat conjunctivitis when dropping eyes and *Candida albicans* infection when used in vaginal topically.

Literature

The only source of natural palmatine is Huangteng. Palmatine is mainly extracted from the dried cane, and it is one of the main medicinal ingredients of Huangteng. Huangteng (*Daemonorops margaritae (Hance) Becc*) is one of the families of Palmae, *Daemonorops* Bl, the dry rattan of which named Huanglian rattan, Big Huangteng, Gold key, Tu Huanglian, and so on. It is concentrated in the tropical and subtropical regions and common in the mountain valley, the altitude of which is 180–1000 m. In China, it is mainly distributed in Hong Kong, Hainan, southeastern of Guangdong, and southwest of Guangxi. It is also distributed in other countries adjacent to China such as Myanmar, Vietnam, Laos, and Cambodia.

Huangteng is rich in alkaloids, lactones, and other active natural compounds. Its medicinal value was found in the *Compendium of Materia Medica*. Residents in

Yunnan folk often use the root, stems, and leaves of Huangteng to treat a variety of infectious diseases, such as skin diseases, falciparum malaria, and so on because of its detoxified, diuretic, antibacterial, and anti-inflammatory function.

History of R&D

In the middle of the last century, Garrison health team of China and Honghe state hospital interviewed to explore the medicinal use of Huangteng in the screen edge of ethnic minority areas firstly and found its detoxification and anti-inflammatory effect of antibacterial. Then, it was promoted to be used in Yunnan province after the initial clinical validation. Soon it was included in "Pharmacopoeia of the People's Republic of China" (1977) for the treatment of gynecological inflammation, surgical infection, bacillary dysentery, enteritis, conjunctivitis, and respiratory and urinary tract infections [1]. Further studies had shown that the active ingredient was palmatine, also included in "Pharmacopoeia of the People's Republic of China" as active pharmaceutical ingredient (API) [2]. And now, common film, sustained-release tablets, dispersible tablets, hard capsules, soft capsules, injection needle, powder, and other series of preparations are developed relying on palmatine API platform.

Palmatine on the market is mostly natural, which is mainly obtained from the extraction of dry rattan of Huangteng, one plant of Menispermaceae. The palmatine is abundant in the dry rattan of Huangteng with more than 2% palmatine chloride [3]. It is an important part of technological research to obtain natural palmatine with high quality and high purity.

Honghe state in Yunnan province isolated and extracted palmatine from Huangteng firstly and then produced needle, tablets, ointment and topical solution, and other dosage forms for the treatment of various suppurative infections. It is now believed that the optimum conditions for the extraction of palmatine should be acid-water decocting [4], firstly extracted from the rattan of Huangteng and then adjusted to pH 9-10, and NaCl salting out [5, 6], which is obtained by precipitation purification step by step, while ultrasonic can be used as an auxiliary means of extraction.

The growth period of Huangteng is very long so that the extraction of natural palmatine cannot meet the market demand. Therefore, in order to alleviate the contradiction of resources, researchers began to explore the method of chemical synthesis of palmatine and its active derivatives in place of natural palmatine. At present, the improved synthetic method using berberine as raw material is the one with high yield, the total yield of which is 54%, and the purity of palmatine is 99% [7].

Pharmacology

Palmatine is a detoxification agent with a variety of pharmacological effects.

Broad-spectrum antimicrobial antiviral effect: It can inhibit the activity of West Nile virus NS2B-NS3 protein, the Asian influenza A virus, dengue virus, and yellow fever virus [8]. Also, it has an inhibitory effect on a variety of Gram-positive and Gram-negative bacteria, especially on fungi. Palmatine inhibits the proliferation of the *Cordyceps* epidermis and other 12 kinds of fungi [9] to different degrees and has a good effect on shallow or deep infection caused by *Candida albicans*.

Enhance the ability of leukocyte phagocytosis: 0.3 mL/100 g, once/day, a total of 10 days of intraperitoneal injection can enhance the macrophage phagocytosis function. Palmatine has a different degree of prophylactic or therapeutic effect on rat adjuvant arthritis when orally administered with suspension prepared by 0.5% carboxymethylcellulose for 9 days on dose of 0.5 mg/100 g. Palmatine can partially counter allergy caused by trichosanthin, and its anti-inflammatory effects may be related to the immune mechanism [10]. Studies have shown that palmatine can improve cellular immunity, humoral immunity, and non-specific immune function.

Insect resistance: Palmatine has a significant anti-*Trichomonas vaginalis* activity in vitro; the anti-vaginal trichomonas activity is equivalent to metronidazole, with seven different concentrations (20, 10, 5, 2.5, 1.25, 0.625, and 0.3125 μg/mL) and different times (2–48 h).

Cardiovascular system: Palmatine can protect myocardial infarction, having a slight excitement on the frog heart and antiarrhythmic effect. It can also lower blood pressure when injecting intravenously anesthetized rabbits.

In addition, the palmatine also showed an inhibition of central nervous function. However, the mechanism of pharmacological mechanism of brassinone is still not clear and need further studies.

Clinical Application

Palmatine has an effect of detoxification; therefore, it is used mainly for gynecological inflammation, such as acute and chronic pelvic inflammatory disease, acute and chronic annex inflammation, cervical erosion, endometritis, mycoplasma vaginitis, puerperal infection, bacillary dysentery, enteritis, urinary tract infection, surgical infection, conjunctivitis, respiratory tract infections, and so on.

At present, palmatine has achieved some curative effect in combination therapy, but the small amount of adverse reactions of palmatine preparations in clinical treatment cannot be ignored, such as allergic reactions and symptoms of catarrh.

Discussion

Palmatine is a purely natural botanical drug developed by China. It is mainly extracted from Huangteng and has obvious antibacterial and anti-inflammatory effects. It can also treat skin diseases and falciparum malaria and other diseases. Now it is produced as anti-inflammatory preparations to treat various inflammations, especially gynecological inflammation. Although a large number of researchers carried out a series of studies on the Huangteng, the mechanism of its action and the principle of adverse reactions are still unknown. It is believed that with the continuous deepening of pharmacological research and the continuous development of modern pharmacy technology, palmatine will have a broad application prospects in clinical treatment.

Fig. 1 *Fibraurea recisa*

Fig. 2 The structure of palmatine

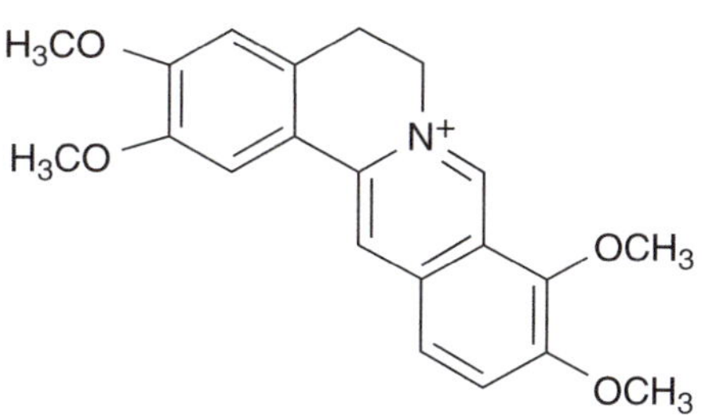

References

1. Chinese Pharmacopoeia Commission. Pharmacopoeia of the People's Republic of China (1997). Beijing: People's Medical Publishing House; 1997. p. 521–4.
2. Chinese Pharmacopoeia Commission. Pharmacopoeia of the People's Republic of China (2005). Beijing: Chemical Industry Press; 2005. p. 215–6.
3. Lin QS. Chemical composition of Chinese herbal medicine. Beijing: Science Press; 1977. p. 222.
4. Ma YS. Effects of different experimental conditions on the extraction rate of palmatine from Huangteng. J Yunnan Univ Tradit Chin Med. 1998;21(l):21–2.
5. Tang AL, Liu XF, Feng DM. A comparative study on the extraction of barmartine from Huangteng by boiling method and cold leaching method. Acta Med Sin. 2003;16(1):75–6.
6. Zheng YH, Jin ZJ, Pu HS, et al. Exploration on extraction conditions of palmatine from Huangteng. J Med Sci Yanbian Univ. 1999;22(3):176–8.
7. Chen TJ, Xia WP, Zhou JY. The synthesis of palmatine. Chin J Pharmaceuticals. 2010;41(7):494–5.
8. Jia F, Zou G, Fan J, et al. Identification of palmatine as an inhibiter of West Nile virus. Arch Virol. 2010;155(8):1325–9.
9. Cong KJ, Xin TC, Guo EL. Anti-fungal experiment and clinical observation of alkaloids from Huangteng. Chin Tradit Herb Drug. 1980;11(12):558–9.
10. Research collaboration group of Huangteng. A preliminary study on the pharmacological action of Huangteng. New Med Newsl. 1978;1:26–32.

Quinine

Lu Wen, Yu-He Yuan, Ling-Lei Kong, and Nai-Hong Chen

Abstract Quinine commonly known as cinchona alkaloid belongs to the alkaloids from natural compounds, extracted from the bark of Rubiaceae plant and cinchona tree. Quinine is the antidote to malaria; the discovery and application of it have saved millions of malaria patients. Biological activity study proved that quinine could kill all kinds of parasite schizonts; it mainly affects the life cycle of the parasite by effectively controlling the developmental conditions. It is mainly used for clinical treatment of not only falciparum malaria caused by chloroquine-resistant strains of plasmodium falciparum but also plasmodium vivax malaria.

Keywords Quinine · Cinchona · Malaria

Alias: quinin, quinine
Origin: Cinchona (Fig. 1)
Chemical name (Fig. 2)

(8S,9R)-6′-methoxycinchonan-9-ol

Molecular formula, $C_{20}H_{24}N_2O_2$; **MW**, 324.42; **CAS**, 130-95-0
Derivatives (Fig. 3)

L. Wen · Y.-H. Yuan · N.-H. Chen (✉)
Institute of Materia Medica, Chinese Academy of Medical Sciences and Peking Union Medical College, Beijing, China
e-mail: chennh@imm.ac.cn

L.-L. Kong
Beijing Key Laboratory of Drug Targets Identification and Drug Screening, Centre for Pharmaceutical Screening, Institute of Materia Medica, Chinese Academy of Medical Sciences and Peking Union Medical College, Beijing, China

G.-H. Du, *Natural Small Molecule Drugs from Plants*,
https://doi.org/10.1007/978-981-10-8022-7_100

Properties

Appearance: white granular or microcrystalline powder. No smell, slightly bitter. *Solubility*: easily dissolved in ethanol, chloroform, and ethyl. Slightly soluble in water and glycerol. *Melting point*: 173–175 °C. *Specific optical rotation*: −172° (ETOH, C = 1).

Dosage Forms and Indications

The product of sulfate was recorded in the *Pharmacopoeia of the People's Republic of China* (2015), the *Japanese Pharmacopoeia* (17th ed.), the *British Pharmacopoeia* (2017), the *Pharmacopeia* (40), the *European Pharmacopoeia* (9.0th ed.); the *International Pharmacopoeia* (5th ed.).

The current clinical dosage forms are as follows: quinine sulfate, quinine sulfate pills, two quinine hydrochloride injection, and quinine compound injection. It is mainly used for clinical treatment of not only falciparum malaria caused by chloroquine-resistant strains of plasmodium falciparum but also plasmodium vivax malaria.

Literature

Quinine commonly known as cinchona alkaloid belongs to the alkaloids from natural compounds, extracted from the bark of Rubiaceae plant cinchona tree. Quinine is the antidote to malaria; the discovery and application of it have saved millions of malaria patients.

The real origin of quinine is no empirical, but Indians use cinchona bark bubble water to treat fever and high fever. About 400 years ago, many European colonists infected with severe malaria died in America because they did not adapt to the local climatic conditions. In history, the Spanish Ambassador's wife Anna unfortunately got infected with malaria; an Indian girl took a risk sending the powder made of cinchona bark, eventually saving Anna's life.

Scientific research shows that the cinchona bark, roots, branches, and stem contain more than 25 kinds of alkaloids, especially the content of alkaloids in the bark is highest; dried bark contains 7–10% of alkaloids, 70% of whom are quinine. In 1817, the French chemists Caventou and Pelletier first obtained quinine monomer isolating from cinchona bark, intended to treat malaria; later quinine was proved to be the anti-malaria ingredient of cinchona bark. The natural source of quinine is limited, only found in South America and Southeast Asia, far beyond affording the folk healing required. According to statistics, when there was no other better antimalarial drug, southern soldiers who died of malaria were more than the injuries to death along the American Civil War. At the end of the nineteenth century, quinine was transported from Europe to China, known as cinchona frost, and was very rare at that time.

History of R&D

After getting quinine monomer in 1817, German chemist Rabe used chemical degradation method to get the flat structure of quinine in 1907 [1]. However, the stereochemical structure of quinine was not determined until the 1940s.

Although quinine molecule is not big, and also is not a particularly complicated structure, there are four chiral centers and has a special complex three-dimensional structure, which make it difficult to chemical synthesize. In 1918, German chemists Rabe and Kindler declared that they found the method which made quinine from dextral quinotoxine by three steps chemical reaction. These three steps chemical reactions were relatively simple but laid a good foundation for the following synthetic study [2]. Announced in 1943, Swiss chemist Prelog found a synthesis method of quinotoxine, which also provided a very important help for its later research [3].

In 1944, American chemist Woodward and Doering finished the total synthesis of quinine. They used 7-hydroxyisoquinoline as initial material to synthesize (d, l)-quinotoxine and applied Professor Rabe and Kindler's following method to complete total synthesis of quinine. Woodward and Doering were regarded as pioneers in definition stereo selectivity reaction and application in synthesis. This study was considered a milestone in the history of organic synthesis. But this kind of quinine synthetic method is too complex, which is confined to the laboratory experiment instead of applying to industrial production.

In 1970, in the United States, Uskokovic team in Roche Pharmaceuticals Company also announced the total synthesis method of quinine. During this period, there are also several chemists to complete the synthesis of quinine. In 2001, a team led by a Stork professor at Columbia University in the United States planned to complete the total synthesis of quinine, chiral compounds (S)-beta-vinyl-gamma butyrolactone for initial and subsequent reaction all around the chiral center, in the reaction using silicon protection, electrophilic substitution, nucleophilic addition, redox reaction, and so on. Since then, several scientists finished quinine total synthesis, such as Igarashi professor at Harvard University in the United States, and planned to use asymmetric catalytic reaction to synthesize quinine; the synthesis method is also regarded as a perfect example of integrating application of modern organic synthesis theory and technology and classical organic synthetic objects. At present, the main source of quinine was extracted from plants or semisynthetic; it is not synthesized in industrialized production according to the total synthetic routes studied by chemists [4].

Quinidine is the right optical isomer of quinine. It is ineffective in treating malaria but has a prominent anti-arrhythmic effect, which was used to treat arrhythmia in clinical.

Pharmacology

Biological activity study showed that quinine mainly effect in the life cycle of the parasite, through a variety of ways to kill all kinds of parasites schizont, effectively controls symptoms [5]. Quinine could gather together and combine with H^+ in parasite lysosome, thus improving lysosome pH value and finally influencing the growth and reproduction of parasite. Quinine can inhibit the heme polymerase activity, prevent heme transferring from lysosome to cytoplasm, make free hemoglobin in the plasmodium, and the heme attack the membrane system; therefore it leads to the death of parasites; quinine can reduce parasite oxygen consumption, inhibit the activity of malaria parasite phosphorylase, and interfere its sugar metabolism. Quinine can form complex with plasmodium DNA and inhibit DNA replication and RNA transcription, thus inhibiting the parasite protein synthesis. Quinine, however, cannot cure benign malaria; a long course of treatment can cure a malignant malaria but has no direct effect on the ligand of malignant malaria, so it can not interrupt transmission. Adults: (1) Take quinine sulfate for the falciparum malaria caused by chloroquine-resistant strain, three times per day, 0.3–0.6 g each time, 14 days for a course of treatment. (2) Severe cases (cerebral type) can be used quinine dihydrochloride, according to the weight 5–10 mg/kg (maximum 500 mg), add it into 500 ml normal saline injection by intravenous drip, drip off 4 h, 12 h after repeat, and change to oral after getting better (generally 3 days).

Children: the treatment of falciparum malaria caused by chloroquine-resistant strains of plasmodium falciparum, for children less than 1 year old, quinine sulfate, 0.1–0.2 g each time, two to three times per day; 1–3 years old, 0.2–0.3 g each time; 4–6 years old, 0.3–0.5 g per day; 7–11, 0.5–1 g per day; 10 days is a course of treatment. Severe patients use quinine dihydrochloride injection; the doses are the same as adults.

Clinical Application

It is mainly used for clinical treatment of not only falciparum malaria caused by chloroquine-resistant strains of plasmodium falciparum but also plasmodium vivax malaria. In addition to the antimalarial, quinine could inhibit the abnormal immune response [6] and can be used in the treatment of immune disorder diseases, such as lupus erythematosus and rheumatoid arthritis. Researchers have found that it had effect on Kaposi's sarcoma disease. The US FDA approved quinine in the treatment of throat diseases and cancer in 2004 [7].

Discussion

Although the anti-malaria drug chloroquine was synthesized along the Second World War, and the more excellent natural antimalarial drug artemisinin was found in the 1970s, the discovery, application, and total synthesis of quinine, both in the human history and in scientific research history, are definitely a brilliant record. Creatures of nature synthesized numerous complex secondary metabolites in its long process of evolution, the diversity of these natural secondary metabolite structures not only greatly enrich the molecular database, broaden the scientist's horizons, but provide molecular biologists with molecular probes and tools to explore the microorganism world, greatly promote the continuous development of chemistry and life science. The development of quinine also provides useful implications for the research and development of new drugs.

Fig. 1 Cinchona

Fig. 2 The chemical structure of quinine

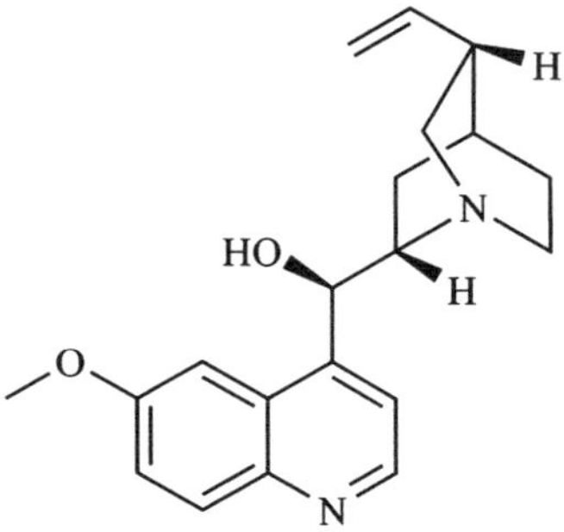

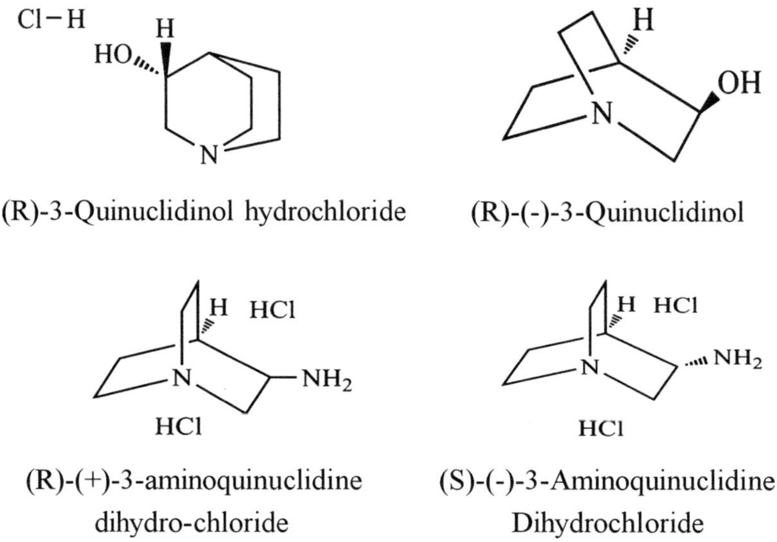

Fig. 3 The structures of quinine derivatives

References

1. der Kenntnis RPZ. Chinaalkaloide. VII. Mitteilung: Über ein neues oxydationsprodukt des Cinchonins. Ber Dtsch Chem Ges. 1907;40(3):3655–8.
2. Rabe P, Kindler K. Cinchona alkaloids. XIX. Partial synthesis of quinine. Ber Dtsch Chem Ges. 1918;51:466–7.
3. Prostenik M, Prelog V. Synthetic experiments in the series of the cinchona alkaloids. IV. Homomeroquinene and the partial synthesis of quinotoxine. Helv Chim Acta. 1943;26:1965–71.
4. Rui-Xia G. Historical story on natural medicinal chemistry: isolation, structure, and total synthesis of quinine. Chin Tradit Herb Drug. 2014;45(19):2737–41.
5. Achan J, Talisuna AO, Erhart A, et al. Quinine, an old anti-malarial drug in a modern world: role in the treatment of malaria. Malar J. 2011;10(144):1–12.
6. Ruocco V, Ruocco E, Schwartz RA, et al. Kaposi sarcoma and quinine: a potentially overlooked triggering factor in millions of Africans. J Am Acad Dermatol. 2011;64(2):434–6.
7. Dias DA, Urban S, Roessner U. A historical overview of natural products in drug discovery. Metabolites. 2012;2(2):303–36.

Santonin

De-Zhi Yang, Li-Da Du, and Yang Lu

Abstract Santonin is a lactone compound extracted from the flower bud of *Artemisia cina* Berg in Compositae, crown daisy chrysanthemum or other plants in Artemisia. It has been recorded in ancient China. Its English name is santonica wormseed. There are more than 100 plus in sagebrush, in which about 30 plus live in China, and parts of them can be used medicinally. Santonin has been used for a long time as a kind of deworming drug. Santonin has obvious central nervous system toxicity. A small amount can cause colour deficiency, and a large amount can cause epileptiform, excessive excitement turning into severe repression, and even coma. Santonin is a highly toxic substance. Children's lethal dose is 0.15 g; adults' lethal dose is about 1 g. Santonin has already been phased out for the toxicity.

Keywords Santonin · Wormseed · Lactone · Deworming · Toxicity

Alias: Santonine
Origin: Santonica wormseed (Fig. 1)
Chemical name (Fig. 2)

3a,5,5a,9b-Tetrahydro-3,5a,9-trimethylnap-hthol[1,2-b]-furan-2,8(3H,4H)-dione

Molecular formula, $C_{15}H_{18}O_3$; **MW**, 246.3; **CAS**, 481-06-1

D.-Z. Yang · Y. Lu (✉)
Beijing Key Laboratory of Polymorphic Drugs, Research Center of Polymorphic Drugs, Institute of Materia Medica, Chinese Academy of Medical Sciences and Peking Union Medical College, Beijing, China
e-mail: luy@imm.ac.cn

L.-D. Du
School of Biomedical Sciences, Chinese University of Hong Kong, Hong Kong, China

G.-H. Du, *Natural Small Molecule Drugs from Plants*,
https://doi.org/10.1007/978-981-10-8022-7_101

Properties

Appearance: a colourless prism crystal or white crystalline powder, odourless, a little taste bitter and turning yellow in the sunlight. *Solubility*: hardly soluble in water, sparingly soluble in ethanol, freely soluble in boiling ethanol and chloroform and slightly soluble in diethyl ether. *Melting point*: 170–174 °C. *Specific optical rotation*: −170 to −175°, determined on 1% ethanol solution at 25 °C.

Santonin is a kind of ketolide with two double bonds. It forms a salt by the ring-opening when dissolving in alkaline solutions and reforms ketolide by acidification.

Dosage Forms

Santonin was recorded in the *Pharmacopoeia of the People's Republic of China* (1963). It was formulated into different drug products, such as santonin tablet, santonin and calomel tablet and santonin and phenolphthalein tablet.

Indications

As a kind of deworming drug, santonin was effective for the treatment of human roundworm infection, but it is no longer in use at present because of the development of more effective medicines.

Literature

Santonin is a lactone compound extracted from the flower bud of *Artemisia cina* Berg in Compositae, crown daisy chrysanthemum or other plants in Artemisia. It has been recorded in ancient China [1]. Its English name is santonica wormseed, Chinese alias is 'quhaosu' and Latin name is *Seriphidium cinum* (Berg ex Poljak.) Pol-jak. [*Artemisia cina* Berg]. Santonica wormseed, a variant name for *Artemisia cina* Berg, is the leaf and anthotaxy of Artemisia cina Berg in sagebrush plants in Compositae. There are more than 100 plus in sagebrush, in which about 30 plus live in China, and parts of them can be used medicinally.

Santonica wormseed, used as raw material to extract and manufacture santonin, lives in cold, dry sandy loam. The place of origin is the Southern Central Asia in the former Soviet Union. China introduced the earliest cultivation in Xinjiang, and now it is introduced and cultivated in the North, Northeast and Northwest China. Santonica wormseed mainly contains hispidulin, quercetin and caffeic acid, and its flower contains 1–3.5% α-santonin, 1–3% artemisinin and volatile oil (mainly is 1, 8-cineole) and so on. Leaf also contains a small amount of α-santonin. In addition, it also contains β-cintonin, 3,4,5,7-tetrahydroxy-3-methoxyflavone7-glucoside and aglucone.

Santonin, used clinically in modern times, is a sesquiterpene lactone compound extracted from flower buds of *Artemisia maritima* L. and *A. cina* Berg. Besides santonin, ψ-santonin and artemisinin are also contained in these plants. β-Santonin is contained in Seriphidium finitum. All these eudesmane derivatives have roundworm-expelling effect, among which α-santonin shows the strongest effect [2] (Fig. 3).

History of R&D

The study on the chemical properties and structure of santonin was carried out at the end of the nineteenth century [3] all the earliest research works were from the Italian scholars, such as Cannizzaro, Andreocci, Gucci, Francesconi, etc. The chemical structures determined before 1910 were all not exactly correct. Until 1929–1930, British scholars, Clemo, Haworth and Walton, finally determined the exact structure of santonin for the transformation from synthesized santonin into desmotroposantonin and santonous acid. However, until 1940, there were no reports on structural configuration. Chinese scholars, Huang Minglong et al., basically completed the research work on the structural configuration of santonin in 1951, and the results were proved by Japanese scholars, Abe, Y. and Harukawa et al. in 1954.

Shandong Xinhua Pharmaceutical Company introduced santonica wormseed from the former Soviet Union and succeeded in planting. For the next 5 years, the area of fields was expanded. The experiment of extracting santonin from these plants finally succeeded, and santonin was put in production. By scientific research, santonin can be synthesized and used as deworming drug. Due to the toxic effects, such as xanthopsia, santonin use was stopped in China.

The earliest record of α-santonin was in page 18 in the Chinese Pharmacopoeia, 1953 edition [4]. It was subsequently recorded in page 511 in the Chinese Pharmacopoeia second part, 1963 edition [5], and in the page 36 in the Chinese Pharmacopoeia second part, 1977 edition [6]. It was deleted in the later edition. Its chemical properties are the same as mentioned above.

There are many ways to extract santonin [7]. In China, the main methods of production of medicinal santonin were calcium salt method and benzene extract method. In the early years, santonin is produced in large scale abroad. Due to high prevalence, the excellent roundworm-expelling effect and convenience to use, a large amount of santonin was needed in China, which was mainly dependent on imports. Introduction and cultivation of santonica wormseed had made great achievements in 1957, which made it possible of mass production. However, santonica wormseed cultivated in China was different in species from others in other countries.

Therefore, the research team led by researcher Fu Fengyong devote to study on extracting process of α-santonin from santonica wormseed (Fig. 4).

Pharmacology

Santonin has a roundworm-expelling effect. It can excite worm's ganglion which leads the worm not to be adsorbed in the intestinal wall, and then the worm is excreted out of the body after using laxatives. The effect is limited to the roundworm, and it has little effect on other helminth. Excessive application can cause toxic effect. When it is excreted in urine, it can make the urine dark yellow or pink. It can be used in the treatment of ascariasis. During the medication, grease should be avoided, and laxative needs to be used such as salts. Patients with hepatosis, nephrosis and acute gastritis must contraindicate its use. The lethal dose 50 (LD_{50}) injected under the skin of mice is 250–400 mg/kg.

Clinical Application

Santonin has been used for a long time as a kind of deworming drug. Its mechanism is between inhibitory effect on γ-GABA and excitatory effect of cholinergic function. It also acts on the human central nervous system and can cause some adverse effects, such as dizziness, leipopsychia, headache, epilepsy, xanthopsia, paresthesia and so on.

Santonin is easily dissolved and absorbed in the intestine due to alkaline intestinal fluid and solvent effect of bile salts. Especially, it can cause severe toxic reaction more easily for increasing absorption because intake of fatty food promotes bile production, secretion and release.

Santonin has obvious central nervous system toxicity. A small amount can cause colour deficiency, and a large amount can cause epileptiform, excessive excitement turning into severe repression and even coma. Santonin is a highly toxic substance. Children's lethal dose is 0.15 g; adults' lethal dose is about 1 g. Santonin has already been phased out for the toxicity.

Discussion

Santonin has a certain toxic effect. As a kind of deworming drug, santonin has provided help for lots of patients and has made a great contribution to human health and also provides a strong theoretical and practical basis for the development of antiparasitic drug. It represents gradual achievement of medical science history.

Fig. 1 Santonica wormseed

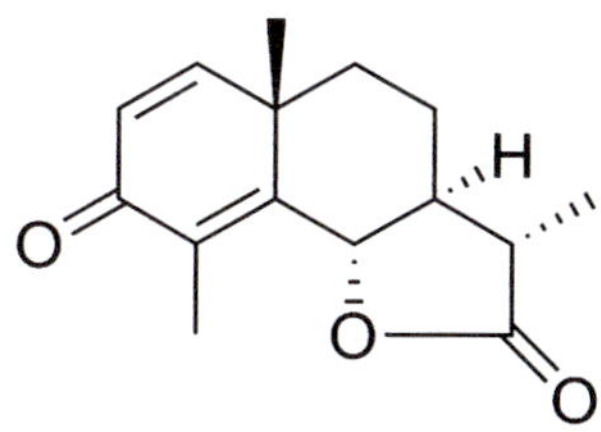

Fig. 2 The structure of santonin

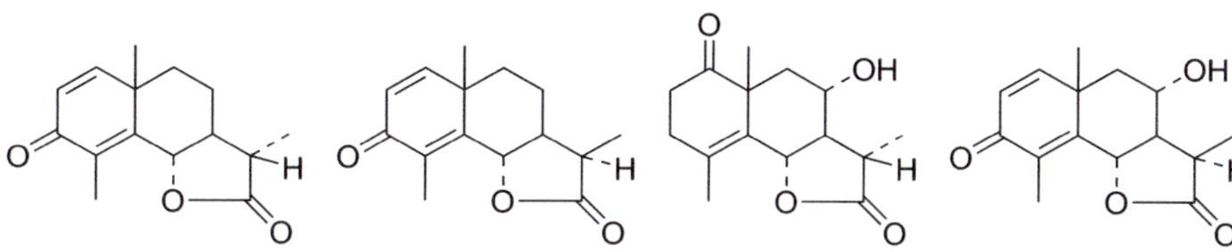

α-santonin β-santonin ψ-santonin artemisin

Fig. 3 Main components in *Artemisia maritima* L. and *A. cina* Berg

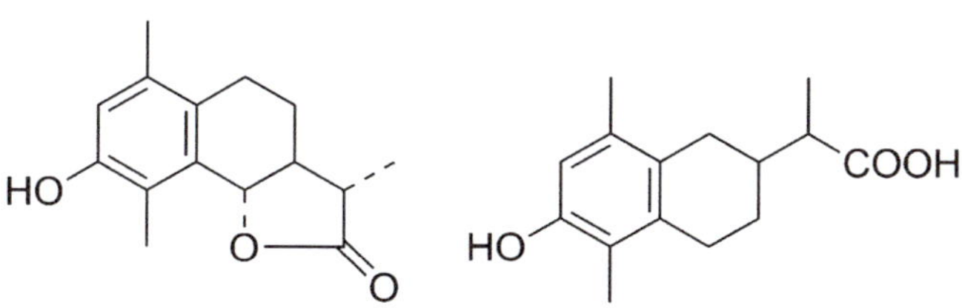

desmotroposantonin santonous acid

Fig. 4 Chemical structure of desmotroposantonin and santonous acid

References

1. The Second Military Medical University. Chinese medicinal plant map. Shanghai: Shanghai Education Publishing; 1960.
2. Beijing Medical College. Chemical ingredients of Chinese traditional medicine. Beijing: The Commercial Press, People's Medical Publishing House; 1980.
3. Minglong H, Shanwei Z. The stereochemistry of santonin and its related compounds. Acta Pharm Sin. 1956;4(1):73–95.
4. China Pharmacopoeia Committee. Pharmacopoeia of the People's Republic of China (1953). Beijing: The Commercial Press., 1953.
5. China Pharmacopoeia Committee. Pharmacopoeia of the People's Republic of China. Beijing: The Commercial Press, People's Medical Publishing House; 1964. 1964
6. China Pharmacopoeia Committee. Pharmacopoeia of the People's Republic of China. Beijing: The Commercial Press, People's Medical Publishing House; 1977. 1977
7. Fengyong F, Cheng Y, Tianmin S. Isolation of santonin from Artemisia Incana Keller. Acta Pharm Sin. 1963;10(3):140–6.

Part VI
Vitamins

Overview

Are vitamins considered as drugs? This is a frequently asked question and even pondered by many people. Vitamins are available in most supermarkets all over the world as dietary supplements or health-care products with various dosages. However, the answer is quite simple, that is, "whether to treat vitamin deficiency-related disease." If the answer is yes, then the vitamin would be considered as a drug. In case of preventing vitamin deficiency, it either can be deemed as a drug or as a health-care product. This is related to the content of pharmaceutical and nutritional theories; we will not expand the discussion here.

Vitamins are essential for the human health, but cannot be produced by the body. The most part of vitamins are obtained from food. The body will suffer diseases when lacking vitamins, such as scurvy caused by vitamin C deficiency, beriberi caused by vitamin B1 deficiency, and rickets caused by vitamin D deficiency. The pharmacological research provides a theoretical basis for the therapy of vitamin deficiency diseases.

The recognization of vitamins is a great progress of life science in the twentieth century. From Hopkins and Eijkman postulated the concept of vitamin in 1906 to the last vitamin (vitamin B12) being identified in 1948, the identification process of the 13 known vitamins took only 42 years. In the course of vitamin discovery, there are 20 scientists who shared 11 Nobel Prizes.

In the path of vitamin discovery, researchers made some detours. When the vitamin deficiency disease was found, at the same time, the microbiology was developing rapidly. Researchers regarded the vitamin deficiency disease as a bacterial infectious disease and used in therapy strategy for a long-term. Through experience accumulation and the successful establishment of pathological animal models, the vitamin discovery process transitioned from empirical research to experimental science research, which has greatly promoted the discovery of vitamins.

Thirteen vitamins are recognized currently, including vitamin A, vitamin B1, vitamin B2, vitamin B6, vitamin B12, vitamin C, vitamin D, vitamin E, vitamin K, nicotinic acid, folic acid, biotin, and pantothenic acid. The 13 vitamins are categorized into two major classes—fat-soluble and water-soluble vitamins. Fat-soluble vitamins are A, D, and E. The others are water-soluble vitamins. Information about vitamins from plant sources as well as the vitamin deficiencies that can cause obvious diseases are collected in this chapter. For more details please refer to the related monographs.

Vitamin A

Xiang-Ying Kong, Li-Da Du, and Guan-Hua Du

Abstract Human vitamin A is derived from the natural food. Natural vitamin A exists in different forms. Vitamin A not only significantly affects visual function but also has a greater physiological impact than visual function, especially for reproduction and embryonic development, immune function, and maintaining healthy skin. Vitamin A in clinical applications includes two aspects: one is to treat vitamin A deficiency disease, and the other is to prevent vitamin A deficiency. Therefore, vitamin A will protect the health of the healthy people and relieve the pain of vitamin A-deficient people. The process of vitamin A discovery provides a very valuable experience and effective drugs for the medical sciences. And the understanding of vitamin A provides a reliable material and rational understanding for the prevention of human disease.

Keywords Vitamin A · Retinol · Night blindness

Alias: Vitamin A, retinol, antiophthalmic factor
Chemical name

(2E,4E,6E,8E)-3,7-dimethyl-9-(2,6,6-trimethyl-1-cyclohexen-1-yl)-2,4,6,8-nonatetra-en-1-ol (Fig. 1)

X.-Y. Kong
Institute of Chinese Materia Medica, China Academy of Chinese Medical Sciences, Beijing, China

L.-D. Du
School of Biomedical Sciences, Chinese University of Hong Kong, Hong Kong, China

G.-H. Du (✉)
Beijing Key Laboratory of Drug Targets Identification and Drug Screening, Centre for Pharmaceutical Screening, Institute of Materia Medica, Chinese Academy of Medical Sciences and Peking Union Medical College, Beijing, China
e-mail: dugh@imm.ac.cn

G.-H. Du, *Natural Small Molecule Drugs from Plants*,
https://doi.org/10.1007/978-981-10-8022-7_102

Properties

Vitamin A1 (VA1), **Molecular formula**, $C_{20}H_{30}O$; **MW**, 286.45; **CAS**, 68-26-8. *Melting point*: 62–64 °C. *Boiling point*: 137–138 °C.

VA2, **Molecular formula**, $C_{20}H_{28}O$; **MW**, 284.44; *Melting point*: 17–19 °C.

Dosage Forms and Indications

Vitamin A was contained in a group of pharmacopoeias, including *Pharmacopoeia of the People's Republic of China* (2015), *British Pharmacopoeia* (2017), *United States Pharmacopeia* (40), the *Japanese Pharmacopoeia* (17th ed.), and the *European Pharmacopoeia* (9.0th ed.).

Formulations of vitamin A are mainly capsules and drops. Clinically they were used for the treatment of diseases caused by vitamin A deficiency and prevention of vitamin A deficiency.

Literature

Human vitamin A is derived from the natural food. Natural vitamin A exists in different forms. In animal tissues, vitamin A is present in the form of retinoids. However, in plants, the form of vitamin A called carotenoids is contained in the green, orange, and yellow plant tissue. Vitamin A compounds such as retinol, retinal, carotene, and so on from these foods can be converted to vitamin A in the human body. Therefore, food is the main source of vitamin A [1].

As early as 1000 years ago, the *Qian Jin Yao Fang* written by Sun Simiao in Tang Dynasty recorded that animal liver can cure night blindness. This is the early recognition in vitamin A supplementation. The traditional Chinese medicine books also recorded that nourishing the liver can improve eyesight. Researches on the therapy of vitamin A deficient-diseases are mainly related to nourishing the liver and kidney, supplementing essence and blood, and activating qi.

History of R&D

The vitamin research is the great achievement in the development of life sciences, while human beings only took half a century to discover and understand vitamins. However, everything is still very difficult for scientists in the early stage of vitamin discovery. From 1913 to 1915, Elmer McCollum and Marguerite Davis indicated that the growth rate was maintained by at least two different kinds of growth factors: one can be separated from eggs or butter, and the other one which prevented multiple neuritis of chicks and pigeons can be extracted by water; thus they were named fat-soluble vitamin A and water-soluble vitamin B.

In 1919, the researchers demonstrated that fat-soluble vitamin A not only supported the rate of growth but also prevented eye dryness and night blindness in the process of property study. In 1920, Dr. J.C. Drummond named this active lipid as vitamin A. It exists in cod liver oil and prevents the occurrence of eye dryness and night blindness.

Pharmacology

Intake of vitamin A precursors, such as carotenoids, retinyl esters, retinol, and retinal, can maintain the epithelial cell differentiation, normal proliferation, and visual function. All of these substances can be metabolized into retinol, retinal, and retinoic acid. But unlike retinol and retinal, retinoic acid cannot be reduced to retinol and retinal. Intake of retinoic acid can only maintain the systemic function of vitamin A.

Visual and vitamin A. 11-cis-retinal plays an important role as a photographic group of retinal cones and visual pigments in rod cells. 11-cis-retinal would be transformed into all-trans-retinal form under the light induction. The dissociation of all-trans retinal and opsin was coupled with the nerve stimulation of the brain's visual center. By a series of biochemical processes, nerve impulses format in the rod cells at the end of synapse, and then the optic nerve conducts the nerve impulses along. The visual process is a component renewable cycle, and all-trans-retinal can be enzymatically modified to 11-cis form in dark conditions.

The systemic effects of vitamin A. Vitamin A not only significantly affects visual function but also has a greater physiological impact than visual function. Vitamin A deficiency destroys the visual cycle, leads to dark adaptation damage (night blindness or nyctalopia), and destroys systemic function which is necessary to maintain life (e.g., corneal injury, infection, and hypoplasia). Vitamin A deficiency can lead to animal death.

Vitamin A functions in reproduction and embryonic development. Vitamin A plays an important role in the reproductive process of sperm production and ovulation, but its biochemical basis is unclear. Vitamin A plays a key role in the development of embryos and organism and maintenance of tissue function. The main organs affected by vitamin A deficiency are the heart, eye tissue, circulatory system, genitourinary system, and respiratory system. Vitamin A is necessary for embryonic development.

Vitamin A functions on immune function. The lymphoid organs, cell distribution, histology, lymphocytes, and other characteristics will change when the animals lack vitamin A. Vitamin A deficiency can lead to immune function decrease, induce inflammation, and exacerbate inflammatory symptoms.

Vitamin A functions in dermatology. Vitamin A plays an important role in maintaining healthy skin. Vitamin A deficiency disrupts human keratin cell terminal differentiation and makes the skin rough, dry, scaly, and clogged and the pores enlarged.

It is reported that vitamin A can degrade malignant melanoma and T-cell lymphoma epidermal transfer, reduce the oil secretion of the common acne and the number of bacteria in the epidermis and capillaries, and inhibit immune response of monocytes and neutrophils.

Vitamin A plays an important role as an important function material in the body system, such as hematopoietic function, bone development, tumor prevention, and so on. Therefore, supplement of vitamin A is necessary for health requirements.

Clinical Application

Vitamin A in clinical applications includes two aspects: one is to treat vitamin A deficiency disease, and the other is to prevent vitamin A deficiency. For the former, vitamin A is a drug which has significant effects and no side effect, and it is easy to be done, but the key issue is the accurate diagnosis. For the latter, it is not necessary to add vitamin A in the normal diet and food balance and adequate situation. And if there is a vitamin A deficiency diet for a short time, the body also has a certain vitamin A reserve to ensure the body supply. There is no report that too much vitamin A supplement is useful.

Vitamin A deficiency can cause many diseases, such as night blindness and dry eye. Night blindness is retinal dark adaptation dysfunction, which begins to develop after lacking of vitamin A in the diet for a year, and the health can be restored after vitamin A supplement. Dry eye is a permanent change of the front of the eye morphology, cannot be corrected, and the early intervention is very important, and the progression of the disease can be blocked before the permanent blindness. In addition, vitamin A deficiency can be seen in the following: systemic diseases such as loss of appetite, dysplasia, membrane drying and keratosis, infection, and death; rough skin and hair; muscle weakness; periosteal hyperplasia, cranial cavity, and spinal cord limitations; nephritis cerebrospinal fluid pressure increases and ataxia; sperm-free, vaginal horniness, stillbirth, and embryo absorption.

It could achieve the desired results by moderately taking vitamin A supplement for the abovementioned diseases caused by vitamin A deficiency.

Discussion

The process of vitamin A discovery provides a very valuable experience and effective drugs for the medical sciences. Vitamin A is the ideal drug which is effective and safe for vitamin A deficiency disease. And the understanding of vitamin A provides a reliable material and rational understanding for the prevention of human disease. Vitamin A will protect the health of the healthy people and relieve the pain of vitamin A-deficient people.

Fig. 1 Chemical structure of vitamin A

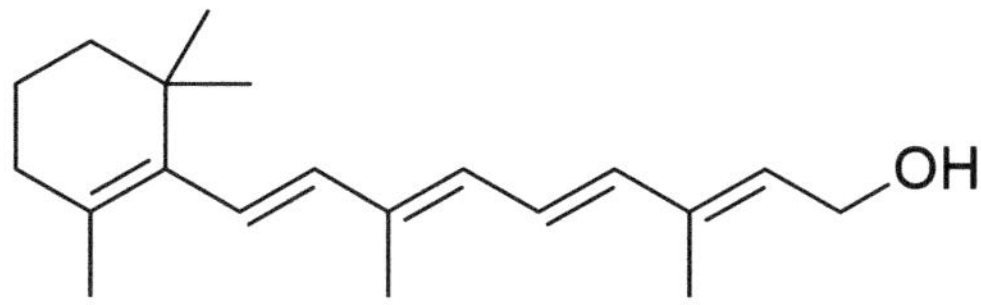

Vitamin A1

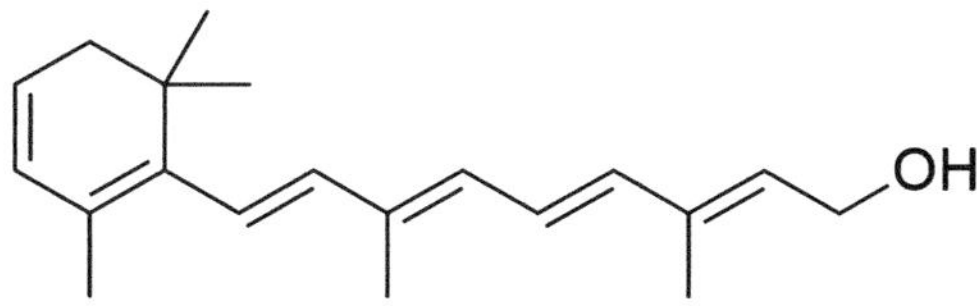

Vitamin A2

Reference

1. Zhang DS, Du GH, translation. The vitamins: fundamental aspects in nutrition and health. Beijing: Science Press; 2009.

Vitamin B

You-Wen Zhang, Li-Da Du, and Guan-Hua Du

Abstract Vitamin B, which is classified as water-soluble B by McCollum in early period, is the indispensable material *in vivo* to convert saccharide, fat and protein to heat. Vitamin B includes thiamin, vitamin B_2, niacin, vitamin B_6, pantothenic acid, biotin, folate, and vitamin B_{12}. They can be used as drugs and play an important role in the treatment of diseases, like neuritis, beriberi, angular stomatitis, fetal anomaly, etc. This type of vitamins has totally different chemical structures, but most of them are from plant extracts. Vitamin B could promote oxygen utilization *in vivo* to improve oxygen levels in many tissues, and it also could help organism in removing toxins. Vitamin B deficiency could cause indigestion, constipation, halitosis, and so on.

Keywords Vitamin B · Vitamin B deficiency · Glycometabolism

Vitamin B is one type of vitamin, which is classified as water-soluble B by McCollum in early period [1]. Most of these vitamins can be used as drugs and play an important role in the treatment of disease, but they are not entirely medicinal. This type of vitamins has totally different chemical structures, but most of them are from plant extracts. In accordance with the requirements of this book, we make a summary of vitamin B below.

Y.-W. Zhang
Beijing Key Laboratory of Anti-infective Agents, Institute of Medicinal Biotechnology, Chinese Academy of Medical Sciences and Peking Union Medical College, Beijing, China

L.-D. Du
School of Biomedical Sciences, Chinese University of Hong Kong, Hong Kong, China

G.-H. Du (✉)
Beijing Key Laboratory of Drug Targets Identification and Drug Screening, Centre for Pharmaceutical Screening, Institute of Materia Medica, Chinese Academy of Medical Sciences and Peking Union Medical College, Beijing, China
e-mail: dugh@imm.ac.cn

G.-H. Du, *Natural Small Molecule Drugs from Plants*,
https://doi.org/10.1007/978-981-10-8022-7_103

Vitamin Terminology

The vitamin terminology evolved non-systematically during the course of their discovery. Current standards for vitamin nomenclature policy were established by the International Union of Nutritional Sciences in 1978 [2].

Vitamin B includes thiamine (vitamin B_1, antiberiberi vitamin), vitamin B2 (riboflavin), niacin (vitamin PP, niacin, vitamin B_3, antipellagra vitamin), vitamin B_6 (pyridoxine, antidermatitis vitamin), pantothenic acid (nicotinic acid, vitamin B_5), biotin (vitamin B_7, vitamin H), folate (vitamin B_9, pteroylglutamic acid), and vitamin B_{12} (cobalamin, anti-anemia vitamin).

Thiamine

Alias: Vitamin B_1, aneurine, thiamine
Chemical name

3-[(4-amino-2-methyl-5-pyrimidinyl)methyl]-5-(2-hydroxyethyl)-4-methylthia zolium

Molecular formula, $C_{12}H_{17}N_4OS+$; **MW**, 265.355; **CAS**, 70-16–6 (Fig. 1)

Properties

Free thiamin is unstable because of its quaternary nitrogen; in water, it is cleaved to the thiol form. For this reason, the hydrochloride and mononitrate forms are used in commerce. Thiamin hydrochloride (actually, thiamin chloride hydrochloride) is a colorless crystal that is very soluble in water (1 g/ml, thus making it a very suitable form for parenteral administration), soluble in methanol and glycerol, but practically insoluble in acetone, ether, chloroform, and benzene. The mononitrate form is more stable than the hydrochloride form, but it is less soluble in water. It is used in food/feed supplementation and in dry pharmaceutical preparations. Free thiamin is easily oxidized to thiamin disulfide and other derivatives, including thiochrome, a yellow biologically inactive product with strong blue fluorescence that can be used for the quantitative determination of thiamin. Its metabolically active form is thiamin diphosphate, also called thiamin pyrophosphate.

Derivatives

Thiamine (free base), thiamin pyrophosphate, thiochrome

Dosage Forms

Injecta, tablets, capsule

Riboflavin

Alias: Vitamin B_2, Vitamin G

The metabolically active forms are commonly called flavin mononucleotide (FMN) and flavin adenine dinucleotide (FAD). More properly, these compounds should be called riboflavin monophosphate and riboflavin adenine diphosphate, respectively.

Chemical name

7,8-dimethyl-10-(1′-D-ribityl)isoalloxazine

Molecular formula, $C_{17}H_20N_4O_6$; **MW**, 376.369; **CAS**, 83-88-5 (Fig. 2)**:**

Properties

Riboflavin is moderately soluble in water (10–13 mg/dl) and ethanol but insoluble in ether, chloroform, and acetone. It is soluble but unstable under alkaline conditions.

The catalytic functions of riboflavin are carried out primarily at positions N-1, N-5, and C-4 of the isoalloxazine nucleus. In addition, the methyl group at C-8 participates in covalent bonding with enzyme proteins. The flavin coenzymes are highly versatile redox cofactors because they can participate in either one- or two-electron redox reactions.

Riboflavin antagonists include analogs of the isoalloxazine ring (e.g., diethylriboflavin, dichlororiboflavin) and the ribityl side chain (e.g., d-araboflavin, d-galactoflavin, 7-ethylriboflavin).

Niacin

Alias: Vitamin B_3
Chemical name

Niacin is the generic descriptor for pyridine 3-carboxylic acid and derivatives exhibiting qualitatively the biological activity of nicotinamide (Fig. 3).

Properties

Nicotinic acid and nicotinamide are colorless crystalline substances. Each is insoluble or only sparingly soluble in organic solvents. Nicotinic acid is slightly soluble in water and ethanol; nicotinamide is very soluble in water and moderately soluble in ethanol.

Nicotinic acid is amphoteric and forms salts with acids as well as bases. Its carboxyl group can form esters and anhydrides and can be reduced. Both nicotinic acid and nicotinamide are very stable in dry form, but in solution nicotinamide is hydrolyzed by acids and bases to yield nicotinic acid.

The coenzyme forms of niacin are the pyridine nucleotides, NAD(H) and NADP(H). In each of these compounds, the electron-withdrawing effect of the N-1 atom and the amide group of the oxidized pyridine nucleus enables the pyridine C-4 atom to react with many nucleophilic agents (e.g., sulfite, cyanide, and hydride ions). It is the reaction with hydride ions (H^-) that is the basis of the enzymatic hydrogen transfer by the pyridine nucleotides; the reaction involves the transfer of two electrons in a single step.

Several substituted pyridines are antagonists of niacin in biological systems: pyridine-3-sulfonic acid, 3-acetylpyridine, isonicotinic acid hydrazine, 17 and 6-aminonicotinamide.

Pantothenic Acid

Alias: Vitamin B_5, pantoyl-β-alanine

Pantothenic acid is the trivial designation for the compound dihydroxy-β,β-dimethylbutyryl-β-alanine. It has two metabolically active forms: coenzyme A, in which the vitamin is linked via a phosphodiester group with adenosine-3′,5′-diphosphate, and acyl-carrier protein, in which it is linked via a phosphodiester to a serinyl residue of the protein.

Chemical name

3-(2,4-dihydroxy-3,3-dimethylbutanamido)propanoic acid (Fig. 4)

Properties

Pantothenic acid is composed of β-alanine joined to 2,4-dihydroxy-3,3-dimethylbutyric acid via an amide linkage. The molecule has an asymmetric center, and only the R-enantiomer, usually called D-(+)pantothenic acid, is biologically active and occurs naturally. *Appearance*: pantothenic acid is a yellow, viscous oil. Its calcium and other salts, however, are colorless crystalline substances; calcium pantothenate is the main product of commerce. *Solubility*: neither form is soluble in organic solvents, but each is soluble in water and ethanol. *Stability*: aqueous solutions of pantothenic acid are unstable when heated under acidic or alkaline conditions.

Biotin

Alias: Vitamin H, Coenzyme R
Chemical name

cis-hexahydro-2-oxo-1*H*-thieno[3,4-d]imidazole- 4-pentanoic acid

Molecular formula, $C_{10}H_{16}N_2O_3S$; **MW**, 244.31; **CAS**, 58-85-5 (Fig. 5)

Properties

Appearance: biotin is a white crystalline substance. *Stability*: it is fairly stable to air, heat, and light in dry form. In solution, however, it tends to degrade under strongly acidic or basic conditions. In the eight possible stereoisomers of biotin, only the (+)-isomer (called d-biotin) has biological activity. Biotin is covalently bound to its enzymes by an amide bond to the ε-amino group of a lysine residue and C-2 of the thiophane nucleus. The biotin molecule is activated by polarization of the O and N-1'atoms of the ureido nucleus. This leads to increased nucleophilicity at N-1′, which promotes the formation of a covalent bond between the electrophilic carbonyl phosphate formed from bicarbonate and ATP, and allows biotin to serve as a transport agent for CO_2.

Folate

Alias: Vitamin B_9

Chemical name

Folate is the generic descriptor for folic acid (pteroylmonoglutamic acid), and related compounds exhibit the biological activity of folic acid (Fig. 6). The terms folacin, folic acids, and folates are used only as general terms for this group of heterocyclic compounds based on the N-[(6-pteridinyl)methyl]-p-aminobenzoic acid skeleton conjugated with one or more l-glutamic acid residues. Folates can consist of a mono- or polyglutamic conjugate; these are named for the number of glutamic residues (n), using such notations as $PteGlu_n$. The reduced compound tetrahydropteroylglutamic acid is called tetrahydrofolic acid.

Properties

The folates include a large number of chemically related species, each differing with respect to the various substituents possible at three sites on the pteroylglutamic acid basic structure. More than 170 different folates are theoretically possible. Not all of these occur in nature; but it has been estimated that as many as 100 different forms are found in animals. The folates from most natural sources usually have a single carbon unit at N-5 and/or N-10; these forms participate in the metabolism of the *single-carbon pool*.

Folic acid (pteroylmonoglutamic acid) is an orange-yellow crystalline substance that is soluble in water but insoluble in ethanol or less polar organic solvents. It is unstable to light, to acidic or alkaline conditions, to reducing agents, and, except in dry form, to heat. It is reduced *in vivo* enzymatically (or in vitro with a reductant such as dithionite) first to 7,8-dihydrofolic acid (FH_2) and then to FH_4; both of these compounds are unstable in aerobic environments and must be protected by the presence of an antioxidant (e.g., ascorbic acid, 2-mercaptoethanol).

Two derivatives of folic acid, each having an amino group in the place of the hydroxyl at C-4, are folate antagonists of biomedical use: aminopterin (4-aminofolic acid) and methotrexate (4-amino-N^{10}-methylfolic acid). Aminopterin is used as a rodenticide; methotrexate is an antineoplastic agent.

Dosage Forms

Tablets, injecta, capsule

Vitamin B_{12}

Alias: Mecobalamin, Adenosylcobalamin, Hydroxocobalamin (Vitamin B_{12b})
Chemical name

Vitamin B_{12} is the generic descriptor for all corrinoids (i.e., compounds containing the corrin nucleus) exhibiting the qualitative biological activity of cyanocobalamin (Fig. 7). Cyanocobalamin is the trivial designation of the vitamin B_{12}-active corrinoid (also called cobalamin) with a cyano ligand (CN^-) at the β position of the cobalt atom.

Properties

Vitamin B_{12} is an octahedral cobalt complex consisting of a porphyrin-like, cobalt-centered macroring (called a corrin ring or nucleus). The corrinoids are red, red-orange, or yellow crystalline substances that show intense absorption spectra above 300 nm owing to the π–π transitions of the corrin nucleus. They are soluble in water and are fairly stable to heat but decompose at temperatures above ~210 °C without melting.

Vitamin B_{12} reacts with ascorbic acid, resulting in the reduction and subsequent degradation of the former, which releases its cobalt atom as the free ion. Cobalamins are more stable in the presence of ascorbic acid but unstable to light.

Dosage Forms

Injecta, tablets, capsule

Clinical Applications and Functions of Vitamin B

Vitamin B is the indispensable material *in vivo* to convert saccharide, fat, and protein to heat. It can be used in the treatment of neuritis, beriberi, angular stomatitis, fetal anomaly, etc. The specific functions of vitamin B have been listed below:

1. It is the critical substance in glycometabolism process.
2. It is closely related to the glycometabolism, lipid metabolism, and amino acid metabolism.

3. It can be used to treat seborrheic dermatitis, pimples, and acne caused by lipid metabolism disorder.
4. Vitamin B deficiency could cause indigestion, constipation, halitosis, and so on.
5. Due to being an integral part of coenzymes, vitamin B_3 is required for the normal metabolism of fatty acids, amino acids, and carbohydrates. Severe vitamin B_3 deficiency could cause pellagra, known as three D symptom (dermatitis, diarrhea, and dementia), which could lead to neural, dermatic, and gastrointestinal lesions.
6. Vitamin B could promote oxygen utilization *in vivo* to improve oxygen levels in the skin, nail, and hair tissues and then relieve dandruff symptom of patients.
7. Vitamin B could help organism in removing toxins, such as alcohol and nicotine, to relieve headache symptom and protect liver.
8. Vitamin B_{11} or vitamin B_{12} deficiency will affect the synthesis of thymine, purine, etc. It will lead to arrest of erythropoiesis prior to the late stages of differentiation, resulting in apoptotic reduction of cells surviving to postmitotic, and terminal stages in the condition called megaloblastic anemia.
9. The lack of folate in the first 3 months of pregnancy can lead to fetal neural tube defects (NTDs), which can increase the incidence of split brain and anencephalus.

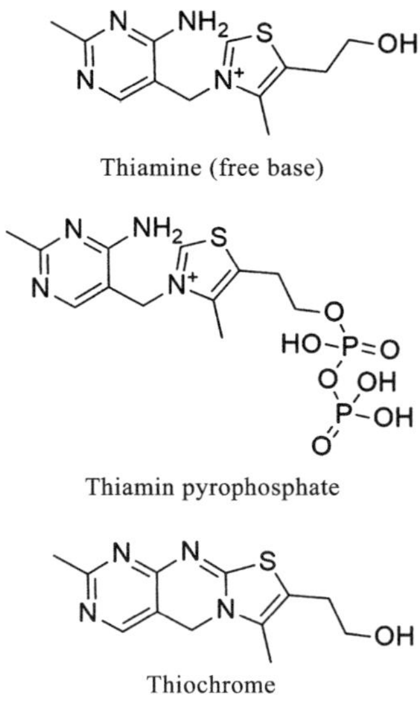

Fig. 1 Chemical structure of thiamin

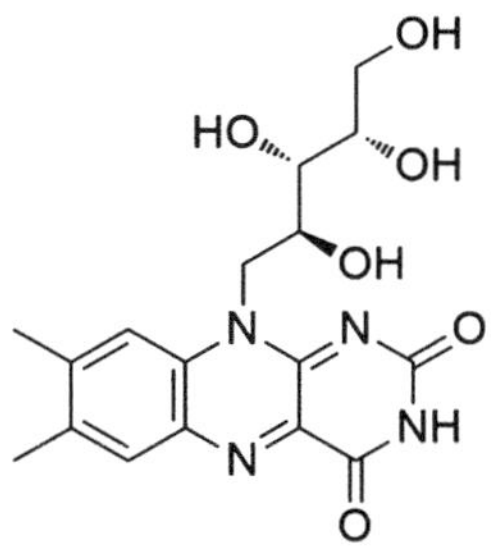

Riboflavin

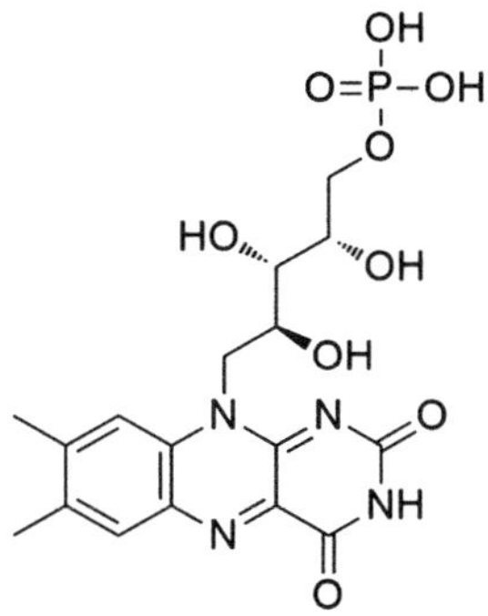

Flavin mononucleotide (FMN)

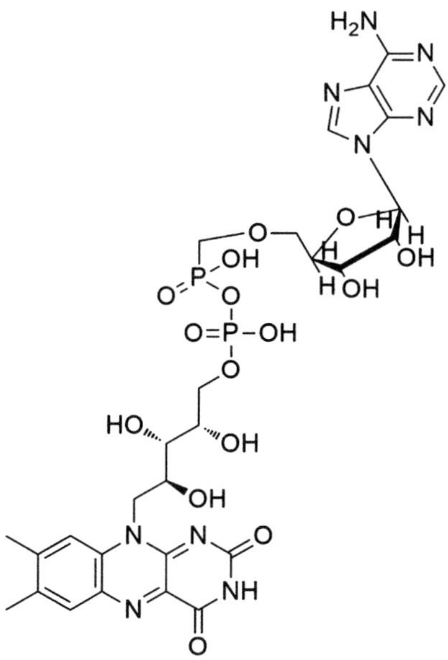

Flavin adenine dinucleotide (FAD)

Fig. 2 Chemical structures of riboflavin and its nucleotide forms

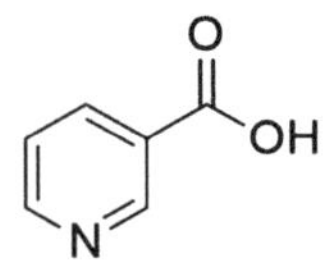

Nicotinic acid

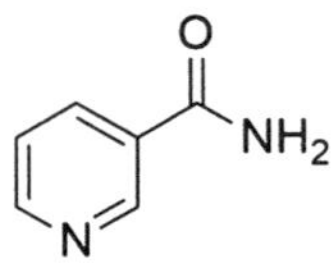

Nicotinamide

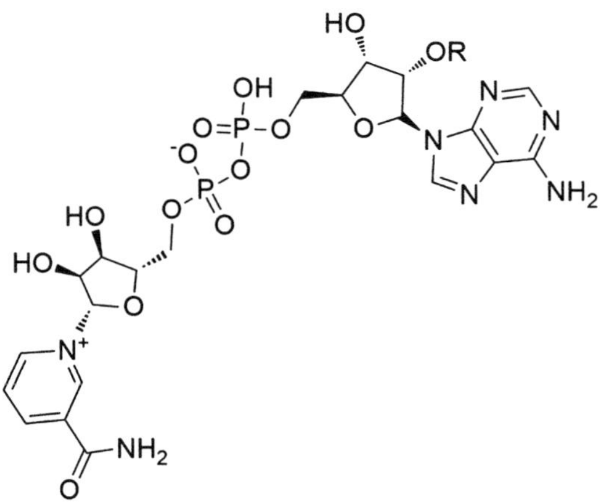

Nicotinamide adenine dinucleotide (NAD^+), R=H

Nicotinamide adenine dinucleotide phosphate ($NADP^+$), R=PO_3H_2

Fig. 3 Chemical structures of niacin

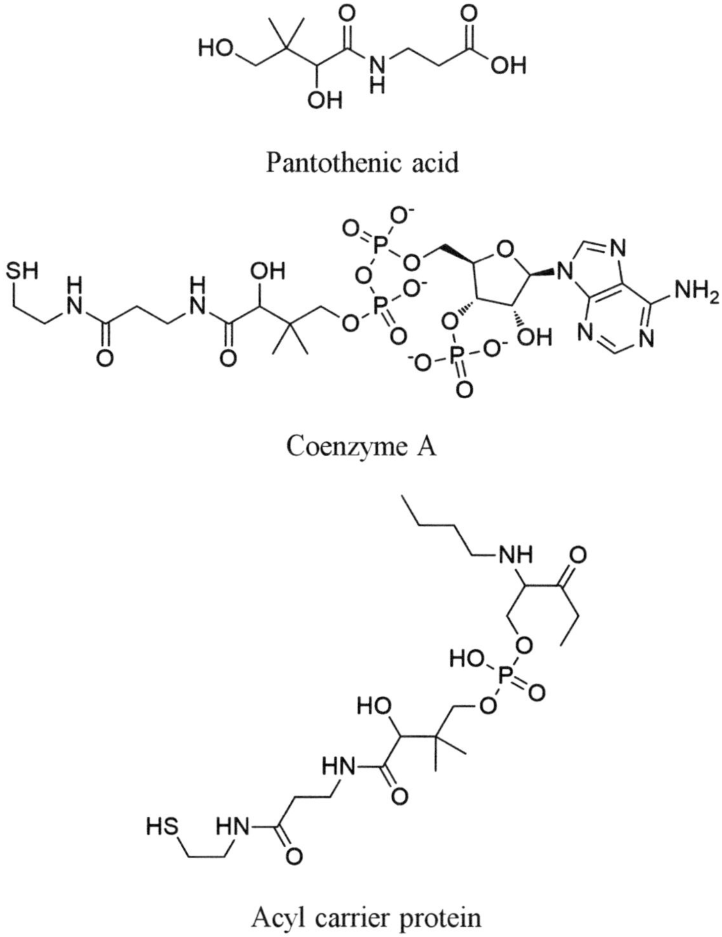

Fig. 4 Chemical structure of pantothenic acid

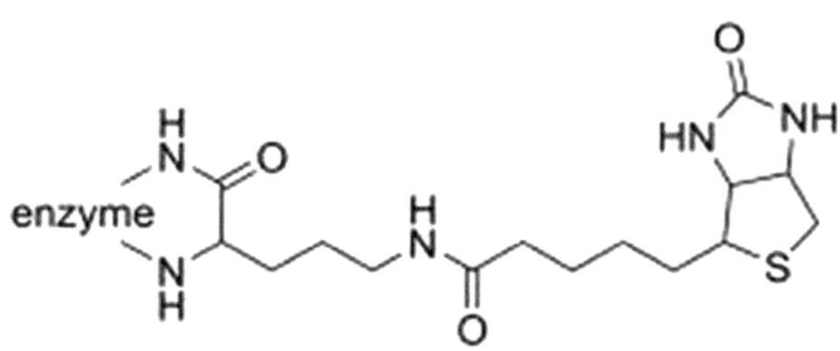

Fig. 5 The structure formula of enzyme-bound biotin (i.e., *biocytin*)

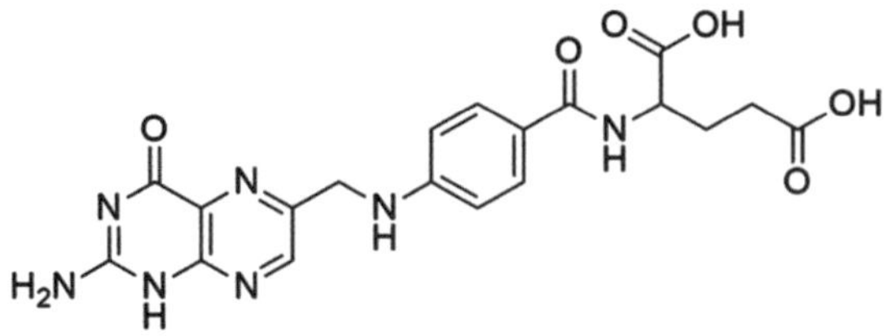

Pteroylglutamic acid

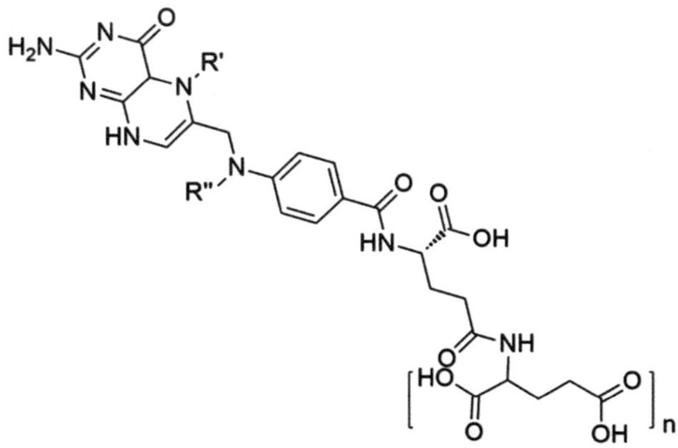

Leucovorin (and its derivatives)

Fig. 6 Chemical structures of the folate group

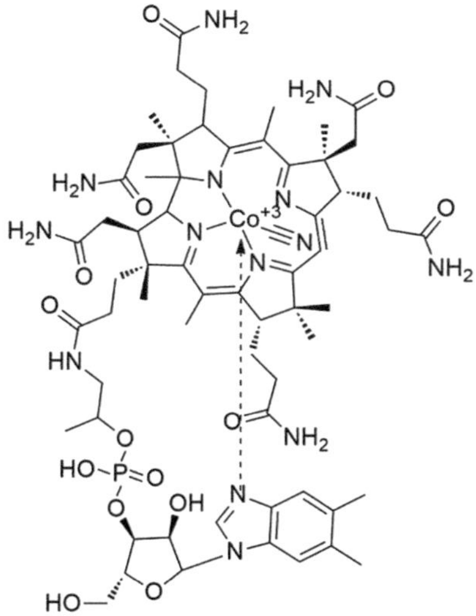

Cyanocobalamine

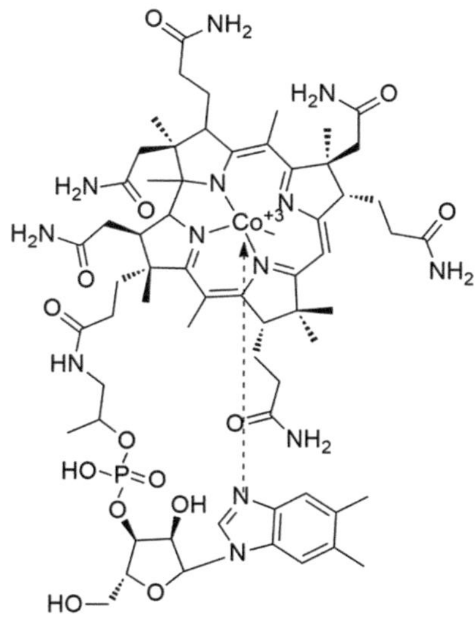

Methylcobalamin

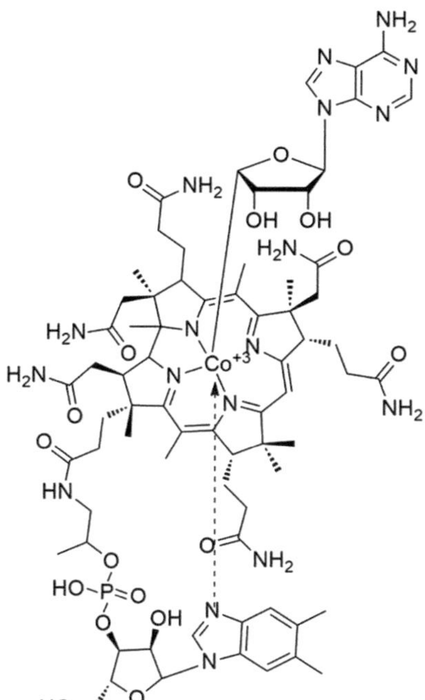

5 '-Deoxyadenosylcobalamin

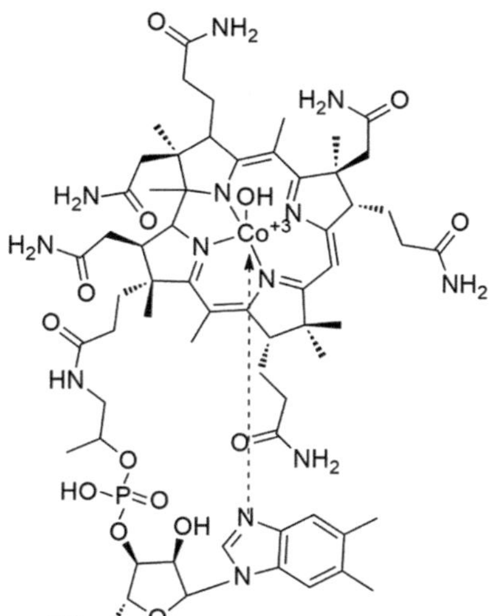

Hydroxocobalamin

Fig. 7 Chemical structures of vitamin B_{12}

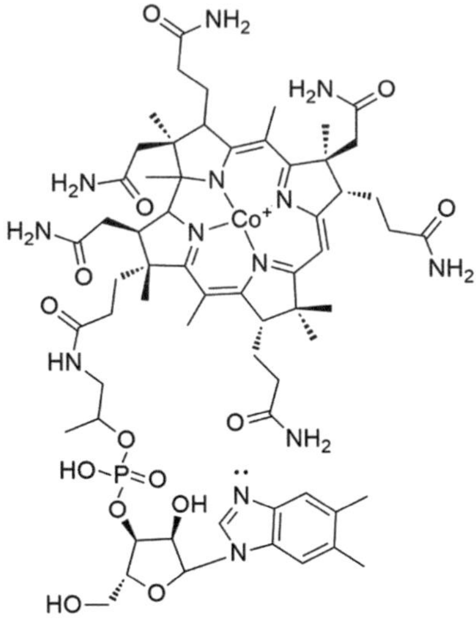

Cob(I)alamin

Fig. 7 (continued)

References

1. Danshen Zhang, Guanhua Du. The vitamins: fundamental aspects in nutrition and health. Beijing: Science Press, 2009 (in Chinese).
2. Combs GF. The vitamins: fundamental aspects in nutrition and health. San Diego: Elsevier Academic Press; 2008.

Vitamin B_6

Li-Da Du, You-Wen Zhang, and Guan-Hua Du

Abstract Vitamin B_6 is the generic descriptor for all 3-hydroxy-2-methylpyridine derivatives exhibiting qualitatively the biological activity of pyridoxine [3-hydroxy-4,5-bis(hydroxymethyl)-2-methylpyridine]. The metabolically active form of vitamin B_6 is pyridoxal phosphate, which functions as a coenzyme for reactions involving amino acids. Otherwise, vitamin B_6 has two main usages. One is for the treatment of vitamin B_6 deficiency which manifested as dermatologic, circulatory, and neurologic changes. The other is for the treatment of related diseases by using higher doses. The correct clinical application of vitamin B_6 can achieve significant clinical therapeutic efficacy, which is of great significance for the maintenance of human health.

Keywords Vitamin B_6 · Pyridoxal phosphate · Vitamin B_6 deficiency

Alias: Pyridoxine
Chemical name (Fig. 1)

Pyridoxol (6-methyl-5-hydroxy-3, 4 pyridine dimethanol hydrochloride), pyridoxal, pyridoxamine

Molecular formula, $C_8H_{11}NO_3$; **MW**, 169.18; **CAS**, 65-23-6

L.-D. Du
School of Biomedical Sciences, Chinese University of Hong Kong, Hong Kong, China

Y.-W. Zhang
Beijing Key Laboratory of Anti-infective Agents, Institute of Medicinal Biotechnology, Chinese Academy of Medical Sciences and Peking Union Medical College, Beijing, China

G.-H. Du (✉)
Beijing Key Laboratory of Drug Targets Identification and Drug Screening, Centre for Pharmaceutical Screening, Institute of Materia Medica, Chinese Academy of Medical Sciences and Peking Union Medical College, Beijing, China
e-mail: dugh@imm.ac.cn

G.-H. Du, *Natural Small Molecule Drugs from Plants*,
https://doi.org/10.1007/978-981-10-8022-7_104

Properties

It is one kind of B vitamins, containing pyridoxine or pyridoxal or pyridoxamine. *Appearance*: colorless crystals at room temperature. *Solubility*: soluble in water and ethanol. *Stability*: stable in acid liquor but easily destroyed in alkali liquor. Pyridoxol is resistant to high temperature, but pyridoxal and pyridoxamine are not.

Dosage Forms

Injecta, capsule, tablet

Indications

Vitamin B_6 deficiency

History of R&D

The discovery of vitamin was tortuous and legendary. After fat-soluble A and water-soluble B were discovered by the year of 1915, the discovery of vitamins entered into a rapid developed period. In separation process of riboflavin by Kuhn and his colleagues, they noticed the unusual relationship between growth-promoting activity and fluorescence of extracts. Then they supposed that the existence of no-fluorescent substances were very necessary for growth-promoting activity of riboflavin. And they considered this phenomenon as the evidence of a second chemical existence in the thermostable complex. At last, they named this substance as vitamin B_6 [1].

In 1934, Szent-Györgyi made a definition for vitamin B_6 activity. He found it could prevent symmetrical scaling dermatitis which previously called acrodynia or pellagra of rats. Its definition contributes to the effective separation of symptoms in vitamin B_6 deficiency from that in riboflavin deficiency. Two years later, his team obtained a partial purification of vitamin B_6. Until 1938, the separation of crystalline form of vitamin B_6 was accomplished, and its chemical structure was also quickly clarified as 3-hydroxy-4,5-hydroxymethyl-2-methylpyridine. In 1939, Folkers completed the synthesis of vitamin B_6, which was called pyridoxine by Szent-Györgyi [2].

Vitamin B_6 is widely distributed in foods, including meats, whole-grain products (especially wheat), vegetables, and nuts. In the cereal grains, vitamin B_6 is concentrated primarily in the germ and aleuronic layer. Thus, the refining of grains in the production of flours, which removes much of these fractions, results in substantial reductions of vitamin B_6 content. The chemical forms of vitamin B_6 tend to vary among foods between plant and animal origin: plant tissues contain most pyridoxine (the free alcohol form, pyridoxol), whereas animal tissues contain most pyridoxal and pyridoxamine.

Pharmacology

The metabolically active form of vitamin B_6 is pyridoxal phosphate, which serves as a coenzyme of numerous enzymes, most of which are involved in the metabolism of amino acids. Vitamin B_6 functions through the following general mechanisms: decarboxylation, transamination, racemization, elimination, replacement reactions, and β-group interconversions.

Pyridoxal phosphate is practically involved in all amino acid metabolism reactions, such as transaminations, transsulfuration, and selenoamino acid metabolism, in both their biosynthesis and their catabolism. Vitamin B_6 also plays an important role in the tryptophan–niacin conversion, histamine synthesis, neurotransmitter synthesis, and hemoglobin synthesis.

Vitamin B_6 has two roles in gluconeogenesis, transaminations and glycogen utilization. It is required for the utilization of glycogen to release glucose by serving as a coenzyme of glycogen phosphorylase.

Vitamin B_6 plays the role of transamination and glycogen heterogenesis, as a coenzyme of glycogen phosphorylase to participate in glycogen releasing glucose. Changes in lipid metabolism also have been reported in response to vitamin B_6 deprivation. It means that vitamin B_6 has a close relationship with lipid levels in vivo.

Due to involvement in a wide range of metabolic process, vitamin B_6 has a close relationship with many diseases, such as homocystinemia. Homocystinemia has been associated with increased risks to occlusive vascular disease, total and cardiovascular disease-related mortality, stroke, dementia, Alzheimer's disease, fracture, and chronic heart failure. Vitamin B_6 has been shown to affect the postnatal development of a glutamate receptor in the brain that is thought to have an important role in learning. Vitamin B_6 deficiency in mother during pregnancy and lactation period will affect neurodevelopment of fetuses and infants. Moreover, some studies suggested that vitamin B_6 has relationships with the immune function and cancer.

Clinical Application

Vitamin B_6 has two main usages. One is for the treatment of vitamin B_6 deficiency. The other is for the treatment of related diseases by using higher doses.

Lack of vitamin B_6 is usually characterized by weakness, sleeplessness, nervous disorders (peripheral neuropathies), cheilosis, glossitis, stomatitis, and impaired cell-mediated immunity. For these disorders, vitamin B_6 supplements can quickly relieve symptoms and achieve satisfactory results.

Several genetic vitamin B_6 metabolic disorders have been proven to be caused by the lack or dysfunction of pyridoxine-dependent enzymes, such as homocystinuria, cystathionuria, GABA deficiency, pyridoxine-responsive seizures, hyperoxaluria, sideroblastic anemia, etc. These patients all need supplement of vitamin B_6.

According to the experience of clinical practice, vitamin B_6 is used for treating a range of related diseases in large doses, such as sideroblastic anemia, sickle cell anemia, iron storage disease, inhibition of lactation, schizophrenia, asthma, and herpes, etc. In addition, it is also used for carpal tunnel syndrome, Chinese restaurant syndrome, premenstrual syndrome, and morning sickness, etc.

Discussion

Vitamin B_6 is involved in a wide range of metabolic processes and has a close relationship with many diseases. At the same time, the correct clinical application can also achieve significant clinical therapeutic efficacy, which is of great significance for the maintenance of human health.

Fig. 1 Chemical structure of vitamin B_6

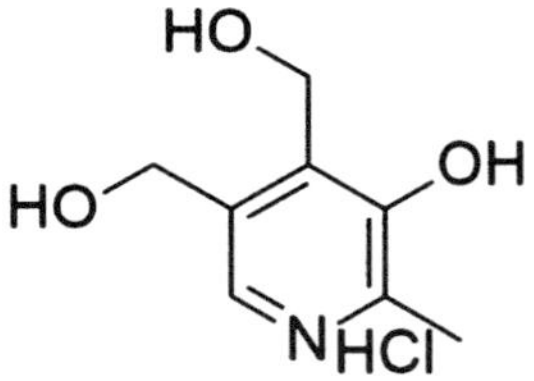

References

1. Zhang D, Guanhua D. [the vitamins: fundamental aspects in nutrition and health](in Chinese). Beijing: Science Press; 2009.
2. Combs GF. The vitamins: fundamental aspects in nutrition and health. San Diego: Elsevier Academic Press; 2008.

Vitamin C

Li-Da Du, Xiang-Ying Kong, and Guan-Hua Du

Abstract Vitamin C is a general term for compounds having ascorbic acid activity, including ascorbic acid, dehydroascorbic acid, and its isomers. A few hundred years ago, researchers found that scurvy could be prevented by eating green vegetables or fruits and obtained the crystal from different plants and identified it as ascorbic acid vitamin C. Vitamin C is considered as a classical enzyme cofactor or antioxidant but also as a transition material in metal ion reaction. And all of these functions of vitamin C are related to the property of oxidation reduction. Vitamin C plays an important role in maintaining human health, such as electron transfer, antioxidant effect, co-substrate function of enzyme, antihistamine effect, and immunomodulatory effects. Therefore, vitamin C not only exerts good healthcare function but also produces therapeutic effect, which emerges with wide prospects in the clinic.

Keywords Vitamin C · Ascorbic acid · Antioxidant

Alias: *L*-ascorbic acid
Chemical name

L-(+)-*threo*-2,3,4,5,6-pentahydroxy-2-hexenoic acid-4-lactone (Fig. 1)

Molecular formula, $C_6H_8O_6$; **MW**, 176.13; **CAS**, 50-81-7

L.-D. Du
School of Biomedical Sciences, Chinese University of Hong Kong, Hong Kong, China

X.-Y. Kong
Institute of Chinese Materia Medica, China Academy of Chinese Medical Sciences, Beijing, China

G.-H. Du (✉)
Beijing Key Laboratory of Drug Targets Identification and Drug Screening, Centre for Pharmaceutical Screening, Institute of Materia Medica, Chinese Academy of Medical Sciences and Peking Union Medical College, Beijing, China
e-mail: dugh@imm.ac.cn

G.-H. Du, *Natural Small Molecule Drugs from Plants*,
https://doi.org/10.1007/978-981-10-8022-7_105

Properties

Appearance: white crystal or crystalline powder, and it is odorless and flavors sour. The color changes yellowish when exposed in the air for a long time. Its aqueous solution is acidic reaction. *Solubility*: vitamin C is soluble in water, slightly soluble in ethanol, and insoluble in chloroform or ether. *Melting point*: 190–192 °C. It would decompose when it melts. *Specific optical rotation*: +20.5 to +21.5°. Ascorbic acid is two-base acid (the pKa is 4.1 and 11.8). It occurs mainly in the form of sodium salt and calcium salt, and its aqueous solution is strongly acidic reaction. Ascorbic acid is a strong reducing agent. It can be oxidized to dehydroascorbic acid under the mild oxidative condition (pKa = −0.45). Meanwhile it is a good free radical quencher.

Dosage Forms and Indications

It was included in *Pharmacopoeia of the People's Republic of China (2015)* and United States Pharmacopeia (40).

Main clinical application forms are tablets, injections, and granules. It can be used for treating the diseases caused by vitamin deficiency such as scurvy.

History of R&D

Vitamin C is a general term for compounds having ascorbic acid activity, including ascorbic acid, dehydroascorbic acid, and its isomers.

The understanding of vitamin C has gone through a long and painful process. A few hundred years ago, people have found that scurvy can be prevented by eating green vegetables or fruits. The Ebers Papyrus (about 1150 BC) and the work of Hippocrates (about 420 BC) also described the relevant cases. In the middle ages, scurvy was common in Nordic residents and crew. Portuguese explorer da Gama revealed that more than 60% of the crewmen lost their lives because of scurvy in their voyage of the Cape of Good Hope in 1498. French explorer Jacques Cartier reported that only three of the 103 crew members had no scurvy symptoms (25 deaths) during the second Newfoundland expedition in 1535. In 1593, British Admiral Richard Hawkins recorded that he had seen nearly 10,000 seafarers die of the disease in their navy career. In the case of food shortages, scurvy would continue to erupt. For example, the scurvy broke out during the California Gold Rush period, in the British prison, in the Crimean War army, in the United States of America North and South war prisoners, as well as in the public of Paris surrounded by war (1871).

Although the relationship between scurvy and stored food is obvious, but the treatments of this disease have been misguided. By 1601, British armed Captain James Lancaster discovered the disease on the ship of the East India Company and regarded the scurvy as "rot," which could be made tissue alkaline. The famous London College of physicians indicated that acids can be used as substitutes. At that time, the surgeons of British ship were all equipped with sulfate, and the patients with scurvy were treated by acetic acid or dilute sulfuric acid.

Until 1746, the British naval surgeon (later known as the founder of modern naval hygiene) James Linde conducted the first control clinical trials, and he made a comparison of every kind of recommended treatments. He paired the 12 seafarers with scurvy for food therapy for 2 weeks, including lemons and oranges, dilute sulfuric acid, vinegar, or other presumptive drugs. The conclusion is clear: seafarers treated with lemon and orange recovered in 6 days, while other therapies did not bring any improvement. In 1753, he published the *A Treatise on the Scurvy* which has been regarded as monumental classic. The publication of this book had a great impact on medical ideology.

Linde didn't believe it was right that "a kind of soap" contained with citrus is conducive to clear the ocean air-clogged skin to perspire, and the significance of the conclusion is to determine the role of fresh fruit in the scurvy treatment. So the daily distribution of lemon juice to seafarers in the British Navy had became an established practice by 1804. At the early stage of the nineteenth century, the understanding and treatment of scurvy had developed to a right approach. The exposition of scurvy etiology and metabolic theory took more than a century.

By the early stage of the twentieth century, inspired by the animal model of beriberi, researchers in the Christchurch Oslo University discovered one animal that could suffer scurvy accidentally and then established a valuable scurvy animal model. This experiment demonstrated that the extract isolated from lemon had anti-scurvy activity. Until 1932, many research groups obtained the anti-scurvy crystal from different plants and identified the crystal as ascorbic acid vitamin C. Next year, the chemical structure of ascorbic acid was elucidated, and then its artificial synthesis was accomplished.

Pharmacology

Vitamin C is considered as a classical enzyme cofactor or antioxidant but also as a transition material in metal ion reaction. And all of these functions of vitamin C are related to the property of antioxidation.

Electron Transfer

Ascorbic acid has a reversible monovalent oxidation that makes it easy to lose electrons, which plays an important role in the biochemical oxidation-reduction system, participating in many electronic transport reactions, such as biosynthesis of collagen, the degradation of the 4-hydroxybenzoic acid, the synthesis of norepinephrine, and the desaturation of saturated fatty acids.

Antioxidant Function

Ascorbic acid can react with the free radicals in the body to form an intermediate with low reactive. And during the process, the active oxygen free radicals with toxic reaction have been reduced.

Co-substrate Function of Enzyme

As a collaborative substrate for enzymes, ascorbic acid becomes the substrate of the enzyme by binding to an oxygen atom, like monooxygenase. At present, at least eight enzymes are known to interact with vitamin C.

In addition, vitamin C is also involved in the synthesis of a variety of substances which play important roles *in vivo*, such as the promotion of collagen synthesis, promoting catecholamine biosynthesis, and being involved in peptide biosynthesis and carnitine synthesis. Vitamin C is also involved in a lot of *in vivo* metabolic processes, including steroid metabolism, tyrosine metabolism, and metal ion metabolism.

In many aspects, vitamin C plays an important role in maintaining human health. Vitamin C plays an antihistamine effect and regulates immune function to achieve the effect in the treatment of disease. In the infectious diseases, vitamin C can play a protective role including common cold, *Helicobacter pylori* infection, and herpes, as same as the treatment of inflammation, cardiovascular disease, diabetes, and cataract. On the other hand, vitamin C is also closely related to exercise tolerance, neurological function, lung function, pregnancy, skin health, dental health, environmental stress, and so on.

Clinical Application

Vitamin C, mainly used for complementary treatment in Vitamin C deficiency patients, can produce ideal results in the clinical treatment. Currently, there is fewer and fewer people falling ill due to the deficiency of vitamin C. Vitamin C deficiency affects normal metabolism in vivo and becomes an important factor in health hazard; therefore, supplement of vitamin C is of vital importance in health maintenance.

Symptoms of Vitamin C Deficiency in Human

Adults will suffer from typical scurvy when they lack vitamin C. The disease mainly occurs in interstitial tissue, manifested as poor wound healing, edema, and bleeding in the skin, mucous membrane, viscera, or muscle. In addition, gingival swelling and bleeding caused by weakened collagen structure of the bone, cartilage, teeth, and connective tissue may lead to periodontal disease. On the other hand, lethargy, fatigue, leg rheumatism, muscle atrophy, skin disease, large area of the thigh hematoma and congestion, multiple organ bleeding, and ecchymosis are also the symptoms of the lack of vitamin C.

In childhood, this syndrome is called Moeller–Barlow disease. Non-breastfed infants usually have these disorders at 6 months (maternal production and preservation of vitamin C have been depleted), which is characterized by width of the interface between the bone and the cartilage, especially at ribs, and the pressure at the bone cartilage, thus leading to the intense pain at joint and recurrent anemia fever. Children with scurvy usually limp or walk hard, with lower limb tenderness, gum bleeding, and ecchymosis bleeding. However, supplementary vitamin C has a significant effect in scurvy treatment. The clinical symptoms of Moeller–Barlow disease will be improved significantly within 1 week by vitamin C treatment.

Epidemiological data suggest that the reduction of plasma vitamin C concentration increases the occurrence of ischemic heart disease or hypertension.

Pharmacological Treatment

According to the characteristics of vitamin C, it is recommended that a relatively large dose of vitamin C can play an adjuvant therapy in the treatment of infectious diseases and other possible concurrent diseases [1].

Discussion

As a classic vitamin drug, vitamin C not only exerts good healthcare function but also produces therapeutic effect, which emerges with wide prospects in the clinic.

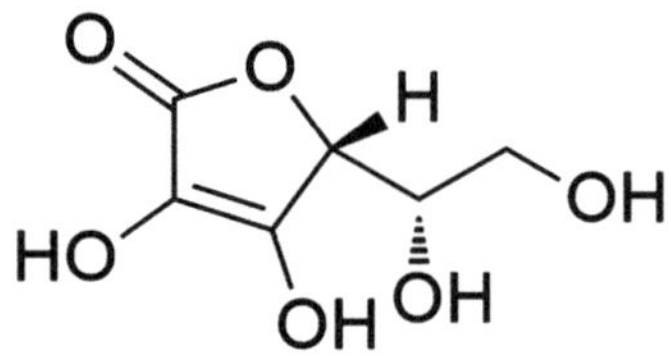

Fig. 1 Chemical structure of vitamin C

Reference

1. Zhang DS, Du GH, translation. The vitamins fundamental aspects in nutrition and health. Beijing: Science Press, 2009.

Vitamin K

Li-Da Du, You-Wen Zhang, and Guan-Hua Du

Abstract Vitamin K is the generic descriptor for 2-methyl-1,4-naphthoquinone and all its derivatives exhibiting qualitatively biological activity of phylloquinone. K are side-chain homologs; each is hydrophobic and insoluble in aqueous environments as plasma, interstitial fluids, and cytoplasm. The 1,4-naphthoquinone ring system of vitamin K renders it susceptible to metabolic reduction. The typical role of vitamin K is maintaining normal blood coagulation function. Thus, deficiencies of vitamin K have a narrow clinical spectrum: hemorrhagic disorders. The most common cause of vitamin K deficiency is that the microfloral production or absorption of the vitamin is interfered.

Keywords Vitamin K · Hemorrhagic disorders · Blood coagulation

Alias: 2-Methyl-1,4-naphthoquinone and all its derivatives

Chemical Name

Vitamin K is the generic descriptor for 2-methyl-1,4-naphthoquinone and all of its derivatives exhibiting qualitatively the biological (antihemorrhagic) activity of phylloquinone (Fig. 1). The phylloquinone group includes forms with phytyl side

L.-D. Du
School of Biomedical Sciences, Chinese University of Hong Kong, Hong Kong, China

Y.-W. Zhang
Beijing Key Laboratory of Drug Targets Identification and Drug Screening, Centre for Pharmaceutical Screening, Institute of Materia Medica, Chinese Academy of Medical Sciences and Peking Union Medical College, Beijing, China

G.-H. Du (✉)
Beijing Key Laboratory of Drug Targets Identification and Drug Screening, Centre for Pharmaceutical Screening, Institute of Materia Medica, Chinese Academy of Medical Sciences and Peking Union Medical College, Beijing, China
e-mail: dugh@imm.ac.cn

G.-H. Du, *Natural Small Molecule Drugs from Plants*,
https://doi.org/10.1007/978-981-10-8022-7_106

chains and side chains which were alkylated, thus consisting of several saturated isoprenoid units. Menadione is the compound of commerce; it is made in several forms (e.g., menadione sodium bisulfite complex, menadione dimethylpyrimidinol bisulfite) [1].

Properties

Appearance: phylloquinone is a yellow oil at room temperature, but the other vitamers K are yellow crystals. *Solubility*: the vitamers K and MK and most forms of menadione are insoluble in water, slightly soluble in ethanol, and readily soluble in ether, chloroform, fats, and oils. *Stability*: the vitamers K are sensitive to light and alkali, but are relatively stable to heat and oxidizing environments.

Menadione, the formal parent compound of the menaquinone series does not occur naturally but is a common synthetic form called menadione (2-methyl-1,4-naphthoquinone). This compound forms a water-soluble sodium bisulfite addition product, menadione sodium bisulfite, whose practical utility is limited by its instability in complex matrices such as feeds. However, in the presence of excess sodium bisulfite, it crystallizes as a complex with an additional mole of sodium bisulfite (i.e., menadione sodium bisulfite complex), which has greater stability, therefore, is used widely as a supplement to poultry feeds. A third water-soluble compound is menadione pyridinol bisulfite (MPB).

Derivatives (Fig. 2 and Table 1 [2])

Dosage Forms

Injecta, Tablet

Table 1 [2] Systems of vitamin K nomenclature

Chemical name	IUPAC[a] system[b]	IUNS[c] system	Traditional
2-Methyl-3-phytyl-1,4-naphthoquinone	Phylloquinone (K)	Phytylmenaquinone (PMQ)	K_1
2-Methyl-3-multiprenyl-1,4-naphthoquinone (n)	Menaquinone-n (MK-n)	Prenylmenaquinone-n (MQ-n)	$K_{2(n)}$
2-Methyl-1,4-naphthoquinone	Menadione	Menaquinone	K_3

[a]International Union of Pure and Applied Chemists
[b]Preferred system
[c]International Union of Nutritional Sciences

Indications

Vitamin K deficiency

Vitamin K Biopotency

The biopotency of vitamin K depends on both the nature and length of their isoprenoid side chains. Generally, the menaquinones have greater biopotencies than the corresponding phylloquinone analogs. The reported biopotencies of the menadiones tend to be variable. Menadione has the lowest biopotency. The biopotencies of other menadiones tend to be 100 times higher than menadione.

Pharmacology

The typical role of vitamin K is to maintain normal blood coagulation function. Its role is related to the metabolic processes.

Blood Clotting

Blood clotting is initiated by injury tissues through the release of collagen fibers and tissue factors, the cell surface proteins, and they interact with vitamin K-dependent Gla proteins in the circulation. These signals are amplified via the clotting pathway ultimately to form the clot.

Vitamin K-Dependent γ-Carboxylations

Vitamin K is the cofactor of a specific microsomal carboxylase that uses the oxygenation of vitamin K hydroquinone to drive the γ-carboxylation of peptide-bound glutamic acid residues. The vitamin K-dependent γ-carboxylation of specific glutamyl residues on the zymogen precursors of each blood clotting factor occurs post-translationally at the N-terminus of the nascent polypeptide. Carboxylation confers Ca^{2+}-binding capacities to these proteins. This facilitates the formation of Ca^{2+} bridges between the clotting factors and phospholipids on membrane surfaces of blood platelets and endothelial and vascular cells, as well as between Gla residues to form internal Gla-Gla linkages.

Vitamin K-Dependent Gla Proteins

Vitamin K functions in the posttranslational modification at least a dozen proteins via carboxylation of specific glutamate residues to γ-carboxyglutamate (Gla) residues.

Osteocalcin and Calcified Tissues

The best characterized vitamin K-dependent protein of calcified tissues is osteocalcin (also called bone GLA protein), a protein found in rapidly growing regions of bones, which can be used as a marker of bone formation.

Vitamin K in Atherosclerosis

Vitamin K-dependent Gla proteins may play roles in atherogenesis, which involves thrombus-induced coagulation as well as intimal calcification. It is possible that increased dietary intake of the vitamin K may be useful in reducing atherosclerosis risk.

Vitamin K and Cancer

Vitamin K could reduce the incidence of cancer by attributing to the reactive oxygen species generation and deplete glutathione due to its arylation effect.

Clinical Application

Coagulopathy

The predominant clinical sign of vitamin K deficiency is hemorrhage. Several congenital disorders of vitamin K-dependent proteins have been identified in human patients: at least a dozen forms of congenital dysprothrombinemia. Patients with these disorders show coagulopathies and lower plasma vitamin K level.

Undercarboxylated Proteins

A more sensitive indicator of low vitamin K status is the presence in plasma of under-γ-carboxylated vitamin K-dependent proteins. It is a useful marker for sub-clinical vitamin K deficiency. Another marker is under-γ-carboxylated osteocalcin. Studies have shown that increases concentrations of these factors in the serum are more sensitive to minidose warfarin therapy than decreases in prothrombin or osteocalcin.

Risk Factors for Vitamin K Deficiency

The most common cause of vitamin K deficiency is that the microfloral production or absorption of the vitamin is interfered.

Discussion

Although vitamin K deficiency mainly leads to the symptom of coagulopathy, the functions of vitamin K are actually very extensive. The factors that cause vitamin K deficiency are complex, and gut microbes play an inignorable role in it. Therefore, prompt supplement of vitamin K is an effective treatment for vitamin K-deficient patients.

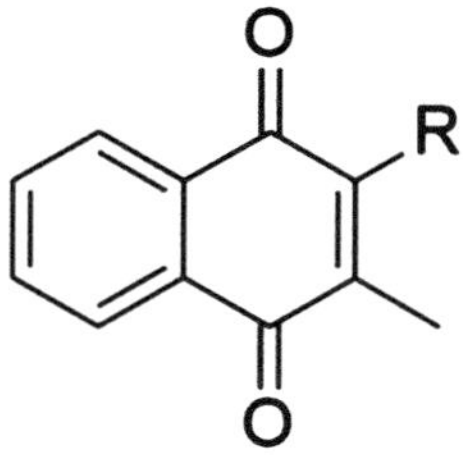

Fig. 1 Derivative of 2-methyl-1,4-naphthoquinone; ring structure has a hydrophobic constituent at the 3-position and can be alkylated with an isoprenoid side chain

Fig. 2 Chemical structures of the vitamin K group

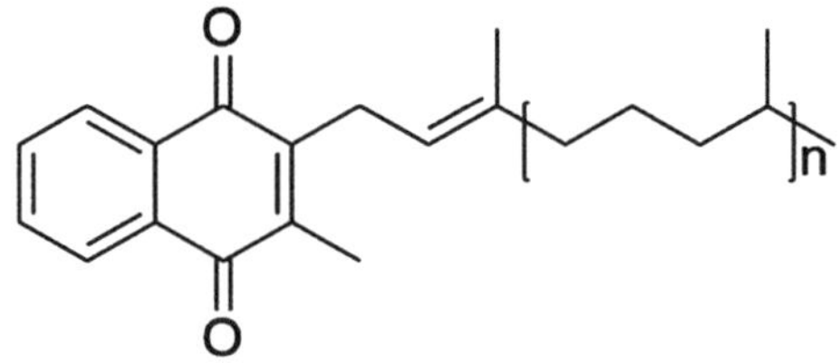

Phylloquinones

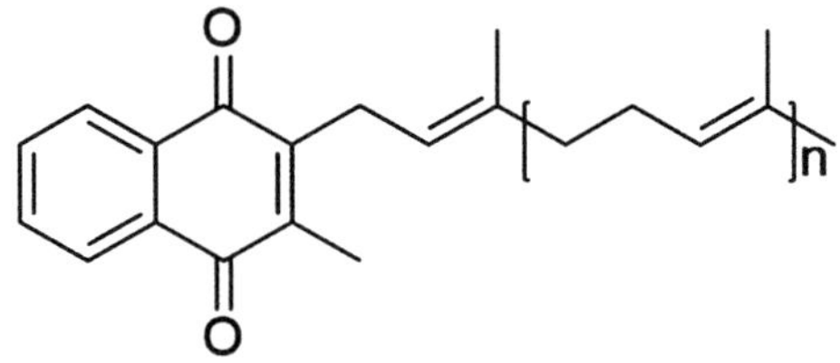

Menaquinones

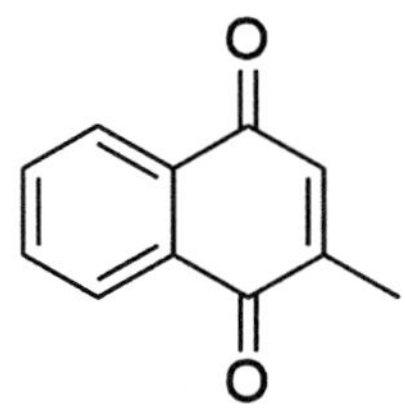

Menadione

References

1. Zhang D, Guanhua D. The vitamins: fundamental aspects in nutrition and health. Beijing: Science Press; 2009. (in Chinese)
2. Combs GF. The vitamins: fundamental aspects in nutrition and health. San Diego: Elsevier Academic Press; 2008.

Part VII
Natural Product Compounds in Pharmacopeia

Overview

This chapter includes some small natural product compounds. These compounds were recorded in the pharmacopeia not as drugs but as controls for drug detection, indicating they are of certain importance. Compounds such as curcumin, luteolin, and caffeic acid are active pharmaceutical ingredients and they are used as controls to detect certain chemicals in natural extracts, while compounds such as aristolochic acid are used to quantify the toxic ingredients in compound drugs.

Most of the compounds that act as active ingredients are validated as compounds with certain activities by long-term researches. Though not used as drugs alone, the bioactivity or pharmacological action of these compounds is approved. By overcoming the shortage of monomeric compounds and identifying appropriate indications, these compounds could be used as drugs to treat certain diseases.

With the research and exploration of natural drugs, more natural drugs would be used in the clinical practice. The quality control of these compound drugs needs more active compounds as controls. Therefore, the research of these active compounds is of equal significance.

Arbutin

Xiao-Xiu Li, Li-Da Du, and Guan-Hua Du

Abstract Arbutin is extracted from the bearberry plant or other plants. Arbutin could effectively inhibit the activity of tyrosinase in skin cells and block the formation of melanin. Arbutin is therefore used as a skin-lightening agent. In the history of arbutin R&D, it is found that arbutin is a hydroquinone compound with two epimers, α and β arbutin. Alpha arbutin is safer and has a stronger inhibitory effect on tyrosinase. Moreover, arbutin is a major component of medicine for treating burn and scald, characterized by rapid elimination of pain and swelling and fast healing, leaving no scars. Arbutin can also be used as raw materials for intestinal anti-inflammatory drug, with sterilization, anti-inflammatory effect, and nontoxic side effects.

Keywords Arbutin · Whitening effects · Cosmetics

Origin: *Bergenia purpurascens* (yanbaicai) (Fig. 1)
Chemical name (Fig. 2)

4-hydroxyphenyl-beta-D-glucopyranoside

Molecular formula, $C_{12}H_{16}O_7$; **MW**, 272.25; **CAS**, 497-76-7

X.-X. Li
Liaoning Key Laboratory of Behavior and Cognitive Neuroscience, Department of Pharmacology, Shenyang Medical College, Shenyang, China

L.-D. Du
School of Biomedical Sciences, Chinese University of Hong Kong, Hong Kong, China

G.-H. Du (✉)
Beijing Key Laboratory of Drug Targets Identification and Drug Screening, Centre for Pharmaceutical Screening, Institute of Materia Medica, Chinese Academy of Medical Sciences and Peking Union Medical College, Beijing, China
e-mail: dugh@imm.ac.cn

G.-H. Du, *Natural Small Molecule Drugs from Plants*,
https://doi.org/10.1007/978-981-10-8022-7_107

Properties

Appearance: white powder. *Solubility*: soluble in hot water. *Melting point*: 198–201 °C

Dosage Forms

Arbutin is recorded in *British Pharmacopoeia* (2017) and *European Pharmacopoeia* (9.0th ed.).

Coating agent, cream, and liposomes

Indications

Arbutin has bactericidal, anti-inflammatory, and whitening effects and is mainly used in whitening cosmetics.

Literature

Arbutin is a hydroquinone compound derived from *houyeyanbaicai* 厚叶岩白菜 (*Bergenia crassifolia* (L.) Fritsch.). It is also distributed in *wufanshu* (*Vaccinium bracteatum* Thunb.), and *xiongguo (bearberry)*, Pear tree, *jishutiao* (*Viburnum opulus* Linn. var.*calvescens* (Rehd.) *Hara f. calvescens*), *xingxiangtuerfeng* (*Ainsliaea fragrans* Champ.), *heiguoxianleihuaqiu* (*Aronia melanocarpa*), *nvzhen* (*Ligustrum lucidum* Ait.), *juanbai* (*Selaginella tamariscina* (Beauv.) Spring), *maodadingcao* (*Gerbera piloselloides* (L.) Cass), *payanhong* (*Veronicastrum axillare* (Sieb. and Zucc.) *Yamazaki*), and *yueju (Vaccinium vitis-idaea* Linn.) also contain this compound [1]. In addition, it can be obtained from biotransformation, enzyme synthesis, and organic synthesis [2].

History of R&D

Arbutin is a hydroquinone compound with two epimers, α and β arbutin. The sources of α-arbutin and β-arbutin are completely different. β-arbutin can be prepared by plant extraction, plant cell culture, and artificial synthesis. Arbutin can relieve cough and asthma and has whitening effect.

The Japanese cosmetics company Shiseido developed the arbutin as a whitening agent in the 1990s. Arbutin can not only reduce skin freckles, senile plaques, and chloasma but also relieve acne and improve healing after skin burns. Arbutin is the epimer of β-arbutin, and the spatial orientation of their glycosidic bonds is just the opposite. Alpha arbutin is generally prepared by different microbial enzymes. A molecule of glucose and a molecule of hydroquinone combine to form a molecule α-arbutin [3]. Alpha arbutin improves ultraviolet burn scar. α-Arbutin can be used in a variety of skin whitening cosmetics since it is chemically stable.

Pharmacology

Arbutin could effectively inhibit the activity of tyrosinase in skin cells and block the formation of melanin without affecting cell proliferation [4]. Furthermore, it could accelerate the decomposition and excretion of melanin and thereby reduce skin pigmentation and eliminate freckles. In addition, arbutin shows no toxicity, irritation, sensitization, and other side effects [5]. Alpha arbutin is safer and has a stronger inhibitory effect on tyrosinase [6, 7]. At present the whitening cosmetics market in the developed countries has been almost monopolized by arbutin.

Clinical Application

Arbutin is mainly used in high-level cosmetics and has been formulated into skin cream, freckle cream, and senior pearl cream. Arbutin is a major component of medicine for treating burn and scald, characterized by rapid elimination of pain and swelling and fast healing, leaving no scars. Arbutin can also be used as raw materials for intestinal anti-inflammatory drug, with sterilization, anti-inflammatory effect, and nontoxic side effects.

Discussion

Arbutin may be made synthetically or derived from bearberry (uva-ursi) plant or other plants and exhibits antioxidant activity. Most notably, arbutin has an increasing amount of research attesting to its effectiveness for brightening an uneven skin tone. Research has shown the synthetic version, chemically known as deoxyarbutin, is more effective for brightening than the natural derivative. Arbutin has very minor risks and side effects. Note that it is very important to get the right concentration to prevent any side effects and risks.

Fig. 1 *Bergenia purpurascens* (yanbaicai)

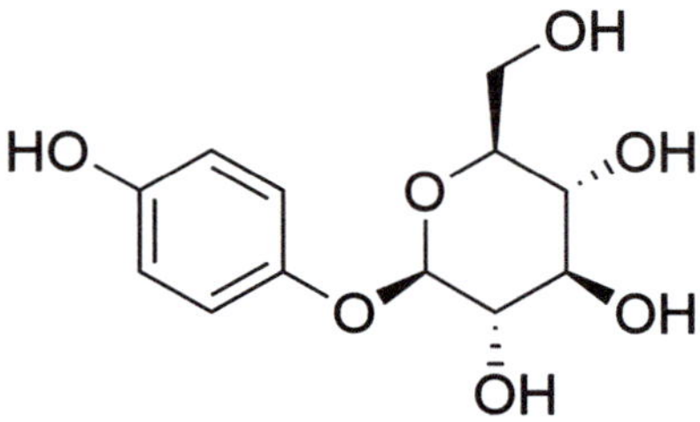

Fig. 2 Structure of arbutin

References

1. Yan XY, Tang XF, Wang X. Research progress and application of arbutin. Tradit Chin Med Inf. 2007;24(4):18–22.
2. Seo DH, Jung JH, Lee JE, et al. Biotechnological production of arbutins (α- and β-arbutins), skin-lightening agents, and their derivatives. Appl Microbiol Biotechnol. 2012;95(6):1417–25.
3. Satoshi K, Hiroshi S. α-D-glucosyl transfer to phenolic compounds by sucrose phosphorylase from leuconostoc mesenteroides and production of α-arbutin. Biochemist. 1994;58(1):38.
4. Hu ZM, Zhou Q, Lei TC, et al. Effects of hydroquinone and its glucoside derivatives on melanogenesis and antioxidation: biosafety as skin whitening agents. J Dermatol Sci. 2009;55(3):179–84.
5. Liang K, Xu K, Bessarab D, et al. Arbutin encapsulated micelles improved transdermal delivery and suppression of cellular melanin production. BMC Res Notes. 2016;9(1):254.
6. Lim YJ, Lee EH, Kang TH, et al. Inhibitory effects of arbutin on melanin biosynthesis of alpha-melanocyte stimulating hormone-induced hyperpigmentation in cultured brownish guinea pig skin tissues. Arch Pharm Res. 2009;32(3):367.
7. Funayama M, Arakawa H, Yamamoto R, et al. Effects of alpha and beta arbutin on activity of tyrosinases from mushroom and mouse melanoma. Biosci Biotechnol Biochem. 1995;59(1):143.

Aristolochic Acid

Rong Yan, Li Li, and Guan-Hua Du

Abstract Aristolochic acid first appeared in Chinese medicine in the fifth century AD, while it was used to treat kidney and urinary diseases, as well as gout, snakebites, and a variety of other ailments. Balkan endemic nephropathy (BEN) was later found to be also the result of aristolochic acid (AA) consumption. *Aristolochia* species are nephrotoxic and carcinogenic, and FDA has issued warnings regarding consumption of AA-containing supplements.

Keywords Aristolochic acid · *Aristolochia* · Balkan endemic nephropathy

Alias: Aristinic acid, *Aristolochia* yellow, Aristolochic acid A, Aristolochin, Aristolochine, Descresept, Tardolyt, TR 1736

Origin: Aristolochic (Fig. 1)

Chemical name (Fig. 2)

3,4-Methylenedioxy-8-methoxy-10-nitro-1-phenanthrenecarboxylic acid

Molecular formula, $C_{17}H_{11}NO_7$; **MW**, 341.29; **CAS**, 313-67-7

Properties

Appearance: brown sheet crystal or yellow powder. *Solubility*: practically insoluble in water and soluble in ethanol, chloroform, ethyl ether, acetone, glacial acetic acid, and aniline. *Melting point*: 260–265 °C (500–509 °F; 533–538 K).

R. Yan · L. Li · G.-H. Du (✉)
Beijing Key Laboratory of Drug Targets Identification and Drug Screening, Centre for Pharmaceutical Screening, Institute of Materia Medica, Chinese Academy of Medical Sciences and Peking Union Medical College, Beijing, China
e-mail: dugh@imm.ac.cn

G.-H. Du, *Natural Small Molecule Drugs from Plants*,
https://doi.org/10.1007/978-981-10-8022-7_108

Dosage Forms

Aristolochic acid (AA) is contained in the *Japanese Pharmacopoeia* (17th ed.) and the *British Pharmacopoeia* (2017).

Indications

Due to the widely associated kidney problems and urothelial cancers, the FDA has issued warnings regarding consumption of AA-containing supplements.

Literature

Aristolochic acid first appeared in Chinese medicine in the fifth century AD, while it was used to treat kidney and urinary diseases, as well as gout, snakebites, and a variety of other ailments. In many of these cases, aristolochic acid was just one of the main components of the salves. In the first century, aristolochic acid was first described as a composition of ingested medicine to treat symptoms such as asthma, hiccups, pains, and spasms.

History of R&D

At a clinic in Brussels, Belgium, a group of women who had all taken the same weight loss supplement, *Aristolochia fangchi*, which contained aristolochic acid, was first diagnosed with aristolochic acid poisoning. Balkan endemic nephropathy (BEN) was later found to be also the result of aristolochic acid (AA) consumption. Balkan endemic nephropathy is likely caused by low-level AA exposure, possibly from the contamination of wheat flour seeds by *Aristolochia clematitis*. BEN falls under the umbrella of what is now known as aristolochic acid nephropathy, the prevalent symptom of AA poisoning [1].

Pharmacology

In 1992, some cases of women present with a rapidly progressive renal failure after having a slimming regimen including powdered extracts of Chinese herbal preparations. This outbreak of renal failure eventually resulted in about 100 cases in 1998, 70% of them being in end-stage renal disease (ESRD) [2]. Recent research has confirmed that the main reason leading to renal injury is aristolochic acid found in many Chinese herbs [3, 4]. Aristolochic acid, a potent human carcinogen from *Aristolochia* plants, is associated with the incidence of urothelial carcinoma in the upper urinary tract (UUC). After the metabolic activation, aristolochic acid reacted with DNA to form aristolactam (AL)-DNA adducts. Mainly pathological changes are in renal cortex, where they serve as a specific biomarker, and are also found in urothelium, where they lead to a unique mutational signature in the TP53 tumor-suppressor gene. The conclusion is that exposure to aristolochic acid gives rise to the incidence of UUC, a finding with significant implications for the global public health [5].

Clinical Application

Aristolochia species are nephrotoxic and carcinogenic, and FDA has issued warnings regarding consumption of AA-containing supplements.

Discussion

Recently, the pharmacology research of the *Aristolochia* species concentrate on the nephrotoxic aristolochic acids responsible for a tragic disease of Chinese herbal preparations recognized in 1992. It may thus be of more than academic interest to examine the remaining *Aristolochia* plants for their presence of aristolochic acid to prevent the issues like Chinese herb nephropathy and Balkan endemic nephropathy.

Fig. 1 *Aristolochia* (马兜铃 madouling)

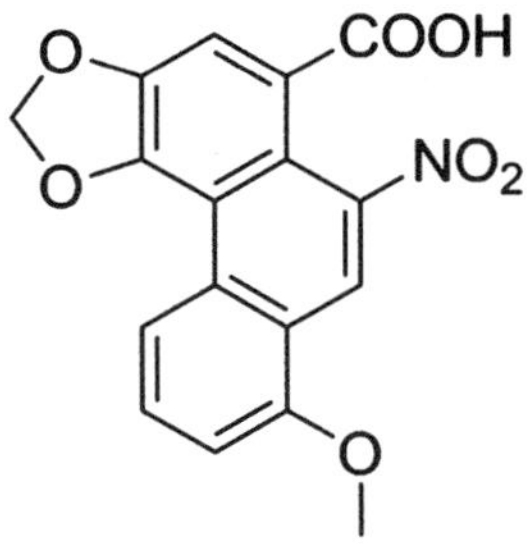

Fig. 2 Chemical structure of aristolochic acid

References

1. Kuo PC, Li YC, Wu TS. Chemical constituents and pharmacology of the Aristolochia (madou ling) species. J Tradit Complement Med. 2012;2(4):249–66.
2. Vanherweghem JL, et al. Rapidly progressive interstitial renal fibrosis in young women: association with slimming regimen including Chinese herbs. Lancet. 1993;341(8842):387–91.
3. Balachandran P, et al. Structure activity relationships of aristolochic acid analogues: toxicity in cultured renal epithelial cells. Kidney Int. 2005;67(5):1797–805.
4. Nedelko T, et al. TP53 mutation signature supports involvement of aristolochic acid in the aetiology of endemic nephropathy-associated tumours. Int J Cancer. 2009;124(4):987–90.
5. Chen CH, et al. Aristolochic acid-associated urothelial cancer in Taiwan. Proc Natl Acad Sci U S A. 2012;109(21):8241–6.

Bergapten

Xiao-Bo Wang, Ling-Lei Kong, and Guan-Hua Du

Abstract Bergapten is extracted from the roots, stems, and leaves of bergamot, of which has a photosensitive activity. It has a significant effect on the treatment of various skin diseases. But, there is no clinical report on it up to now. And, bergapten can treat the diseases such as vitiligo and different types of psoriasis and has an anti-HIV activity. It also has certain inhibitory activity on tumor and cell proliferation. Due to the long synthetic route and the harsh reaction conditions, synthetic bergamot cannot meet the application requirements at present. We have to find more ways to improve the research on bergapten.

Keywords Bergapten · Autofluorescence · Photosensitive activity · Skin diseases

Alias: 5-Methoxypsoralen; Bergapten; Bergaptene; Majudin; Psoraderm; Bergaptol methyl ether; Geralen; Heraclin
Origin: Bergamot (Fig. 1)
Chemical name (Fig. 2)

4-Methoxy-7H-furo[3,2-g]chromen-7-one

Molecular formula, $C_{12}H_8O_4$; **MW**, 216.19; **CAS**, 484-20-8

X.-B. Wang
Guangdong Pharmaceutical University, Guangzhou, China

L.-L. Kong · G.-H. Du (✉)
Beijing Key Laboratory of Drug Targets Identification and Drug Screening, Centre for Pharmaceutical Screening, Institute of Materia Medica, Chinese Academy of Medical Sciences and Peking Union Medical College, Beijing, China
e-mail: dugh@imm.ac.cn

G.-H. Du, *Natural Small Molecule Drugs from Plants*,
https://doi.org/10.1007/978-981-10-8022-7_109

Properties

Appearance: Crystalline state is mercerizing white needled crystal. Autofluorescence can be observed under the excitation of UV at 432 nm. By directly adding bergapten into sulfuric acid, the solution became golden. Because of the structure of coumarin lactone, hydroxamic acid iron reaction [1] can be occurred. *Solubility*: It is soluble in chloroform; slightly soluble in benzene, ethyl acetate, and ethanol; and insoluble in water. It has a photosensitive activity. Crystalline state is mercerizing white needled crystal. *Melting point*: 181–183 °C.

Derivatives

This product is contained in the *British Pharmacopoeia* (2017) and *European Pharmacopoeia* (9.0th ed.).

Indications

No clinical report

Literature

Bergapten exists in roots, stems, and leaves of a variety of Rutaceae, Umbelliferae, and leguminous plants such as ephedra, fennel, and vetch.

History of R&D

It was first isolated from a plant called *Ammi majus L* in 1947. It was first synthesized by JEAN and JACQUES [2] in 1982. The efficacy of psoralen in the treatment of psoriasis and other skin diseases has been studied in modern medicine, and more and more studies have proven that the extracts and derivatives of psoralen have significant effect on the treatment of various skin diseases. Bergapten used in the treatment of PUVA (psoralen plus ultraviolet) and treatment of vitiligo has more than 20 years of history [3]. 8-Methoxypsoralen in the treatment of psoriasis frequently has a problem of phototoxicity and drug intolerance. Therefore, researchers have studied 5-MOP (5-methoxypsoralen) on this basis. The results showed that the 5-methoxypsoralen has a close effect and fewer side effects of the treatment companies with the 8-methoxypsoralen with a double dose.

Pharmacology

Bergapten could be used to treat vitiligo mainly because it can be embedded between the thymine base pairs of DNA in darkness. Under UV light (365 nm) double furan ring or pyran ring on the irradiation can occur [2+2] halo reaction showed the structure and biological activity of thymine. The curative effect of high-dose 5-MOP (1.2–1.6 mg/kg) in the treatment of UVA is better and has a less cure time [4].

As the main component of the drug, it can be used to treat different types of psoriasis.

The study of Shikishima et al. [5] showed that the bergapten has an anti-HIV activity, which was mainly affected by HIV reverse transcriptase, protease, and integrase, and the IC50 value was 24.8 mg/L.

Modern pharmacological studies have shown that it has a certain inhibitory activity on tumor and cell proliferation [6].

Clinical Application

It has the optical activity to the skin, has the function of killing the soft animal, has the anticoagulant function and the hemostatic effect against the heparin, and has the antimicrobial activity. In a trial for chemical therapy in the treatment of psoriasis patients, the result has shown that the drug can stimulate the melanin and increase the skin pigmentation, which is essential for the treatment of vitiligo [7].

Discussion

At present, bergamot is mainly extracted from bergamot (*Citrus bergamia*). With the amount of the lemon tree getting less [8], the extraction of furan coumarins would decrease and make cost getting higher. Due to the long synthetic route and the harsh reaction conditions, synthetic bergamot cannot meet the application requirements at present. Improvement on the synthetic bergamot is urgent and would accelerate the bergapten research.

Fig. 1 Bergamot

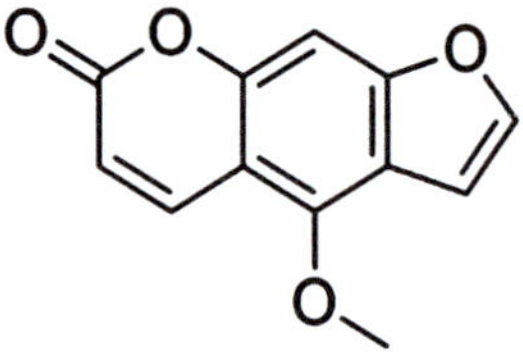

Fig. 2 Structure of bergapten

References

1. Liu X. Chemistry of natural product. 1st ed. Beijing: Chemical Industry Press; 2003.
2. Goupil J J. 5-methoxy-psoralene as a new medicament and process for the synthesis[J]. 1982.
3. Shao CG. The progress in etiology and treatment of psoriasis. Foreign Med Sci Sect Dermatol Venereol. 1994;20(2):105.
4. Hönigsmann H, Jaschke E, Gschnait F, et al. 5-Methoxypsoralen (bergapten) in photochemotherapy of psoriasis. Br J Dermatol. 1979;101(4):369–78.
5. Shikishima Y, Takalshi Y, Honda G, et al. Chemical constituents of Prangos tschimganica, structure elucidation and absolute configuration of coumarin and furanocoumarin derivatives with anti-HIV activity. Chart Pharm Bull. 2001;49(7):877–80.
6. Ji L, Xu ZL, et al. A review of the chemical constituents of the psoralen[J]. China J Chin Materia Medica. 1995;20(2):120–2.
7. Dong F, Liu HZ, Sun Y, et al. The preliminary determination of anti tumor activity in vitro isolation and identification of bergapten in Radix Glehniae. J Plant Resour Environ. 2010;19(1):95–6.
8. Milesi S, Massot B, Gontier E, et al. Ruta graveolens, L.: a promising species for the production of furanocoumarins. Plant Sci. 2001;161(1):189–99.

Curcumin

Tian-yi Yuan and Guan-Hua Du

Abstract Curcumin is the principal curcuminoid of turmeric (*Curcuma longa*). It is sold as an herbal supplement, cosmetic ingredient, food flavoring, and food coloring. In China, curcumin is mainly used in food, while in Western countries, it has been regarded as a health-care product. Curcumin has been proved to have multiple pharmacological effects including anti-fibrosis, antitumor, anticancer effects, and so on. As its broad biological activities, it is applied in a lot of diseases such as hyperlipidemia, infection, and cancer. However, curcumin still needs research to confirm its effects and mechanisms and find its exact indications. There is still a long way to go to make curcumin better applied in clinical practice in the future.

Keywords Curcumin · Food additives · Wide application

Alias: curcuma; C.I. 75300; C.I. Natural Yellow 3; diferuloylmethane; curcumin I
Origin: *Curcuma longa* (Fig. 1)
Chemical name (Fig. 2)

1,7-bis(4-Hydroxy-3-methoxyphenyl)hepta-1,6-diene-3,5-dione

Molecular formula, $C_{21}H_{20}O_6$; **MW**, 368.38; **CAS**, 458-37-7

Properties

Appearance: orange-brown crystalline powder and tastes a little bitter. It will turn into reddish brown in alkaline solution and yellow in neutral and acidic solution. It has strong stability against the reducing agent. It has excellent pigmentation which

T.-y. Yuan · G.-H. Du (✉)
Beijing Key Laboratory of Drug Targets Identification and Drug Screening, Centre for Pharmaceutical Screening, Institute of Materia Medica, Chinese Academy of Medical Sciences and Peking Union Medical College, Beijing, China
e-mail: dugh@imm.ac.cn

G.-H. Du, *Natural Small Molecule Drugs from Plants*,
https://doi.org/10.1007/978-981-10-8022-7_110

is not easy to fade. It is sensitive to light, heat, and iron ion. When PH is greater than 8, curcumin turns from yellow to red, which can be used as a pH indicator.

Solubility: insoluble in water or diethyl ether and soluble in ethanol, propylene glycol, acetic acid, and alkali solution.

Melting point: about 183 °C.

Dosage Forms and Indications

This product is contained in the *British Pharmacopoeia* (2017), *United States Pharmacopeia* (40), and *European Pharmacopoeia* (8.7th ed.).

Curcumin is mainly used in food in China at present (E100), such as coloring agent for sausage products, canned food, and pickled products. It is a kind of natural pigment and high-quality food additive. In the United States, curcumin has been listed as the third generation of cancer chemoprevention agent by the US National Cancer Institute and was included in the *United States Pharmacopeia* in 2000. It has been used as a health-care product, including capsules, tablets, and so on. In the treatment of cancer and related diseases, curcumin has been tested in phase I and II clinical trials in several research centers across the world and has been approved by the US FDA into the phase III clinical trial [1].

Literature

The main source of curcumin is the root of Zingiberaceae *Curcuma aromatica*, rhizome of *Curcuma longa* (Jiang Huang), *Curcuma zedoaria*, and *Acorus calamus*. Among them, Jiang Huang contains about 3–6% curcumin. The traditional Chinese medicine, Jiang Huang, is the root tuber of perennial herbaceous plant *Curcuma longa* L. of family Zingiberaceae. It was firstly recorded in the "Tang materia medica" (Xin Xiu Ben Cao). It is pungent, bitter, and warm and enters the liver and spleen meridians. It activates the blood, moves qi, dredges meridians, and alleviates pain. In India and other Asian countries, Jiang Huang has more than 6000 years of application history. In Japan, Jiang Huang has a long history of health care, and the people of Okinawa Island regarded Jiang Huang as a holy tribute to the emperor.

Jiang Huang mainly comes from Taiwan, Fujian, Guangdong, Guangxi, Yunnan, and Tibet of China and other regions in East Asia and Southeast Asia. It grows in warm and humid climate and sunny environment with abundant rainfall and fears cold frost, drought, and flood. At present, *Chinese Pharmacopoeia* only included Jiang Huang and Yu Jin which contains curcumin, while curcumin is not included.

History of R&D

Curcumin is one such agent that was described about two centuries ago as the yellow coloring matter from the rhizomes of *Curcuma longa*. Besides curcumin, more than 300 different components, including phenolics and terpenoids, have been identified in turmeric, but curcumin is one of the most important active components [2]. Pure curcumin was prepared in 1842 by Vogel Jr. After 1870, the possible structure of curcumin was reported by several chemists in the subsequent decades. The chemical structure of curcumin as diferuloylmethane or 1,6-heptadiene-3,5-dione-1,7-bis (4-hydroxy-3-methoxyphenyl)-(1E, 6E) was reported by Milobedzka et al. (1910). Lampe and Milobedzka (1913) reported the synthesis of curcumin. However, Srinivasan (1953) for the first time used chromatography to separate and quantify the components of curcumin [3].

Jiang Huang has been used for more than 6000 years; it is also well known for its medicinal value and active ingredients. But it was not until the middle of the twentieth century that scientists conducted a systematic study on their pharmacological effects. In 1949, Schraufstatter and Bernt found that curcumin has a variety of antibacterial effects against *Streptococcus*, *Salmonella*, *Mucor*, *Mycobacterium* and so on [4]. In the 1970s, the study also found that it has lipid-lowering, anti-inflammatory, antioxidant, and antidiabetic effects. In 1980s, it was found to have antitumor effects. In the last 30 years, there are many reports about the clinical and pharmacological effects of curcumin.

At present, more than 65 human clinical trials have been completed, and more than 35 clinical trials are in progress. In addition, the study of curcumin derivatives has also become a hot topic in recent years.

Pharmacology

1. Anti-fibrosis effects: curcumin has the effect of anti-fibrosis in the lung, liver, kidney, and so on. It could inhibit the release of various inflammatory factors and reduce the expression of collagen, laminin, hyaluronic acid, and other extracellular matrix content. It could also reduce the transforming growth factors such as TGF-β to inhibit cell proliferation [5].
2. Antitumor effects: the antitumor effect of curcumin is currently the most studied pharmacological effects and attracts a lot of attention worldwide. Curcumin has been proved to inhibit the proliferation of a variety of tumor cells through regulating a variety of transcription factors (NF-κB, AP-1, etc.), mitogen-activated protein kinase (MAPK), growth factor receptor kinase (PDGFR, VEGFR, etc.), and cyclooxygenase. It plays an important role in the cell cycle and further to inhibit proliferation. Curcumin can also inhibit the migration of tumor cells by activating caspase and inducing tumor cell apoptosis [6].

3. Anti-inflammatory effects: curcumin has a strong inhibitory effect on different kinds of inflammation. The mechanism might relate to the reduction of the expression of prostaglandins and leukotriene to decrease the release of various inflammatory factors. The anti-inflammatory effect of curcumin is close to that of nonsteroidal anti-inflammatory drugs and glucocorticoids, but it has higher safety and lower side effects [7].
4. Antimicrobial effects: curcumin has a strong inhibitory effect on bacteria, viruses, fungi, and parasites [8]. Researchers believe that curcumin may play a role in inhibiting microbial survival and reproduction by destroying microbial cell membranes, inducing their genetic changes, and so on.
5. Hypolipidemic effect: many researchers believe that curcumin will become a hypolipidemic drug with a good prospect. It can lower the levels of total blood cholesterol and triglyceride levels, increase apolipoprotein A level, promote low-density lipoprotein (LDL) metabolism, and increase LDL excretion to reduce LDL body content [9].
6. Drug metabolism: rats were treated with a single dose of refined curcumin orally, 60–65% of which was absorbed by the gastrointestinal tract. Within 5 days, 40% of curcumin were excreted from the feces. The plasma concentration reached the peak after 3 days. The transformation of curcumin happened in the process of hepato-enteral circulation [10].

Clinical Application

1. Cholagogic effect could promote bile formation and secretion.
2. Hypolipidemic effect could reduce the level of cholesterol in the blood and prevent atherosclerosis.
3. Antibacterial and antiviral effect could inhibit *Staphylococcus aureus* and HIV.
4. Liver protection.
5. Anticancer and antitumor effect.
6. Help with the prevention of dementia.
7. Anti-inflammation and treatment of acne and dermatitis.
8. There are no reports of adverse effect of curcumin till now.

Discussion

As a natural small molecule from plant, curcumin has hypolipidemic, antitumor, anti-inflammation, cholagogic, and antioxidant effects. Curcumin has a wide medicinal resource and it also can be artificially synthesized. Therefore, the price of curcumin is quite low. Although curcumin has a wide range of pharmacological effects and no obvious side effects and it also has many reports, it still has not become a drug yet on the market. This suggests that curcumin still has great research value and potential. To find the exact indications and related mechanisms of curcumin will be helpful for its better application in clinical practice in the future.

Fig. 1 *Curcuma longa* (Jiang Huang, 姜黄)

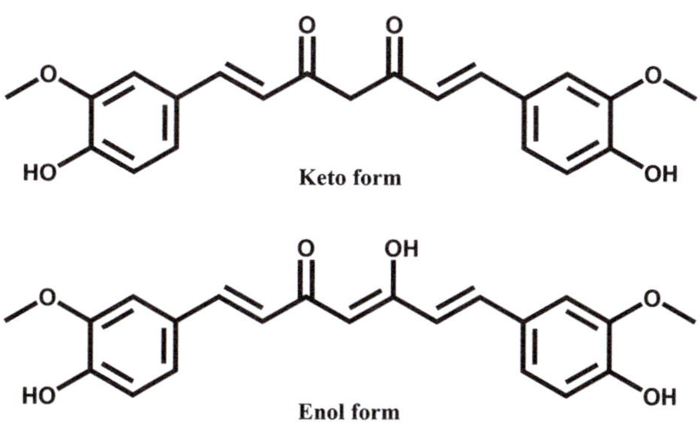

Fig. 2 Chemical structure of curcumin

References

1. Gupta SC, Patchva S, Aggarwal BB. Therapeutic roles of curcumin: lessons learned from clinical trials. AAPS J. 2013;15(1):195–218.
2. Gupta SC, Kismali G, Aggarwal BB. Curcumin, a component of turmeric: from farm to pharmacy. Biofactors. 2013;39(1):2–13.
3. Prasad S, Gupta SC, Tyagi AK, Aggarwal BB. Curcumin, a component of golden spice: from bedside to bench and back. Biotechnol Adv. 2014;32(6):1053–64.
4. Schraufstatter E, Bernt H. Antibacterial action of curcumin and related compounds. Nature. 1949;164(4167):456.
5. Wu HC, Zhang B. The mechanism progress of anti-fibrosis by curcumin. Zhongguo Zhong Xi Yi Jie He Za Zhi. 2013;33(1):135–7.
6. Vallianou NG, Evangelopoulos A, Schizas N, Kazazis C. Potential anticancer properties and mechanisms of action of curcumin. Anticancer Res. 2015;35(2):645–51.
7. He Y, Yue Y, Zheng X, Zhang K, Chen S, Du Z. Curcumin, inflammation, and chronic diseases: how are they linked? Molecules. 2015;20(5):9183–213.
8. Moghadamtousi SZ, Kadir HA, Hassandarvish P, Tajik H, Abubakar S, Zandi K. A review on antibacterial, antiviral, and antifungal activity of curcumin. Biomed Res Int. 2014;2014:186864.
9. Sahebkar A. A systematic review and meta-analysis of randomized controlled trials investigating the effects of curcumin on blood lipid levels. Clin Nutr. 2014;33(3):406–14. eng.
10. Wang Q, Wang K. Metabolism of curcumin. Chin Pharmacol Bull. 2003;10:1097–101.

Cytisine

Xiao-Ming Zhu, Li-Da Du, and Guan-Hua Du

Abstract Cytisine is an alkaloid, which is isolated from *Laburnum* and *Cytisus* of the family *Fabaceae*. Cytisine has a similar chemical structure to that of nicotine, and it possesses selective binding affinity for neuronal nicotinic acetylcholine receptors, which is even superior to that of nicotine and many other agonists. In 1960s, cytisine was firstly used as a smoking cessation aid in Bulgaria. However, cytisine remains relatively unknown outside Eastern Europe, and further clinical studies for its long-term therapeutic effects are also essential.

Keywords Cytisine · Alkaloid · Nicotine · Smoking cessation

Alias: (−)-Cytisine; Baptitoxine; Laburnin; Sophorine
Origin: *Thermopsis lanceolata* R.Br (Fig. 1)
Chemical name (Fig. 2)

(1R,5S)-1,2,3,4,5,6-Hexahydro-1,5-methanopyrido[1,2-a][1,5] diazocin-8-one

Molecular formula, $C_{11}H_{14}N_2O$; **MW**, 190.24; **CAS**, 485-35-8

X.-M. Zhu
State Key Laboratory of Quality Research in Chinese Medicine, Macau University of Science and Technology, Taipa, Macau, China

L.-D. Du
School of Biomedical Sciences, Chinese University of Hong Kong, Hong Kong, China

G.-H. Du (✉)
Beijing Key Laboratory of Drug Targets Identification and Drug Screening, Centre for Pharmaceutical Screening, Institute of Materia Medica, Chinese Academy of Medical Sciences and Peking Union Medical College, Beijing, China
e-mail: dugh@imm.ac.cn

G.-H. Du, *Natural Small Molecule Drugs from Plants*,
https://doi.org/10.1007/978-981-10-8022-7_111

Properties

Appearance: white or yellow powder. *Solubility*: soluble in water, pyridine, acetone, ethanol, and methanol and only slightly soluble in benzene but insoluble in petroleum ether. *Melting point*: 154–156 °C. *Specific optical rotation*: −114.0 to −116.0° ($C = 0.1$, H_2O). Toxic, stable under room temperature, and should be kept in a dry and dark place.

Derivatives (Fig. 3)

Dosage Forms

Tablet, each containing 1.5 mg of cytisine.

Indications

It is the first effective medicine in the world for smoking cessation, which does not contain nicotine and antidepressants. It is administered perorally according to the following schedule:

First 3 days: one tablet six times daily (every 2 h) with a parallel reduction of the number of cigarettes smoked.

If the result is unsatisfactory, the treatment is discontinued, and a new therapy can be resumed after 2–3 months. In case of good effect, the treatment should continue according to the following schedule:

Fourth to 12th day: one tablet every 2 and 1/2 h (five tablets daily)
Thirteenth to 16th day: one tablet every 3 h (four tablets daily)
Seventeenth to 20th day: one tablet every 5 h (three tablets daily)
Twenty-first to 25th day: one to two tablets daily

Literature

It is an alkaloid that occurs naturally in several plant genera, such as *Laburnum* and *Cytisus* of the family *Fabaceae*. Pi-Zhen-Ye-Huang-Hua批针叶黄华 (*Thermopsis lanceolata* R.Br) contains large amount of cytisine, and it is distributed in the northeast, north, and northwest of China. This herb has been traditionally used for treating cough in China and was recorded in "Yue-Wang-Yao-Zhen," "Jing-Zhu-Ben-Cao," "Zhong-Hua-Ben-Cao," and "Ning-Xia-Zhong-Cao-Yao-Shou-Ce."

History of R&D

The largest amount of cytisine (about 1–5%) is also found in the seeds of common garden decorative plant Du-Dou 毒豆 (*Laburnum anagyroides*). In the early twentieth century, cytisine was identified as the toxic component of this plant. The extracts of *Laburnum* seeds or flowers have been used in traditional medicine for hundreds of years. In the Second World War, the leaves of *Laburnum anagyroides* were used as a tobacco substitute by soldiers. In 1960s, cytisine was firstly used as a smoking cessation aid in Bulgaria. The pharmacological and clinical studies of cytisine were also conducted in other Eastern European countries such as Poland and Russia. These results were promising. Stoyanov S. and Yanachkova M. studied 70 volunteers with a long experience in smoking. They found that 57% volunteers stopped smoking and 31.4% volunteers decreased cigarettes: reduction of the cigarettes smoked from 20–30 to 3–4 a day. Cytisine was developed and marked as Tabex® (Sopharma, Bulgaria), which has been widely distributed in the Eastern Europe [1].

Pharmacology

Tobacco smoking is one of the main threats to human health. Nicotine is responsible for the dependence of tobacco, and it is commonly accepted that nicotine addiction results from its interaction with neuronal nicotinic acetylcholine receptors (nAChRs). Cytisine has a similar chemical structure to that of nicotine, and in vitro binding assays showed that cytisine possesses selective binding affinity for nAChRs, which is superior to that of nicotine and many other nAChR agonists. In vivo studies indicated that cytisine moderately increased the dopamine level in the mesolimbic system and attenuate the withdrawal symptoms.

Clinical Application

Cytisine has been used in the Eastern European countries for many years because of its proven benefits and low cost as compared with other cessation medications. A recent clinical trial in New Zealand found that at 1 month, continuous abstinence from smoking was reported for 40% of participants receiving cytisine (264 of 655) and 31% of participants receiving nicotine replacement therapy (203 of 655) [2]. The available data indicates that cytisine is generally well-tolerated in human. However, high dosages of cytisine can be toxic.

Discussion

The smoking cessation application of cytisine is highly dependent on its high affinity to the neuronal nicotinic acetylcholine receptors. However, cytisine remains relatively unknown outside Eastern Europe, and further clinical studies for its long-term therapeutic effects are also essential.

Fig. 1 *Thermopsis lanceolata* R.Br

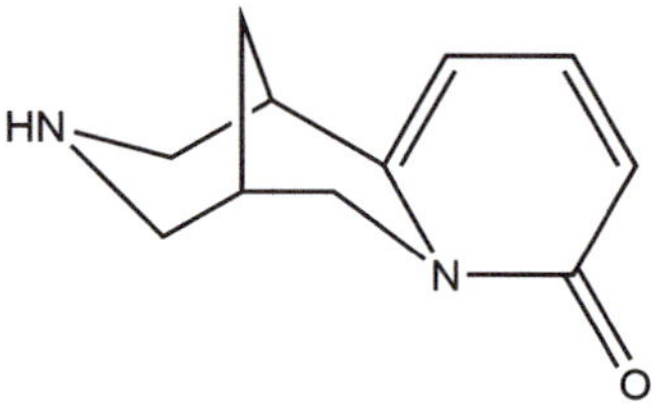

Fig. 2 Chemical structure of cytisine

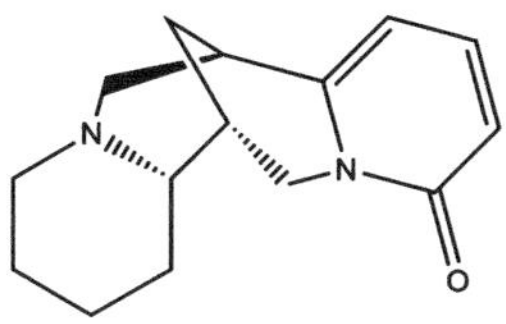

Thermopsine

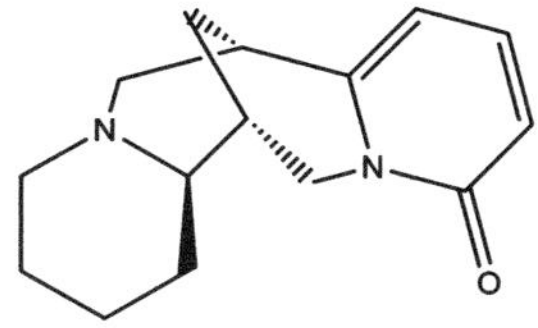

Anagyrine

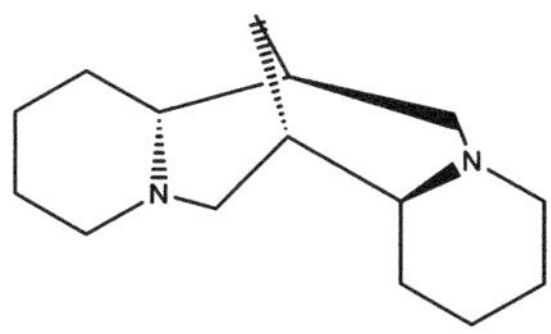

Sparteine

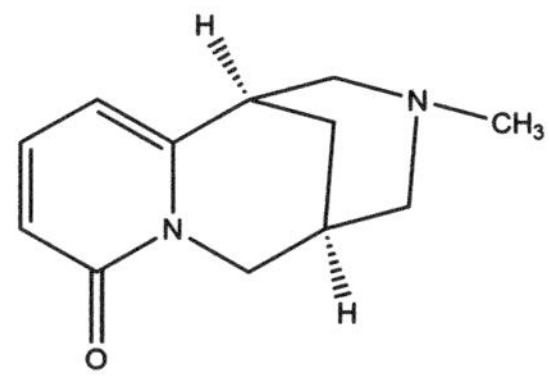

N-Methylcytisine

Fig. 3 Chemical structures of cytisine derivatives

References

1. Tutka P, Zatonski W. Cytisine for the treatment of nicotine addiction: from a molecule to therapeutic efficacy. Pharmacol Rep. 2006;58(6):777–98.
2. Walker N, Howe C, Glover M, McRobbie H, Barnes J, Nosa V, Parag V, Bassett B, Bullen C. Cytisine versus nicotine for smoking cessation. N Engl J Med. 2014;371(25):2353–62.

Dauricine

Xi Chen, Xiu-Ying Yang, and Guan-Hua Du

Abstract Dauricine is a kind of phenolic aromatic compound isolated from Rhizoma Menispermi which is a type of traditional Chinese medicine. Alkaloids are the main chemical components of Rhizoma Menispermi. Besides dauricine, there are several other alkaloid in it such as dauriciline, *O*-methyldauricine, and daurisoline. Dauricine was first separated and structure-confirmed by Fukuda Mayo. The main pharmacological actions of dauricine are protecting against anti-arrhythmia, anti-myocardial ischemia, anti-cerebral ischemia and anti-cerebral ischemia/reperfusion injury. Also, dauricine has antitumor activity. Due to its strong lipid solubility and easy across into cells, it can rapidly and widely distribute in the body, and the drug concentration in organ are significantly higher than that in plasma. Though there has no medicine in the market using dauricine as the main component till now, it still need to be further developed according to its optimistic pharmacological results.

Keywords Dauricine · Phenolic aromatic · Cardiovascular protection · Antitumor

Origin: *Menispermum dauricum* (Fig. 1)
Chemical name (Fig. 2)

4-[[(1R)-6,7-dimethoxy-2-methyl-3,4-dihydro-1H-isoquinolin-1-yl]methyl]-2-[4-[[(1R)-6,7-dimethoxy-2-methyl-3,4-dihydro-1H-isoquinolin-1-yl]methyl]phenoxy]phenol

Molecular formula, $C_{38}H_{44}N_2O_6$; **MW**, 624.77; **CAS**, 524-17-4

Dauricine derivatives (Fig. 3)

X. Chen · X.-Y. Yang · G.-H. Du (✉)
Beijing Key Laboratory of Drug Targets Identification and Drug Screening, Centre for Pharmaceutical Screening, Institute of Materia Medica, Chinese Academy of Medical Sciences and Peking Union Medical College, Beijing, China
e-mail: dugh@imm.ac.cn

G.-H. Du, *Natural Small Molecule Drugs from Plants*,
https://doi.org/10.1007/978-981-10-8022-7_112

Properties

Appearance: yellowish-white powder or crystal. *Solubility*: soluble in ethanol, methanol, chloroform, acetone, benzene and slightly soluble in ether. *Density*: 1.185 g/cm^3. *Melting point*: 115 °C. *Boiling point*: 712.3 °C at 760 mmHg atmospheric pressure. *Specific optical rotation*: MeOH-139.

Dosage Forms and Indications

Drugs whose main component is dauricine are still in clinical trials. Several formulations of dauricine have currently been developed, including tablets, capsules, injections, and so on. The main indications of dauricine in clinical trials are antiplatelet, antihypertensive and anti-arrhythmic.

Literature

Dauricine is a kind of phenolic aromatic compound which is isolated from Menispermaceae plants such as Asian *Menispermum dauricum* or Canadian *Menispermum dauricum* [1] and can be classified as isoquinoline alkaloids [2].

Rhizoma Menispermi is the dry rhizomc of Mcnispermaceae plant *Menispermum dauricum. Menispermum dauricum* mainly distributes in the northern part of East Asia and is mainly distributed in the north, east and northeast regions of China. The usage of Rhizoma Menispermi as a kind of traditional Chinese medicine was firstly recorded in *Kaibao Herbs* with its function of detoxifying many drugs, relieving pains, reducing swelling and sore, curing fever and cough, and killing parasites. According to the *Chinese Pharmacopoeia*, Rhizoma Menispermi in the north of China is different from the herbs used in the south of China known as the same name. Based on the traditional Chinese Medicine theory, Rhizoma Menispermi has the feature of bitterness and coldness, with the meridian tropism in the lung, stomach, and intestine. The main effects of Rhizoma Menispermi are heat-clearing, detoxicating, wind-dispelling and pain-relieving according to its vital function lies in the therapy of swollen sore throat, enteritis diarrhea and rheumatic arthralgia. Alkaloid is the main chemical component of Rhizoma Menispermi, and besides dauricine, there are also other kinds of alkaloids such as dauriciline, O-methyldauricine, and daurisoline [3] in it.

History of R&D

Menispermum dauricum has been long used as a traditional Chinese medicine, but the research of its effective components has been slow. Dauricine was firstly separated and structure-confirmed by Fukuda Mayo using alumina column chromatography separation assay in 1964 [4]. At the same year, dauricine was first synthesized by two Japanese scientists, Tetsuji Tamayama and Rtani Rikazu, by means of Arndt-Eistert reaction and Bischler-Napieralski reaction [5]. Thereafter, lots of scientists from various countries successively separated daurinoline, *Menispermum* Xinlin alkali, *Menispermum* succinylcholine alkali, chelilanthifoline, stepholidine, and many other monomeric components of *Menispermum dauricum*.

The study of dauricine in China started from the Cultural Revolution Period. Under the guidance of integrated traditional Chinese and Western medicine, a large number of effective components in traditional Chinese herbs were extracted, separated and applied to clinic. Dauricine was firstly reported on the mainland as an active ingredient in herbs which can clear away heat and toxic materials. Moreover, dauricine was also used as an antineoplastic and muscle relaxant. There were many medicines whose principal component was dauricine had been under clinical research for its role of cardiovascular protection since the 1980s, most of these medicines were stagnating in phase II clinical trials [6].

Pharmacology

The main pharmacological actions of dauricine are as follows;

Anti-arrhythmia: Dauricine can slow the heart rate and prolong conduction time in sinus rhythm, therefore reducing the incidence of arrhythmia [7].

Anti-myocardial ischemia: Dauricine has protective effect on myocardial ischemia and can improve hemodynamic disturbances during ischemia which is supposed to produce effect via mechanism of improving myocardial metabolism and anti-oxidation.

Anti-cerebral ischemia: Dauricine has an effective regulation on gene expression of Bcl-2 and Bax in cortical neurons after the ischemic brain injury, protecting against cerebral ischemia [8].

Anti-cerebral ischemia/reperfusion injury: Dauricine can inhibit apoptosis induced by cerebral ischemia/reperfusion injury by upregulating Bcl-2 expression and downregulating Bax expression. Also dauricine has the effect of anti-oxidation.

Antitumor activity: Dauricine can affect TGF-beta signaling pathway via upregulating gene expression of DPC4, and then the gene expression of P53 and P16 in tumor tissue, and finally reduce the gene expression of hFGF in tumor tissue. Thereby, the growth of various tumor cells could be inhibited [9].

The investigations of metabolism in vivo of dauricine indicate that it is rapidly and widely distributed in body and the drug concentration in organs is significantly higher than that in plasma, and this is possibly related to its strong lipid solubility and easy access to cells.

Clinical Application

No medicine in the market takes dauricine as the main component up till now. Many animal experiments and clinical studies suggest that dauricine can reverse the effect of MDR as verapamil. Since the clinical application of the classical drug verapamil is limited for the serious toxic side effects, the moderate pharmacological effect of dauricine has a vital clinical significance. The main side effect of dauricine by oral absorption is gastrointestinal reaction, such as diarrhea, nausea, abdominal distension, and so on. In addition, dauricine can accumulate in the body, leading to hepatorenal toxicity and central nervous system toxicity [10].

Discussion

The separation of dauricine is a successful achievement in the research of effective components in traditional Chinese medicine. Both preclinical and clinical studies show favorable results for dauricine's functions. However, the poor molecular stability of dauricine due to its double benzylisoquinoline class structure makes it difficult to be developed as a clinical drug. Additionally, dauricine has certain toxic effects, and its safe dose range is difficult to control. All these disadvantages slow down its clinical research progress. Therefore, dauricine remains to be further optimized and developed.

Fig. 1 *Menispermum dauricum* DC

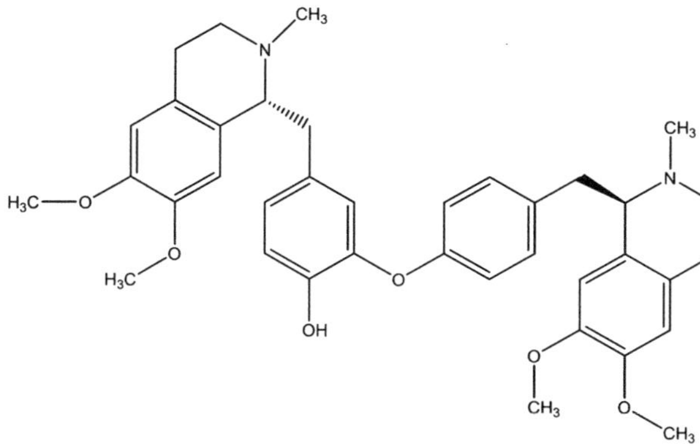

Fig. 2 Dauricine

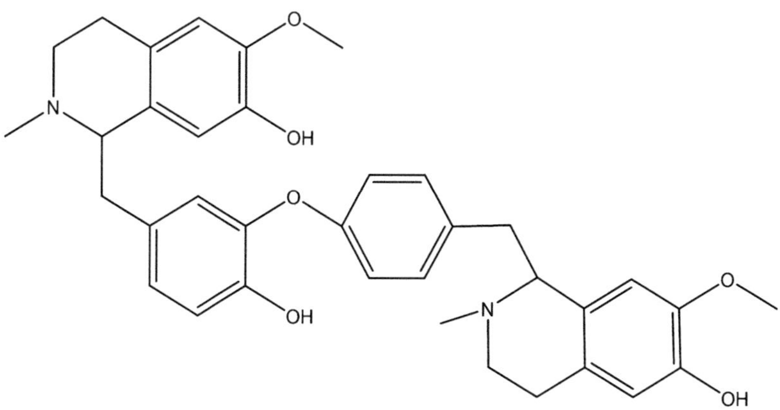

Dauriciline

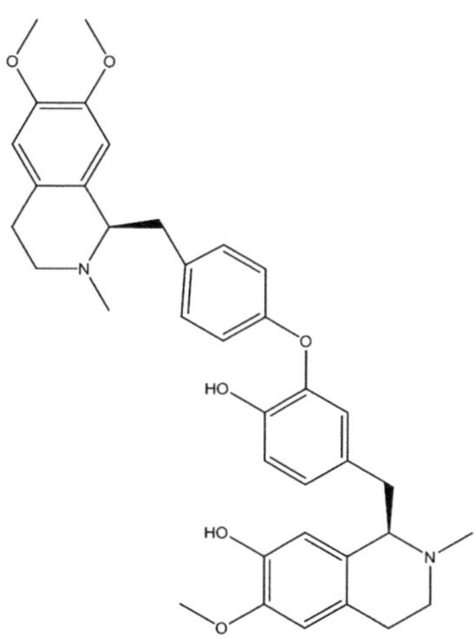

O-Methyldauricine

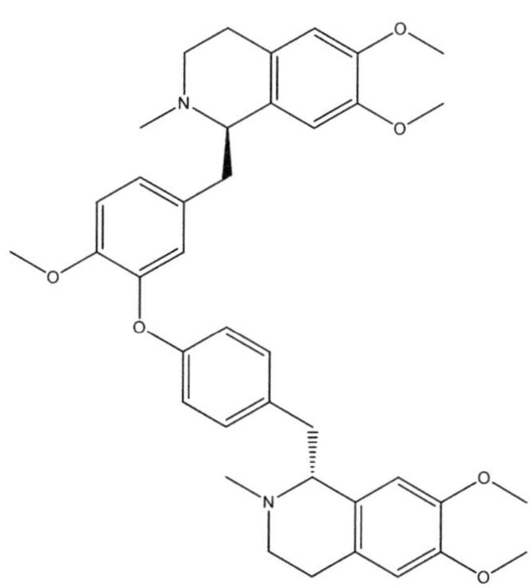

Daurisoline

Fig. 3 The structure of dauricine derivatives

References

1. Kametani T, Fukumoto K. Total synthesis of (±)-dauricine. Tetrahedron Lett. 1964;5(38):2771.
2. "CHEBI: 4331" – dauricine. ChEBI.
3. Wei L. The quality of traditional Chinese medicine (TCM) Rhizoma Menispermi and application research. Liaoning: Liaoning University of Chinese Medicine; 2006.
4. Xiongtai G, Wang W, Xiaoying S, et al. Subprostrate sophora alkali extraction and separation of research. J Wuhan Med Coll. 1983;02:195–6.
5. Manske RHF. The alkaloids: chemistry and physiology V9. New York: Academic; 1967. p. 141. ISBN:9780080865331
6. Jiequan Z. A new antiarrhythmic drug, identification of dauricine in Wuhan. New Drugs Clin Rem. 1985;06:376.
7. Yang Z, Li C, Wang X, Zhai C, Yi Z, Wang L, Liu B, Du B, Wu H, Guo X, Liu M, Li D, Luo J. Dauricine induces apoptosis, inhibits proliferation and invasion through inhibiting NF-kappaB signaling pathway in colon cancer cells. J Cell Physiol. 2010;225(1):266–75.
8. Jiayin X, Yunming S, Lixin G, et al. Effect of PAMD on Bcl-2 mRNA and Bax mRNA expression following cerebral ischemia. J Harbin Commer Univ (Nat Sci Ed). 2008;24(1):16–9.
9. Jin H, Dai J, Chen X, Liu J, Zhong D, Gu Y, Zheng J. Pulmonary toxicity and metabolic activation of Dauricine in CD-1 mice. J Pharmacol Exp Ther. 2009;332(3):738–46.
10. Qian Z, Ruomin J, Guangtao Y. Preliminary study on hepatocytes and nephrocytes toxicity induced by dauricine. Chin J Pharmacovigilance. 2012;10:580–3.

Hyperoside

Jun Zhang, Ling-Lei Kong, and Guan-Hua Du

Abstract Hyperoside is isolated from redosier dogwood. Hyperoside has a protective effect against myocardial ischemia and cerebral ischemia. However, hyperoside has a significant local analgesic and anti-inflammatory effect and a strong cough effect. Hyperoside has an obvious protective effect on liver tissue and gastric mucosa. Hyperoside significantly enhances immune function. In addition, hyperoside has hypolipidemic and antidepressant pharmacological effects. In recent years, the protective effect of hyperoside on ischemic organs, especially the protective effect of cerebral ischemia, has been of interest, and it was also found to have antidepressant effects. The mechanism of pharmacological effects of hyperoside is related to the action and anti-free-radical production and inhibition of calcium influx.

Keywords Hyperoside · Anti-ischemia · Analgesic · Anti-inflammatory

Alias: Begonia due to hypericin, begonia flavonoids, terrestrial glycosides, quercetin-3-galactoside, quercetin-3-D-galactoside
Origin: Chinese St. Johnswort (Fig. 1)
Chemical name (Fig. 2)

2-(3,4-Dihydroxyphenyl)-3-(beta-D-galactopyranosyloxy)-5,7-dihydroxy-4H-1-benzopyran-4-one

Molecular formula, $C_{21}H_{20}O_{12}$; **MW**, 282.34; **CAS**, 482-36-0

Z. Jun · L.-L. Kong · G.-H. Du (✉)
Beijing Key Laboratory of Drug Target Identification and Drug Screening, Centre for Pharmaceutical Screening, Institute of Materia Medica, Chinese Academy of Medical Sciences and Peking Union Medical College, Beijing, China
e-mail: dugh@imm.ac.cn

G.-H. Du, *Natural Small Molecule Drugs from Plants*,
https://doi.org/10.1007/978-981-10-8022-7_113

Properties

Appearance: yellowish needlelike crystals. *Melting point*: 227–229 °C. *Specific optical rotation*: −83° (c = 0.2, pyridine). *Solubility*: soluble in ethanol, methanol, acetone, and pyridine. Hydrochloric acid–magnesium powder reaction yielded formation of cherry red; ferric chloride reaction was green; α-naphthol reaction was positive.

Indications

This product is available in the British Pharmacopoeia (2017) and the European Pharmacopoeia (9.0th ed.).

Monomeric compounds are currently used clinically. In clinical practice, hypericin is the main active ingredient of proprietary Chinese medicines, such as *Acanthopanax* capsules, of which *Acanthopanax* stem and leaf extract are the raw materials for the preparation. Xin'an capsules, prepared with hawthorn leaf extract, are rich in flavonoids, in which hyperoside, a kind of flavonoid, is one of the main components. Qi yue lipid-lowering tablets are prepared from the effective parts of Chinese medicinal herbs such as hawthorn (Nucleation) and *Astragalus membranaceus*. Flavonoid is one of the main active ingredients in hawthorn, in which hyperoside content is higher. Xinxuening tablets, containing ursolic acid, vitexin rhamnoside, hyperoside, citric acid, and others, is a preparation made from hawthorn and pueraria and other traditional Chinese medicines, in which hawthorn enhances the actions of the medicine. Clinical indications of Xinxuening tablets are coronary heart disease, angina pectoris, chest tightness, palpitations, high blood pressure, arrhythmia, hyperlipidemia, and mental depression.

Literature

Hypericin, which is widely found in various plants, such as *Hypericum*, Rosaceae, Campanulaceae, Labiatae, *Rhododendron*, Asteraceae Kwai, Garciniaceae, Leguminosae, *Euonymus*, and other fruits and whole grass, is a flavonoid compound.

History of R&D

In 1960, Nair et al. [1] isolated hyperin from redosier dogwood, in which its content is 0.075%. Flavonoids in the treatment of cardiovascular and cerebrovascular diseases play a pivotal role, among which rutin is a typical representative. Isoquetin mainly presents in the leaves of kumquat flowers and the oleander plant kenaf, and it can also be obtained by chemical synthesis. It shows an anti-inflammatory effect through the capillary permeability test and other animal experiments. It also has a toxic effect on the larvae of Chilohabala larvae. It is one of the active ingredients of *Hypericum japonicum*, which is used in the treatment of hepatitis because of its inhibition of the activities of hepatic enzyme.

Pharmacology

Hyperoside has a protective effect against myocardial ischemia. It achieves its protective effect on the myocardium by reducing the apoptosis rate of myocardial cells, inhibiting myocardial cell lactate dehydrogenase release, enhancing anti-free-radical effect [2, 3], and inhibiting calcium influx so as to [4].

Hyperoside has a protective effect against cerebral ischemia. Hyperoside can significantly lower oxygen-free radicals, reduce the content of malondialdehyde and NO in brain tissue, inhibit the decrease of the activity of LDH, SOD and glutathione peroxidase, and hence result in reduced brain energy, and improve the anti-anoxic ability [5]. At the same time, hyperoside can reduce the degree of cerebral edema in ischemia-reperfusion rats [6].

Hypericin has a significant local analgesic effect that is weaker than that of morphine and stronger than that of aspirin and does not create dependency; it is a new type of local analgesic. Studies have shown that the analgesic effect of hyperoside occurs by reducing the Ca^{2+} of the pain nerve, thereby inhibiting high potassium-induced Ca^{2+} influx, which distinguishes it from morphine and aspirin [7].

Hyperoside has a significant anti-inflammatory effect and a strong cough effect and inhibits the role of eye aldose reductase, which may be beneficial for the prevention of diabetic cataracts. Hypericin has an obvious protective effect on liver tissue [8] and gastric mucosa [9], and its mechanism is related to antioxidant activity and the promotion of a normal return of NO level and improvement of SOD activity.

Hyperoside significantly enhances immune function. In vivo, hyperoside has a significant inhibitory effect on spleen B, T lymphocyte proliferation and peritoneal macrophage phagocytosis, and mouse thymus index; in vitro hyperoside, at a concentration of 6.25–100 μg/mL, significantly enhanced B, T lymphocyte proliferation and promoted the production of interleukin 2 in T lymphocytes [10].

In addition, hypericin still has hypolipidemic and antidepressant pharmacological effects. Its derivative rutin has anti-inflammatory and antiviral effects. Another derivative, quercetin, is effective at inhibiting expectorant, cough, and asthma for use in the treatment of chronic bronchitis, coronary heart disease, and hypertension.

Discussion

Research on hyperoside has mainly focused on its pharmacological effects. In the 1990s, scientists began to study its analgesic effects more. In recent years, the protective effect of hyperoside against ischemic organs, especially the protective effect against cerebral ischemia, has been revealed, and its antidepressant effects were also discovered [11].

Fig. 1 Chinese St. Johnswort

Fig. 2 Structure of hyperoside

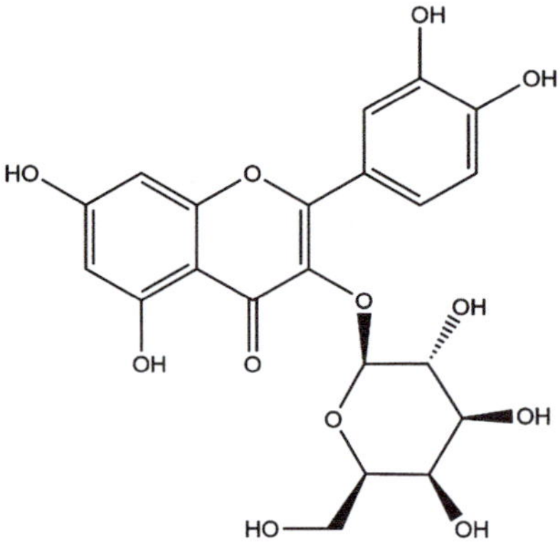

References

1. Meng X-y, Liang GUO, Yu-xin LI, et al. Source, use and extraction and purification of rutin. J Changchun College Tradit Chin Med. 2003;19(2):61–4.
2. Qinglin L, Gang M, Zhiwu C, et al. Protective effect of hypericron on myocardial ischemic injury. J Anhui Med Univ. 2001;36(1):15–8.
3. Khan for the group, Yan Z, Zhilian F, et al. Hypericum test on myocardial lipid peroxidation. Chin J Pharmacol. 1995;11(2):123–5.
4. Wang W-q, Chuan-gang MA, Shu-yun XU. Effects of hyperoside on myocardial ischemia and reperfusion injury. Chin J Pharmacol. 1996;17(4):341–4.
5. Jiasheng Z, Zhiyu C, Ma C, et al. Effects of hypericum perforatum on oxygen and white nitric oxide in ischemic brain injury in rats. Chin J Tradit Chin Med. 1999;24(7):431–3.
6. Hongyan C, Wang J, et al. Protective effects of hyperoside on cerebral ischemia-reperfusion injury in rats. J Integr Tradit Chin West Med. 2006;4(5):5265–9.
7. Minfang L, Hui L, Wang X. Advances in the pharmacological effects of hyperoside. Chin J Inform TCM. 2008;15(4):102–4.
8. Yu-mei H, Xiaoman Z, Jinping Y. Hyperoside and vitamin E on liver tissue protection. J Jiangxi Med College. 2002;42(1):37–8.
9. Wei-zhong Z, Zhi-wu C, Bi-wei S, et al. Protective effect and mechanism of Hyp on gastric mucosal lesion in mice. J Anhui Med Univ. 1999;34(3):178–81.
10. Ligang G, Min Y, Yu-chen Y, et al. Effects of hypothyroid in vivo and in vivo on immune cell function in mice. Chin J Tradit Chin Med Pharm. 2001;8(11):42–4.
11. Butterweck V, Hegger M, Winterhoff H. Flavonoids of St. John's wort reduce HPA axis function in the rat. Plants Med. 2004;70(10):1008–11.

Luteolin

Guo-Rong He, Shou-Bao Wang, and Guan-Hua Du

Abstract Luteolin is extracted from *Honeysuckle*金银花 (*Lonicera japonica Thunb*), which is the most important flavone in *Honeysuckle*. Plants rich in luteolin have been used for treating various diseases such as hypertension, inflammatory disorders, and cancer. The biological effects of luteolin could be functionally related to each other. For instance, the anti-inflammatory activity may be linked to its anti-cancer property. Luteolin can selectively inhibit the fatty acid synthase activity in prostate cancer and breast cancer cells, which is related to the inhibitory effect of luteolin on tumor cell growth and apoptosis. Luteolin can enhance the transfer of synapses in the hippocampus dentate gyrus, causing long-term potentiation. Most studies on luteolin were carried out in vitro, lacking in vivo experimental results. Therefore, it is necessary to further explore the pharmacological activity and pharmacokinetics of luteolin in vivo, so as to provide a solid theoretical basis for its clinical application.

Keywords Luteolin · Flavone · Antitumor · Anti-inflammatory · Neuroprotection

Alias: 2-(3,4-Dihydroxyphenyl)- 5,7-dihydroxy-4-chromenone, 3′,4′,5,7-tetrahydroxy-flavon, digitoflavone, luteoline
Origin: *Reseda odorata L* (Fig. 1)
Chemical name (Fig. 2)

2-(3,4-Dihydroxyphenyl)- 5,7-dihydroxy-4-chromenone

Molecular formula, $C_{15}H_{10}O_6$; **MW**, 286.23; **CAS**, 491-70-3

G.-R. He · S.-B. Wang · G.-H. Du (✉)
Beijing Key Laboratory of Drug Targets Identification and Drug Screening, Centre for Pharmaceutical Screening, Institute of Materia Medica, Chinese Academy of Medical Sciences and Peking Union Medical College, Beijing, China
e-mail: dugh@imm.ac.cn

G.-H. Du, *Natural Small Molecule Drugs from Plants*,
https://doi.org/10.1007/978-981-10-8022-7_114

Properties

Appearance: yellow needle crystal. *Solubility*: slightly soluble in water and soluble in alkaline solution (monohydrate). *Density*, 1.654 g/cm^3. *Melting point*, 330 °C. *Boiling point*, 616.1 °C (760 mmHg). *Flash point*, 239.5 °C. *Vapor pressure*, 9.03E-16 mmHg (25 °C). *Acidity*, weak acid.

Dosage Forms

Luteolin is contained in British Pharmacopoeia (2017) and European Pharmacopoeia (9.0th ed.).

Luteolin is the main active component of medicinal plants, and it is not used alone as a drug.

Indications

Luteolin compound prescription is mainly used for relieving cough, eliminating phlegm, diminishing inflammation, treating cardiovascular diseases, and treating amyotrophic lateral sclerosis, severe acute respiratory syndrome (SARS), hepatitis, etc.

Literature

Luteolin is a flavone derived from *Honeysuckle*金银花 (*Lonicera japonica Thunb*). Luteolin is widely distributed in nature. It can be isolated from a variety of natural herbs, vegetables, and fruits. At present, luteolin is found mainly in honeysuckle, chrysanthemum, *Schizonepeta*, *Ajuga*, artichokes, *Scutellaria*, and *Callicarpa nudiflora* natural herbs. Dietary sources include celery, broccoli, green pepper, parsley, thyme, dandelion, perilla, chamomile tea, carrots, olive oil, peppermint, rosemary, navel oranges, and oregano. It can also be found in the seeds of the palm *Aiphanes aculeata* [1].

History of R&D

At present, luteolin does not have the application of the proprietary medicine, but as one of the main active ingredients of medicinal plants, there is a long history of application. In northern and southern dynasties, honeysuckle with sweet taste, nontoxic, can treat swelling, lose weight, and prolong life under long-term use. In Tang dynasty, honeysuckle was used for treatment of abdominal distension, hot toxic blood dysentery, and water dysentery. It was showed that the clinical application of honeysuckle had made significant progress.

During the Song and Yuan dynasties, honeysuckle was widely used for the treatment of diseases such as sore and ulcer. To the Ming dynasty, there were many treatises about honeysuckle. For example, it is said in *Compendium of Materia Medica*: honeysuckle cure all rheumatism QI and all sorts of swollen poison, ulcer, scab, and heat dissipation detoxify. The prescription has also expanded its scope of application. Up to the Qing dynasty, the application of honeysuckle can not only inherit the theory of the predecessors but also put forward some original ideas and innovation in some respects.

In recent years, through the in-depth study of pharmacological effects, it is found that luteolin has significant effects on antitumor, cardioprotection, neuroprotection, respiratory system, immune regulation, anti-inflammatory, spasmolysis, expectorant, anti-allergic, enzyme activities, antioxidant, diuretic, and other aspects.

Pharmacology

Luteolin can selectively inhibit the fatty acid synthase activity in prostate cancer and breast cancer cells, which is related to the inhibitory effect of luteolin on tumor cell growth and apoptosis. Luteolin can significantly reduce the incidence of colon cancer and the size of tumor caused by dimethylhydrazine, which may be related to the regulation of lipid peroxidation, antioxidation, and antiproliferative effect. Luteolin dose dependently inhibits the invasion and motility of ovarian cancer cell HO8910PM in vitro, which is related to the inhibitory effect of luteolin on the secretion of matrix metalloproteinase-9 (MMP-9) and downregulation of the expression of extracellular signal-regulated kinase 2 (ERK2) [2–4].

The anti-inflammatory activity of luteolin is related to the inhibition of nitric oxide (NO) and other inflammatory cytokines such as tumor necrosis factor alpha (TNF-α) and interleukin-6 (IL-6) generation and inhibition of protein tyrosine phosphorylation and nuclear transcription factor KB (NF-KB)-mediated gene expression.

Luteolin can enhance the transfer of synapses in the hippocampus dentate gyrus, causing long-term potentiation. Moreover, in chronic hypoperfusion injury caused by vascular occlusion, luteolin can still protect synapses, causing long-term potentiation, and reduce the escape latency in the Morris water maze test in rats. Western blot analysis showed that luteolin can not only activate the normal cAMP response element binding protein (CREB) in the hippocampus of rats but also improve the expression of phosphorylated CREB in chronic perfusion injury in rat hippocampus, which may be the reason why luteolin promotes long-term potentiation and improves memory capacity.

Luteolin can reduce the degree of hepatic fibrosis, hydroxyproline level in liver tissues (HYP), malondialdehyde (MDA) content, and mRNA expression of procollagen type I and inhibit hepatic stellate cell (HSC) proliferation and collagen synthesis in vitro. Luteolin can also improve the histological changes of pulmonary fibrosis induced by bleomycin, reduce the lung weight index, significantly reduce the increase in MDA and HYP, and inhibit the level of mRNA of transforming growth factor beta 1 (TGF-β1) in lung tissue. Luteolin in vitro can inhibit the proliferation of human embryonic lung fibroblast cells and promote apoptosis [5–8].

Clinical Application

The natural extract containing luteolin has been used in clinical treatment of many diseases. *Lamiophlomis rotata* Kudo capsule is made from traditional Chinese medicine *Lamiophlomis rotata* Kudo, which consists of the medicinal components such as flavonoids, saponins, sterols, amino acids, and many trace elements. Among these components, luteolin content is not less than 0.80 mg/g. This capsule is mainly used for a variety of surgical incision pain, postoperative bleeding, fracture, sprain of muscles, rheumatic pain, dysmenorrhea, uterine bleeding, gingival swelling, and bleeding.

MaiShu capsule is the processed capsule of groundnut, whose medicinal ingredient is flavonoid compound mainly consisting of luteolin. MaiShu capsule is used for the treatment of hyperlipidemia.

Botrychium ternatum oral liquid is consisted of *Botrychium ternatum* and other medicinal herbs. The main active ingredient is luteolin. It is used for the treatment of respiratory diseases.

Discussion

Luteolin, as a natural flavonoid, has a wide range of pharmacological activities and widely exists in various vegetables and fruits. So people can easily prevent and reduce disease by changing their diet [8]. Most studies on luteolin were carried out in vitro, lacking in vivo experimental results. Therefore, it is necessary to further explore the pharmacological activity and pharmacokinetics of luteolin in vivo, so as to provide a solid theoretical basis for its clinical application.

Fig. 1 *Reseda odorata L*

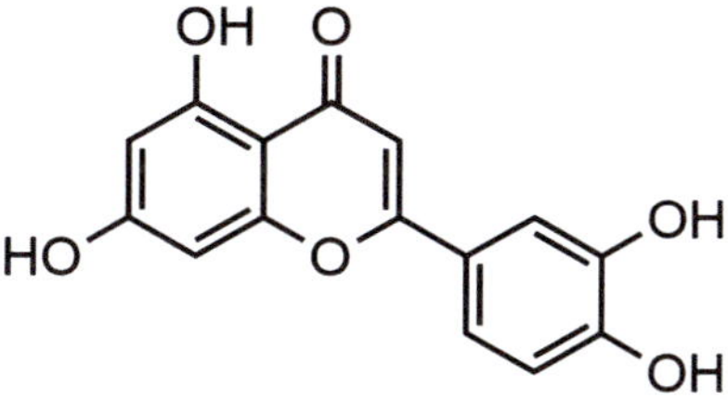

Fig. 2 The structure of luteolin

References

1. López-Lázaro M. Distribution and biological activities of the flavonoid luteolin. Mini Rev Med Chem. 2009;9(1):31–59.
2. Lin Y, Shi R, Wang X, et al. Luteolin, a flavonoid with potential for cancer prevention and therapy. Curr Cancer Drug Targets. 2008;8(7):634–46.
3. Rao PS. Luteolin induces apoptosis in multidrug resistant cancer cells without affecting the drug transporter function: involvement of cell line-specific apoptotic mechanisms. Int J Cancer. 2012;130(11):2703–14.
4. Dajas F, Rivera-Megret F, Blasina F, et al. Neuroprotection by flavonoids. Braz J Med Biol Res. 2003;36:1612–20.
5. Wang J, He Y, Zhang W, et al. Advances in studies on pharmacological effects of luteolin. Chin Bull Life Sci. 2013;25(6):560–5.
6. Han W, Xing Y, Kang T. Advances in studies on biological activities of Luteolin. Yunnan J Tradit Chin Med Matera Med. 2010;31(4):60–2.
7. Ulubelen A, Miski M, Neuman P, et al. Flavonoids of Salvia tomentosa (Labiatae). J Nat Prod. 1979;42(4):261–3.
8. Zhao G, Qin GW, Wang J, et al. Functional activation of monoamine transporters by luteolin and apigenin isolated from the fruit of Perilla frutescens (L.) Britt. Neurochem Int. 2010;56(1):168–76.

Magnolol and Honokiol

Hai-Guang Yang and Guan-Hua Du

Abstract Magnolol is derived from root bark and branch bark of magnolia or *Magnolia officinalis*. Magnolol shows many pharmacological effects, such as prolonging of central muscle relaxation, inhibition of the central nervous, anti-inflammatory, antibacterial, antimicrobial, antiulcer, antioxidant, antitumor, hormone regulation, and antidiabetes. They are still in the stage of preclinical evaluation.

Keywords Magnolol · Honokiol · Pharmacological effects

Magnolol

Origin: Magnolia bark (Fig. 1)
Chemical name (Fig. 2)

5,5′-Diallyl-2,2′-biphenyldiol

Molecular formula, $C_{18}H_{18}O_2$; **MW**, 266.33; **CAS**, 528–43-8

Honokiol

Chemical name (Fig. 3)

5,3′-Diallyl-2,4′-dihydroxybiphenyl

Molecular formula, $C_{18}H_{18}O_2$; **MW**, 266.33; **CAS**, 35354-74-6

H.-G. Yang · G.-H. Du (✉)
Beijing Key Laboratory of Drug Targets Identification and Drug Screening, Centre for Pharmaceutical Screening, Institute of Materia Medica, Chinese Academy of Medical Sciences and Peking Union Medical College, Beijing, China
e-mail: dugh@imm.ac.cn

G.-H. Du, *Natural Small Molecule Drugs from Plants*,
https://doi.org/10.1007/978-981-10-8022-7_115

Properties

Appearance: white fine powder. *Solubility*: practically insoluble in water; very soluble in benzene, ethyl ether, chloroform, and acetone. *Melting point*: magnolol is 102 °C. Honokiol is 87.5 °C.

Dosage Forms and Indications

These compounds are recorded in the *British Pharmacopoeia* (2017) and *European Pharmacopoeia* (8.7th ed.). They are mainly used as antibacterial and antifungal agents.

Literature

Magnolol is derived from the root bark and branch bark of magnolia or *Magnolia officinalis. Magnolia officinalis* was first written in the 《ben jing》, it prefers cool and humid climate and well-drained acid soil. *Cortex magnoliae officinalis* (hou po) is mainly produced in Sichuan (Wan quan, Shizhu, Guanxian), Hubei (Shien, Yichang, Lichuan), Zhejiang (Longquan, Anhui), and other places. It is also distributed in Fujian, Jiangxi, Hunan, Guangxi, Yunnan, Guizhou, Shanxi, Gansu, and other places. *Cortex magnoliae officinalis* from Sichuan and Hubei is called "chuan pu," with better quality, while in Zhejiang, it is called "warm park," with a larger output [1, 2].

History of R&D

In 1930, magnolol was first isolated from the bark of *Cortex magnoliae officinalis* in China by Japanese Sugii, and it was also isolated from *Cortex magnoliae officinalis* in Japan. In 1973, magnolol and its isomers honokiol were isolated from *Cortex magnoliae officinalis* in China and Japan by Japanese Fujita [3].

The drug research and development of magnolol and honokiol remain in pharmacological activity and preclinical research. Kyung Hee University from Korea declared a preclinical study in 2003, in which allergy, anxiety, angina, and heart failure are indications for honokiol.

Pharmacology

Magnolol and honokiol have the pharmacological effects of prolonging central muscle relaxation, inhibition of the central nervous, anti-inflammatory, antibacterial, antimicrobial, antiulcer, antioxidant, antitumor, hormone regulation, and antidiabetes [4].

Discussion

It is found that magnolol and honokiol have many pharmacological activities, and they are still in the stage of preclinical activity evaluation.

Fig. 1 Magnolia bark

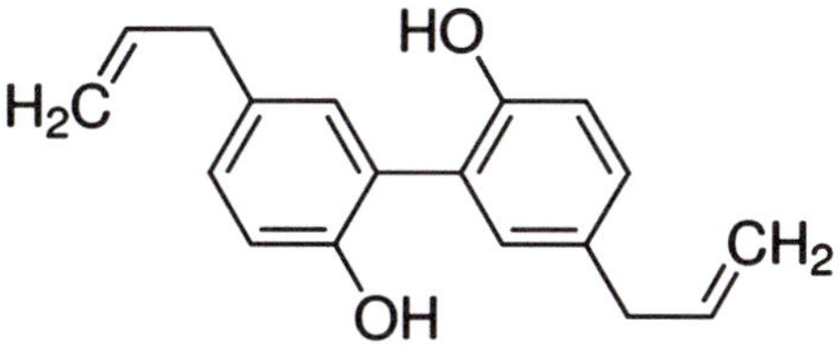

Fig. 2 The structure of magnolol

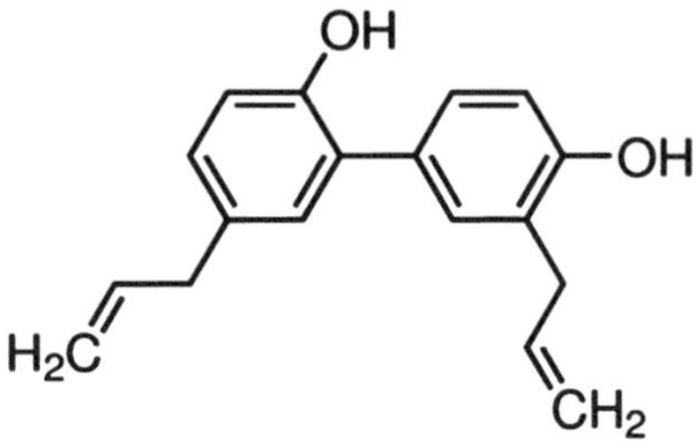

Fig. 3 The structure of honokiol

References

1. Song LR. Chinese Materia Medica. Editorial Committee of Chinese Materia Medica, vol. 6. Shanghai: Shanghai Science & Technology Press; 2011. p. 880–7.
2. Chinese Pharmacopoeia Commission. Chinese pharmacopoeia. Beijing: China Medical Science and Technology Press; 2010. p. 51.
3. Fujita M, Itokawa H, Sashida Y. Studies on the components of Magnolia obovata Thunb. 3. Occurrence of magnolol and honokiol in M. obovata and other allied plants. Yakugaku Zasshi. 1973;93(4):429–34.
4. Zhang Y, Tang F. Advance in latest studies on pharmacological effects of magnolol. Chin J Chin Mater Med. 2012;37(23):3526–30.

Oleuropein

Xue Zhang, Li-Da Du, and Yang Lu

Abstract Oleuropein is a phenolic compound extracted from olive (*Canarium album* (Lour.) Raeusch., Gan Lan) leaves mainly. Clinical and experiment statistics demonstrated oleuropein is beneficial to human health, discovering its activities of antioxidation, anti-inflammatory, antitumor, atherosclerosis prevention, blood glucose hypoglycemia reduction, and nerve protection. In recent years, the modern pharmacological research provides a theoretical foundation for its clinical application. At the same time, there are some new pharmacological effects discovered by scientists. It is believed that it is beneficial and it has antitumor effect and effects on cardiovascular diseases and metabolic diseases.

Keywords Oleuropein · Olive · Skin care · Dietary supplement

Alias: Oleuropin
Origin: Olive (Fig. 1)
Chemical name (Fig. 2)

2-(3,4-Dihydroxyphenyl)ethyl[(2S,3E,4S)-3-ethylidene-2-(b-D-glucopyranosyloxy)-5-(methoxycarbonyl)-3,4-dihydro-2H-pyran-4-yl]acetate

Molecular formula, $C_{25}H_{32}O_{13}$; **MW**, 540.51; **CAS**, 32619-42-4

X. Zhang · Y. Lu (✉)
Beijing Key Laboratory of Polymorphic Drug, Institute of Materia Medica, Chinese Academy of Medical Sciences and Peking Union Medical College, Beijing, China
e-mail: luy@imm.ac.cn

L.-D. Du
School of Biomedical Sciences, Chinese University of Hong Kong, Hong Kong, China

G.-H. Du, *Natural Small Molecule Drugs from Plants*,
https://doi.org/10.1007/978-981-10-8022-7_116

Properties

Appearance: crystal in ethyl acetate. *Solubility*: very soluble in ethanol, acetone, glacial acetic acid, and 5% NaOH solution; soluble in water, butanol, ethyl acetate, and butyl acetate; insoluble in ether, petroleum ether, chloroform, and carbon tetrachloride. *Melting point*: 87–89 °C. *Specific optical rotation*: −147° in water/ethanol/propane with a concentration of 1 mol/L; able to convert to −127° in water in 9 h.

Dosage Forms

This product is included in *British Pharmacopoeia* (2017) and *European Pharmacopoeia* (9.0th ed.).

Olive leaf extract tablet.

Indications

Dietary supplement

Literature

Oleuropein is a phenolic compound extracted from olive (*Canarium album* (Lour.) Raeusch., Gan Lan) leaves mainly. The olive tree belongs to family Oleaceae *Olea* genus of evergreen trees, and it is the world famous fruit and woody species. There is highly edible value for cultivating trees. Affiliated to famous economic forest and subtropical fruit trees, they are rich in olive oil which is considered as high-quality edible oil [1]. Broad leaf is opposite to single leaf. There are four crack corolla, female and male, two rooms in the ovary, two ovules in each room, and oiled drupe in the ellipse olive. There are more than 500 kinds of olive, 140 kinds of which are planted broadly. China is not only the hometown of olives, but also there are numerous olives cultivated here among the world. It distributes in the province of Fujian mainly, and it is also cultivated in Guangdong, Guangxi, Zhejiang, Sichuan, Taiwan province, and other provinces. In Thailand, Vietnam, Burma, Laos, the Philippines, India, and Malaysia, the plant is also cultivated widely. Adoring drought and light, olive grows rapidly in the Mediterranean countries in Europe and the United States. Olives are produced mainly in Greece, Italy, Tunisia, and Spain. And now olive is introduced into the countries all over the world. In the 1950s, little of olive is introduced in northwest of China, such as in Gansu, Shanxi, Sichuan, Chongqing, Fujian,

Guangxi, Guangdong, and Hunan province. The original olive still can be seen in the mountain area of Sichuan, Guizhou, and Hunan province.

History of R&D

The usage of olive ramification (including olive oil) in human health dates back to several centuries ago. For centuries, olive oil has been added to cosmetics and pharmacological agent. However, the unrefined mixture is applied primarily, and later studies have found that the main component of it which can be absorbed is olive polyphenols [1, 2]. Olive polyphenol, which gives a unique bitter taste, is mainly from the seeds, leaves, and immature bark of the olive (14% of dry weight). The content and purity are affected by fruit ripening stage, production, and extraction technology. In 1959, oleuropein was isolated, and the chemical structure was identified [1]. In addition, oleuropein can be found in other kinds of plants, such as *Lauraceae, Syringa, Ligustrum, Hibiscus and Jasminum*. Up to now, the amount of Oleaceae containing oleuropein is 25 at least, including jasmine, cloves, lobular lilac, etc. Based on the molecular structure, pharmacological effects have been studied further. Clinical and experiment statistics demonstrated it is beneficial to human health. People gradually realized the function of its antioxidative, anti-inflammatory, and antitumor effects, atherosclerosis prevention, blood glucose hypoglycemia reduction, and nerve protection.

Pharmacology

According to the study, the pharmacological effects including antiviral, antitumor, antioxidative, and antimicrobial and cardiovascular protection are investigated. In cardiovascular system, olive leaf extract can reduce the discomfort caused by inadequate arterial blood flow induced by angina and intermittent claudication. Oleuropein can decrease the degree of LDL oxidation [3] and protect from coronary heart disease. The intensity of vascular smooth muscle can be decreased, and the ability of decreasing blood pressure can be achieved. Oleuropein regulates enzymes which participates in specific metabolism of protein, carbohydrate, and lipid and activates pepsin, trypsin, lipase, glycerol dehydrogenase, and GAPDH [2]. The activity of other enzymes is also inhibited by oleuropein.

Oleuropein plays a role in resisting the virus; it can suppress VHSV and acts as a specific HIV inhibitor. It also shows significant activity against respiratory syncytial virus [4].

Oleuropein destroys the structure of actin filaments in cellular and noncellular detection and inhibits the proliferation and migration of tumor cells. The replication, movement, and invasion of tumor cells can be prevented. However, all of the above are reversible in normal cells [5].

In recent years, with the further study on oleuropein, pharmacological effect attracts people's attention, and oleuropein is applied broadly.

Clinical Application

In the United States and other European countries, olive leaf extract serves as a dietary supplement, and its recommended dose is 50–100 mg as an immunomodulator. FDA also approves olive leaf extracts to be used as antioxidants in foods. In addition, 80% of oleuropein are used for skin care products mainly, and it can protect skin cells from UV. At the same time, it maintains soft and elastic skin effectively to protect skin. With light color and high content, oleuropein is very suitable for cosmetic formula design. In the cream and liquid formula, it is added at a concentration of 0.5–1%. The other pharmacological effects are still in preclinical study.

Discussion

In recent years, the modern pharmacological research provides a theoretical foundation for its clinical application. At the same time, there are some new pharmacological effects discovered by scientists. It is believed that it is beneficial and it has antitumor effect and effects on cardiovascular diseases and metabolic diseases. Its high bioavailability and low toxicity have been demonstrated with animal experiment [6]. The results reveal that there is rosy prospect of applying for clinical.

Fig. 1 Olive

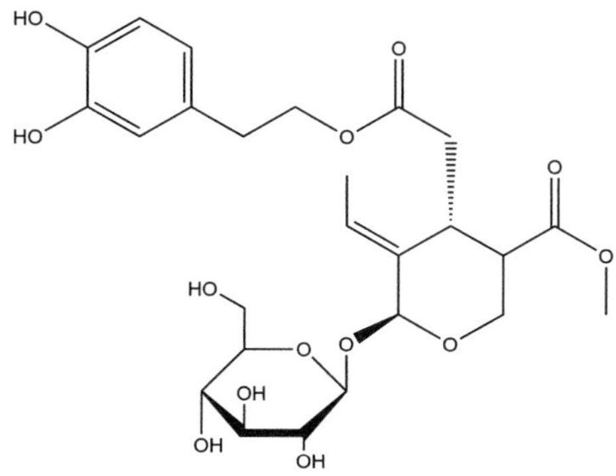

Fig. 2 Chemical structure of oleuropein

References

1. Petkov V, Manolov P. Pharmacological analysis of the iridoid oleuropein. Arzneimittelforschung. 1972;22:1476–86.
2. Barbaro B, Toietta G, Maggio R, et al. Effects of the olive-derived polyphenol oleuropein on human health. Int J Mol Sci. 2014;15:18508–24.
3. Xie P, Huang L, Zhang C, et al. Preparation and anti-oxidative activity of hydroxytyrosol by alkaline hydrolysis of oleuropein. Chin Tradit Herb Drug. 2013;44(15):2075–8.
4. Xiang Z, Chen H, He C. Advancement in research on the chemical constituents and pharmacological effects of Canarium album. Lihizhen Med Materia Medica Res. 2006;17(11):2299–300.
5. Hassan ZK, Elamin MH, Daghestani MH, et al. Oleuropein induces anti-metastatic effects in breast cancer. Asian Pac J Cancer Prev. 2012;13:4555–9.
6. Yi Y, Su Z, Liu Y, et al. Studies on pharmacokinetics of oleuropein in rats. Shizhen Guoyi Guoyao. 2012;8:1896–8.

Osthole

Xiao-Xiu Li, Li-Da Du, and Guan-Hua Du

Abstract Osthole is derived from Umbelliferae and Rutaceae plants. It is a Chinese medicine which is used widely by traditional medicine doctors. Cnidium has been recorded in almost all of the pharmaceutical monograph. Osthole exhibits a variety of pharmacological effects including neuroprotective, osteogenic, immunomodulatory, anticancer, hepatoprotective, cardiovascular protective, and antimicrobial activities. It is expected to be used in the treatment of cardiovascular system, central nervous system, and immune system diseases. Further studies are needed to enrich the knowledge of the pharmacological effects and address the safety issues of osthole in order to develop this natural product and its derivatives as an agent for disease prevention and therapeutics in human.

Keywords Osthole · Anticancer · Neuroprotective · Cardiovascular protective

Alias: Osthol
Origin: Cnidium (Fig. 1)
Chemical name (Fig. 2)

7-Methoxy-8-(3-methylbut-2-enyl)-2-chromenone

Molecular formula, $C_{15}H_{16}O_3$; **MW**, 244.29; **CAS**, 484-12-8

X.-X. Li
Department of Pharmacology, Liaoning Key Laboratory of Behavior and Cognitive Neuroscience, Shenyang Medical College, Shenyang, China

L.-D. Du
School of Biomedical Sciences, Chinese University of Hong Kong, Hong Kong, China

G.-H. Du (✉)
Beijing Key Laboratory of Drug Targets Identification and Drug Screening, Centre for Pharmaceutical Screening, Institute of Materia Medica, Chinese Academy of Medical Sciences and Peking Union Medical College, Beijing, China
e-mail: dugh@imm.ac.cn

G.-H. Du, *Natural Small Molecule Drugs from Plants*,
https://doi.org/10.1007/978-981-10-8022-7_117

Properties

Appearance: white needlelike crystalline powder. *Solubility*: soluble in methanol. *Melting point*: 83–84 °C.

Dosage Forms

Osthole is recorded in *British Pharmacopoeia* (2017) and *European Pharmacopoeia* (9.0th ed.).

Injections, encapsulants, ointments, gels

Indications

Impotence, infertility, uterine cold, wet nose with low back pain, vulva eczema, woman vulvae, trichomonas vaginitis

Literature

Cnidium is a Chinese medicine which is used widely by traditional medicine doctors. Cnidium has been recorded in almost all of the pharmaceutical monographs [1–3], including *Shen Nong's Classic Materia Medica*, *New Materia Medica*, *Compendium of Materia Medica*, etc. Osthole is a major bioactivity compound of the Cnidium, which has the effect of dampness dispelling, insecticidal antipruritic, and aphrodisiac and is used for the treatment for pruritus and eczema, lumbago, kidney impotence, and infertility [4].

Osthole is derived from Umbelliferae and Rutaceae plants. Osthole is also distributed in Asteraceae and Leguminosae plants. Osthole is present in 14 genera of Umbelliferae plants: danggui (*Angelica* L.), gudanggui (*Archangelica* Hoffm.), mianguoqin (*Cachrys* L.), shechuang (*Cnidium* Cusson), awei (*Ferula* L.), duhuo (*Heracleum* L.), yanfeng (*Libanotis* Hill), ouqin (*Petroselinum* Hill), qianhu (*Peucedanum* L.), huiqin (*Pimpinella* L.), liangshechuang (*Selinum* L.). Shuanchiqin (Prangos Lindl), xifengqin (Seseli L.), oufangfeng (Pastinaca L.). Osthole is present in 17 genera of Rutaceae plants: ganju (*Citrus* L.), huangpi (*Clausena* Burm. F.), xiangju (*Feronia* Correa), jupanmu (*Flindersia* R. Br.), niyunxiang (*Haplophyllum* A. Juss.), mizhuyu (*Melicope* J. R. and G. Forst.), xiaoyunmu (*Micromelum* Bl.), jiulixiang (*Murraya* Koenig ex L.), zhi (Poncirus Raf.), yinyu (Skimmia Thunb.), Limonia, Myrtopsis, Pentaceras, Phebalium, Pilocarpus, Thamnosma, Ticorea) [5].

History of R&D

Cnidium is a Chinese medicine which is used widely by traditional medicine doctors. Osthole was isolated in 1950 [6] and then its chemical structure was determined [7]. Osthole can be extracted by solvent method, acid-base method, ultrasound, and supercritical and other methods. The main methods for the separation of osthole are recrystallization, chromatography, and so on. With the development of modern pharmacological experimental methods and techniques, the pharmacological effects of osthole have been discovered and developed. It is found that osthole has many pharmacological activities such as anti-osteoporosis, antiplatelet aggregation, antiviral, anti-mutation, anti-allergic reaction, and so on. In recent years, it has been found that osthole has obvious antitumor activity. In addition, osthole also has strong insecticidal and antifungal effects. In the future, osthole is expected to be developed as a new kind of low-toxicity, high-efficiency, and low-residue pesticide.

Pharmacology [8]

1. Osthole can enhance the non-specific immune function of mice.
2. Osthole exhibits protective effect on the heart and circulatory system. Osthole can dilate blood vessels, reduce blood pressure, reduce blood lipid, and inhibit the formation of venous thrombosis.
3. Osthole has a central sedative effect and can improve the learning and memory function of mice.
4. Osthole presents the effect of relieving asthma by regulating the function of hypothalamus-pituitary-adrenal cortex system.
5. Osthole has antiaging and antioxidant effects, which may be related to the direct scavenging of oxygen free radicals.
6. Osthole reduces osteoporosis and bone resorption in ovariectomized rats and increases bone mass. Osthole can effectively prevent osteoporosis in ovariectomized rats, prevent bone loss, and maintain bone metabolism balance.
7. Osthole can relieve itching, which may be related to blocking the Ca^{2+} channel of mast cells, decreasing the flow of Ca^{2+}, and inhibiting the release of histamine from mast cells.
8. Osthole is a noncompetitive antagonist of Ca^{2+}, which can inhibit the smooth muscle of the uterus. In addition, osthole has a certain androgen-like, estrogen-like effect.
9. Osthole has anti-inflammatory effects on acute and chronic inflammation models, and its anti-inflammatory mechanism is not through the pituitary-adrenal cortex system, but also with the synthesis of PGE. It may be possible to exert anti-inflammatory effects by enhancing non-specific immune function and/or anti-allergic reactions [9].

10. Accumulating experimental evidences have shown that osthole exhibited antiproliferative properties and induced apoptosis in various kinds of tumor cells.

Besides, osthole has broad-spectrum antibacterial and insecticidal effects.

Clinical Application

In vitro and in vivo experimental results have revealed that osthole demonstrates multiple pharmacological actions including neuroprotective, osteogenic, immunomodulatory, anticancer, hepatoprotective, cardiovascular protective, and antimicrobial activities. Osthole is expected to be used in the treatment of cardiovascular system, central nervous system, and immune system diseases.

Discussion

Osthole is one of the active ingredients of dicoumarol in the plant. It exhibits a variety of pharmacological effects including neuroprotection, osteogenesis, immunomodulation, and anticancer properties, making it a potential multitarget complementary medicine. Further studies are needed to enrich the knowledge of the pharmacological effects and address the safety issues of osthole in order to develop this natural product and its derivatives as an agent for disease prevention and therapeutics in human.

Fig. 1 Cnidium

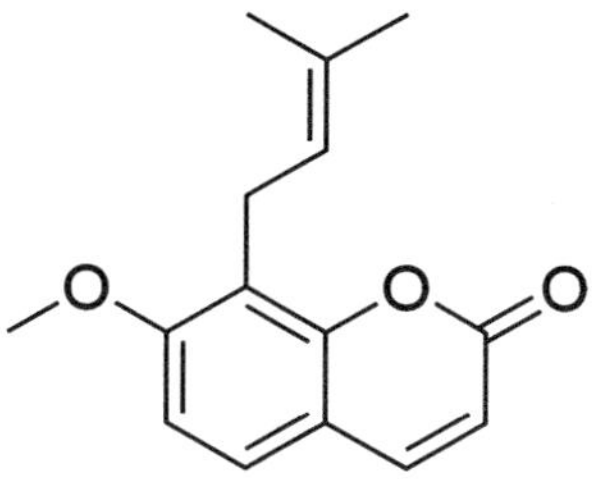

Fig. 2 Chemical structure of osthole

References

1. Liu CQ. Chinese medicine male subjects recommended. Beijing: Huaxia Publishing House; 1990.
2. Shang ZJ. New materia medica copy of the second edition. An Hui: Anhui Science and Technology Press; 1996.
3. Qing CC. Compendium of materia medica. Tai Yuan: Shanxi Science and Technology Press; 1999.
4. Chinese Pharmacopoeia Commission. Chinese pharmacopoeia, vol. 2015. Beijing: China Medical Science and Technology Press; 2015. p. 315–6.
5. Zhang QY, Zheng HC, Qin LP. Distribution and pharmacological activity of osthole in plant. Foreign Med Bot Drug. 2002;17(1):16–8.
6. Mao YD, Parks LM. Identification of osthole in Hseh Tsuang seed. J Am Pharm Assoc. 1950;39(2):107.
7. Jamwal KS, Anand KK, Chopra IC. Pharmacological properties of a crystalline substance (Osthol) isolated from Prangos pabularia Lindl. Arch Int Pharmacodyn Ther. 1962;138:400–11.
8. Li WN, Xie JX. Advances in pharmacological effects of osthol. Lishizhen Med Mater Med Res. 2005;16:6.
9. Liang HJ, Suk FM, Wang CK, et al. Osthole, a potential antidiabetic agent, alleviates hyperglycemia in db/db mice. Chem Biol Interact. 2009;181(3):309–15.

Quercetin

Li Li, Li Zhang, and Guan-Hua Du

Abstract Quercetin, a plant polyphenol from the flavonoid group, is widely found in flowers, leaves, and fruits of various plants. Experimental studies showed that quercetin had antitumor, anti-inflammatory, anti-oxidation, hypoglycemic, anti-obesity, antidepressant, and other effects. However, there is still insufficient evidence shown that quercetin as a single component has a significant effect on the treatment of the disease in clinic. Threevein *Aster*, having quercetin as one of the main active ingredients, has been used for the domestic clinical treatment of chronic bronchitis in China. It is still unable to determine the recommended dietary intake of quercetin, due to the large structure of quercetin derivatives, the complexity of chemical properties, and the diversity of distribution in food. Therefore, a well-designed research scheme for the appropriate measurement and analysis methods in the study of quercetin's food content, dietary intake, relative bioavailability, and metabolic process is of great significance to its application.

Keywords Quercetin · Polyphenol · Flavonoid · Threevein *Aster*

Alias: Sophoretin, Meletin, Quercetine, Xanthaurine, Quercetol, Quercitin, Quertine, Flavin meletin
Origin: *Aster tataricus* (Fig. 1)
Chemical name (Fig. 2)

2-(3, 4-Dihydroxyphenyl)-3, 5, 7-trihydroxy-4H-1-chromen-4-one

Molecular formula, $C_{15}H_{10}O_7$; **MW**, 302.24; **CAS**, 117-39-5

L. Li · L. Zhang · G.-H. Du (✉)
Beijing Key Laboratory of Drug Targets Identification and Drug Screening, Centre for Pharmaceutical Screening, Institute of Materia Medica, Chinese Academy of Medical Sciences and Peking Union Medical College, Beijing, China
e-mail: dugh@imm.ac.cn

G.-H. Du, *Natural Small Molecule Drugs from Plants*,
https://doi.org/10.1007/978-981-10-8022-7_118

Properties

Appearance: yellow needle-like crystalline powder. *Solubility*: slightly soluble in water; soluble in ethanol, acetone, pyridine, and acetic acid; easily soluble in ether and methanol. *Melting point*: 314–317 °C.

Derivatives (Fig. 3)

Dosage Forms

Quercetin is recorded in the *British Pharmacopoeia* (2017), *European Pharmacopoeia* (8.7th ed.), and *United States Pharmacopeia* (40). Currently there is no single component preparation of quercetin, which is contained as a reference in the *British Pharmacopoeia*, *European Pharmacopoeia*, and *United States Pharmacopeia*. At present, quercetin as one of the main active ingredients of Threevein *Aster* in tablet is used in China.

Indications

It is mainly used for the treatment of chronic bronchitis.

Literature

The name of quercetin has been used since 1857, which is derived from quercetum (oak forest) after *Quercus* [1]. Quercetin is widely found in flowers, leaves, and fruits of various plants. Vegetables (such as onions, ginger, celery, etc.), fruits (such as apples, strawberries, etc.), beverages (such as tea, coffee, red wine, fruit juice, etc.), and more than 100 kinds of Chinese herbal medicines (such as Threevein *Aster*, mountain white chrysanthemum, Huai rice, *Apocynum*, *Ginkgo biloba*, etc.) contain this ingredient.

Threevein *Aster* is a Chinese herbal medicine and used in Jiangxi province, China, for more than 30 years. Its plant name is three veins Mala, Compositae. It is rich in drug sources, which can be found in Southern provinces of China. Clinical practice proved that it has significant anti-inflammatory and expectorant effects, and it is a good prescription for the treatment of elderly chronic bronchitis.

History of R&D

In 1936, Szent-Györgyi firstly reported the separation and identification of biological activity of quercetin [2]. Usually quercetin is presented in the form of glycosides such as lutin, quercitrin, and mycoside, which can be hydrolyzed to get the quercetin. Quercetin has a polyphenol hydroxyl structure, which is of weak lipophilicity and poor hydrophilicity, resulting in its low bioavailability and limiting its clinical application. The synthesis of phenolic derivatives improves its bioavailability, which are lipid-soluble quercetin derivatives such as 3-*O*-methylquercetin, hydrophilic quercetin derivatives such as 3′-*ON*-carboxymethylformamide quercetin, and quercetin glycosides.

Threevein *Aster*, having quercetin as one of the main active ingredients, has been used for the domestic clinical treatment for chronic bronchitis in China since 1971.

Pharmacology

Experimental studies showed that quercetin had antitumor, anti-inflammatory, anti-oxidation, hypoglycemic, anti-obesity, antidepressant, and other effects [3]. *In vitro* cell experiments and *in vivo* animal experiments have shown that quercetin could inhibit the growth of various malignant tumor cells such as human ovarian cancer, breast cancer, gastrointestinal tumor cells, and leukemia, and it could induce cancer cell apoptosis and had a reversal of tumor multidrug resistance (MDR) effect, while, combined with other anticancer drugs, it could enhance the effect of anticancer drugs.

Quercetin could alleviate the inflammatory response that was aggravated by the activation of the central granulocytes. In the experimental study on the treatment of non-bacterial prostatitis and acute gouty arthritis, quercetin also showed a good anti-inflammatory effect.

The experimental results showed that quercetin had a good direct scavenging effect on free radicals and exhibited antioxidant activity. In addition, it also had the anti-hepatic fibrosis, pulmonary fibrosis, keloid hyperplasia and glaucoma filtering bubble scarring and other effects, its mechanism involving the inhibition of fibroblast proliferation, inhibition of collagen synthesis, preventing oxidative damage and so on. Moreover, studies have shown that quercetin also had antibacterial, anti-aging, antidepressant, antileukemia, antidiabetic, and other pharmacological effects.

Clinical Application

Since the first clinical phase I trial of quercetin in 1996 found that it had antitumor activity [4], quercetin has also been reported in early clinical trials of cardiovascular disease, diabetes, and other diseases. However, there is still insufficient evidence shown that quercetin has a significant effect on the treatment of the disease in clinic. The US FDA has issued a warning [5], emphasizing that quercetin is not a definite nutrient, unable to determine its content in the diet, nor can it be used as a drug.

China's Threevein *Aster* consists of a single Chinese herb, which was released by the *Pharmacopoeia of the People's Republic of China* (1977) Part I [6]. One of the main active ingredients obtained following the hydrolysis of Threevein *Aster* is quercetin, which has the function of relieving cough and eliminating phlegm and can be used for the treatment of chronic bronchitis. The anti-inflammatory effect of Threevein *Aster* is poor. Side effects after use include stomach discomfort, dizziness, and abdominal pain, while withdrawal can make them disappear.

Discussion

Although quercetin is widely distributed in nature and its pharmacological effects are extensive, the medicinal development of quercetin is extremely slow, and there is no relevant drug at the moment. Due to the large structure of quercetin derivatives, the complexity of chemical properties, and the diversity of distribution in food, it is difficult to establish appropriate measurement and analysis methods in the study of food content, dietary intake, relative bioavailability, and metabolic process, resulting in the current inability to determine the recommended dietary intake [7]. Therefore, a well-designed research scheme for quercetin is of great significance to its application.

Fig. 1 *Aster tataricus*

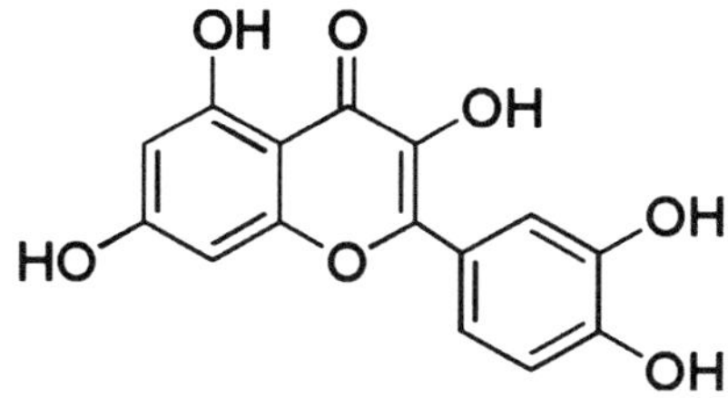

Fig. 2 Chemical structure of quercetin

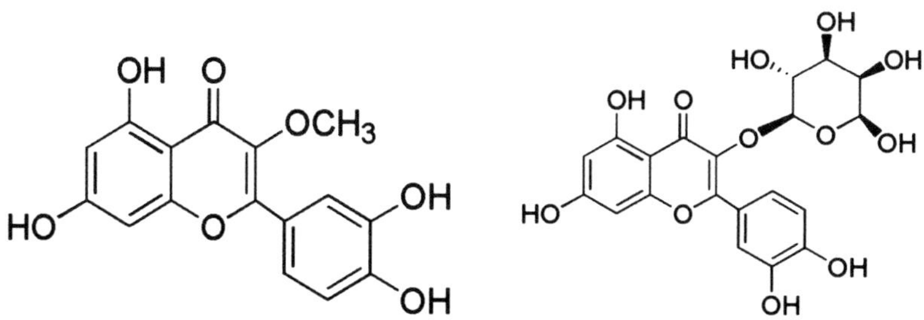

3-O-methyl quercetin 3-D-galactoside quercetin

Fig. 3 Chemical structures of quercetin derivatives

References

1. Quercetin. The free encyclopedia. In Wikipedia, from https://en.wikipedia.org/wiki/Quercetin. Retrieved 20:34, 6 May 2017.
2. Harborne JB. Flavonoids in the environment: structure activity relationships. In Progress in clinical and biological research. New York: Alan R. Liss, Inc. 1986.
3. Zhang ZQ, Zhu SX. The pharmacological activity of quercetin and the progress of clinical application. Chin J Pharmaceut Res. 2013;32(7):400–3. (in Chinese)
4. Ferry DR, Smith A, Malkhandi J, et al. Phase I clinical trial of the flavonoid quercetin: pharmacokinetics and evidence for in vivo tyrosine kinase inhibition. Clin Cancer Res. 1996;2(4):659–68.
5. "River Hills Harvest dba Elderberrylife". Adams, AM, Inspections, Compliance, Enforcement, and Criminal Investigations, US FDA. 22 April 2014. Retrieved 5 November 2014, from https://www.fda.gov/ICECI/EnforcementActions/WarningLetters/2014/ucm394486.htm
6. Pharmacopoeia Commission. Pharmacopoeia of the People's Republic of China (1977) part one. Beijing: People's Medical Publishing House; 1978. (in Chinese)
7. D'Andrea G. Quercetin: a flavonol with multifaceted therapeutic applications? Fitoterapia. 2015;106:256–71.

Shikonin

Wen Sun, Li-Da Du, Shou-Bao Wang, Xiu-Ping Chen, and Guan-Hua Du

Abstract Shikonin is a naphthoquinone isolated from *Arnebia* species such as *Arnebia euchroma*. The dried root of *A.euchroma* is a traditional Chinese medicine with anti-inflammatory and antiviral activities. Though there are about 500 publications on shikonin, no clinical trials for shikonin and its derivatives have been reported yet because of inherent properties such as poor solubility in water and low bioavailability. Therefore, it is of great importance to conquer these problems to exploit the clinical potential of shikonin and its derivatives.

Keywords Shikonin · Anti-inflammatory · *Arnebia euchroma* (Royle) Johnst

Origin: *Arnebia euchroma* (Royle) Johnst (Fig. 1)
Chemical name (Fig. 2)

5,8-Dihydroxy-2-[(1R)-1-hydroxy-4-methylpent-3-enyl]naphthalene-1,4-dione

Molecular formula, $C_{16}H_{16}O_5$; **MW**, 288.31; **CAS**, 517-89-5
Derivatives (Fig. 3)

W. Sun · X.-P. Chen (✉)
State Key Laboratory of Quality Research in Chinese Medicine, Institute of Chinese Medical Sciences, University of Macau, Macao, China
e-mail: xpchen@umac.mo

L.-D. Du
School of Biomedical Sciences, Chinese University of Hong Kong, Hong Kong, China

S.-B. Wang · G.-H. Du
Beijing Key Laboratory of Drug Targets Identification and Drug Screening, Centre for Pharmaceutical Screening, Institute of Materia Medica, Chinese Academy of Medical Sciences and Peking Union Medical College, Beijing, China

G.-H. Du, *Natural Small Molecule Drugs from Plants*,
https://doi.org/10.1007/978-981-10-8022-7_119

Properties

Appearance: purple lamellae or crystalline powder. *Melting point:* 147 °C. *Specific optical rotation*: +138°(benzene). *Solubility*: soluble in ethanol, vegetable oils, and other organic solvents.

Indications

No clinical usage till now.

Literature

Shikonin could be found in the plant of *Lithospermum erythrorhizon* Sieb. et. Zucc and *Arnebia euchroma* (Royle) Johnst. It could also be produced using plant cell culture techniques. Shikonin is recorded in *British Pharmacopoeia* as a reference compound.

Zicao, a traditional Chinese medicine firstly recorded in *Shen Nong's Herbal Classic*, has long been used medically in history [1]. *Chinese Pharmacopoeia* (Edition 2015) recorded the root of *Arnebia euchroma* (Royle) Johnst and *Arnebia guttata* Bge as Zicao, which is mainly used for inflammatory diseases such as macular eruptions, measles, sore throat, carbuncles, and burns.

History of R&D

Kuroda and Majima firstly identified acetyl shikonin from *L. erythrorhizon* in 1922 [2], followed by the discovery of other shikonin derivatives, including shikonin. The chemical structure of shikonin was not precisely identified till 1936 for its high physicochemical similarity with naphthazarin [3]. There have been about 500 publications on shikonin up to now. Great attention has also been paid on the biosynthesis of shikonin and its derivatives, and an increasing number of shikonin derivatives have been designed and synthesized for exploring their antitumor effect. There are two types of derivatives: one is modifications of 1′-OH with parent nucleus naphthazarin remained, as shown in Fig. 3a, b, and the other is modifications of both 1′-OH and parent nucleus naphthazarin, as shown in Fig. 3c, d [4].

Pharmacology

Shikonin possesses anti-inflammatory, antioxidant, antiviral, cardiovascular protective, and antitumor activities.

Shikonin reduces inflammation by inhibiting the biosynthesis of leukotrienes and 5-hydroxyeicosatetraenoic acid and thus reduces synthesis of inflammation-related active molecules, which selectively block chemokine binding to CC chemokine receptor 1 [5, 6].

Shikonin shows free radical scavenging and antioxidant (especially toward superoxide anion and DPPH) activities. It significantly inhibits autoxidation caused by β-carotene and linoleic acid [7]. Its anti-HCV effects have been reported recently with an EC_{50} at 25 ng/mL, which is lower than that of ribavirin (2.6 μg/mL) [8].

Recent studies also reveal shikonin possesses cardiovascular protective effects. Shikonin inhibits the activity of TNF-α promoter, revealing its transcriptional antagonism to pro-inflammatory cytokine [9]. Shikonin also shows antitumor potentials by inducing apoptosis and/or necrosis, inhibiting DNA topoisomerase activity and angiogenesis, and regulating proliferative signaling pathways (including MAPK, VEGF, and PTKs [4]). In addition, shikonin circumvents cancer drug resistance by induction of necroptotic cell death [10].

Clinical Application

Shikonin and its derivatives have not been approved for clinical use yet. Studies are confined to cellular and animal experiments. Its original plant Zicao has a long history of medical use both orally and externally in China. Various dosage forms of Zicao, including tablets, injections, oils, creams, tinctures, plastics, and pastes, have been developed for different clinical applications especially in dermatology, gynecology, pediatrics, ophthalmology, and otorhinolaryngology. Among them, puccoon oil and lithospermum cream are the most widely used forms.

Discussion

Zicao has long been used as medical herb with shikonin and its derivatives as the main active components. Although various pharmacological effects of shikonin, including antitumor, anti-angiogenesis, and antiproliferation, have been demonstrated since 1990, no clinical trials for shikonin and its derivatives have been reported yet. Some inherent properties such as poor solubility in water, low bioavailability, poor heat stability, and ready oxidation and multimerization dramatically limit its clinical application. Capsules and targeted drug delivery system with biodegradable polysaccharide may help overcome these shortcomings. Though the anti-inflammatory and antitumor effects of shikonin have been confirmed, further studies are still needed to explore its clinical potential.

Fig. 1 *Arnebia euchroma* (Royle) Johnst

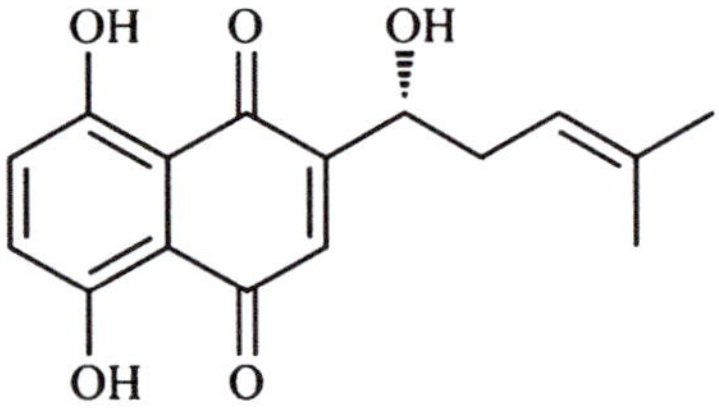

Fig. 2 Chemical structure of shikonin

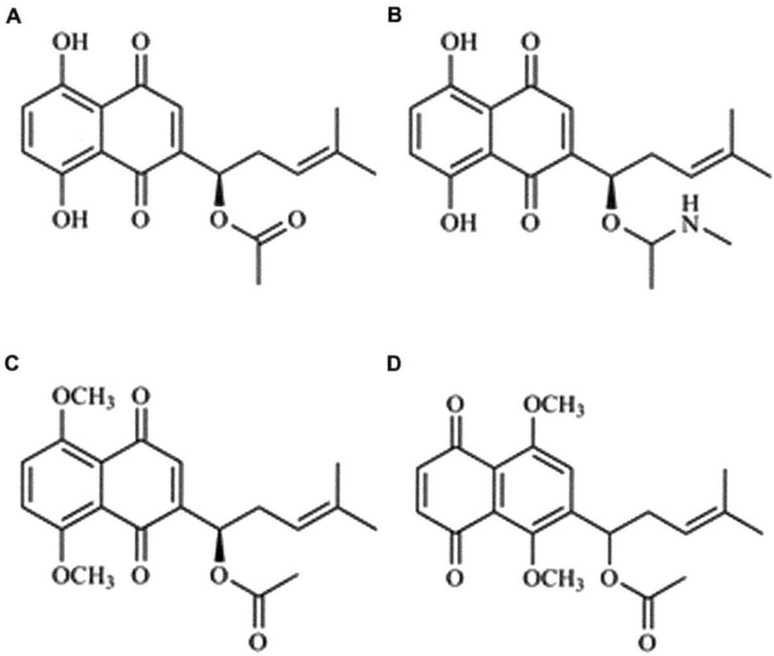

Fig. 3 Typical shikonin derivatives. (**a**) Derivative of acetyl shikonin; (**b**) Derivative of carbamate; (**c**) Derivative of 5, 8-*O*-dimethyl shikonin; (**d**) Derivative of 6-epimers 5, 8-*O*-dimethyl shikonin

References

1. Tao HJ. Bencaojing Jizhu. Beijing: People's Medical Publishing House; 1994.
2. Mamima R, Kuroda S. The coloring matter of *Lithospermumerythrorhizon*. Acta Phytochim. 1922;1:43–65.
3. Brockmann H. Die konstitution des alkannins, shikonins and alkannans. Ann Chem. 1936;521(1):1–47.
4. Wang RB, Yin RT, Zhou W, et al. Shikonin and its derivatives: a patent review. Expert Opin Ther Pat. 2012;22(9):977–97.
5. Wang WJ, Bai JY, Liu DP, et al. The anti-inflammatory activity of shikonin and its inhibitory effect on leukotriene B4 biosynthesis. Acta Pharm Sin. 1994;29(3):161–5.
6. Chen X. Shikonin, a component of anti-inflammatory Chinese herbal medicine, selectively blocks chemokine binding to CC chemokinereceptor-1. Int Immunopharmacol. 2001;1(2):229–36.
7. Zhang GP, Yang JX, Zhu YA, et al. Study on antioxidant activity of *Lithospermum erythrorhizon* in vitro. J Wuhan Bot Res. 2007;25(5):490–3.
8. Li HM, Tang YL, Zhang ZH, et al. Compounds from *Arnebiaeuchroma* and their related anti-HCV and antibacterial activities. Planta Med. 2012;78(1):39–45.

9. Zhang ZQ, Cao XC, Zhu WL. Inhibitory effects of shikonin on vascular smooth muscle cells and macrophage tumor necrosis factor-α promoter activity. Chin J Pathophysiol. 2007;23(7):1378–81.
10. Han WD, Li L, Qiu S, et al. Shikonin circumvents cancer drug resistance by induction of a necroptotic death. Mol Cancer Ther. 2007;6(5):1641–9.

Toosendanin

Rui Zhao, Li-Da Du, and Guan-Hua Du

Abstract *Fructus Meliae Toosendan* is a traditional Chinese medicine which is still used in internal medicine. Toosendanin is a tetracyclic triterpene compound extracted from *fructus toosendan*. The structure was identified in 1975. A synthetic strategy is put forth in 2009. It is an ascaridole agent and showed anti-botulinum, antibacterial, anticancer, and anti-inflammatory effects. It is also a neuromuscular blocking agent.

Keywords Toosendanin · Tetracyclic triterpene compound · Insecticide

Alias: Chuanliansu, Azedarachin
Origin: Fructus Meliae Toosendan (Fig. 1)
Chemical name (Fig. 2)

24-Norchola-20,22-diene-4-carboxaldehyde,3,12-bis(acetyloxy)-14,15:21,23-diepoxy-1,7,19-trihydroxy-4,8-dimethyl-11-oxo-,cyclic4,19-hemiacetal

Molecular formula, $C_{30}H_{38}O_{11}$; **MW**, 574.62; **CAS**, 58812-37-6

R. Zhao
Beijing Chao-Yang Hospital, Capital Medical University, Beijing, China

L.-D. Du
School of Biomedical Sciences, Chinese University of Hong Kong, Hong Kong, China

G.-H. Du (✉)
Beijing Key Laboratory of Drug Targets Identification and Drug Screening, Centre for Pharmaceutical Screening, Institute of Materia Medica, Chinese Academy of Medical Sciences and Peking Union Medical College, Beijing, China
e-mail: dugh@imm.ac.cn

G.-H. Du, *Natural Small Molecule Drugs from Plants*,
https://doi.org/10.1007/978-981-10-8022-7_120

Properties

Appearance: white acicular crystal, colorless, bitter, and odorless. *Solubility*: soluble in ethanol, methanol, ethyl acetate, acetone, pyridine, etc., and slightly soluble in hot water, chloroform, benzene, and ether and insoluble in petroleum ether.

Dosage Forms

There is no medicine for the preparation of toosendanin.

Indications

Insecticide

Literature

Fructus Meliae Toosendan (chuan lian zi), the fruit of *Melia toosendan Sieb. Et Zucc*, is mainly produced in Sichuan, Hubei, Henan, and Hunan province in China [1].

Toosendanin is a tetracyclic triterpene compound extracted from fructus toosendan. It is an ascaridole agent and showed anti-botulinum, antibacterial, anticancer, and anti-inflammatory effects. It is also a neuromuscular blocking agent.

History of R&D

Toosendanin is the main chemical composition of the fructus toosendan. Researches on this compound could date back to 1953. Toosendanin has been named in 1955, while its structure was identified 20 years later by mass spectrometry. Toosendanin is a triterpenoid derivative containing a β-substituted furan ring [2]. A new synthetic strategy is put forth allowing access to a 4-acetoxy CD fragment analogue of toosendanin, which was achieved from mesityl oxide and acetylacetone in 14 steps in 2009 [3].

The anti-parasite of toosendanin was reported in 1971. In 1980, it was identified as a selective neuromuscular blocking agent that acts on neuromuscular junction [4]. The anti-botulinum effect of toosendanin was discovered in 1982 [5]. In recent years, it is also observed that toosendanin can induce differentiation and apoptosis [6]. Toosendanin could inhibit the growth of feeding larvae and insects. It has been used as pollution-free, no residue of botanical insecticides for the production of fruits and vegetables in China [7].

Pharmacology

By interfering with neurotransmitter release and causing an initial facilitated diffusion toosendanin eventually blocks synaptic transmission at both the neuromuscular junction and central synapses.

Toosendanin is sharing many similar actions with botulinum neurotoxin on blocking neuromuscular transmission, but it has very low biological activity compared with botulinum neurotoxin. Studies suggest that the anti-botulinum effect is achieved by preventing botulinum neurotoxin from approaching its enzymatic substrate, SNARE protein.

Toosendanin can induce differentiation and apoptosis in several cell lines and suppress proliferation of various human cancer cells. It inhibits K^+ channel and facilitates L-type Ca^{2+} channel, which is associated with parts of its biological activities [8].

In a word, activities of toosendanin include selective neuromuscular blocking, effectively antagonizing botulism, inducing cell differentiation and apoptosis and inhibiting proliferation of various human cancer cells, inhibiting feeding and development in insects, and modifying K^+ and Ca^{2+} channel activity.

Clinical Application

Fructus Meliae Toosendan is a traditional Chinese medicine which is still used in internal medicine.

Discussion

Fructus Meliae Toosendan (chuan lian zi), the fruit of *Melia toosendan Sieb. Et Zucc*, could be used as digestive tract parasiticide was recorded about two thousand years ago. Toosendanin, a triterpenoid derivative, was extracted from the bark of *Melia toosendan Sieb. Et Zucc* by Chinese scientists in the 1950s. Studies have demonstrated that toosendanin possesses special biological actions as well as considerable various values in scientific research, clinic medicine, and agriculture [9].

Fig. 1 Fructus Meliae Toosendan

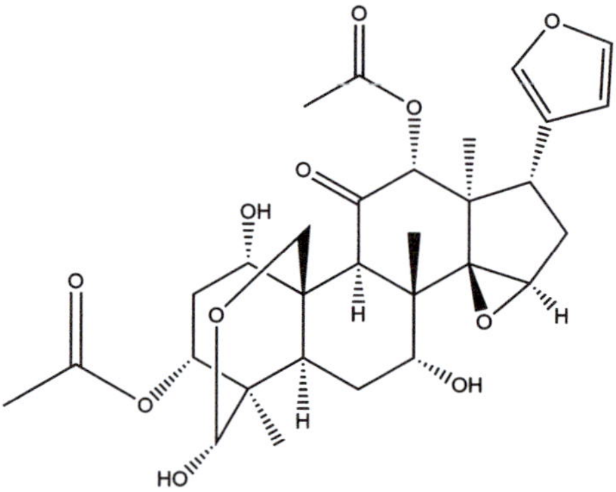

Fig. 2 Chemical structure of toosendanin

References

1. Chinese Academy of Medical Sciences. Zhong Yao Zhi. Beijing: People's Medical Publishing House; 1961. p. 162–5.
2. Zhong CC, Xi JX. The structure of Toosendanin. Acta Chim Sin. 1975;33(1):35–47.
3. Nakai Y, Tepp WH. Function-oriented synthesis applied to the anti-botulinum natural product toosendanin. Bioorg Med Chem. 2009;17(3):1152–7.
4. Shi YL, Wei NS. Toosendanin- a presynaptic neuromuscular junction transfer blocking agent. Acta Physiol Sin. 1980;32(3):293–7.
5. Shi YL, Xu K. Botulinum resistant effect of toosendanin. Chin Sci Bull. 1983;28(14):885–7.

6. Shi YL, Wang WP. Biological effects of toosendanin, an active ingredient of herbal vermifuge in Chinese traditional medicine. Acta Physiol Sin. 2006;58(05):397–406.
7. Zhang X, Wang XL. The development of botanical insecticides- toosendanin. Acta Northwest Univ Agric Sin. 1993;21(1):1–5.
8. Shi YL, Wang WP. Toosendanin modifies K+-and Ca2+-channel activity and intracellular Ca2+ concentration. Prog Biochem Biophys. 2007;34(02):132–7.
9. Wang XJ, Liu YR. Review on anti-tumor effects and mechanisms of toosendanin. Sci Technol Eng. 2011;11(02):281–5.

MIX
Papier aus verantwortungsvollen Quellen
Paper from responsible sources
FSC® C105338

If you have any concerns about our products,
you can contact us on
ProductSafety@springernature.com

In case Publisher is established outside the EU,
the EU authorized representative is:
Springer Nature Customer Service Center GmbH
Europaplatz 3, 69115 Heidelberg, Germany

Printed by Libri Plureos GmbH
in Hamburg, Germany